Lie Groups and Lie Algebras

Mathematics and Its Applications

Managing Editor:

M. HAZEWINKEL

Centre for Mathematics and Computer Science, Amsterdam, The Netherlands

Volume 433

Lie Groups and Lie Algebras

Their Representations, Generalisations and Applications

Edited by

B. P. Komrakov

International Sophus Lie Center,
Minsk, Belarus

I. S. Krasil'shchik

Moscow Institute for Municipal Economy and Diffiety Institute,
Moscow, Russia

G. L. Litvinov

Institute for New Technologies,
Moscow, Russia

and

A. B. Sossinsky

Institute for Problems in Mechanics,
Russian Academy of Sciences,
Moscow, Russia

KLUWER ACADEMIC PUBLISHERS
DORDRECHT / BOSTON / LONDON

A C.I.P. Catalogue record for this book is available from the Library of Congress.

ISBN 0-7923-4916-4

Published by Kluwer Academic Publishers,
P.O. Box 17, 3300 AA Dordrecht, The Netherlands.

Sold and distributed in the U.S.A. and Canada
by Kluwer Academic Publishers,
101 Philip Drive, Norwell, MA 02061, U.S.A.

In all other countries, sold and distributed
by Kluwer Academic Publishers,
P.O. Box 322, 3300 AH Dordrecht, The Netherlands.

Printed on acid-free paper

Table of Contents

PREFACE

This collection contains papers conceptually related to the classical ideas of Sophus Lie (i.e., to Lie groups and Lie algebras). Obviously, it is impossible to embrace all such topics in a book of reasonable size. The contents of this one reflect the scientific interests of those authors whose activities, to some extent at least, are associated with the *International Sophus Lie Center*.

We have divided the book into five parts in accordance with the basic topics of the papers (although it can be easily seen that some of them may be attributed to several parts simultaneously).

The first part (quantum mathematics) combines the papers related to the methods generated by the concepts of quantization and quantum group. The second part is devoted to the theory of hypergroups and Lie hypergroups, which is one of the most important generalizations of the classical concept of locally compact group and of Lie group. A natural harmonic analysis arises on hypergroups, while any abstract transformation of Fourier type is generated by some hypergroup (commutative or not). Part III contains papers on the geometry of homogeneous spaces, Lie algebras and Lie superalgebras. Classical problems of the representation theory for Lie groups, as well as for topological groups and semigroups, are discussed in the papers of Part IV. Finally, the last part of the collection relates to applications of the ideas of Sophus Lie to differential equations.

We are indebted to Kluwer Academic Publishers (and to Prof. Michiel Hazewinkel and Dr. Paul Roos especially) for their support in the realization of our project. We also want to express our gratitude to Ms. Irina A. Andreeva for her help in the preparation of the manuscript.

The Editors

DUAL QUASITRIANGULAR STRUCTURES RELATED TO THE TEMPERLEY–LIEB ALGEBRA

P. AKUESON AND D. GUREVICH
ISTV, Université de Valenciennes,
59304 Valenciennes, France

Abstract. We consider nonquasiclassical solutions to the quantum Yang–Baxter equation and the corresponding quantum cogroups $\mathrm{Fun}(SL(S))$ constructed earlier in (Gurevich, 1991). We give a criterion of the existence of a dual quasitriangular structure in the algebra $\mathrm{Fun}(SL(S))$ and describe a large class of such objects related to the Temperley–Lieb algebra satisfying this criterion. We show also that this dual quasitriangular structure is in some sense nondegenerate.

Mathematics Subject Classification (1991): 17B37, 81R50.

Key words: quantum Yang–Baxter equation, Poincaré series, quantum group, dual quasitriangular structure, canonical pairing, Temperley–Lieb algebra, Hecke symmetry (of Temperley–Lieb type).

1. Introduction

It became clear after the works of one of the authors (D.G.) that besides the well-known deformational (or quasiclassical) solutions to the quantum Yang–Baxter equation (QYBE) there exists a lot of other solutions that differ drastically from the former ones. Let us explain this in more detail. Let $\mathbf{V}$ be a linear space over the field $K = \mathbb{C}$ or $\mathbb{R}$. We call a *Yang–Baxter operator* a solution $S : \mathbf{V}^{\otimes 2} \to \mathbf{V}^{\otimes 2}$ to the QYBE

$$S^{12}S^{23}S^{12} = S^{23}S^{12}S^{23}, \ S^{12} = S \otimes \mathrm{id}, \ S^{23} = \mathrm{id} \otimes S.$$

A Yang–Baxter operator satisfying a second degree equation

$$(\mathrm{id} + S)(q\,\mathrm{id} - S) = 0$$

1

B. P. Komrakov et al. (eds.), Lie Groups and Lie Algebras, 1–16.

was called in (Gurevich, 1991) a *Hecke symmetry*. The quantum parameter $q \in K$ is assumed to be generic.

It is natural to associate to a Hecke symmetry two algebras defined as follows

$$\wedge_+(\mathbf{V}) = T(\mathbf{V})/\{\mathrm{Im}(q\,\mathrm{id} - S)\}, \quad \wedge_-(\mathbf{V}) = T(\mathbf{V})/\{\mathrm{Im}(\mathrm{id} + S)\}.$$

They are q-counterparts of the symmetric and skew-symmetric algebras of the space $\mathbf{V}$, respectively.

Let us denote $\wedge_\pm^l(\mathbf{V})$ the degree l homogeneous component of the algebra $\wedge_\pm(\mathbf{V})$. It was shown in (Gurevich, 1991) that the *Poincaré series*

$$P_\pm(t) = \sum \dim \wedge_\pm^l(\mathbf{V}) t^l$$

of the algebras $\wedge_\pm(\mathbf{V})$ for a generic q satisfy the standard relation

$$P_+(t)P_-(-t) = 1.$$

Moreover, if the series $P_-(t)$ is a polynomial with leading coefficient 1, it is reciprocal. Hecke symmetries of such a type and the corresponding linear spaces $\mathbf{V}$ are called *even*.

A particular case of an even Hecke symmetry is provided by the quantum groups $U_q(sl(n))$: the operator $S = \sigma \rho^{\otimes 2}(\mathcal{R})$ (where $\mathcal{R}$ is the corresponding universal R-matrix, σ is the flip and $\rho : sl(n) \to \mathrm{End}(\mathbf{V})$ is the fundamental vector representation) is just such a type of solution to the QYBE. In fact, we have a family of operators S_q and we recover the standard flip σ for $q = 1$. Namely, in this sense we call such Yang–Baxter operators (and all related objects) *deformational* or *quasiclassical*.

The Poincaré series for Hecke symmetries of such a type coincide with the classical ones. Thus, in this case we have $P_-(t) = (1 + t)^n$ with $n = \dim \mathbf{V}$ and, therefore, $P_+(t) = (1 - t)^{-n}$.

However, this is no longer true in general case. As shown in (Gurevich, 1991), for any $n = \dim \mathbf{V}$ and any integer p, $2 \leq p \leq n$, there exists a nonempty family of nontrivial even Hecke symmetries such that $\deg P_-(t) = p$. The integer p is called the *rank* of the even Hecke symmetry S or of the corresponding space $\mathbf{V}$.

The classification problem of all even Hecke symmetries of a given rank p is still open. However, all such symmetries of rank $p = 2$ are completely classified.

Let us observe that the case $p = 2$ is related to the Temperley–Lieb (TL) algebra, since the projectors $P_-^i : \mathbf{V}^{\otimes m} \to \mathbf{V}^{\otimes m}$, $1 \leq i \leq m - 1$, defined by

$$P_-^i = \frac{(q\,\mathrm{id} - S^{i,i+1})}{(q + 1)},$$

where $S^{i,i+1}$ is the operator S acting onto the i-th and $(i+1)$-th components of $\mathbf{V}^{\otimes m}$, generate a TL algebra. Let us recall that a TL algebra is the algebra generated by t_i, $1 \leq i \leq n-1$, with the following relations:

$$t_i^2 = t_i, \; t_i t_{i\pm 1} t_i = \lambda t_i \, , t_i t_j = t_j t_i \; |i-j| > 1.$$

(In the case under consideration $\lambda = q(1+q)^{-2}$.) In what follows even rank 2 Hecke symmetries will be called symmetries of the *TL type*.

It is possible to assign to any YB operator S the famous "RTT=TTR" algebra. In the sequel it is called the *quantum matrix algebra* and denoted by $\mathbf{A}(S)$. If S is an even Hecke symmetry of the TL type, in this algebra there exists a so-called *quantum determinant* (it was introduced in (Gurevich, 1991)). If it is a central element, it is possible to define a *quantum cogroup* $\mathrm{Fun}(SL(S))$ looking like the famous quantum function algebra $\mathrm{Fun}_q(SL(n))$ (in (Gurevich, 1991) this algebra was called a quantum group)[1]. Let us note that these quantum cogroups $\mathrm{Fun}(SL(S))$ possess Hopf algebra structures.

However, until now no corepresentation theory of such nonquasiclassical quantum cogroups has been constructed. In particular, it is not clear whether any finite-dimensional $\mathrm{Fun}(SL(S))$-comodule is semisimple for a generic q. Nevertheless this problem seems to be of great interest, since the nonquasiclassical solutions to the QYBE provides us a new type of symmetries, which differ drastically from classical or supersymmetries. (The simplest models possessing such a symmetry of new type corresponding to an involutive S, namely, a "nonquasiclassical harmonic oscillator" was considered in (Gurevich and all, 1992). Let us observe that the partition functions of such models can be expressed in terms of the Poincaré series corresponding to the initial symmetry.)

The present paper is the first in a series aimed at a better understanding the structure of such nonquasiclassical symmetries. More precisely, we discuss here two problems: first, whether the quantum cogroups $\mathrm{Fun}(SL(S))$ have quasitriangular structure and second, what is an explicit description of their dual objects?

It is well known that the notion of a quasitriangular structure was introduced by V. Drinfeld. In fact this notion was motivated by the quantum groups $U_q(\mathfrak{g})$. These objects have an explicit description due to V. Drinfeld and M. Jimbo in terms of deformed Cartan–Weyl system $\{H_\alpha, X_{\pm\alpha}\}$ (cf., e.g., (Chari, Pressley, 1994)). Thus this construction allows one to develop a representation theory of quantum groups.

In the nonquasiclassical case under consideration such a description does not exist. And the problem of an appropriate description of objects dual to

---

[1] Let us note that a subclass of objects of such a type was independently introduced in (Dubois-Violette, Launer, 1990).

the quantum cogroups $\mathrm{Fun}(SL(S))$ is of great interest. (The duality in the present paper is understood in the algebraic sense, i.e., all dual objects are *restricted*).

We attack this problem here by means of the so-called *canonical pairing*. Such a pairing can be defined on any algebra $\mathbf{A}(S)$ (an algebra $\mathbf{A}(S)$ equipped with such a pairing is called *dual quasitriangular*). Nevertheless, only under some additional conditions this pairing can be descended to the quantum cogroup $\mathrm{Fun}(SL(S))$. We show here that this condition is satisfied for the quantum cogroup $\mathrm{Fun}(SL(S))$ related to TL algebras.

Moreover, we show that in this case the canonical pairing is nondegenerate when restricted to the span of the generators of this algebra. This is the main difference between the quasiclassical and nonquasiclassical cases: in the former case this pairing is degenerate. This is a reason why we cannot introduce an object dual to $\mathrm{Fun}_q(SL(n))$ by means of this pairing. Finally, following the paper (Reshetikhin and all, 1990), in this case we must introduce an additional pairing (associated in a similar way to the YB operator S^{-1}). And the above mentioned deformed Cartan–Weyl basis in $U_q(\mathfrak{g})$ can be constructed by means of both pairings (for this construction the reader is referred to (Reshetikhin and all, 1990), cf. also Remark 1).

As for the nonquasiclassical case, we can equip the basic space $\mathbf{V}$ (using nondegeneracy of the canonical pairing in the mentioned sense) with a structure of a $\mathrm{Fun}(SL(S))$-module. Moreover, we can equip any tensor power of the space $\mathbf{V}$ with such a structure. Finally, we get a new tool to study tensor categories generated by such spaces.

The paper is organized as follows. In the next Section we introduce the notion of a dual quasitriangular structure and describe the one connected with the quantum matrix algebras in terms of the canonical pairing. In Section 3 we give the condition ensuring the existence of such a structure on the algebra $\mathrm{Fun}(SL(S))$ mentioned above and in Section 4 we show that this condition is satisfied for a large family of such algebras related to TL algebras. We conclude the paper with a proof of the nondegeneracy of the canonical pairing (in the above sense) for algebras from this family (Section 5) and with a discussion of a hypothetical representation theory for the algebra $\mathrm{Fun}(SL(S))$ (Section 6).

2. Dual quasitriangular structure

The notion of a dual quasitriangular bialgebra (in particular, a Hopf algebra) was introduced by Sh. Majid (see (Majid, 1995) and the references therein) as a dualization of the notion of a quasitriangular bialgebra due to V. Drinfeld. By definition, a dual quasitriangular bialgebra is a bialgebra equipped with some pairing similar to the one defined on the cogroups

$\mathrm{Fun}_q(G)$ by means of the quantum universal R-matrix $\mathcal{R}$:

$$a \otimes b \to \langle\langle a, b \rangle\rangle = \langle a \otimes b, \mathcal{R} \rangle, \ a, \ b \in \mathrm{Fun}_q(G).$$

Here the pairing $\langle \ , \ \rangle$ is that between $\mathrm{Fun}_q(G)$ and $U_q(\mathfrak{g})$ extended to their tensor powers.

More precisely, one says that a bialgebra (or a Hopf algebra) $\mathcal{A}$ is equipped with a *dual quasitriangular structure* and it is called a *dual quasitriangular bialgebra* (or *Hopf algebra*), if it is endowed with a pairing

$$\langle\langle \ , \ \rangle\rangle : \mathcal{A}^{\otimes 2} \to K$$

satisfying the following axioms

$$
\begin{array}{ll}
\text{(i)} & \langle\langle a, bc \rangle\rangle = \langle\langle a_{(1)}, c \rangle\rangle \langle\langle a_{(2)}, b \rangle\rangle, \\
\text{(ii)} & \langle\langle ab, c \rangle\rangle = \langle\langle a, c_{(1)} \rangle\rangle \langle\langle b, c_{(2)} \rangle\rangle, \\
\text{(iii)} & \langle\langle a_{(1)}, b_{(1)} \rangle\rangle a_{(2)} b_{(2)} = b_{(1)} a_{(1)} \langle\langle a_{(2)}, b_{(2)} \rangle\rangle, \\
\text{(iv)} & \langle\langle 1, a \rangle\rangle = \varepsilon(a) = \langle\langle a, 1 \rangle\rangle
\end{array}
$$

for all $a, b, c \in \mathcal{A}$, where $\varepsilon : \mathcal{A} \to K$ is the counit of $\mathcal{A}$ and $\Delta : \mathcal{A} \to \mathcal{A}^{\otimes 2}$, $\Delta(a) = a_{(1)} \otimes a_{(2)}$, is the coproduct. If $\mathcal{A}$ is a Hopf algebra and $\gamma : \mathcal{A} \to \mathcal{A}$ is its antipode, we impose a complementary axiom

$$\text{(v)} \quad \langle\langle a, b \rangle\rangle = \langle\langle \gamma(a), \gamma(b) \rangle\rangle.$$

(If the pairing is invertible in sense of (Majid, 1995), p. 48, the axioms (i)–(iii) imply those (iv) and (v), cf (Majid, 1995).)

Let us note that the axioms (i), (ii), (iv), (v) mean that the product (respectively, the coproduct, the unit, the counit, the antipode) of the algebra $\mathcal{A}$ is dual to the coproduct (respectively, the product, the counit, the unit, the antipode) of the algebra $\mathcal{A}^{op}$, where $\mathcal{A}^{op}$ denotes as usually the bialgebra $\mathcal{A}$ whose product is replaced by the opposite one. So, in fact, we have the pairing of bialgebras (Hopf algebras)

$$\langle\langle \ , \ \rangle\rangle : \mathcal{A} \otimes \mathcal{A}^{op} \to K. \tag{1}$$

In some sense the notion of a dual quantum bialgebra is more fundamental than that of quasitriangular one for the following reason. It is well known that the most popular construction of a quasitriangular Hopf algebra is given by the famous Drinfeld–Jimbo quantum group $U_q(\mathfrak{g})$. Usually it is introduced by means of the Cartan–Weyl generators $\{H_\alpha, X_{\pm\alpha}\}$ and certain relations between them which are quantum (or q-) analogues of the ordinary ones. However, this approach is valid only in the quasiclassical case.

In the general case, including the nonquasiclassical objects, we should first introduce dual quasitriangular bialgebras (or Hopf algebras) and only after that we can proceed to introduce their dual objects. Moreover, an explicit description of the latter objects depends on the properties of the canonical pairing and they are not similar in quasiclassical and nonquasiclassical cases.

Let us describe now a regular way to introduce the dual quasitriangular bialgebras (and Hopf algebras) associated to the YB operators discussed above. Let $\mathbf{V}$ be a linear space equipped with a nontrivial Yang–Baxter operator $S : \mathbf{V}^{\otimes 2} \to \mathbf{V}^{\otimes 2}$. Let us fix a basis $\{x_i\}$ in $\mathbf{V}$ and denote by S_{ij}^{kl} the entries of the operator S ($S(x_i \otimes x_j) = S_{ij}^{kl} x_k \otimes x_l$). From here on summation over repeated indices is assumed.

Let us consider a matrix t with entries t_k^l, $1 \le k, l \le n = \dim \mathbf{V}$. The bialgebra $\mathbf{A}(S)$ of *quantum matrices* associated to S is defined as the algebra generated by 1 and n^2 indeterminates $\{t_k^l\}$ satisfying the following relations

$$S(t \otimes t) = (t \otimes t)S, \text{ or in a basis form } S_{ij}^{mn} t_m^p t_n^q = t_i^u t_j^v S_{uv}^{pq}.$$

This algebra possesses a bialgebra structure, being equipped with the co-matrix coproduct $\Delta(1) = 1 \otimes 1$, $\Delta(t_i^j) = t_i^p \otimes t_p^j$ and the counit $\varepsilon(1) = 1$, $\varepsilon(t_i^j) = \delta_i^j$.

This is just the famous "RTT=TTR" bialgebra introduced in (Reshetikhin and all, 1990). Let us fix $c \in K$, $c \ne 0$, and equip this algebra with a dual quasitriangular structure by setting

$$\langle\langle 1, t_i^k \rangle\rangle_c = \delta_i^k = \langle\langle t_i^k, 1 \rangle\rangle_c \text{ and } \langle\langle t_i^k, t_j^l \rangle\rangle_c = c\, S_{ji}^{kl}$$

and extending the pairing to the whole $\mathbf{A}(S)^{\otimes 2}$ by using the above axioms (i), (ii) and (iv).

We leave to the reader to check that this extension is well defined (here it is precisely the QYBE that plays the crucial role) and, moreover, the axiom (iii) is satisfied as well.

The pairing $\langle\langle \,,\, \rangle\rangle_c$ will be called *canonical*.

Let us note that such a canonical pairing is usually considered with $c = 1$. However, we need this complementary "degree of freedom" to make the pairing $\langle\langle \,,\, \rangle\rangle_c$ compatible with the equation $\det t = 1$ (cf. below). Another (but equivalent) way consists in replacing the Hecke symmetry S by cS. We drop the index c if $c = 1$.

Thus, the bialgebra $\mathbf{A}(S)$ can be *canonically* equipped with a dual quasitriangular structure.

Nevertheless, this bialgebra does not possess any Hopf algebra structure since no antipode is defined in it. To get a Hopf algebra, we must either impose the complementary equation det $t = 1$, i.e., pass to the quotient of the algebra $\mathbf{A}(S)$ by the ideal generated by the element det $t - 1$ (assuming the determinant det t to be well defined)[2] or add to the algebra a new generator $\det^{-1}$. In the latter case we obtain Hopf algebras (quantum cogroup) of GL type (cf. (Gurevich, 1991)).

In any case it is necessary to check that the above dual quasitriangular structure on the algebra $\mathbf{A}(S)$ can be transferred to the final quantum cogroup. As for standard quantum function algebras $\mathrm{Fun}_q(G)$ dual to the quantum groups $U_q(\mathfrak{g})$ (for a classical simple Lie algebra $\mathfrak{g}$) this follows automatically by duality. However, in general this is no longer true. In Section 3 we will give a necessary and sufficient condition ensuring the existence of a dual quasitriangular structure on an SL type quantum cogroup.

Thus, it is possible to associate a dual quasitriangular bialgebra to any YB operator and a dual quasitriangular Hopf algebra (of SL type) to some of them. These algebras look like the function algebras $\mathrm{Fun}(G)$ on a ordinary (semi) group G. This means that we can equip the space $\mathbf{V}$ with a (right to be concrete) comodule structure over the algebra $\mathrm{Fun}(SL(S))$ by

$$\Delta : \mathbf{V} \to \mathbf{V} \otimes \mathbf{A}(S), \quad \Delta(x_i) = x_j \otimes t_i^j.$$

Therefore any tensor power of the space $\mathbf{V}$ also becomes a right $\mathbf{A}(S)$-comodule.

However, these powers are not in general irreducible as comodules over the above coalgebra. Unfortunately, up to now no corepresentation theory of quantum cogroups in nonquasiclassical cases has been constructed yet (such a hypothetical theory in the case connected to the TL algebra is discussed in Section 6). It is worth saying that even in the quasiclassical case it is possible to use quantum cogroups $\mathrm{Fun}_q(G)$ instead of the quantum groups $U_q(\mathfrak{g})$ themselves. However, technically it is more convenient to work with the latter objects.

In the nonquasiclassical case an interesting problem arises: what is an appropriate description of the objects dual to the bialgebras $\mathbf{A}(S)$ or of their quotient of the SL type. If the canonical pairing (1) is nondegenerate, we can consider the bialgebra $\mathbf{A}(S)^{op}$ as the dual object to that $\mathbf{A}(S)$ (and similarly for their quotients of the SL type).

This is just (conjecturally) the case of the Hecke symmetries of TL type. We show that (at least for a large family of such symmetries) the canonical pairing, being restricted to the space $\mathbf{T} = \mathrm{Span}(t_i^j)$, is nondegenerate for a generic q. Nevertheless, this weak version of nondegeneracy is sufficient

---

[2]In the sequel we will restrict ourselves to quantum cogroups of SL type whose construction was suggested in (Gurevich, 1991). Quantum cogroups of SO or Sp type and the corresponding dual quasitriangular structures will be discussed elsewhere.

to equip the initial space $\mathbf{V}$ with a structure of a left $\mathbf{A}(S)^{op}$-module (and therefore, with that of a right $\mathbf{A}(S)$-module).

Having in mind the usual procedure (cf., e.g., (Majid, 1995)) we put

$$t_i^j \triangleright x_k = x_m \langle\!\langle t_k^m, t_i^j \rangle\!\rangle_c = c S_{ik}^{mj} x_m,$$

where $t_i^j \triangleright x_k$ denotes the result of applying the element $t_i^j \in \mathbf{A}(S)^{op}$ to $x_k \in \mathbf{V}$.

Remark 1 Let us observe that if the canonical pairing is degenerate, then the above action is still well defined, but $\mathbf{V}$ becomes reducible as an $\mathbf{A}(S)^{op}$-module since it contains the $\mathbf{A}(S)^{op}$-module $\text{Im}(\triangleright)$, where $\triangleright : \mathbf{A}(S)^{op} \otimes \mathbf{V} \to \mathbf{V}$ is the above map and this module is a proper submodule in $\mathbf{V}$.

It is just the case related to the quantum groups $U_q(\mathfrak{g})$. This is a reasons why one needs a complementary pairing. More precisely, let us introduce (following (Reshetikhin and all, 1990)) two sets of generators $(L^+)_i^j$ and $(L^-)_i^j$ and define the pairing between the spaces $\mathbf{L}^+ = \text{Span}((L^+)_i^j)$, $\mathbf{L}^- = \text{Span}((L^-)_i^j)$ and $\mathbf{T} = \text{Span}(t_i^j)$ as follows

$$\langle\!\langle t_i^k, (L^+)_j^l \rangle\!\rangle = S_{ji}^{kl}, \quad \langle\!\langle t_i^k, (L^-)_j^l \rangle\!\rangle = (S^{-1})_{ji}^{kl}.$$

In fact, in this way we have introduced a pairing between the spaces $\mathbf{L}^+ \oplus \mathbf{L}^-$ and $\mathbf{T}$. Of course, this pairing is degenerate on $\mathbf{L}^+ \oplus \mathbf{L}^-$, but it becomes nondegenerate on $\mathbf{T}$. There exists a natural way to extend the above pairing up to that $(\mathbf{L}^+ \oplus \mathbf{L}^-) \otimes \mathbf{A}(S) \to K$, cf. (Reshetikhin and all, 1990). Thus, the space $\mathbf{L}^+ \oplus \mathbf{L}^-$ is embedded into the algebra $\mathbf{A}(S)^*$ dual to $\mathbf{A}(S)$. The subalgebra of $\mathbf{A}(S)^*$ generated by 1 and the space $\mathbf{L}^+ \oplus \mathbf{L}^-$ is called in (Reshetikhin and all, 1990) the *algebra of regular functions* on $\mathbf{A}(S)$. (Moreover, in (Reshetikhin and all, 1990) the elements $(L^+)_i^j$ are expressed in terms of the generators of the quantum groups $U_q(\mathfrak{g})$.) In a similar way we can define such an algebra in the nonquasiclassical case under consideration, but since the canonical pairing is nondegenerate for a generic q, we restrict ourselves to the generators $(L^+)_i^j$.

3. Dual quasitriangular algebras related to the even Hecke symmetries

First, we recall some facts about the cogroups $\text{Fun}(SL(S))$ introduced in (Gurevich, 1991).

Let us fix an even Hecke symmetry $S : \mathbf{V}^{\otimes 2} \to \mathbf{V}^{\otimes 2}$ of rank $p \geq 2$. Let us denote by $P_-^{(p)}$ the projector of $\mathbf{V}^{\otimes p}$ onto its skew-symmetric component $\wedge_-^p(\mathbf{V})$ (an explicit form of this projector is given in (Gurevich, 1991)).

Then by definition $\dim \operatorname{Im} P_{-}^{(p)} = 1$ and (assuming a base $\{x_i\} \in \mathbf{V}^{\otimes p}$ to be fixed)

$$P_{-}^{(p)} x_{i_1} x_{i_2} \ldots x_{i_p} = u_{i_1 i_2 \ldots i_p} v^{j_1 j_2 \ldots j_p} x_{j_1} x_{j_2} \ldots x_{j_p}$$

with $u_{i_1 i_2 \ldots i_p} v^{i_1 i_2 \ldots i_p} = 1$ (hereafter we drop the sign $\otimes$).

The tensors $U = (u_{i_1 i_2 \ldots i_p})$ and $V = (v^{j_1 j_2 \ldots j_p})$ are quantum analogues of the Levi–Civita ones.

Let us consider the bialgebra $\mathbf{A}(S)$ corresponding to the given even Hecke symmetry S and introduce a distinguished element in it:

$$\det t = u_{i_1 i_2 \ldots i_p} t_{j_1}^{i_1} \ldots t_{j_p}^{i_p} v^{j_1 j_2 \ldots j_p}.$$

In (Gurevich, 1991) it was shown that this element is group-like, i.e.,

$$\Delta\,(\det\,t) = \det t \otimes \det t.$$

It was called a *quantum determinant.*

Under the additional condition that this determinant is central (in general this is not so), we introduce an analogue $\operatorname{Fun}(SL(S))$ of the quantum functional algebra $\operatorname{Fun}_q(SL(n))$ as the quotient algebra of $\mathbf{A}(S)$ over the ideal generated by the element $\det t - 1$. This quotient inherits a bialgebra structure but, moreover, it possesses a Hopf structure (for an explicit description of the antipode, the reader is referred to (Gurevich, 1991)). Our intermediate aim is to study whether it is possible to equip the algebra $\operatorname{Fun}(SL(S))$ with a dual quasitriangular structure.

It is evident that the dual quasitriangular structure on $\mathbf{A}(S)$ defined above can be descended to $\operatorname{Fun}(SL(S))$ iff

$$\langle\langle \det t, a \rangle\rangle_c = \varepsilon(a) = \langle\langle a, \det t \rangle\rangle_c \tag{2}$$

for any $a \in \mathbf{A}(S)$. Using the fact that the quantum determinant is a group-like element, it is possible to show that these relations are valid for any a if they are true for $a = t_i^j$. (As for $a = 1$, relation (2) follows immediately from $u_{i_1 i_2 \ldots i_p} v^{i_1 i_2 \ldots i_p} = 1$.) Moreover, we have

Proposition 1 *We have the following relations*

$$\langle\langle t_k^l, \det t \rangle\rangle_c = c^p(-1)^{p-1} q p_q M_k^l, \quad \langle\langle \det t, t_k^l \rangle\rangle_c = c^p(-1)^{p-1} q p_q N_k^l, \tag{3}$$

where $M_k^l = u_{i_1 i_2 \ldots i_{p-1} k} v^{l i_1 i_2 \ldots i_{p-1}}$, $N_k^l = u_{k i_1 i_2 \ldots i_{p-1}} v^{i_1 i_2 \ldots i_{p-1} l}$ *and* $p_q = 1 + q + \ldots + q^{p-1}$. *(Let us note that the operators $M = (M_k^l)$ and $N = (N_k^l)$ have been introduced in (Gurevich, 1991), p. 816.)*

Proof. By axiom (i) we have

$$\langle\langle t_k^l, \det t\rangle\rangle_c = c^p u_{i_1 i_2 \ldots i_p} v^{j_1 j_2 \ldots j_p} \langle\langle t_k^{m_1}, t_{j_p}^{i_p}\rangle\rangle \langle\langle t_{m_1}^{m_2}, t_{j_{p-1}}^{i_{p-1}}\rangle\rangle \ldots$$

$$\langle\langle t_{m_{p-1}}^l, t_{j_1}^{i_1}\rangle\rangle = c^p u_{i_1 i_2 \ldots i_p} v^{j_1 j_2 \ldots j_p} S_{j_p k}^{m_1 i_p} S_{j_{p-1} m_1}^{m_2 i_{p-1}} \ldots S_{j_1 m_{p-1}}^{l i_1}.$$

The term $u_{i_1 i_2 \ldots i_p} v^{j_1 j_2 \ldots j_p} S_{j_p k}^{m_1 i_p} S_{j_{p-1} m_1}^{m_2 i_{p-1}} \ldots S_{j_1 m_{p-1}}^{l i_1}$ was found in (Gurevich, 1991) while commuting the elements $V = v^{j_1 j_2 \ldots j_p} x_{j_1} x_{j_2} \ldots x_{j_p}$ and x_k (cf. Proposition 5.7 from (Gurevich, 1991)) and is equal to $(-1)^{p-1} q p_q M_k^l$. This proves the first equality. The second one can be proved in the same way using the commutation law for the elements x_k and $V = v^{j_1 j_2 \ldots j_p} x_{j_1} x_{j_2} \ldots x_{j_p}$.
□

Corollary 1 *Equations* (2) *can be satisfied for some $c \in K$ iff the operators M and N are scalar (this property is equivalent by virtue of Proposition 5.9 from (Gurevich, 1991) to the quantum determinant being central) and, moreover, $M = N$. More precisely, if $M = m \, \mathrm{id}$, $N = n \, \mathrm{id}$, $m, n \in K$, and $m = n$, we can satisfy the relations*

$$\langle\langle t_k^l, \det t\rangle\rangle_c = \delta_k^l = \langle\langle \det t, t_k^l\rangle\rangle_c$$

by putting $c^p = (-1)^{p-1} q^{-1} p_q^{-1} m^{-1}$.

Let us note that the operators M and N satisfy the relation $MN = q^{p-1} p_q^{-2} \, \mathrm{id}$ (cf. (Gurevich, 1991)). Thus, if $M = m \, \mathrm{id}$, $N = n \, \mathrm{id}$, the relation $M = N$ is equivalent to

$$m^2 = q^{p-1} p_q^{-2}. \tag{4}$$

Thus, we have reduced the problem of describing the quantum cogroups $\mathrm{Fun}(SL(S))$ allowing a dual quasitriangular structure to the classification problem of all even Hecke symmetries such that the corresponding operator M is scalar, $M = m \, \mathrm{id}$, with m satisfying (4). In the next section we will consider this problem for Hecke symmetries of TL type.

Remark 2 Let us observe that if the operators M and N are not scalar, one cannot define the algebra $\mathrm{Fun}(SL(S))$, but it is possible to define the algebra $\mathrm{Fun}(GL(S))$ by introducing a new generator $\det^{-1}$ satisfying the relations $\det^{-1} \det t = 1$, and the commutation law of $\det^{-1}$ with other generators arising from this relation (cf. (Gurevich, 1991)). Moreover, it is possible to extend the canonical pairing up to that defined on $\mathrm{Fun}(GL(S))$ by setting

$$\langle\langle t_i^p, \det^{-1}\rangle\rangle\langle\langle t_p^j, \det t\rangle\rangle = \delta_i^j, \ \langle\langle \det^{-1}, t_i^p\rangle\rangle\langle\langle \det t, t_p^j\rangle\rangle = \delta_i^j.$$

The details are left to the reader.

Remark 3 Let us observe that if an even Hecke symmetry is of TL type and M is scalar, then $M = N$ since in this case $M = VU$ and $N = UV$ (cf. Section 4). Therefore, if the algebra $\mathrm{Fun}(SL(S))$ is well defined (i.e., the corresponding quantum determinant is central), it automatically has a canonical dual quasitriangular structure. It is not clear whether there exists a Hecke symmetry of rank $p > 2$ such that the algebra $\mathrm{Fun}(SL(S))$ is well defined (i.e., $M = m\,\mathrm{id}$) but the factor m does not satisfy the relation (4) and therefore the corresponding canonical pairing is not compatible with the equation $\det t = 1$.

4. The TL algebra case

Now let us consider the case related to TL algebras. In this case it is possible to give an exhausting classification of the corresponding Hecke symmetries.

Indeed, it is easy to see (cf. (Gurevich, 1991)) that any even Hecke symmetry of TL type can be expressed by means of the Levi–Civita tensors $U = (u_{ij})$ and $V = (v^{kl})$ in the following way

$$S^{kl}_{ij} = q\delta^k_i\delta^l_j - (1 + q)u_{ij}v^{kl}.$$

Then the QYBE and the Hecke second degree relation are equivalent to the system

$$\mathrm{tr}\, UV^t = 1, \quad UVU^tV^t = q(1 + q)^{-2}\,\mathrm{id}. \qquad (5)$$

Hereafter $U \mapsto U^t$ is the transposition operator. Thus, $\mathrm{tr}\, UV^t = u_{ij}v^{ij}$.

Introducing the matrix $Z = (1 + q)VU^t$ ($z^j_i = (1 + q)v^{jk}u_{ik}$) and using the fact that the second relation of (5) can be represented in form $V^tU^tVU = q(1 + q)^{-2}\,\mathrm{id}$, we can reduce the relations (5) to the form

$$(Z^t)^{-1}q = V^{-1}ZV, \quad \mathrm{tr}\, Z = 1 + q. \qquad (6)$$

The family of all solutions to the QYBE over the field $K = \mathbb{C}$ is described by the following

Proposition 2 (Gurevich, 1991) *The pair (Z, V) is a solution of system (6) iff the matrix Z is such that $\mathrm{tr}\, Z = 1 + q$ and for any cell corresponding to an eigenvalue x in the Jordan form, there is an analogous cell with eigenvalue q/x (with the same multiplicity).*

Remark 4 Let us note that U and V are transformed under changes of base as bilinear form matrices, while Z is transformed as an operator matrix (their transformations are coordinated and relations (6) are stable). So, assuming $K = \mathbb{C}$, we can represent the operator Z in Jordan form by an appropriate choice of base. Moreover, we can assume that the cells with eigenvalues x and q/x are in positions symmetric to each other with respect to the center of the matrix Z. Observe that if the number of the cells is odd, the eigenvalue of the middle one is $\pm\sqrt{q}$.

It is not difficult to see that for such a choice of base the tensor V can be taken in the form of a skew-diagonal matrix (i.e., possessing nontrivial terms only at the auxiliary diagonal). Let us fix such a matrix V_0 and note that all other V satisfying (6) are of the form $V = WV_0$, where W commutes with Z. In the sequel we assume that a base possessing these properties is fixed.

Let us observe that in case under consideration we have $M = UV$, $N = VU$. Moreover, relations (4) take the form $m^2 = q(1 + q)^{-2}$. Using the relation $U^t = (1 + q)^{-1}V^{-1}Z$, we can transform the equality $UV = m\,\mathrm{id}$ to

$$Z = (1 + q)mV(V^t)^{-1}. \tag{7}$$

Let us assume that Z has a simple spectrum, i.e., its eigenvalues are pairwise distinct. So, its Jordan form is diagonal: $Z = \mathrm{diag}(z_1, \ldots, z_n)$. The family of diagonal Z, satisfying conditions of Proposition 2 and fulfilling the only relation $\mathrm{tr}\, Z = 1 + q$, can be parametrized by $(z_1, \ldots, z_r)$ with $r = n/2$, if n is even, and with $r = (n - 1)/2$, if n is odd (if n is odd we have also a choice for the value of $z_{r+1} = \pm\sqrt{q}$).

Since Z has a simple spectrum, any W commuting with Z is also diagonal (with arbitrary diagonal entries). This implies that V satisfies (6) iff it is skew-diagonal with arbitrary entries at the auxiliary diagonal. Therefore U is also skew-symmetric. Thus, we have $v^{ij} \neq 0$, $u_{ij} \neq 0$ iff $i + j = n$. For the sake of simplicity, we will use the notation v^i (u_i) instead of $v^{i\,n+1-i}$ ($u_{i\,n+1-i}$). Let us note that $z_i = (1 + q)u_iv^i$ (up to the end of this section there is no summation over repeated indices).

It is easy to see that relation (7) is satisfied iff the entries v_i fulfill the system

$$m(1 + q)v^i/v^{n-i+1} = z_i, \quad 1 \le i \le n. \tag{8}$$

This system is consistent by virtue of the relations

$$z_i z_{n-i+1} = q, \quad m^2 = q(1 + q)^{-2}.$$

Moreover, the family of the solutions of the system (8) can be parametrized by $(v_1, \ldots, v_r)$. Let us note that if n is odd, the value of $z_{r+1} = \pm\sqrt{q}$ depends on that of $m = \pm\sqrt{q}(1 + q)^{-1}$, namely, we have $z_{r+1} = m(1 + q)$.

Thus, we have proved the following

Proposition 3 Let $K = \mathbb{C}$, S be a Hecke symmetry of TL type and Z be the corresponding tensor described in Proposition 2 with a simple spectrum (a parametrization of all such tensors Z was given above). Then the dual quasitriangular structure defined on the algebra $\mathbf{A}(S)$ can be descended on *the quantum cogroup* $\mathrm{Fun}(SL(S))$ *iff V is a skew-diagonal matrix with the entries $v_{i\,n+1-i} = v_i$ satisfying system (8). This system is always compatible and the family of its solutions can be parametrized as above.*

Remark 5 Let us observe that the Hecke symmetries of TL type such that the operator UV is scalar are just those introduced in (Dubois-Violette, Launer, 1990) (the authors of (Dubois-Violette, Launer, 1990) use another normalization of the operator S).

5. Nondegeneracy of the canonical pairing

In the present section we show that when $n = \dim \mathbf{V} > 2$ and q is generic, the canonical pairing $\langle\langle\ ,\ \rangle\rangle_c$ is nondegenerate for those even Hecke symmetries of TL type whose operator Z has a generic simple spectrum. As above, we assume that Z has a diagonal form in a chosen base and therefore the tensors U and V are skew-diagonal. In the sequel we put $c = 1$.

Thus we have the pairing $\langle\langle t_i^j, t_k^l \rangle\rangle = S_{ki}^{jl}$. To show that it is nondegenerate we will compute the Gram determinant, i.e., the determinant of the Gram matrix. The rows and the columns of this matrix are labeled by the bi-index (i, j) running over the set

$$(1,1), \ldots, (1,n), (2,1), \ldots, (2,n), \ldots, (n,1), \ldots, (n,n).$$

So, the term $\langle\langle t_i^j, t_k^l \rangle\rangle = R_{ik}^{jl} = S_{ki}^{jl}$ is situated at the intersection of the (i,j)-row and the (k,l)-column.

Let us note that if S is a Hecke symmetry of TL type, then all the entries of the matrix S_{ki}^{jl} are equal to zero unless either $i = j$, $k = l$ or $i + k = j + l = n + 1$. So, we have just two nonzero elements in the (i,j)-row, namely, R_{ij}^{ji} and $R_{i\,n+1-i}^{j\,n+1-j}$, if $i + j \neq n + 1$, and only one, namely, $R_{i\,n+1-i}^{n+1-i\,i}$, if $i + j = n + 1$. A similar statement is valid for the columns.

This yields that if $i + j \neq n+1$, then the (i,j)- and $(n+1-j, n+1-i)$-rows and (j,i)- and the $(n + 1 - i, n + 1 - j)$-columns possess just four nontrivial elements

$$R_{ij}^{ji},\ R_{i\,n+1-i}^{j\,n+1-j},\ R_{n+1-j\,j}^{n+1-i\,i},\ R_{n+1-j\,n+1-i}^{n+1-i\,n+1-j}$$

situated at their intersections. If $i + j = n + 1$, then two rows (columns) are merged into one and the only nontrivial element $R_{i\,n+1-i}^{n+1-i\,i}$ belongs to the intersection of the $(i, n + 1 - i)$-row and the $(n + 1 - i, i)$-column.

For example, for $n = 3$ we have the following Gram matrix

$$\mathbf{G} = \begin{pmatrix} R_{11}^{11} & 0 & 0 & 0 & 0 & 0 & 0 & 0 & R_{13}^{13} \\ 0 & 0 & 0 & R_{12}^{21} & 0 & 0 & 0 & R_{13}^{22} & 0 \\ 0 & 0 & 0 & 0 & 0 & 0 & R_{13}^{31} & 0 & 0 \\ 0 & R_{21}^{12} & 0 & 0 & 0 & R_{22}^{13} & 0 & 0 & 0 \\ 0 & 0 & 0 & 0 & R_{22}^{22} & 0 & 0 & 0 & 0 \\ 0 & 0 & 0 & R_{22}^{31} & 0 & 0 & 0 & R_{23}^{32} & 0 \\ 0 & 0 & R_{31}^{13} & 0 & 0 & 0 & 0 & 0 & 0 \\ 0 & R_{31}^{22} & 0 & 0 & 0 & R_{32}^{23} & 0 & 0 & 0 \\ R_{31}^{31} & 0 & 0 & 0 & 0 & 0 & 0 & 0 & R_{33}^{33} \end{pmatrix}.$$

By changing the order of the rows and columns, we can reduce this matrix to a block-diagonal form, where all blocks are either one-dimensional and consist of elements $R_{i\,n+1-i}^{n+1-i\,i}$ or two-dimensional and have the following form

$$\begin{pmatrix} R_{ij}^{ji} & R_{i\,n+1-i}^{j\,n+1-j} \\ R_{n+1-j\,j}^{n+1-i\,i} & R_{n+1-j\,n+1-i}^{n+1-i\,n+1-j} \end{pmatrix}.$$

Denote by $I(n)$ the set of indices i, j satisfying $1 \leq i, j \leq n, i + j \neq n + 1$. Then $\det \mathbf{G}$ is equal up to a sign to

$$\prod_{1 \leq i \leq n} R_{i\,n+1-i}^{n+1-i\,i} \sqrt{\left| \prod_{I(n)} \det \begin{pmatrix} R_{ij}^{ji} & R_{i\,n+1-i}^{j\,n+1-j} \\ R_{n+1-j\,j}^{n+1-i\,i} & R_{n+1-j\,n+1-i}^{n+1-i\,n+1-j} \end{pmatrix} \right|}.$$

(The root is motivated by the fact that any factor in the second product is taken two times.)

By straightforward calculations it is not difficult to see that

$$\prod_{1 \leq i \leq n} R_{i\,n+1-i}^{n+1-i\,i} = \prod_{1 \leq i \leq n} (q - z_i)$$

and

$$\prod_{I(n)} \det \begin{pmatrix} R_{ij}^{ji} & R_{i\,n+1-i}^{j\,n+1-j} \\ R_{n+1-j\,j}^{n+1-i\,i} & R_{n+1-j\,n+1-i}^{n+1-i\,n+1-j} \end{pmatrix} = \prod_{I(n)} (q^2 - z_{n+1-i} z_j).$$

Thus, we have proven the following

Proposition 4

$$(\det \mathbf{G})^2 = \prod_{1 \leq i \leq n} (q - z_i)^2 \prod_{I(n)} (q^2 - z_{n+1-i} z_j).$$

It is interesting to observe that the final expression depends only on the matrix Z. This enables us to state the following

Proposition 5 *Let us assume that $S : \mathbf{V}^{\otimes 2} \to \mathbf{V}^{\otimes 2}$ is an even Hecke symmetry of TL type, the corresponding operator Z possesses a simple spectrum and $n = \dim \mathbf{V} \geq 4$. Then for a generic q and for a generic Z (of such type) the canonical pairing is nondegenerate.*

Proof. The set where the determinant $\det \mathbf{G}$ vanishes is an algebraic variety in the space $\mathbb{C}^{n+1}$ generated by the indeterminates z_i and q. It suffices to show that this variety is not contained in that defined by

$$z_i\, z_{n+1-i} = q, \quad \sum z_i = 1 + q.$$

Let us decompose the expression

$$\prod_{I(n)} (q^2 - z_{n+1-i} z_j)$$

into the product of factors with $i = j$ and those with $i \neq j$.

If $i = j$, we have $z_{n+1-i} z_j = q$. Thus, the above product is equal to

$$(q^2 - q)^n \prod_{I(n),\, i \neq j} (q^2 - z_{n+1-i} z_j).$$

So, for q such that $q \neq 0, q \neq 1$, we have $\det \mathbf{G} = 0$ iff $z_i = q$ for some i or $q^2 = z_{n+1-i} z_j$ for some $i \neq j$.

From the above parametrization it is evident that if $n \geq 4$ there exists a matrix $Z = \mathrm{diag}(z_1, \ldots, z_n)$ satisfying the conditions of Proposition 2 and such that $\det \mathbf{G} \neq 0$.

Let us consider the case $n = 3$ separately. In this case the set of such diagonal matrices Z is parametrized by z_1 (after choosing a value of $z_2 = \pm\sqrt{q}$) satisfying the equation $z_1 \pm \sqrt{q} + q/z_1 = 1 + q$. This equation has two solutions for any choice of the value $\pm\sqrt{q}$. It is not difficult to see that for a generic q we have $\det \mathbf{G} \neq 0$ for any of these four values of z_1. $\square$

Let us note that the canonical pairing becomes degenerate if $q = 1$ (this case corresponds to an involutive symmetry $(S^2 = \mathrm{id})$) or if $n = 2$, since in this case the system $z_1 z_2 = q$, $z_1 + z_2 = 1 + q$ has two solutions $z_1 = 1, z_2 = q$ and $z_1 = q, z_2 = 1$ for which the product $(q - z_1)(q - z_2)$ vanishes. And we always have $\det \mathbf{G} = 0$.

This is the principal reason why the latter case $(n = 2)$ which corresponds to a quasiclassical Yang–Baxter operator S (it is in fact the only quasiclassical case related to the TL algebra) differs crucially from the non-quasiclassical ones $(n > 2)$.

Thus, according to the above construction, we can convert the right $\mathrm{Fun}(SL(S))$-comodules $\mathbf{V}^{\otimes m}$ into left $\mathrm{Fun}(SL(S))^{op}$-modules (assuming

S to be an even Hecke symmetry of TL type, q to be generic and Z to have a simple spectrum, also generic) and therefore into right $\mathrm{Fun}(SL(S))$-modules.

6. Discussion of a possible representation theory

Our further aim is to construct some representation theory of the algebra $\mathrm{Fun}(SL(S))$ equipped with the above action. Conjecturally, it looks like that of $SL(2)$. Let us denote by V_m the symmetric component of $\mathbf{V}^{\otimes m}$. (Let us note that in classical and quasiclassical cases $m/2$ is just the spin of the representation $U(sl(2)) \to \mathrm{End}(V_m)$ or $U_q(sl(2)) \to \mathrm{End}(V_m)$.)

It seems very plausible that similarly to the $SL(2)$- or $U_q(sl(2))$-case we have for a generic q the following properties

- the $\mathrm{Fun}(SL(S))$-modules V_m are irreducible,
- any irreducible finite-dimensional $\mathrm{Fun}(SL(S))$-module is isomorphic to one of V_m,
- any finite-dimensional $\mathrm{Fun}(SL(S))$-module is completely reducible,
- we have the classical formula $V_i \otimes V_j = \oplus_{|i-j|\leq k\leq i+j} V_k$.

To motivate the latter formula, let us show that it is satisfied at least "in sense of dimensions", i.e.,

$$\dim V_i \otimes \dim V_j = \sum_{|i-j|\leq k\leq i+j} \dim V_k. \tag{9}$$

Indeed, using the fact that the Poincaré series of the symmetric algebra of the space $\mathbf{V}$ is equal to $(t^2 - nt + 1)^{-1}$, one can see that

$$\dim V_i = \alpha^i + \alpha^{i-2} + \alpha^{i-4} + \ldots + \alpha^{-i},$$

where $\alpha = n/2 + \sqrt{(n/2)^2 - 1}$ is a root of the equation $t^2 - nt + 1 = 0$. Then relation (9) can be established by straightforward calculations. The details are left to the reader.

References

Chari, V. and Pressley A.: *A Guide to Quantum Groups*, Cambridge University Press, 1994.

Dubois-Violette, M. and Launer, G.: The quantum group of a nondegenerate bilinear form, *Physic Letters B* **245** (1990) 175-177.

Gurevich, D.: Algebraic aspects of the quantum Yang–Baxter equation, *Leningrad Math. J.* **2** (1991) 801–828.

Gurevich, D., Rubtsov, V. and Zobin, N.: Quantization of Poisson pairs: R-matrix approach, *JGP* **9** (1992) 25–44.

Majid, Sh.: *Foundations of Quantum Group Theory*, Cambridge University Press, 1995.

Reshetikhin, N., Takhtadzhyan, L. and Faddeev L.: Quantization of Lie groups and Lie algebras, *Leningrad Math. J.* **1** (1990) 193–226.

ON THE QUANTIZATION OF QUADRATIC
POISSON BRACKETS ON A POLYNOMIAL ALGEBRA OF
FOUR VARIABLES

J. DONIN

Department of Mathematics
Bar-Ilan University, 52900 Ramat-Gan, Israel

Abstract. Poisson brackets (P.b) are the natural initial terms for the deformation quantization of commutative algebras. The problem whether any Poisson bracket on the polynomial algebra of n variables can be quantized is open. By the Poincaré–Birkhoff–Witt (PBW) theorem any linear P.b. can be quantized for all n. On the other hand, it is easy to show that in the case $n = 2$ any P.b. is quantizable as well.

Quadratic P.b. appear as the initial terms for the quantization of polynomial algebras as quadratic algebras. The problem of the quantization of quadratic P.b. is also open. L. Makar-Limanov and the author proved that in the case $n = 3$ any quadratic P.b can be quantized. Moreover, this quantization is given as the quotient algebra of tensor algebra of three variables by relations which are similar to those in the PBW theorem.

In this note we consider a certain class of P.b. on the polynomial algebra of four variables that is associated to pairs of quadratic forms. We prove that all such brackets can be quantized but do not admit a PBW type quantization.
Mathematics subject classification (1991): 81R50, 17B37.

Key words: quantization, Poisson brackets, quadratic algebra.

1. Poisson brackets and deformations of commutative algebras

Let A be an associative algebra with unit over a field $\mathbf{k}$ of characteristic zero. We will consider deformations of A over the algebra of formal power series

B. P. Komrakov et al. (eds.), Lie Groups and Lie Algebras, 17–25.

$\mathbf{k}[[\hbar]]$ in the variable $\hbar$. In [1] the general deformation theory of algebras is developed.

By a deformation of A, we mean an algebra $A_\hbar$ over $\mathbf{k}[[\hbar]]$ that is isomorphic to $A[[\hbar]] = A\hat{\otimes}_{\mathbf{k}}\mathbf{k}[[\hbar]]$ as a $\mathbf{k}[[\hbar]]$-module (flatness condition) and $A_\hbar/\hbar A_\hbar = A$ (the symbol $\hat{\otimes}$ denotes the tensor product completed in the $\hbar$-adic topology). We will also denote A by A_0.

If $A'_\hbar$ is another deformation of A, we call the deformations $A_\hbar$ and $A'_\hbar$ equivalent, if there exists a $\mathbf{k}[[\hbar]]$-algebra isomorphism $A_\hbar \to A'_\hbar$ which induces the identity automorphism of A_0.

In other words, $A_\hbar$ consists of elements of the form

$$x = \sum_{i=0}^{\infty} x_i\hbar^i, \quad x_i \in A.$$

The multiplication in $A_\hbar$ is given by a $\mathbf{k}$-bilinear map $F_\hbar : A \times A \to A[[\hbar]]$ written as

$$F_\hbar(x,y) = \sum_{i\geq 0} \hbar^i F_i(x,y), \quad x,y \in A,$$

where $F_0(x,y) = xy$ is the multiplication in A. The terms F_i, $i > 0$, are $\mathbf{k}$-bilinear forms $A \times A \to A$. Associativity means that $F = F_\hbar$ satisfies the following equation

$$F(F(x,y),z) = F(x,F(x,y)).$$

Collecting terms in similar powers of $\hbar$, we obtain

$$\sum_{i+j=n} (F_i(F_j(x,y),z) - F_i(x,F_j(y,z))) = 0, \quad n \geq 0. \qquad (1)$$

If the multiplication in $A'_\hbar$ is given by the bilinear map $F'_\hbar : A \times A \to A[[\hbar]]$, then the equivalence of $A_\hbar$ and $A'_\hbar$ can be given as a power series

$$Q = \mathrm{Id} + \sum_{i\geq 1} \hbar^i Q_i,$$

where Q_i are $\mathbf{k}$-linear maps $Q_i : A \to A$, such that

$$F'(x,y) = Q^{-1}(F(Q(x),Q(y))). \qquad (2)$$

Let us consider the element F_1. From equation (1) for $n = 1$ we get the following relation

$$xF_1(y,z) - F_1(xy,z) + F_1(x,yz) - F_1(x,y)z = 0. \qquad (3)$$

This means that F_1 is a Hochschild 2-cocycle. From (1) for $n = 2$ we have

$$F_1(F_1(x, y), z) - F_1(x, F_1(y, z)) =$$
$$xF_2(y, z) - F_2(xy, z) + F_2(x, yz) - F_2(x, y)z,$$

i.e., the element $F_1(F_1(x, y), z) - F_1(x, F_1(yz))$ is 3-coboundary. If the series F' gives an equivalent deformation of A, then (2) implies

$$F_1'(x, y) - F_1(x, y) = xQ_1(y) - Q_1(xy) + Q_1(x)y,$$

i.e., $F_1'(x, y)$ and $F_1(x, y)$ are cohomologous.

Throughout the remainder of this paper we will be interested in deformations of commutative algebras. Such a deformation is called a (deformation) quantization, [2]. In this case it is easy to see that the form $\tilde{F}_1(x, y) = F_1(y, x)$ is a Hochschild cocycle as well, so the form $(x, y) = F_1(x, y) - F_1(y, x)$ is a skew-symmetric Hochschild cocycle. Furthermore, a straightforward computation shows that a skew-symmetric form is a Hochschild cocycle if and only if it satisfies the Leibnitz rule

$$(xy, z) = x(y, x) + y(x, z). \tag{4}$$

Moreover, an element from the Hochschild two-dimensional cohomology group can be represented by only one skew-symmetric cocycle, so this cocycle depends only on the class of equivalent deformations.

From the associativity of multiplication F it is easy to derive that (x, y) must satisfy the Jacobi identity

$$((x, y), z) + ((y, z), x) + ((z, x), y) = 0. \tag{5}$$

The skew-symmetric form $f : A \otimes A \to A$ satisfying the Leibnitz rule (4) and the Jacobi identity (5) is called a *Poisson bracket* on A. We say that the algebra $A_\hbar$ with the multiplication F is the quantization of the Poisson bracket f if $f(x, y) = F_1(x, y) - F_1(y, x)$.

Let A be an algebra of finite type, i.e., a quotient of the free algebra $T = T(x_1, ..., x_n)$ by relations R, $A = T/R$. One can obtain a deformation of A by taking some deformation of the relations, $R_\hbar$, and forming the algebra $A_\hbar = T[[\hbar]]/R_\hbar$. In this case we need to check the flatness condition, i.e., to verify that $A_\hbar$ is isomorphic to $A[[\hbar]]$ as a $\mathbf{k}[[h]]$-module.

2. Some examples of Poisson brackets and their quantizations

Further we suppose that $A = \mathbf{k}[x_1, ..., x_n]$, the polynomial algebra of n variables. Due to the Leibnitz rule, each Poisson bracket f is defined by a skew-symmetric matrix $\{f_{ij}\}$, where f_{ij} are polynomials such that

$$f(x_i, y_j) = f_{ij}. \tag{6}$$

Then, the Jacobi identity can be written as

$$\sum_r \left(\frac{\partial f_{ij}}{\partial x_r} f_{rk} + \frac{\partial f_{jk}}{\partial x_r} f_{ri} + \frac{\partial f_{ki}}{\partial x_r} f_{rj} \right) = 0.$$

We consider some examples of Poisson brackets on the polynomial algebra and their quantizations. One can find other examples in [5], [6].

1. *Determinant Poisson brackets.* For n polynomials $g_1, ..., g_n$ we denote by $J(g_1, ..., g_n)$ the Jacobi matrix $\{\partial g_i / \partial x_j\}$. Let us fix $n - 2$ polynomials $g_1, ..., g_{n-2}$ and define the bracket on A in the following way

$$f(a, b) = \det J(g_1, ..., g_{n-2}, a, b) \tag{7}$$

for $a, b \in A$. One can easily check that it is a Poisson bracket of rank two, i.e., the Hamiltonian vector fields $f(a, \cdot)$, $a \in A$, form at most a two-dimensional vector space at each point $(x_1, ..., x_n)$. We call it a *determinant Poisson bracket*. It is not known whether each determinant Poisson bracket can be quantized.

2. *Constant Poisson brackets.* Let f be defined by (6), with the f_{ij} being constants. Then the Jacobi identity holds. Denote by $T(V)$ the tensor algebra over the vector space V spanning $x_1, ..., x_n$. Denote by J the ideal in $T(V)[\hbar]$ generated by the elements $x_i x_j - x_j x_i - \hbar f_{ij}$. It follows from the Poincaré–Birkhoff–Witt (PBW) theorem that the algebra $A_\hbar = T(V)[\hbar]/J$ is the quantization of f.

3. *Linear Poisson brackets.* Let f is defined by (6) where f_{ij} are linear combinations of x_k, i.e., $f_{ij} = \sum_r a_{ij}^r x_r$, where a_{ij}^r are constants. If the Jacobi identity holds, we obtain a linear Poisson bracket on A. This bracket transforms the space V spanning x_i into a Lie algebra. Denote by J the ideal in $T(V)[\hbar]$ generated by the elements $x_i x_j - x_j x_i - \hbar f_{ij}$. Again, applying the PBW theorem, we see that the algebra $A_\hbar = T(V)[\hbar]/J$ is the quantization of f.

4. *Quadratic Poisson brackets.* Let f be defined by (6), where $f_{ij} = \sum_{pq} a_{ij}^{pq} x_p x_q$ are quadratic forms. We suppose that $a_{ij}^{pq} = a_{ij}^{qp}$. In this case f is called a *quadratic Poisson bracket*. The quadratic Poisson brackets are very important because they are the initial terms for the deformations of the polynomial algebra as a graded algebra, i.e., if $A_\hbar$ is assumed to be a graded algebra over $\mathbf{k}[[\hbar]]$. It is not known whether any quadratic Poisson bracket on the polynomial algebra of n variables can be quantized. But in [7] we have proved the following

Theorem 2.1 *Any quadratic Poisson bracket f on a polynomial algebra of three variables admits a PBW type quantization.*

We say that f admits a PBW type quantization, if the quantized algebra $A_\hbar$ is as in Examples 2 and 3, i.e., $A_\hbar = T(V)[\hbar]/J$, where the ideal J is generated by the quadratic elements $x_i x_j - x_j x_i - \hbar \sum_{pq} a_{ij}^{pq} x_p x_q$ (here the products $x_p x_q$ are considered in $T(V)$).

In the next section we give examples of quadratic Poisson brackets of the determinant form (7) on a polynomial algebra of four variables, which can be quantized but admit no quantization of PBW type.

Note that such Poisson brackets appear in two cases:

a) in (7) g_1 and g_2 are a linear and a cubic form, respectively;

b) both g_1 and g_2 are quadratic forms.

It is easy to see that the problem of quantizing Poisson brackets in the case a) can be reduced by change of variables to dimension three. So, using Theorem 2.1, we obtain

Proposition 2.2 *In case* a) *the corresponding quadratic Poisson bracket admits a PBW type quantization.*

As we will show in the next section, in case b) for the generic quadratic forms g_1, g_2 the corresponding Poisson brackets can be quantized but does not admit quantizations of PBW type.

3. Quantization of determinant quadratic Poisson brackets of type b)

From now on we assume the field **k** to be algebraically closed. Fix two quadratic forms g_1 and g_2 of four variables in general position. It is known that by a linear change of variables such forms can be simultaneously reduced to diagonal forms:

$$g_1 = \lambda_1 x_1^2 + \lambda_2 x_2^2 + \lambda_3 x_3^2 + \lambda_4 x_4^2, \quad g_1 = \mu_1 x_1^2 + \mu_2 x_2^2 + \mu_3 x_3^2 + \mu_4 x_4^2.$$

According to (7) the Poisson bracket in these coordinates has the form

$$\begin{aligned}
(x_1, x_2) &= 2\lambda_{12} x_3 x_4, & (x_2, x_3) &= 2\lambda_{23} x_1 x_4, \\
(x_1, x_3) &= 2\lambda_{13} x_2 x_4, & (x_2, x_4) &= 2\lambda_{24} x_1 x_3, \\
(x_1, x_4) &= 2\lambda_{14} x_2 x_3, & (x_3, x_4) &= 2\lambda_{34} x_1 x_2,
\end{aligned} \qquad (8)$$

where $\lambda_{ij} = \lambda_i - \mu_j$.

We will consider brackets of the form (8) with the generic λ_{ij} satisfying the condition $\lambda_{ij} = -\lambda_{ji}$. Then the Jacobi identity is equivalent to the condition

$$\lambda_{12}\lambda_{34} + \lambda_{13}\lambda_{42} + \lambda_{14}\lambda_{23} = 0. \qquad (9)$$

Denote $X = \lambda_{12}\lambda_{34}$, $Y = \lambda_{13}\lambda_{42}$, $Z = \lambda_{14}\lambda_{23}$ (in the product $\lambda_{ij}\lambda_{rs}$, the sequence $(ijrs)$ forms an even permutation). Then the Jacobi identity can be rewritten in the form

$$X + Y + Z = 0.$$

Let V be the four-dimensional vector space spanning x_i, $i = 1, ..., 4$. In the tensor algebra $T(V)$, we consider the ideal J generated by the quadratic elements

$$x_1 x_2 - x_2 x_1 - \lambda_{12}(x_3 x_4 + x_4 x_3), ..., x_3 x_4 - x_4 x_3 - \lambda_{34}(x_1 x_2 + x_2 x_1) \quad (10)$$

(each element is associated to the corresponding equality from (8)). Our aim is to determine a condition for λ_{ij} which must be satisfied in order that the homogeneous components of the algebra $T(V)/J$ have the same dimensions as those of the polynomial algebra (when $\lambda_{ij} = 0$ for all i, j). We do not suppose here that (9) holds. A similar condition is contained in [3] for arbitrary quadratic brackets (see also [4], example 8, p. 52). But in our setting we obtain that this condition is reduced only to one equation for λ_{ij} (see (12) below) which can be presented in explicit form.

We have the equalities

$$[x_1, x_2] = \lambda_{12}(x_3 x_4 + x_4 x_3) \mod J, ...$$

From this we obtain modulo J

$$
\begin{aligned}
[[x_1, x_2], x_3] &= \lambda_{12}[x_3 x_4 + x_4 x_3, x_3] = \lambda_{12}(x_3[x_4, x_3] + [x_4, x_3]x_3) = \\
&\lambda_{12}\lambda_{43}(x_3(x_1 x_2 + x_2 x_1) + (x_1 x_2 + x_2 x_1)x_3) = \\
&-X(x_3(x_1 x_2 + x_2 x_1) + (x_1 x_2 + x_2 x_1)x_3).
\end{aligned}
$$

Now it is easy to check the identity

$$
\begin{aligned}
x_3(x_1 x_2 + x_2 x_1) + (x_1 x_2 + x_2 x_1)x_3 = \\
\frac{1}{3}([[x_3, x_1], x_2] + [[x_3, x_2], x_1]) + \frac{2}{3}(x_1 x_2 x_3),
\end{aligned}
$$

where $(x_1 x_2 x_3)$ denotes symmetrization. So, we obtain modulo J

$$[[x_1, x_2], x_3] = -\frac{1}{3}X([[x_3, x_1], x_2] + [[x_3, x_2], x_1]) - \frac{2}{3}X(x_1 x_2 x_3). \quad (11)$$

In a similar way we have modulo J

$$
\begin{aligned}
{[[x_3, x_1], x_2]} &= -\frac{1}{3}Y([[x_2, x_3], x_1] + [[x_2, x_1], x_3]) - \frac{2}{3}Y(x_1 x_2 x_3), \\
{[[x_3, x_2], x_1]} &= \frac{1}{3}Z([[x_1, x_3], x_2] + [[x_1, x_2], x_3]) + \frac{2}{3}Z(x_1 x_2 x_3).
\end{aligned}
$$

Substituting in (11) we get modulo J

$$
\begin{aligned}
{[[x_1, x_2], x_3]} = &\ \tfrac{1}{9}XY([[x_2, x_3], x_1] + [[x_2, x_1], x_3]) - \\
&\ \tfrac{1}{9}XZ([[x_1, x_3], x_2] + [[x_1, x_2], x_3]) + \\
&\ \tfrac{2}{3}((-X + \tfrac{1}{3}XY - \tfrac{1}{3}XZ))(x_1 x_2 x_3).
\end{aligned}
$$

We will suppose that X, Y, Z are free commutative variables. Proceeding by induction up to infinity, we get

$$[[x_1, x_2], x_3] = \frac{2}{3} R_{123}(X, Y, Z)(x_1 x_2 x_3) \mod J,$$

where $R_{123}(X, Y, Z)$ is a power series of the form

$$R_{123}(X, Y, Z) = -X + \frac{1}{3}(XY - XZ) + \cdots + (\frac{1}{3})^{n-1} R^n(X, Y, Z) + \cdots.$$

Let us find the n-th degree homogeneous components $R^n = R^n(X, Y, Z)$ in $R_{123}(X, Y, Z)$. At first, define their combinatorial representations $R_c^n \in T(X, Y, Z)$, where $T(X, Y, Z)$ is the free algebra. We put by definition $R_c^1 = -X$. Let the combinatorial representation R_c^i be already defined for all $i < n$. Let Pa be a monomial from the combinatorial representation R_c^{n-1}, where P is a monomial of degree $n - 2$ and a is equal to one of X, Y, Z. Form a monomial Pab, where b is equal to one of the variables X, Y, Z but different from a. Ascribe a sign to Pab. This sign is equal to sign(Pa), if b is followed by a in the cyclic sequence $X \to Y \to Z \to X$, and to $-$sign(Pa) in the other case. Define $R_c^n \in T(X, Y, Z)$ as the algebraic sum of all monomials which can be obtained from all monomials of R_c^{n-1} in this way (taking into account their signs).

So, we have $R_c^2 = XY - XZ$, $R_c^3 = XYX - XYZ + XZX - XZY$, and so on. Now it is easy to check that the component R^n in $R_{123}(X, Y, Z)$ is equal to R_c^n in which the variables X, Y, Z are considered to be commutative. So, we get $R^2 = XY - XZ$, $R^3 = XYX - XYZ + XZX - XZY = X^2Y + X^2Z - 2XYZ$, and so on.

In $T(V)$ the Jacobi identity holds:

$$[[x_1, x_2], x_3] + [[x_2, x_3], x_1] + [x_3, x_1], x_2] = 0.$$

But one has to be in $T(V)$, so

$$\frac{2}{3}(R_{123} + R_{231} + R_{312})(x_1 x_2 x_3) = 0 \mod J,$$

where the power series $R_{231} = R_{231}(X, Y, Z)$ and $R_{312} = R_{312}(X, Y, Z)$ are constructed as $R_{123} = R_{123}(X, Y, Z)$ above. But this is possible only if

$$R(X, Y, Z) = R_{123}(X, Y, Z) + R_{231}(X, Y, Z) + R_{312}(X, Y, Z) = 0, \quad (12)$$

because $(x_1 x_2 x_3)$ is a symmetric expression and does not belong to J.

One can check that starting from the Jacobi identity for another triple x_i, x_j, x_k we will obtain the same condition (12). Thus, we have

Proposition 3.1 *For generic λ_{ij}, relations* (10) *define the deformation of the polynomial algebra of four variables if and only if condition* (12) *holds.*

One can check that in $R(X, Y, Z)$ only terms of odd degree appear and it has the form

$$R = -(X+Y+Z) + \frac{1}{9}(X^2Y + X^2Z + Y^2X + Y^2Z + \cdots + Z^2Y - 6XYZ) + \cdots .$$

So, R is not divided by $(X + Y + Z)$, and if λ_{ij} satisfy the Jacobi identity then they, in general, do not satisfy (12) (if none of X, Y, Z is equal to zero). Therefore, in the general case the Poisson bracket of the form (8) does not admit the PBW type quantization.

Nevertheless, any Poisson bracket of the form (8) can be quantized. Indeed, let the constants $\lambda_{ij,1}$ define such a bracket. Put $X_1 = \lambda_{12,1}\lambda_{34,1}$, $Y_1 = \lambda_{13,1}\lambda_{42,1}$, $Z_1 = \lambda_{14,1}\lambda_{23,1}$. Using the undetermined coefficients method it is easy to construct power series $X(t) = X_1t + X_3t^3 + \cdots$, $Y(t) = Y_1t + Y_3t^3 + \cdots$, $Z(t) = Z_1t + Z_3t^3 + \cdots$ satisfying (12) (only odd powers of t appear). Setting $t = \hbar^2$, we obtain the power series $X(\hbar) = X_1\hbar^2 + X_3\hbar^6 + \cdots, \ldots$ also satisfying (12). Let $\lambda_{12}(\hbar) = X_1\hbar + X_3\hbar^5 + \cdots$, $\lambda_{34}(\hbar) = \hbar$. Similarly we define all other $\lambda_{ij}(\hbar)$ from $Y(\hbar)$ and $Z(\hbar)$. It is obvious that $\lambda_{12}(\hbar)\lambda_{34}(\hbar) = X(\hbar)$, and so on. So, the power series $\lambda_{ij}(\hbar)$ satisfy condition (12), therefore the relations (10) with $\lambda_{ij} = \lambda_{ij}(\hbar)$ define the quantization of the given Poisson bracket.

Thus, we obtain

Theorem 3.2 *All the quadratic Poisson brackets of the form* (8) *can be quantized. The generic Poisson bracket of the form* (8) *does not admit a PBW type quantization.*

Remark. Only in the case when one of the X, Y, Z is equal to zero (or, what is the same, one of the λ_{ij} is equal to zero) the power series $R(X, Y, Z)$ is divided by $(X + Y + Z)$ and the corresponding Poisson bracket admits a PBW type quantization. Recall that for a determinant Poisson bracket associated to two diagonal quadratic forms we have $\lambda_{ij} = \lambda_i - \mu_j$. Hence, if $\lambda_i \neq \mu_j$ for all i, j this bracket does not admit a PBW type quantization. So, we see that the brackets of type b) corresponding to two quadratic forms with distinct eigenvalues can be quantized but do not admit a PBW type quantizations.

Acknowledgments

I want to thank I. Kantor and S. Shnider for helpful discussions.

References

1. Gerstenhaber, M.: On the deformation of rings and algebras, *Ann. of Math.* **79** (1964) 59–103.
2. Bayen, F., Flato, M., Fronsdal, C., Lichnerowicz, A. and Sternheimer, D.: Deformation theory and quantization, *Ann. Phys.* **111** (1978) 61–110.
3. Drinfeld, V.: On quadratic commutator relations in the quasiclassical case, *Math. Phys. and Functional Analysis*, Kiev, Naukova Dumka, 1986.
4. Manin, Yu.: *Quantum Groups and Non-commutative Geometry*, Les Publications Centre de Récherches Mathématiques, Montréal, 1989.
5. Omori, H., Maeda, Y. and Yoshioka, A.: Deformation quantizations of Poisson algebras, *Contemporary Mathematics* **179** (1994) 229–242.
6. Liu, Z. and Xu, P.: On quadratic Poisson structures, *Letters in Math. Phys.* **26** (1992) 33-42.
7. Donin, J. and Makar-Limanov, L.: Quantization of quadratic Poisson brackets on a polynomial algebra of three variables, *J. of Pure and Applied Alg.* (to appear).

TWO TYPES OF POISSON PENCILS
AND RELATED QUANTUM OBJECTS

D. GUREVICH
ISTV, Université de Valenciennes,
59304 Valenciennes, France

J. DONIN
Department of Mathematics,
Bar-Ilan University, 52900 Ramat-Gan, Israel

AND

V. RUBTSOV
ITEP, Bol.Tcheremushkinskaya 25,
117259 Moscow, Russia

Dedicated to Alain Guichardet with regards and friendship.

Abstract. Two types of Poisson pencils connected to classical R-matrices and their quantum counterparts are considered. A representation theory of the quantum algebras related to some symmetric orbits in $sl(n)^*$ is constructed. A twisted version of quantum mechanics is discussed.

Mathematics Subject Classification (1991): 17B37, 81R50.

Key words: Poisson bracket, Poisson pencil, quantum group, quantum algebra, associative pencil, symmetric orbit, twisted quantum mechanics.

1. Introduction

There are (at least) two reasons for Poisson pencils (P.p.) to be currently of great interest. First, they play a very important role in the theory of integrable systems (in the so-called Magri–Lenart scheme). Second, they

27

B. P. Komrakov et al. (eds.), Lie Groups and Lie Algebras, 27–46.
© 1998 *Kluwer Academic Publishers. Printed in the Netherlands.*

appear as infinitesimal (quasi-classical) objects in the construction of certain quantum homogeneous spaces. Roughly speaking, we can say that the P.p. arising in the framework of the latter construction are of the first type and those connected to integrable systems belong to the second type (while the integrable systems themselves are disregarded).

The main characteristic that joins together these two classes of P.p. is that a classical R-matrix participates in their construction. Let us recall that by a classical R-matrix on a simple Lie algebra $\mathfrak{g}$ one means a skew-symmetric ($R \in \wedge^2(\mathfrak{g})$) solution of the classical Yang–Baxter equation

$$[R^{12}, R^{13}] + [R^{12}, R^{23}] + [R^{13}, R^{23}] = a\varphi, \, a \in k,$$

where φ is a unique (up to a factor) ad-invariant element belonging to $\wedge^3(\mathfrak{g})$). In what follows we deal with the "canonical" classical R-matrix

$$R = \frac{1}{2} \sum X_\alpha \wedge X_{-\alpha}, \tag{1}$$

where $\{H_\alpha, X_\alpha, X_{-\alpha}\}$ is a Cartan–Weyl–Chevalley base of a simple Lie algebra $\mathfrak{g}$ over the field $k = \mathbb{C}$. We assume that a triangular decomposition of the Lie algebra $\mathfrak{g}$ is fixed. The field $k = \mathbb{R}$ is also admitted, but in this case we consider the normal real forms of the Lie algebras corresponding to this triangular decomposition.

Thus, the R-matrix (1) enters the constructions of both type of P.p. under consideration. However, the constructions and properties of these types are completely different. Moreover, the methods of quantizing these two types of P.p. are completely different, as well as the properties of the resulting associative algebras. We will use the term *associative pencils* (a.p.) of the first (second) type for the families of these quantum algebras arising from the first (second) type P.p.

The ground object of the second type P.p. is a quadratic Poisson bracket (P.b.) of Sklyanin type. By this type we mean either the famous Sklyanin bracket[1] or its various analogues: the elliptic Sklyanin algebras in the sense of (Sklyanin, 1982), the so-called second Gelfand–Dikii structures, etc. (Although we restrict ourselves to finite-dimensional Poisson varieties,

---

[1]Let us specify that this bracket is given by

$$\{ \, , \, \}_S = \{ \, , \, \}_l - \{ \, , \, \}_r, \, \{f, g\}_\epsilon = \mu \langle \rho_\epsilon^{\otimes 2}(R), df \otimes dg \rangle, \, \epsilon = l, r,$$

where $\rho_l \, (\rho_r) : \mathfrak{g} \to \text{Vect}(\text{Mat}(n))$ is a representation of the Lie algebra $\mathfrak{g}$ into the space of right (left) invariant vector fields on $\text{Mat}(n)$, R is the classical R-matrix (1) corresponding to $sl(n)$ and $\langle \, , \, \rangle$ is the pairing between vector fields and differentials extended to the their tensor powers. Hereafter μ is the multiplication in the algebra under consideration. (Let us note that for other simple Lie groups G analogous P.b. being extended to $\text{Mat}(n)$ are no longer Poisson.)

we would like to note it is not reasonable to consider these structures as the infinite-dimensional analogues of P.p. on symmetric orbits in $\mathfrak{g}^*$, as it was suggested in (Khoroshkin and all, 1993). The principal aim of the present paper is to make the difference between the two classes of P.p. related to classical R-matrices more transparent.)

All these P.b. (further denoted by $\{\ ,\ \}_2$) are quadratic and their linearization gives rise to another (linear) P.b. (denoted by $\{\ ,\ \}_1$) compatible with the initial one. As a result we have a P.p.

$$\{\ ,\ \}_{a,b} = a\{\ ,\ \}_1 + b\{\ ,\ \}_2 \qquad (2)$$

generated by these two brackets.

Now let us describe the corresponding quantum objects. We construct them in two steps. First, we quantize the Sklyanin bracket and get the well-known quadratic "RTT=TTR" algebra. Then by linearizing the determining quadratic relations (or more precisely, by applying a shift operator to the algebra, cf. below) we get a (second type) a.p. which is the quantum counterpart of the whole P.p. These second type P.p. and their quantum counterparts are considered in Section 2.

Let us note that the shift operator mentioned above does not give rise to any meaningful deformation and it is reduced to a mere change of base.

We are interested in P.p. and their quantum counterparts connected to symmetric orbits in $\mathfrak{g}^*$, where $\mathfrak{g}$ is a simple Lie algebra. It should be pointed out that the latter P.p. are not any reduction of the second type P.p. Let us say a little bit more about the origin of these P.p. on symmetric orbits.

It is well known that any orbit $\mathcal{O}_x$ of any semisimple element in $\mathfrak{g}^*$ can be equipped with the reduced Sklyanin (sometimes called Sklyanin–Drinfeld) bracket. One usually considers the reduction procedure for the compact forms of simple Lie algebras, but it is also valid for normal ones (the R-matrix for compact forms differs from (1) by the factor $\sqrt{-1}$).

It is worth noting that the symmetric orbits in $\mathfrak{g}^*$ admit a complementary nice property: the reduced Sklyanin bracket is compatible with the Kirillov–Kostant–Souriau (KKS) bracket. This fact was shown in (Khoroshkin and all, 1993), (Donin, Gurevich, 1995a) (in (Khoroshkin and all, 1993) it was also shown that only symmetric orbits possess this property).

Moreover, as shown in (Donin, Gurevich, 1995a), both components $\{\ ,\ \}_\epsilon,\ \epsilon = l,\ r$, of the Sklyanin bracket (cf. footnote 1) become Poisson after being reduced to $\mathcal{O}_x$ and one of them (say, $\{\ ,\ \}_r$ coincides up to factor with the KKS one if we identify $\mathcal{O}_x$ with right coset G/H).

Since the brackets $\{\ ,\ \}_\epsilon,\ \epsilon = l, r$, are always compatible, we have a (first type) P.p. (2) on a symmetric orbit $\mathcal{O}_x \in \mathfrak{g}^*$ with $\{\ ,\ \}_1 = \{\ ,\ \}_l^{\text{red}}$ and $\{\ ,\ \}_2 = \{\ ,\ \}_r^{\text{red}}$ ("red" means reduced on $\mathcal{O}_x$)^2.

---

[2]Let us note that similar (first type) P.p. exists on certain nilpotent orbits in $\mathfrak{g}^*$. These

The difference between these two types of P.p. also manifests itself on the quantum level. One usually represents quantum analogues of the Sklyanin bracket reduced to a semisimple orbit in terms of "quantum reduction" by means of pairs of quantum groups or their (restricted) dual objects. However, in this way it is not possible to represent the whole a.p. arising from the first type P.p. under consideration. We discuss here another, more explicit, way to describe the corresponding quantum a.p. Namely, we represent them as certain quotient algebras.

In virtue of the results of the paper (Donin, Shnider, 1995), a formal quantization of the P.p. under consideration exists on any symmetric orbit. However, it is not so easy to describe the quantum objects explicitly or, in other words, to find systems of equations defining them. We consider here two ways to look for such a system in the case when the orbit $\mathcal{O}_x$ is a rank 1 symmetric space in $sl(n)^*$, i.e., that of $SL(n)/S(L(n-1) \times L(1))$ type.

The first way was suggested in (Donin, Gurevich, 1995) (but there the system of equations was given in an inconsistent form). It uses the fact that such an orbit can be described by a system of quadratic equations (cf. Section 3). The second way consists in an attempt to construct a representation theory of the first type quantum algebras (cf. Section 4) and to compute the desired relations in the modules for these algebras. This approach is valid in principle for any symmetric orbit in $\mathfrak{g}^*$ for any simple Lie algebra $\mathfrak{g}$. This enables us to treat this type of quantum algebras from the point of view of twisted quantum mechanics (Section 5). This means that quantum algebras and their representations are objects of a twisted (braided) category. (We consider the term "twisted" as more general, keeping the term "braided" for nonsymmetric categories).

Completing the Introduction, we would like to make two remarks. First, let us note that there is a number of papers devoted to q-analogues of special functions of mathematical physics. In fact they deal with quantum analogues of double cosets. Meanwhile, the problem of finding an explicit description of quantum symmetric spaces (in particular, orbits in $\mathfrak{g}^*$) is disregarded (in some sense the latter problem is more complicated). If the usual q-special functions arise from a one-parameter deformation of the classical objects, our approach enables us to consider their analogues arising

---

pencils are generated by the KKS bracket and the so-called R-matrix bracket defined by

$$\{ \ , \ \}_R = \mu \langle \rho^{\otimes 2}(R), df \otimes dg \rangle, \ \rho = \mathrm{ad}^*.$$

In fact the R-matrix bracket is just that $\{ \ , \ \}_l^{\mathrm{red}}$ for any such orbit but only for the symmetric ones the brackets KKS and $\{ \ , \ \}_r^{\mathrm{red}}$ are proportional to each other. The reader is referred to (Gurevich, Panyushev, 1994) where all orbits in $\mathfrak{g}^*$ possessing the above P.p. are classified.

from the two-parameter deformation. This approach will be developed in a joint paper of one of the author (D.G.) and L. Vainerman.

Second, we would like to emphasize that we consider the present paper as an intermediate review of the papers (Donin, Gurevich, 1995), (Donin and all., 1997), (Gurevich, 1997), where the present topic (i.e., the problem of the explicit description of a.p. arising from quantization of certain P.p. associated with the "canonical" classical R-matrices) was discussed. Meanwhile, this topic is still in progress.

In what follows $U_q(\mathfrak{g})$-Mod stands for the category of all finite-dimensional $U_q(\mathfrak{g})$-modules and their inductive limits. The parameter q is assumed to be generic (or formal, when we speak about flatness of deformation, in fact we do not distinguish these two meanings).

We dedicate this paper to our friend, professor Alain Guichardet. We acknowledge his warm hospitality at Ecole Polytechnique for a long time and benefited a lot from our numerous discussions and conversations.

This paper was finished during the visit of V.R. to the University Lille-1. He is thankful to his colleagues from the group of Mathematical Physics for their hospitality. The work of V.R. was supported partially by the grants RFFI-O1-01011, INTAS-93-2494 and INTAS-1010-CT93-0023.

2. Second type Poisson and associative pencils

Let us fix a simple algebra $\mathfrak{g} = sl(n)$, $n \geq 2$, and consider the corresponding quadratic Sklyanin bracket extended on the space $\mathrm{Fun}(\mathrm{Mat}(n))$ which is the algebra of polynomials on the matrix elements a_i^j. More precisely, this bracket is determined the following multiplication table

$$\{a_k^i, a_k^j\}_2 = a_k^i a_k^j, \quad \{a_i^k, a_j^k\}_2 = a_i^k a_j^k, \quad i < j;$$

$$\{a_i^l, a_k^j\}_2 = 0, \quad \{a_i^j, a_k^l\}_2 = 2\, a_i^l a_k^j, \quad i < k, \, j < l.$$

Let us linearize the bracket $\{\ ,\ \}_2$. To do this we introduce a shift operator Sh_h which sends a_i^j to $a_i^j + h\delta_i^j$ (we extend this operator onto the whole algebra $\mathrm{Fun}(\mathrm{Mat}(n))$ by multiplicativity). Then applying this operator to r.h.s. of the above formulas and taking the linear terms in h, we get a linear bracket with by the following multiplication table

$$\{a_k^i, a_k^j\}_1 = \delta_k^i a_k^j + a_k^i \delta_k^j, \quad \{a_i^k, a_j^k\}_1 = \delta_i^k a_j^k + a_i^k \delta_j^k, \quad i < j;$$

$$\{a_i^l, a_k^j\}_1 = 0, \quad \{a_i^j, a_k^l\}_1 = 2\,(\delta_i^l a_k^j + a_i^l \delta_k^j), \quad i < k, \, j < l.$$

Proposition 1 *The brackets $\{\ ,\ \}_{1,2}$ are compatible.*

Proof. This follows immediately from the following fact: the r.h.s. of the formulas for the bracket $\{\ ,\ \}_2$ does not contain any summand of the form $a_i^i a_j^j$. This implies that the images of r.h.s. elements under the operator Sh_h do not contain any term quadratic in h. The details are left to the reader.

$\square$

Remark 1 It is often more convenient to take, instead of $\{a_i^j\}$, the base consisting of a_i^j, $i \neq j$, $a_1^1 - a_i^i$, $i > 1$, and $a_0 = \sum a_i^i$. Then the mentioned property can be reformulated as follows: the r.h.s. of the above formulas do not contain the term $(a_0)^2$ (with respect to these new generators the operator Sh_h acts nontrivially on a_0) only.

It is easy to see that the bracket $\{\ ,\ \}_1$ can be represented in the form

$$\{a_i^j, a_k^l\}_1 = \{\mathbf{R}(a_i^j), a_k^l\}_{gl} + \{a_i^j, \mathbf{R}(a_k^l)\}_{gl}.$$

Here $\{\ ,\ \}_{gl}$ is the linear bracket corresponding to the Lie algebra $gl(n)$ (namely, $\{a_i^j, a_k^l\}_{gl} = a_i^l \delta_k^j - a_k^j \delta_i^l$) and $\mathbf{R} : W \to W$ is an operator defined in the space $W = \mathrm{Span}(a_i^j)$ as follows $\mathbf{R}(a_i^j) = \mathrm{sign}(j - i)\, a_i^j$ (we assume that $\mathrm{sign}(0) = 0$).

Thus, the space W is equipped with two Lie algebra structures. The first one is $gl(n)$ and the second one corresponds to the Poisson bracket $\{\ ,\ \}_1$ and therefore we have a double Lie algebra structure on the space W according to the terminology of (Semenov-Tian-Shansky, 1983) (our construction coincides with (Semenov-Tian-Shansky, 1983) up to a factor).

Let us consider now the quantum analogue of the above P.p.

Let $U_q(\mathfrak{g})$ be the Drinfeld–Jimbo quantum group corresponding to $\mathfrak{g}$ and $\mathcal{R}$ be the universal quantum R-matrix. Consider the linear space V of vector fundamental representation $\rho : \mathfrak{g} \to \mathrm{End}(V)$ of the initial Lie algebra $\mathfrak{g}$. Denote by ρ_q the corresponding representation of the quantum group $U_q(\mathfrak{g})$ to $\mathrm{End}(V)$ (we assume that $\rho_q \to \rho$ when $q \to 1$). Then the operator $S : V^{\otimes 2} \to V^{\otimes 2}$ defined as follows $S = \sigma \rho^{\otimes 2}(\mathcal{R})$, where σ is the flip $(\sigma(x \otimes y) = y \otimes x)$, satisfies the quantum Yang–Baxter equation (QYBE)

$$S^{12} S^{23} S^{12} = S^{23} S^{12} S^{23}, \quad S^{12} = S \otimes \mathrm{id}, \quad S^{23} = \mathrm{id} \otimes S \qquad (3)$$

and possesses two eigenvalues (we recall that $\mathfrak{g} = sl(n)$).

Let us identify now the space W with $V \otimes V^*$ and equip it with the operator $S_W = S \otimes (S^*)^{-1} : W^{\otimes 2} \to W^{\otimes 2}$, where S^* is defined with respect to the pairing

$$\langle x \otimes y, a \otimes b \rangle = \langle x, a \rangle \langle y, b \rangle, \quad x, y \in V, \ a, b \in V^*.$$

Let us fix a base $\{a_i\} \in V$. Let $\{a^i\} \in V^*$ be the dual base and set $a_i^j = a_i \otimes a^j$. Then we have the following explicit form for the operator S_W:

$$S_W(a_i^k \otimes a_j^l) = S_{ij}^{mn} S^{-1}{}_{pq}^{kl} (a_m^p \otimes a_n^q), \text{ where } S(a_i \otimes a_j) = S_{ij}^{kl} a_k \otimes a_l.$$

It is easy to see that the operator S_W also satisfies the QYBE and possesses 1 as an eigenvalue. Then it is natural to introduce deformed analogues of symmetric and skew-symmetric subspaces of the space $W^{\otimes 2}$ as follows

$$I_-^q = \mathrm{Im}(S_W - \mathrm{id}), \quad I_+^q = \mathrm{Ker}(S_W - \mathrm{id}).$$

Let us describe explicitly the operator S and the spaces $I_\pm$:

$$S(a_i \otimes a_j) = (q-1)\delta_{i,j} a_i \otimes a_j + a_j \otimes a_i + \sum_{i<j}(q - q^{-1}) a_i \otimes a_j,$$

$$I_-^q = \mathrm{Span}(a_k^i a_k^j - q a_k^j a_k^i, \; a_i^k a_j^k - q a_j^k a_i^k, \; i < j;$$

$$a_i^l a_k^j - a_k^j a_i^l, \; a_i^j a_k^l - a_k^l a_i^j - (q - q^{-1}) a_k^j a_i^l, \; i < k, j < l)$$

and

$$I_+^q = \mathrm{Span}((a_i^k)^2, \; q a_k^i a_k^j + a_k^j a_k^i, \; q a_i^k a_j^k + a_j^k a_i^k, \; i < j;$$

$$a_i^j a_k^l + a_k^l a_i^j, \; a_i^l a_k^j + a_k^j a_i^l + (q - q^{-1}) a_i^j a_k^l, \; i < k, j < l).$$

Let us introduce also the quotient algebra $A_{0,q} = T(W)/\{I_-^q\}$. In what follows $T(W)$ denotes the free tensor algebra of the space W and $\{I\}$ denotes the two-sided ideal generated by the set $I \subset T(W)$ (here $I = I_-^q$ is a subspace of the space $W^{\otimes 2}$).

It is well known (cf. (Reshetikhin and all, 1990)) that the quantum analogue $\mathrm{Fun}_q(SL(n))$ of the space $\mathrm{Fun}(SL(n))$ is the quotient algebra of $A_{0,q}$ over the ideal generated by the element $\det_q -1$ ($\det_q$ is the so-called "quantum determinant"). Moreover, the latter quotient algebra can be endowed with a Hopf structure. However, we are interested rather in the algebra $A_{0,q}$ itself. This algebra is quadratic and is a flat deformation of the classical counterpart, namely, of the symmetric algebra $\mathrm{Sym}(W)$ of the space W (cf. (Donin, Shnider, 1996) for details). Let us note that the skew-symmetric algebra of W has also an evident quantum analogue $T(W)/\{I_+^q\}$.

Now we want to introduce a two-parameter deformation of the algebra $\mathrm{Sym}(W)$ quantizing the whole P.p. under consideration. This deformation can be realized by means of the same shift operator as above. Applying this operator to I_-^q we obtain the following space

$$J_{h,q} = \mathrm{Span}(a_k^i a_k^j - q a_k^j a_k^i - h(\delta_k^i a_k^j + a_k^i \delta_k^j),$$

$$a_i^k a_j^k - q a_j^k a_i^k - h(\delta_i^k a_j^k + a_i^k \delta_j^k), \ i < j; \ a_i^l a_k^j - a_k^j a_i^l,$$

$$a_i^j a_k^l - a_k^l a_i^j - (q - q^{-1}) a_k^j a_i^l - hm \, (\delta_i^l a_k^j + a_i^l \delta_k^j), \ i < k, j < l),$$

where $m = 1 + q^{-1}$ (we have realized here a substitution $h(q - 1) \to h$).

Let us introduce the algebra $A_{h,q} = T(W)/\{J_{h,q}\}$. It is evident that this algebra is a flat deformation of the initial commutative algebra $A_{0,1}$ since the passage from the algebra $A_{0,q}$ to that $A_{h,q}$ is trivial from the deformation point of view and reduces to a change of a base. Thus, we have constructed the a.p. $A_{h,q}$, which is a quantum analogue of the P.p. under consideration (it is easy to see that quasi-classical term of this two-parameter deformation is just the above second type P.p.).

A similar method can be applied to quantize the P.p. generated by the elliptic Sklyanin P.b (Sklyanin, 1982). This bracket (denoted also by $\{\ ,\ \}_2$) is determined by the following multiplication table

$$\{S_1 \, S_0\} = 2J_{23}S_2S_3, \quad \{S_1 \, S_2\} = -2S_0S_1$$

and their cyclic permutations with respect to the indices $(1, 2, 3)$ with some elliptic functions J_{ij} (cf. (Sklyanin, 1982)). The linearization of this bracket defined by the shift operator $S_0 \to S_0 + h$, $S_i \to S_i, i \neq 0$ (S_0 plays here the role of the element a_0 from Remark 1) gives rise to the linear one corresponding to the Lie algebra $\mathfrak{g} = so(3) \oplus k$. The brackets are compatible (by the same reason as above).

To quantize the P.p. generated by them, it suffices to quantize the bracket $\{\ ,\ \}_2$ and apply the shift operator. It is well known (cf. (Sklyanin, 1982)) that a quantum analogue of the bracket $\{\ ,\ \}_2$ is the quotient algebra $T(W)/\{I\}$, where $W = \mathrm{Span}(S_0, ..., S_3)$ and $\{I\}$ is the two-sided ideal in $T(W)$ generated by the elements

$$S_1S_0 - S_0S_1 + iJ_{23}(S_2S_3 + S_3S_2), \ S_1S_2 - S_2S_1 - i(S_0S_3 + S_3S_0)$$

and their cyclic permutations with some elliptic functions J_{ij}.

We leave to the reader the explicit description of the resulting two-parameter quantum a.p.

The algebra $T(W)/\{I\}$ originally defined by E. Sklyanin is an elliptic analogue of the symmetric algebra of the space W. Unfortunately, we do not know any natural elliptic analogue of the skew-symmetric one (it seems very plausible that such analogue does not exist).

Remark 2 There are other quantum analogues of symmetric and skew-symmetric algebras of the space W defined by the so-called reflection equations (RE)

$$Su_1Su_1 = u_1Su_1S, \ u_1 = u \otimes 1, \ u = (u_i^j),$$

where S is a solution of (3). Conjecturally this "RE algebra" is a flat deformation of its classical counterpart (assuming S to be of Hecke type). At least a necessary condition for flatness of a deformation, i.e., the existence of a P.b. as a quasiclassical term of the deformation, is satisfied. The mentioned bracket is also quadratic and admits a linearization. These two brackets are also compatible. We get a quantization of the P.p. generated by them by applying to the above algebra a shift operator (for details the reader is referred to (Isaev, 1995), (Isaev, Pyatov, 1995), (Gurevich, 1997)).

3. Rank 1 quantum orbits

In what follows we consider the P.p. and their quantum counterparts connected to symmetric orbits. These P.p. were described in the Introduction: they are generated by the KKS bracket and by the so-called R-matrix bracket.

We want to describe the corresponding quantum a.p. explicitly by means of a system of equations. In the present section we recall the method suggested in (Donin, Gurevich, 1995) to look for such systems. Unfortunately, it is valid only for rank 1 symmetric spaces, namely, those of $SL(n)/S(L(n-1) \times L(1))$ type.

Let us fix a Lie algebra $\mathfrak{g} = sl(n)$ and an element $x = \omega \in \mathfrak{h}^*$, where $\mathfrak{h} \subset \mathfrak{g} = sl(n)$ is the Cartan subalgebra (a triangular decomposition of $\mathfrak{g}$ is assumed to be fixed). We consider ω as an element of $\mathfrak{g}^*$ extending it by zero to the nilpotent subalgebras. Let $\mathcal{O}_\omega$ be its orbit in $\mathfrak{g}^*$.

It is well known that this orbit is symmetric iff $\omega(h_i) = 0$ for all $i \neq i_0$, where $\{h_i = e_{i,i} - e_{i+1,i+1}, i = 0, ..., n-1\}$ is the standard basis in $\mathfrak{h}$. The rank of this symmetric space is $\mathrm{rk}(\mathcal{O}_\omega) = \min(i_0, n - i_0)$. We assume here that $rk(\mathcal{O}_\omega) = 1$. So we have $i_0 = 1$ or $i_0 = n - 1$. For the sake of concreteness we set $i_0 = 1$.

Thus, we have a family $\mathcal{O}_\omega$ of such orbits parametrized by the value $\omega(h_1) \in k$ ($k = \mathbb{C}$ or $k = \mathbb{R}$ corresponding to the case under question). Let us represent the algebra $\mathrm{Fun}(\mathcal{O}_\omega)$ as a quotient algebra of $\mathrm{Sym}(\mathfrak{g}) = \mathrm{Fun}(\mathfrak{g}^*)$. It is not difficult to show that this algebra can be described by a system of equations quadratic in generators of the space $\mathfrak{g}$.

Let us describe this system explicitly. Consider the space $V = \mathfrak{g}$ as an object of the category $\mathfrak{g}$-Mod. Let $V^{\otimes 2} = \oplus V_\beta$ be a decomposition of $V^{\otimes 2}$ into a direct sum of isotypic $\mathfrak{g}$-modules, where β denotes the highest weight (h.w.) of the corresponding module. Let α be h.w. of $\mathfrak{g}$ as a $\mathfrak{g}$-module.

Then in the above decomposition there is a trivial module V_0 ($\beta = 0$), a module $V_{2\alpha}$ of the h.w. $\beta = 2\alpha$, and two irreducible modules isomorphic to $\mathfrak{g}$ itself. One of them belongs to the space $I_+ \subset V^{\otimes 2}$ of symmetric tensors and the other one to the space $I_- \subset V^{\otimes 2}$ of skew-symmetric tensors. We will use the notation V_+ (V_-) for the former (latter) of them.

Note that a similar decomposition is valid for other simple Lie algebras, but for them there exists only one module isomorphic to $\mathfrak{g}$. It belongs to the skew-symmetric part and we keep the notation V_- for it.

Let us denote by C the generator (called *Casimir*) of the module V_0. Then the orbit $\mathcal{O}_\omega$ is given by the following system of equations

$$I_- = 0; V_\beta = 0 \; \forall \, V_\beta \subset I_+, \; V_\beta \notin \{V_0, V_+, V_{2\alpha}\}; \; C = c_0, \; V_+ = c_1\mathfrak{g} \qquad (4)$$

with some factors c_0 and c_1 (the latter relation is a symbolic way to write the system which arises if we choose some bases in $\mathfrak{g}$ and V_+ and equate the highest weight element of V_+ to that one of $\mathfrak{g}$ times a factor c_1 and consider all the relations that follow).

It is not difficult to find the values of the factors c_i, $i = 0, 1$. It suffices to substitute "the point" ω to this system. (Let us note that in (Donin, Gurevich, 1995) this system was given in a inconsistent form, where the condition $c_1 = 0$ was assumed.)

A similar procedure can be realized in the category $U_q(\mathfrak{g})$-Mod. More precisely, we equip the space V with the structure of a $U_q(\mathfrak{g})$-module deforming the initial $\mathfrak{g}$-module structure on V (there exists a regular way to convert any finite-dimensional $\mathfrak{g}$-module into a $U_q(\mathfrak{g})$ one, cf., i.e., (Chari, Pressley, 1994)). Using the comultiplication in the quantum algebra $U_q(\mathfrak{g})$, we equip the space $V^{\otimes 2}$ with a $U_q(\mathfrak{g})$-module structure and decompose it as above into a direct sum of irreducible $U_q(\mathfrak{g})$-modules. We will denote these components by V_β^q. By C_q we denote a generator of the module V_0^q (we call C_q a *braided Casimir*).

Then we can impose a similar system of equations by replacing all V_β participating in the system (4) by their quantum (or q-) analogues V_β^q and C by C_q. The only problem is: what are the proper values of the factors $c_i(q)$, $i = 0, 1$, which now depend on q.

In (Donin, Gurevich, 1995) the following way to find out a proper system of equations was suggested. Let us consider the data (V, I^q, ν_0, ν_1), where $I^q = V^{\otimes 2} \setminus V_{2\alpha}^q$ and ν_i, $i = 0, 1$ are two $U_q(\mathfrak{g})$-morphisms defined as follows $\nu_0 : V_0^q \to k$ $(\nu_0(C_q) = c_0)$ and $\nu_1 : V_+^q \to c_1\mathfrak{g}$ (all other components of $V^{\otimes 2} \setminus V_{2\alpha}^q$ are sent by ν_i to zero).

Let us also consider the graded quadratic algebra $T(V)/\{I^q\}$ and its filtered analogue $T(V)/\{I_{c_0,c_1}^q\}$, where $\{I_{c_0,c_1}^q\}$ is the ideal generated by the elements

$$I_-^q, \; V_\beta \subset I_+^q, \; V_\beta \notin \{V_0, V_+, V_{2\alpha}\}, \; C - c_0(q), \; V_+^q - c_1(q)\mathfrak{g}$$

(see the end of this Section for the definition of the spaces $I_\pm^q$).

Let us note that the algebra $T(V)/\{I^q\}$ is the quantum analogue of the function space on the cone and it is a flat deformation of its classical counterpart. The flatness can be shown, e.g., in the way suggested in (Donin, Gurevich, 1995a), where an intertwining operator of the initial commutative product and the deformed one is given. Moreover, the algebra $T(V)/\{I^q\}$ is Koszul for a generic q since it is so for the case $q = 1$ by virtue of (Bezrukavnikov, 1995).

Let us assume now that the above data satisfy the following system:

$$\mathrm{Im}(\nu_1 \otimes \mathrm{id} - \mathrm{id} \otimes \nu_1)(I \otimes V \bigcap V \otimes I) \subset I,$$

$$(\nu_1(\nu_1 \otimes \mathrm{id} - \mathrm{id} \otimes \nu_1) + \nu_0 \otimes \mathrm{id} - \mathrm{id} \otimes \nu_0)(I \otimes V \bigcap V \otimes I) = 0,$$

$$\nu_0(\nu_1 \otimes \mathrm{id} - \mathrm{id} \otimes \nu_1)(I \otimes V \bigcap V \otimes I) = 0.$$

Then by virtue of the PBW theorem in the form of (Braverman, Gaitsgory, 1994) we can conclude that the algebra $T(V)/\{I^q_{c_0,c_1}\}$ is a flat deformation of the algebra $T(V)/\{I^q\}$.

Let us note that the above conditions represent a more general form of the Jacobi relation connected to deformation theory. Thus, if the above form of the Jacobi identity is fulfilled, the algebra $T(V)/\{I^q_{c_0,c_1}\}$ is a flat deformation of the orbit $\mathcal{O}_\omega$ (more precisely, of the function algebra on it).

So, the proper quantities $c_i(q)$, $i = 0, 1$, if they exist, can be found from the above equations. However, a priori it is not clear why such quantities exist. Let us assume here that they exist and denote by $A_{0,q}$ the corresponding algebra $T(V)/\{I^q_{c_0,c_1}\}$. We are interested in its further deformation.

To do this, we begin by discussing the following question: what is the deformational quantization of the KKS bracket? We want to represent the latter quantum object also as a quotient of the enveloping algebra $U(sl(n))_h$. The index h here means that we have introduced a factor h in the bracket in the definition of the enveloping algebra, i.e.,

$$U(\mathfrak{g})_h = T(\mathfrak{g})/\{xy - yx - h[x, y]\}.$$

There are some factors $c_i(h)$, $i = 0, 1$, now depending on h such that the quotient algebra $A_{h,0} = U(sl(n))_h/\{J\}$, where the ideal $\{J\}$ is generated by the family of elements from (4) lying in I_+ but with new $c_i(h)$, is a flat deformation of the initial algebra corresponding to the case $h = 0$. These factors can be also found by means of the above form of Jacobi identity.

This approach can also be applied in order to get the quantum algebras corresponding to the whole P.p. under consideration, since these algebras are quadratic as well. The only problem is: what is the proper quantum analogue of the algebra $U(sl(n))_h$? Or, in other words, what is a consistent way to introduce a quantum analogue of the Lie bracket?

We will introduce a q-generalization of the ordinary Lie bracket by means of the following

Definition 1 Let $\mathfrak{g}$ be a simple Lie algebra equipped with a $U_q(\mathfrak{g})$-module structure and V_β^q be the irreducible modules in the category $U_q(\mathfrak{g})$-Mod entering its tensor square. We call *q-Lie bracket* the operator $[\ ,\ ]_q : V^{\otimes 2} \to V$ defined as follows

$$[\ ,\ ]_q|_{V_\beta^q} = 0 \text{ for all } V_\beta^q \neq V_-^q$$

and $[\ ,\ ]_q : V_-^q \to V$ is a $U_q(\mathfrak{g})$-isomorphism.

Let us observe that in this way the q-Lie bracket is defined up to a factor. We fix this factor and introduce the q-analogue of the algebra $U(\mathfrak{g})_h$ as follows

$$U(\mathfrak{g})_{h,q} = T(V)/\{\operatorname{Im}(\operatorname{id} - h[\ ,\ ]_q)I_-^q\}.$$

The space $V = \mathfrak{g}$ equipped with this bracket will be denoted by $\bar{\mathfrak{g}}$ and called *braided Lie algebra*[3]. We will also use the notation $U(\bar{\mathfrak{g}})$ for the algebras $U(\mathfrak{g})_{1,q}$. Let us note that this is another, as compared with the quantum group $U_q(\mathfrak{g})$, q-analogue of the enveloping algebra $U(\mathfrak{g})$, but the deformation $U(\mathfrak{g}) \to U(\mathfrak{g})_{h,q}$ is not flat except for the $sl(2)$ case. However, some quotient algebras of this algebra are flat deformations of their classical counterparts.

Let us return to the case $\mathfrak{g} = sl(n)$ and introduce the first type a.p. $A_{h,q}$ as the quotient algebra of $U(sl(n))_{h,q}$ by the ideal

$$\{C_q - c_0,\ V_+^q - c_1V,\ I_+^q \setminus (V_{2\alpha}^q \oplus V_0^q \oplus V_+^q)\}$$

with some factors $c_i(h,q)$, $i = 0, 1$.

In order to look for the consistent factors $c_i(h,q)$, we must only modify the above morphism ν_1 on the "q-skew-symmetric" subspace I_-^q by setting $\nu_1 : V_-^q \to h\mathfrak{g}$ (all other components are still sent by ν_1 to zero) and verify the above form of Jacobi identity. This provides us with the factors $c_i(h,q)$ ensuring the flatness of the deformation of the function algebra on the orbit $\mathcal{O}_\omega$ under consideration.

Let us emphasize that the existence of the proper factors $c_i(h,q)$, as well as that of the above factors $c_i(q)$, can be deduced from the paper (Donin, Shnider, 1995), where a formal quantization of the P.p. under question was considered. In the next Section we discuss another way to look for appropriate factors $c_i(h,q)$.

---

[3] Other definitions of q-counterparts of Lie algebras have been suggested recently in (Delius, Huffmann, 1995) and (Lyubashenko, Sudbery, 1995). It is very plausible that they are equivalent to ours.

It remains only to note that the quasiclassical term of two-parameter deformation $\mathrm{Fun}(\mathcal{O}_\omega) \to A_{h,q}$ is just the above P.p. on the orbit $\mathcal{O}_\omega$ (cf. (Donin, Gurevich, 1995)).

It is worth to note that the product in the resulting quantum algebras $A_{h,q}$ is $U_q(sl(n))$-invariant. This means that the following property

$$X\mu(a \otimes b) = \mu(X_1(a) \otimes X_2(b)), \quad a, b \in A_{h,q}, \ X \in U_q(sl(n))$$

is satisfied. (Hereafter we use Sweedler's notation $X_1 \otimes X_2$ for $\Delta(X)$.) This property follows immediately from the construction of the algebras. This algebra is the inductive limit of finite-dimensional $U_q(sl(n))$-modules (moreover, it is multiplicity free, i.e., the multiplicity of each irreducible $U_q(\mathfrak{g})$-module is ≤ 1). Thus, this algebra belongs to the category $U_q(sl(n))$-Mod.

Let us note that a particular case of the a.p. $A_{h,q}$ is the algebra $A_{0,q}$ arising as the result of quantization of the only R-matrix bracket. This algebra is commutative in the category $U_q(sl(n))$-Mod in the following sense. This category is balanced (see (Chari, Pressley, 1994) for the definition). Moreover, for any two finite-dimensional objects U and V of this category there exists an involutive operator $\widetilde{S} : U \otimes V \to V \otimes U$ such that it is a $U_q(\mathfrak{g})$-morphism, it commutes with S and it is a deformation of the flip.

Then $I_\pm^q = \mathrm{Im}(\mathrm{id} \pm \widetilde{S})$ with $\widetilde{S} : V^{\otimes 2} \to V^{\otimes 2}$. Using the fact that the algebra $A_{h,q}$ can be decomposed into a direct sum of finite-dimensional objects of the category $U_q(sl(n))$-Mod we can extend the operator $\widetilde{S}$ onto $A_{h,q}^{\otimes 2}$. Then the above mentioned commutativity of the algebra $A_{0,q}$ means that the multiplication operator μ satisfies the following relation: $\mu = \mu\widetilde{S}$. A proof of this fact can be obtained from (Donin, Shnider, 1995). Strictly speaking, just the algebra $A_{0,q}$ is the quantum analogue of the orbit under consideration.

Let us note that the above method can be also applied to find the equations describing quantum algebras arising from similar P.p. on certain nilpotent orbits in $\mathfrak{g}^*$ (see the Introduction). Thus, such a.p. related to the highest weight element orbits in $\mathfrak{g}^*$ can be defined by means of the above ideal I_{c_0,c_1}^q, but with $c_0(h,q) = 0$ and with a suitable $c_1(h,q)$.

4. Modules for first type quantum algebras

In the previous section we discussed the first step of quantization procedure. The resulting object of this step is an a.p. represented as a quotient algebras over some suitable ideal. Now we want to consider the second step of quantization, consisting in an attempt to represent the above algebras in certain linear spaces. At this step the difference between two types of quantum algebras (a.p.) under question becomes clearer.

Moreover, this step provides us with another way to look for consistent factors $c_i(h,q)$. Briefly speaking, this method reduces to the computation of the factor $c_i(h,q)$ on the image of the first type algebras into the space $\mathrm{End}(V)$.

To do this, we recall a natural way to equip the space of endomorphisms of a $U_q(\mathfrak{g})$-module with a $U_q(\mathfrak{g})$-module structure. Let U^q be a (finite-dimensional) $U_q(\mathfrak{g})$-module and $\rho_q : U_q(\mathfrak{g}) \to \mathrm{End}(U^q)$ be the corresponding representation. Let us introduce the representation $\rho_q^{\mathrm{End}} : U_q(\mathfrak{g}) \to \mathrm{End}(\mathrm{End}(U^q))$ by putting

$$\rho_q^{\mathrm{End}}(a)M = \rho(a_1) \circ M \circ \rho(\gamma(a_2)), \ a \in U_q(\mathfrak{g}), \ M \in \mathrm{End}(U^q).$$

We denote the matrix product by $\circ$, while γ is the antipode in $U_q(\mathfrak{g})$.

We deal with a coordinate representation of module elements. We consider the endomorphisms as matrices and their action is the left multiplication by these matrices.

Let us note that this way of equipping $\mathrm{End}(U^q)$ with a $U_q(\mathfrak{g})$-module structure is compatible with the matrix product in it in the following sense:

$$\rho_q^{\mathrm{End}}(a)(M_1 \circ M_2) = \rho_q^{\mathrm{End}}(a_1)M_1 \circ \rho_q^{\mathrm{End}}(a_2)M_2.$$

This means that $\circ : \mathrm{End}(U^q)^{\otimes 2} \to \mathrm{End}(U^q)$ is a $U_q(\mathfrak{g})$-morphism.

Now we will introduce a useful notion for constructing a representation theory of the algebras under consideration.

Definition 2 Let $\bar{\mathfrak{g}}$ be a braided Lie algebra. We say that a $U_q(\mathfrak{g})$-module U^q is a *braided module* or, more precisely, a *braided $\bar{\mathfrak{g}}$-module*, if it can be equipped with a structure of a $U(\bar{\mathfrak{g}})$-module and the representation $\rho : U(\bar{\mathfrak{g}}) \to \mathrm{End}(U^q)$ is a $U_q(\mathfrak{g})$-morphism. We also say that the classical counterpart $U = U^1$ of the $U_q(\mathfrak{g})$-module U^q allows braiding.

A natural way to construct braided modules is given by the following

Proposition 2 Let U^q be a $U_q(\mathfrak{g})$-module. If the decomposition $\mathrm{End}(U^q) = \oplus V_\gamma^q$ of the $U_q(\mathfrak{g})$-module $\mathrm{End}(U^q)$ into the direct sum of irreducible $U_q(\mathfrak{g})$-modules is such that

1. *it does not contain modules isomorphic to $V_\beta^q \subset I_-^q$ apart from those isomorphic to V^q, where by V^q we denote $\mathfrak{g} = V$ equipped with the $U_q(\mathfrak{g})$-module structure,*

2. *the multiplicity of the module V^q is 1,*
then U^q *can be equipped with a braided module structure (briefly, braided structure).*

Proof. By the assumption, there exists a unique $U_q(\mathfrak{g})$-submodule in $\mathrm{End}(U^q)$ isomorphic to V^q. Consider a $U_q(\mathfrak{g})$-morphism defined up to a factor

$$\rho : V^q \to \mathrm{End}(U^q). \tag{5}$$

This map is an almost representation of the braided Lie algebra $\bar{\mathfrak{g}}$ in the sense of the following

Definition 3 We say that a map (5) is an *almost representation* of the braided Lie algebra $\bar{\mathfrak{g}}$ if it is a $U_q(\mathfrak{g})$-morphism and the following properties are satisfied

1. $\circ \rho^{\otimes 2}(V_\beta^q) = 0$ for all $V_\beta^q \subset I_-^q$ apart from that V_-^q,
2. $\circ \rho^{\otimes 2} V_-^q = \nu \rho[\ ,\ ]_q V_-^q$ with some $\nu \neq 0$.

In a more explicit form these conditions can be reformulated as follows. If the elements $\{b_{k,\beta}^{i,j} u_i u_j, 1 \leq k \leq \dim V_\beta^q\}$ form a basis of the space $V_\beta^q \subset I_-^q$ and similarly the elements $\{b_{k,-}^{i,j} u_i u_j, 1 \leq k \leq \dim V_-^q\}$ form a basis of the space V_-^q then

$$b_{k,\beta}^{i,j} \rho(u_i)\rho(u_j) = 0, \text{ if } V_\beta^q \neq V_-^q \text{ and } b_{k,-}^{i,j} \rho(u_i)\rho(u_j) = \nu b_{k,-}^{i,j} \rho([u_i, u_j]_q).$$

Let us complete the proof. The image of the composed map $\circ \rho^{\otimes 2}(V_-^q)$ is isomorphic to the $U_q(\mathfrak{g})$-module V^q since ρ and $\circ$ are $U_q(\mathfrak{g})$-morphisms. Such a module in the decomposition of $\text{End}(U^q)$ is unique, therefore the image of the space $[\ ,\ ]_q V_-^q = V^q$ with respect to the morphism ρ coincides with the previous one (it suffices to show that the images of the highest weight element of the module V^q with respect to both operators coincide up to a factor). This gives the second property of Definition 2 (the property that $\nu \neq 0$ for a generic q follows from the fact that this is so for $q = 1$).

The first property of the Definition follows from the fact that $\text{End}(U^q)$ does not contain any modules isomorphic to $V_\beta^q \subset I_-^q$, $V_\beta^q \neq V_-^q$. Finally, changing the scale, i.e., considering the map $\rho_\nu = \nu^{-1}\rho$ instead of ρ, we get a representation of the algebra $U(\bar{\mathfrak{g}})$. This completes the proof. $\square$

A natural question arises: how many $U_q(\mathfrak{g})$-modules can be converted into braided ones or, in other words, how many $\mathfrak{g}$-modules allow braiding? The answer to this question for the $sl(n)$-case is given by the following proposition, which can be proved by straightforward computations using Young diagram techniques.

Proposition 3 *Let ω be a fundamental weight of the Lie algebra $sl(n)$. Then the $sl(n)$-modules $V_{k\omega}$ (for any nonnegative integer k) allow braiding. In other words, their q-analogues $V_{k\omega}^q$ are braided modules.*

For other simple Lie algebras, $\mathfrak{g}$ it seems very plausible that a similar statement is valid for fundamental weights ω such that their orbits in $\mathfrak{g}^*$ are symmetric (in the $sl(n)$-case all fundamental weights satisfy this condition). Note that all orbits of such type have been classified by E. Cartan (cf. (Khoroshkin and all, 1993)).

Let us now discuss how the braided modules can be used to find the above factors $c_i(h, q)$.

Once more we set $\mathfrak{g} = sl(n)$ and we take as fundamental the weight ω such that one $\omega(h_1) = 1$, $\omega(h_i) = 0$, $i > 1$. Then we consider the $U(sl(n))$-module $V_{k\omega}^q$ (which is in fact $V_{k\omega}$ equipped with a representation $\rho_q : U_q(\mathfrak{g}) \to \operatorname{End}(V_{k\omega}^q)$). Let us realize a braiding of the modules, i.e., construct a $U_q(\mathfrak{g})$-morphism

$$\rho : U(\mathfrak{g})_{h,q} \to \operatorname{End}(V_{k\omega}^q)$$

in the way described above for the algebra $U(\bar{\mathfrak{g}}) = U(\mathfrak{g})_{1,q}$.

Proposition 4 *The representation ρ is factorized to a representation of the algebras $A_{h,q}$ with certain $c_i(h,q)$. (This means that $\rho(V_\beta^q) = 0$ if $V_\beta^q \subset I_+^q \setminus (V_{2\alpha}^q \oplus V_0^q \oplus V_+^q)\}$, $\rho(C_q) = c_0(h,q)\mathrm{id}$, and $\rho(V_+^q) = c_1(h,q)\rho(V^q)$.)*

Proof. It suffices to show that the modules belonging to $I_+^q \setminus (V_{2\alpha}^q \oplus V_0^q \oplus V_+^q)\}$ are not represented in $\operatorname{End}(V_{k\omega}^q)$ for any k and that multiplicity of the modules V_0^q and V_+^q in $\operatorname{End}(V_{k\omega}^q)$ is 1. This can be done by straightforward calculations by means of Young diagram techniques.

Finally, the factors $c_i(h,q)$ are defined by these relations. Now if we want to find the relations defining the "quantum orbits" (i.e., algebras corresponding to the case $h = 0$), we pass to the limits $k \to \infty$ and $h \to 0$ in such a way that $c_0(h,q)$ has a limit (denoted by $c_0(q)$). By virtue of (Donin, Shnider, 1995) $c_1(h,q)$ also has a limit (denoted by $c_1(q)$). These two constants are just the factors that we are looking for. $\square$

It is interesting to emphasize that to find the system of equations describing a commutative algebra in the category $U_q(\mathfrak{g})$-Mod in the framework of this approach, we are looking first for the system corresponding to the noncommutative algebra $A_{h,q}$ ($h \neq 0$).

5. On "twisted quantum mechanics"

In this section we discuss the so-called twisted quantum mechanics. Roughly speaking, it is a quantum mechanics in twisted categories. Our aim is to consider the above quantum algebras from this point of view.

It is well known (De Wilde–Lecomte–Fedosov) that any symplectic Poisson structure can be quantized and the result of quantization can be treated as an operator algebra $\operatorname{End}(V)$ in a complex Hilbert space V. The space $\operatorname{End}(V)$ is equipped in this case with a trace and a conjugation (involution) possessing the usual properties

$$\operatorname{tr}(A \circ B) = \operatorname{tr}(B \circ A), \ \operatorname{tr} A^* = \operatorname{tr} A, \ (A \circ B)^* = B^* \circ A^*.$$

The quantum observables are identified with Hermitian (self-adjoint) operators in this space and they form a linear space over $\mathbb{R}$ closed with respect to the bracket $i[\ ,\ ]$.

If we consider a super-version of quantum mechanics, this bracket must be replaced by its super-analogue. The notions of ordinary trace and conjugation operators must be also replaced by their super-counterparts; this leads, in particular, to a modification of the notion of self-adjoint operators. Meanwhile, the observables that play the role of Hamiltonians must be even (since only for an even operator H the equation $dA/dt = i[H, A]$ is consistent) and such an operator is self-adjoint in the usual sense iff it is super-self-adjoint. Moreover, the super-bracket with such an operator becomes the ordinary Lie bracket.

What is a proper analogue of this scheme in a twisted category? If a category in question is symmetric, i.e., the Yang–Baxter twist S is involutive ($S^2 = \mathrm{id}$) in it, the corresponding generalization was suggested in (Gurevich and all, 1992). In this case "S-analogues" of Lie bracket, trace and conjugation operators introduced there satisfy the following version of the above relations

$$\mathrm{tr}(A \circ B) = \mathrm{tr} \circ S(A \otimes B), \ \mathrm{tr}\, A^* = \mathrm{tr}\, A, \ (A \circ B)^* = \circ(* \otimes *)S(A \otimes B). \quad (6)$$

Let us recall a construction of a conjugation operator in this case (assuming V to be finite-dimensional). Let us suppose that we can identify V and V^* (this means that there exists a pairing $\langle\ ,\ \rangle : V^{\otimes 2} \to k$, which is a morphism of the category). Then $\mathrm{End}(V)$ can be identified with $V^{\otimes 2}$ and the conjugation $* : \mathrm{End}(V) \to \mathrm{End}(V)$ is just the image of the operator $S : V^{\otimes 2} \to V^{\otimes 2}$ under this identification. Therefore the conjugation satisfies the following relation: $(* \otimes \mathrm{id})S = S(\mathrm{id} \otimes *)$.

More precisely, we consider the space $\mathfrak{g} = V$ over the field $k = \mathbb{R}$ and assume all matrix elements of S to be real. This implies that if we introduce an S-Lie bracket in $\mathrm{End}(V)$ by

$$[A, B] = A \circ B - \circ S(A \otimes B),$$

its structural constants are real as well.

Let us extend this bracket to the space $V_{\mathbb{C}} = V \otimes \mathbb{C}$ by linearity. Let $* : V_{\mathbb{C}} \to V_{\mathbb{C}}$ be a conjugation, i.e., an involutive ($*^2 = \mathrm{id}$) operator such that $(\lambda z)^* = \bar{\lambda} z^*$, $\lambda \in \mathbb{C}$, $z \in V_{\mathbb{C}}$. Assume also that the relation $(* \otimes \mathrm{id})S = S(\mathrm{id} \otimes *)$ and the third relation (6) are satisfied. Then it is easy to see that this conjugation is compatible with the S-Lie bracket in the following sense

$$* [A, B] = -[A^*, B^*]. \quad (7)$$

It is not so evident what is the proper definition of a conjugation operator, say, in the algebra $\mathrm{End}(V)$, where V is an object of a twisted but nonsymmetric category. Now possessing a q-analogue of the Lie bracket, we can try to define the compatibility of such an operator with the twisted structure by means of relation (7). (Let us note that the operator S does not enter explicitly in it and formula (7) is in some sense universal.)

Let us consider a space $V_{\mathbb{C}}$ equipped with a bracket $[\,,\,]: V_{\mathbb{C}}^{\otimes 2} \to V_{\mathbb{C}}$.

Definition 4 We say that a conjugation $*$ is *compatible* with this bracket if relation (7) is satisfied.

Proposition 5 *The odd elements with respect to this involution (i.e., elements such that $z^* = -z$) form a subalgebra, i.e., the element $[a, b]$ is odd if a and b are.*

Proof. It is obvious. □

Therefore the space of "$*$-even" operators is closed with respect to the bracket $i[\,,\,]$.

In (Donin and all., 1997), we have classified all such conjugations for $U_q(sl(2))$-case. Let us reproduce the final result here, but first represent the multiplication table for the q-Lie bracket in some base $\{u, v, w,\}$:

$$[u, u] = 0, \ [u, v] = -q^2 Mu, \ [u, w] = (q + q^{-1})^{-1} Mv,$$

$$[v, u] = Mu, \ [v, v] = (1 - q^2) Mv, \ [v, w] = -q^2 Mw,$$

$$[w, u] = -(q + q^{-1})^{-1} Mv, \ [w, v] = Mw, \ [w, w] = 0,$$

Here M is an arbitrary real factor (cf. (Gurevich, 1997) for details).

Proposition 6 *For a real $q \neq 1$ there exist only two conjugations in the space $V_{\mathbb{C}}$ compatible with the q-Lie bracket, namely, $a^* = -\bar{a}$ for any $a \in V_{\mathbb{C}}$ and the following one: $u^* = u, \ v^* = -v, \ w^* = w$.*

Although these conjugations are rather trivial they are, together with quantum traces (their construction in End(V), where V is an object of a rigid category, is well known), the ingredients of twisted quantum mechanics in the sense of the following informal definition.

Definition 5 We say that an associative algebra is a *subject of twisted quantum mechanics*, if it belongs to a twisted category, it is represented in the space End(V) equipped with a twisted Lie bracket, a trace and a conjugation as above and the representation map is a morphism in this category.

Unfortunately, we cannot give a final axiom system of twisted quantum mechanics for a noninvolutive S (though it seems very plausible that the above relations between the operators under consideration are still valid in the category $U_q(\mathfrak{g})$-Mod, if we replace S by $\tilde{S}$). However, we want to point out the principal difference between the twisted version of quantum mechanics and its classical version: the twisted (quantum) trace must occur in calculations of partition functions.

In the above sense the second type quantum algebras represented in some linear spaces in spirit of the paper (Vaksman, Soibelman, 1988) are not subjects of twisted quantum mechanics. Though the algebra $A_{0,q} = \text{Sym}(W)$ itself belongs to the twisted category generated by the space W, its modules constructed in (Vaksman, Soibelman, 1988) for $su(2)$ case do not belong to this category.

As for the second type algebras $A_{h,q}$ with $h \neq 0$, they differ nonessentially from the algebras $A_{0,q}$.

This is not the case for the first type algebra $A_{h,q}$. The representation theory of the algebras $A_{h,q}$ for generic h and that for the case $h = 0$ are completely different. From this point of view $h = 0$ is a singular point and we disregard it.

Let us note that our first type quantum algebras are subjects of twisted quantum mechanics arising from the quantization of Poisson brackets. However, it is possible to consider similar objects connected to nondeformational solutions of the QYBE (cf. (Gurevich, 1991), (Gurevich and all, 1992)).

Concluding the paper, we would like to formulate two problems: that of calculating the above factors $c_i(h, q)$ (or more generally, giving an exact description of the two parameter deformation of all symmetric orbits in $\mathfrak{g}^*$ for any simple Lie algebra $\mathfrak{g}$) and the problem of generalizing our approach to infinite-dimensional Lie algebras.

References

Bezrukavnikov, R. (1995): Koszul property and Frobenius splitting of Schubert Varieties, *alg-geom/9502021*.

Braverman, A. and Gaitsgory, D. (1994): Poincaré–Birkhoff–Witt theorem for quadratic algebras of Koszul type, *hep-th/9411113*.

Chari, V. and Pressley, A. (1994): A Guide to Quantum Groups, *Cambridge University Press*, 1994.

Delius, G. and Hüffmann, A. (1995): On quantum algebras and quantum root systems, *q-alg/9506017*.

Donin, J. and Gurevich, D. (1995): Quantum orbits of R-matrix type, *Lett. Math. Phys.* **35** (1995) 263–276.

Donin, J. and Gurevich, D. (1995a): Some Poisson structures associated to Drinfeld–Jimbo R-matrices and their quantization, *Israel J. Math.* **92** (1995) 23–32.

Donin, J., Gurevich, D. and Majid, S. (1993): R-matrix brackets and their quantization, *Ann. Inst. Henri Poincaré* **58** (1993) 235–246.

Donin, J., Gurevich, D. and Rubtsov, V.: (1997): Quantum hyperboloid and braided modules, in: *Contacts Franco–Belges à Reims, 1995*, Publ. de SMF, série Colloques et Séminaires, 1997.

Donin, J. and Shnider, S. (1995): Quantum symmetric spaces, *J. of Pure and Applied Algebra* **100** (1995) 103–115.

Donin, J. and Shnider, S. (1996): Quasiassociativity and flatness criteria for quadratic algebra deformation, *Israel J. Math.* (to appear).

Gurevich, D. (1991): Algebraic aspects of the quantum Yang–Baxter equation, *Leningrad Math. J.* **2** (1991) 801–828.

Gurevich, D. (1997): Braided modules and reflection equations, in: *Quantum Groups and Quantum Spaces*, Banach Center Publ. **40**, Inst. of Math., Polish Academy of Sciences, Warszawa 1997.

Gurevich, D. and Panyushev, D. (1994): On Poisson pairs associated to modified R-matrices, *Duke Math. J.* **73** (1994) 249–255.

Gurevich, D. and Rubtsov, V. (1995): Quantization of Poisson pencils and generalized Lie algebras, *Teor. i Mat. Fiz.* **103** (1995) 476–488.

Gurevich, D., Rubtsov, V. and Zobin, N. (1992): Quantization of Poisson pairs: R-matrix approach, *J. Geom. and Phys.* **9** (1992) 25–44.

Isaev, A. (1995): Interrelation between quantum groups and reflection equation (braided) algebras, *Lett. Math. Phys.* **34** (1995) 333–341.

Isaev, A. and Pyatov, P. (1995): Covariant differential complex on quantum linear groups, *J. Phys. A: Math. Gen.* **28** (1995) 2227–2246.

Khoroshkin, S., Radul, A. and Rubtsov, V. (1993): Families of Poisson structures on Hermitean Symmetric Spaces, *Comm. Math. Phys.* **153** (1993) 299–315.

Lyubashenko, V. and Sudbery, A. (1995): Quantum Lie algebras of type A_n, q-alg/9510004.

Reshetikhin, N., Takhtadzhyan, L. and Faddeev, L. (1990): Quantization of Lie groups and Lie algebras, *Leningrad Math. J.* **1** (1990) 193–226.

Semenov-Tian-Shansky, M. (1983): What is classical r-matrix? *Funct. Anal. Appl.* **17** (1983) 259–272.

Sklyanin, E. (1982): Some algebraic structures connected with the Yang–Baxter equation, *Funct. Anal. Appl.* **16** (1982) 263–270.

Vaksman, L. and Soibelman, Y. (1988): Algebra of functions on the quantum group SU(2), *Funct. Anal. Appl.* **22** (1988) 170-181.

WAVE PACKET TRANSFORM IN SYMPLECTIC GEOMETRY AND ASYMPTOTIC QUANTIZATION

VLADIMIR NAZAIKINSKII AND BORIS STERNIN
Department of Physics,
Department of Computer Mathematics and Cybernetics,
Moscow State University, 119899 Moscow, Russia.
E-mail: nve@qs.phys.msu.su, boris@sternin.msk.su

Abstract. We discuss a universal quantization procedure based on an integral transform that takes functions on the configuration space to functions on the phase space and is closely related to the Bargmann transform. In the leading term this procedure yields Schrödinger's quantization of observables, Maslov's quantization of Lagrangian modules, and Fock's quantization of canonical transforms.

Mathematics Subject Classification (1991): Primary 53B30, 53C80, 83C57, 83C75.

Key words: quantization, Bargmann transform, wave packet, intertwining operators, Lagrangian manifold, Maslov's canonical operator, Fourier integral operators, anti-Wick symbol, Hamiltonian.

Introduction

This text is an abridged version of the preprint [1], where the ideas of [2] were developed. We omit some proofs, which can be found in [1]. We deal with *asymptotic, or semi-classical, quantization*. Let us first explain this notion in some detail.

1. By quantization of classical mechanics physicists mean the assignment of quantum objects to the corresponding classical ones. The main classes of objects are *states* and *observables*. We recall that in classical mechanics the state of a system is determined by a point (q, p) in the phase space (the

B. P. Komrakov et al. (eds.), Lie Groups and Lie Algebras, 47–69.
© 1998 *Kluwer Academic Publishers. Printed in the Netherlands.*

space of coordinates and momenta), and observables are functions $f(q,p)$ on this space. In quantum mechanics, the states of the system are described by ψ-*functions* (or *wave functions*) $\psi(x)$ and observables are described by linear operators in the state space.

For the Schrödinger quantization, the correspondence between the classical and the quantum *observables* is given by the rule

$$q \mapsto \hat{q} = x, \qquad p \mapsto \hat{p} = -ih\frac{\partial}{\partial x},$$

so that the (pseudodifferential) operator corresponding to an observable $f(q,p)$ has the form

$$\hat{f} = f(\overset{2}{\hat{q}}, \overset{1}{\hat{p}}) \tag{1}$$

(the numbers 1 and 2 indicate the order of action of the operators $\hat{q}$ and $\hat{p}$ [3]; different orderings, such as $f(\overset{1}{\hat{q}}, \overset{2}{\hat{p}})$, the Weyl ordering $f(\overset{2}{\hat{q}} + \overset{3}{\hat{q}}, \overset{1}{\hat{p}})$, or the Jordan ordering $(1/2)(f(\overset{1}{\hat{q}}, \overset{2}{\hat{p}}) + f(\overset{2}{\hat{q}}, \overset{1}{\hat{p}}))$, give the same result with the accuracy $O(h)$.)

The correspondence between classical and quantum *states* is not so simple. Although simultaneous measurement of the coordinates and the momenta is impossible, it makes sense to speak of the *joint probability density of the coordinates and the momenta* for a quantum particle in a state $\psi(x)$: the mean value of an arbitrary observable $f(q,p)$ in the state ψ is

$$\langle f \rangle_\psi = (\psi, \hat{f}\psi)_{L_2} = \operatorname{tr}(\hat{f}\hat{P}_\psi) = \int_{\mathbb{R}^{2n}} f(q,p)\rho(q,p)\,dq\,dp,$$

where $\hat{P}_\psi$ is the orthogonal projection on ψ and

$$\rho(q,p) = \psi(q)\tilde{\psi}(p)e^{ipx/h}, \quad \tilde{\psi}(p) = (2\pi h)^{-n/2}\int e^{ipx/h}\psi(x)\,dx.$$

The function $\rho(q,p)$ is the desired joint probability density (it is known as the *density function* corresponding to $\psi(x)$). In the semiclassical limit (i.e., $h \to 0$) the density function vanishes for some classical states (p,q); if the support of the limit density is a manifold and if a certain additional condition is satisfied, then this manifold is necessary an *isotropic* submanifold of the phase space, that is, a submanifold on which the Cartan form $p\,dq$ is closed.

From the viewpoint of a quantum particle, this submanifold is the *oscillation front* of the ψ-function. Quantization must assign a ψ-function (more precisely, a class of ψ-functions) to a given isotropic manifold in such a way that the oscillation fronts of these functions lie in this isotropic manifold.

2. Let us now present a mathematical treatment of the above physical considerations. A similar discussion can be found in [4]. It will be convenient to use the language of category theory.

Let us fix a phase space, for example, the cotangent space of a smooth real manifold M with the canonical symplectic structure $dp \wedge dq$.

i) Consider the category $\mathcal{C}$ whose *objects* are the modules $C^\infty(\Lambda)$ of smooth complex functions on compact Lagrangian manifolds Λ over the ring of classical observables. Thus, an object in this category is the abelian group $C^\infty(\Lambda)$ for some Lagrangian manifold Λ with the following action of the ring $C^\infty(T^*M)$ of classical observables:

$$f \cdot \varphi = f(p, q)\,|_\Lambda\,\varphi,$$

where the usual pointwise multiplication is used on the right-hand side.

Morphisms in this category are induced by *symplectic transforms* of the phase space. Namely, let Λ_1 and Λ_2 be Lagrangian manifolds. If

$$g : T^*M \to T^*M \tag{2}$$

is a symplectic transform such that $\Lambda_2 = g(\Lambda_1)$, then to (2) we assign the module homomorphism

$$g^* : C^\infty(\Lambda_2) \to C^\infty(\Lambda_1)$$

over the ring homomorphism

$$g^* : C^\infty(T^*M) \to C^\infty(T^*M).$$

Note that since Λ_1 and Λ_2 are compact, we can always assume that g is well-behaved at infinity (i.e., all derivatives of g are uniformly bounded).

ii) Consider the category $\mathcal{Q}$ whose *objects* are the spaces $C_h^\infty(M, \Lambda)$ of smooth functions $\psi(x, h)$ depending on the parameter $h \in (0, 1]$ with oscillation fronts in Λ. The space $C_h^\infty(M, \Lambda)$ is viewed as a module over the ring $\mathrm{PSD}(M)$ of quantum observables (that is, pseudodifferential operators). Note that pseudodifferential operators preserve oscillation fronts and hence the module structure is well defined.

Morphisms in this category are given by invertible mappings

$$T : C^\infty(M) \to C^\infty(M)$$

such that

$$T : C_h^\infty(M, \Lambda_1) \to C_h^\infty(M, \Lambda_2) \tag{3}$$

and the operator $T\hat{H}T^{-1}$ is a pseudodifferential operator for any pseudodifferential operator $\hat{H}$. Note that (3) is a module homomorphism over the ring homomorphism $\hat{H} \to T\hat{H}T^{-1}$.

iii) A *semi-classical quantization* is a projective contravariant functor[1]:

$$\mathcal{F} : \mathcal{C} \to \mathcal{Q} \quad (\mathrm{mod}\, O(h))$$

such that the module of smooth functions on M with oscillation front in a given Lagrangian manifold is assigned to the module of smooth functions on this Lagrangian manifold together with the $\mathcal{F}$-functorial mappings

$$\mu : C^\infty(T^*M) \to \mathrm{PSD}(M)$$

and

$$K_\sigma : C_0^\infty(\Lambda) \to C_h^\infty(M, \Lambda)$$

for each Lagrangian manifold Λ equipped with a measure σ. (By saying that these mappings are $\mathcal{F}$-functorial, we mean that they are naturally included in projectively commutative diagrams involving morphisms in $\mathcal{C}$ and $\mathcal{Q}$ related by $\mathcal{F}$.)

Such a functor exists and can be explicitly constructed on the basis of a certain integral transform, which we call the *wave packet transform*. This is an *invertible* transform taking functions $f(x, h)$ determined on the configuration space $\mathbb{R}^n$ to some subspace of functions $\tilde{f}(q, p, h)$ determined on the phase space $T^*\mathbb{R}^n$. By using such a transform, one can carry out a unified construction of quantization of all classical objects.

Moreover, this procedure coincides in the leading term with the Schrödinger quantization[2] [6] for observables, with the Fock quantization [7] for canonical (symplectic) transforms, and with the Maslov quantization [8] for Lagrangian modules.

Implementing this construction, we obtain $1/h$-pseudodifferential operators as quantization of observables, Fourier integral operators as quantization of symplectic transforms, and Maslov's canonical operator as quantization of Lagrangian modules (the reader can find the notions used here, for example, in [9]).

3. We conclude these preliminary considerations with some remarks. To obtain the correspondence between classical and quantum objects, that is, to construct the quantization procedure, we try to decompose any quantum state $\psi(x, h)$ into a sum of elements corresponding to points (q, p) of the phase space $T^*\mathbb{R}^n$, that is, to classical states. Such a decomposition is a *microlocalization procedure*. Various versions of this procedure were widely

---

[1]By a *projective* functor we mean a mapping between categories such that the composition of morphisms is preserved up to a unimodular factor; this makes sense if the sets of morphisms have the structure of vector spaces over $\mathbb{C}$.

[2]We actually use the anti-Wick quantization (see, e. g., [5]), which coincides with the Schrödinger quantization in the leading term.

presented in the literature (e.g., see [10]). The version in this paper leads to a transform providing the direct quantization procedure.

The localization of a function $f(x, h)$ in the phase space can be accomplished by localization along the fibers of $T^*\mathbb{R}^n$ followed by localization along the base. The localization along the base uses the "integral partition of unity" of the form

$$1 = \left(\frac{1}{2\pi h}\right)^{n/2} \int e^{-\frac{1}{2h}(x-x_0)^2} dx.$$

Localization along the base is therefore obtained with the help of multiplication by

$$\delta_h(x - x_0) = \left(\frac{1}{2\pi h}\right)^{n/2} e^{-\frac{1}{2h}(x-x_0)^2};$$

note that

$$\delta_h(x - x_0) \to \delta(x - x_0)$$

as $h \to 0$.

Localization along the fibers can be done with the help of the quantum Fourier transform, in other words, by means of a p-representation at the point x_0:

$$F_{x \to p_0}[f] = \left(\frac{1}{2\pi ih}\right)^{n/2} \int e^{-\frac{i}{h}p_0(x-x_0)} f(x) \, dx.$$

By composition of these two localizations we obtain the *microlocal element* corresponding to the function $f(x, h)$ in the form

$$
\begin{aligned}
f_{(x_0,p_0)} &= F_{x \to p_0}\{\delta_h(x - x_0) f(x, h)\} = \left(\frac{1}{2\pi h}\right)^{n/2} \left(\frac{1}{2\pi ih}\right)^{n/2} \times \\
&\times \int \exp\left\{\frac{i}{h}\left[-p_0(x - x_0) + \frac{i}{2}(x - x_0)^2\right]\right\} f(x) \, dx.
\end{aligned}
$$

The latter formula determines an integral transform, which we call the *wave packet transform* of the function $f(x, h)$. The inverse transform is given by

$$\tilde{f}(x_0, p_0) \mapsto f(x) = \left(\frac{i}{2\pi h}\right)^{n/2} \int e^{\frac{i}{h}[p_0(x-x_0) + \frac{i}{2}(x-x_0)^2]} \tilde{f}(x_0, p_0) \, dx_0 dp_0.$$

It is convenient to renormalize the obtained transform is such a way that the Parseval identity takes place. This normalization is used below.

1. Definition and basic properties of the wave packet transform

1. In 1961 V. Bargmann [11] introduced a remarkable integral transform relating the "harmonic oscillator representation" of the creation–annihilation operators for Bose particles in quantum field theory to the Fock representation of these operators [12, 13]. Let us briefly recall these results. In the harmonic oscillator representation, the creation and annihilation operators act in the space $L^2(\mathbb{R}_q^n)$ of square integrable functions of the variables $q = (q_1, \ldots, q_n)$ and have the form[3] $(\hat{p}_i = -i\partial/\partial q_i)$:

$$a_i^* = \frac{1}{\sqrt{2}}(q_i - i\hat{p}_i) = \frac{1}{\sqrt{2}}\left(q_i - \frac{\partial}{\partial q_i}\right), \quad i = 1, \ldots, n \quad \text{(creation}$$
$$\text{operators)}; \tag{4}$$
$$a_i = \frac{1}{\sqrt{2}}(q_i + i\hat{p}_i) = \frac{1}{\sqrt{2}}\left(q_i + \frac{\partial}{\partial q_i}\right), \quad i = 1, \ldots, n \quad \text{(annihilation}$$
$$\text{operators)}.$$

The operators a_i and a_i^* are adjoints of each other with respect to the inner product on $L^2(\mathbb{R}_q^n)$ and satisfy the commutation relations

$$[a_i, a_j] = [a_i^*, a_j^*] = 0, \quad [a_i, a_j^*] = \delta_{ij}, \quad i, j = 1, \ldots, n. \tag{5}$$

Fock introduced a different solution of the commutation relations (5), namely,

$$\tilde{a}_i = \frac{\partial}{\partial z_i}, \quad \tilde{a}_i^* = z_i, \quad i = 1, \ldots, n. \tag{6}$$

Here it is required, in analogy with (5), that the operators $\tilde{a}_i$ and $\tilde{a}_i^*$ be mutually adjoint in some Hilbert space of functions of $z = (z_1, \ldots, z_n)$. One can achieve this by assuming that the z_i are *complex variables*, $z_i = x_i + iy_i$. Then the operators (6) are pairwise adjoint in the Hilbert space $\mathcal{F}_n$ of entire analytic functions $f(z)$ with the inner product

$$(f, g) = \frac{1}{\pi^n} \int_{\mathbb{C}^n} \bar{f}(z)\, g(z) e^{-\bar{z}z} dx\, dy = \sum_{|\alpha|=0}^{\infty} \alpha! \bar{f}_\alpha g_\alpha \tag{7}$$

(here $dx\, dy$ is the standard Lebesgue measure in $\mathbb{C}^n$,

$$\bar{z}z = \sum_{i=1}^{n} \bar{z}_i z_i,$$

$\alpha = (\alpha_1, \ldots, \alpha_n)$ is a multi-index, and f_α and g_α are the Taylor coefficients of f and g: $f = \sum_{|\alpha|=0}^{\infty} f_\alpha z^\alpha$, and similarly for g).

---

[3]In the system of units in which $h = 1$.

The Bargmann transform A_n acts from $L^2(\mathbb{R}_q^n)$ into $\mathcal{F}_n$ according to the formula

$$(A_n\psi)(z) = \int_{\mathbb{R}_q^n} A_n(z,q)\,\psi(q)\,dq, \quad \psi \in L^2(\mathbb{R}_q^n), \tag{8}$$

where the kernel has the form

$$A_n(z,q) = \frac{1}{\pi^{n/4}} \exp\left\{-\frac{1}{2}(z^2 + q^2) + \sqrt{2}zq\right\}. \tag{9}$$

The main properties of the Bargmann transform are given by the following theorem.

Theorem 1 i) *The transform*

$$A_n : L^2(\mathbb{R}_q^n) \to \mathcal{F}_n$$

is an isometric isomorphism (that is, a unitary operator).
  ii) *The inverse transform is given by the formula*

$$(A_n^{-1}f)(q) = \lim_{\lambda \to 1} \frac{1}{\pi^n} \int_{\mathbb{C}^n} \overline{A_n(z,q)} f(\lambda z) e^{-\bar{z}z} dx\,dy, \tag{10}$$

where $\lambda \to 1$ from below, and the limit is understood in the strong sense in $L^2(\mathbb{R}_q^n)$.
  iii) *The transform A_n is an intertwining operator for the representations (4) and (6) of the commutation relations (5), that is,*

$$A_n \cdot a_i = \tilde{a}_i \cdot A_n, \quad A_n \cdot a_i^* = \tilde{a}_i^* \cdot A_n, \quad i = 1,\ldots,n. \tag{11}$$

The comparison of the formulas

$$a_i^* = \frac{1}{\sqrt{2}}(q_i - i\hat{p}_i) \text{ and } \tilde{a}_i^* = z_i$$

suggests that it might be useful to identify the complex space $\mathbb{C}^n$ on which the elements of $\mathcal{F}_n$ are defined with the phase space $\mathbb{R}_q^n \oplus \mathbb{R}_p^n$ according to the formula

$$z = \frac{1}{\sqrt{2}}(q - ip).$$

In the "exact" theory, this identification, as Bargmann noted, is of limited applicability since q_k and $\hat{p}_k$ do not commute. However, it is quite adequate in the asymptotic theory (as $h \to 0$), but we need to consider a different transform.

2. Consider the *Gaussian wave packet*

$$G_h(q,p;x) = \exp\left\{\frac{i}{h}(x-q)p - \frac{1}{2h}(x-q)^2\right\}, \quad x \in \mathbb{R}^n, \tag{12}$$

where $h > 0$, $q \in \mathbb{R}^n$, and $p \in \mathbb{R}^n$ are parameters. We use the function (12) as a kernel to define an integral transform U acting on $L^2(\mathbb{R}_x^n)$ as follows:

$$U[f](q,p) = 2^{-\frac{n}{2}}(\pi h)^{-\frac{3n}{4}} \int_{\mathbb{R}_x^n} \overline{G_h(q,p;x)}\, f(x)\, dx, \quad f \in L^2(\mathbb{R}_x^n), \tag{13}$$

where the bar denotes complex conjugation and $dx = dx_1 \cdot dx_2 \cdot \ldots \cdot dx_n$ is the standard Lebesgue measure on $\mathbb{R}_x^n$. The integral on the right-hand side in (13) is obviously well defined, since $G_h(q,p;x)$ belongs to $L^2(\mathbb{R}_x^n)$ for any fixed h, q, and p.

Definition 1 The integral transform U defined in (13) will be called the *wave packet transform*.

Remark 1 In [2] this transform was called the "Fourier–Gauss transform," but we prefer the present name since this transform is the symplectic analog of the wave packet transform considered by Cordoba and Fefferman [14] (see also [15]).

Theorem 2 (i) *The wave packet transform is a bounded operator in the spaces*

$$U : L^2(\mathbb{R}_x^n) \to L^2(\mathbb{R}_{q,p}^{2n}) \tag{14}$$

and satisfies the Parseval identity

$$(Uf, Ug)_{L^2} = (f,g)_{L^2}. \tag{15}$$

(ii) *The adjoint operator*

$$U^* : L^2(\mathbb{R}_{q,p}^{2n}) \to L^2(\mathbb{R}_x^n) \tag{16}$$

is given by the formula

$$U^*[\psi](x) = 2^{-n/2}(\pi h)^{-\frac{3n}{4}} \int_{\mathbb{R}_{q,p}^{2n}} G_h(q,p;x)\psi(q,p)\, dq\, dp, \tag{17}$$

where the integral on the right-hand side (which is not absolutely convergent at infinity in general) is understood as the limit of the similar integrals with $\psi(q,p)$ replaces by $\psi_k(x,p)$, where $\{\psi_k\}$ is a sequence of compactly supported functions convergent to ψ in $L^2(\mathbb{R}_{q,p}^{2n})$.

(iii) *One has the inversion formula*

$$U^*U = 1. \tag{18}$$

(iv) *The range of U is the closed subspace $\mathcal{F}^2(\mathbb{R}^{2n}_{q,p}) \subset L^2(\mathbb{R}^{2n}_{q,p})$ of functions $F(q,p)$ that satisfy the equations*

$$\left[h\frac{\partial}{\partial q_j} - ih\frac{\partial}{\partial p_j} - ip_j \right] F(q,p) = 0, \quad j = 1, \ldots, n. \tag{19}$$

Remark 2 Obviously, condition (19) is equivalent to saying that

$$\exp\{p^2/(2h)\} F(q,p)$$

is an analytic function of the variables $q - ip = (q_1 - ip_1, \ldots, q_n - ip_n)$.

The following statement is quite obvious.

Theorem 3 *Set*

$$U^{-1} = U^*|_{\mathcal{F}^2(\mathbb{R}^{2n}_{q,p})}. \tag{20}$$

Then

$$U^{-1}U = 1, \quad UU^{-1} = 1, \tag{21}$$

but

$$U^*U = 1, \quad UU^* = P, \tag{22}$$

where P is the operator of orthogonal projection in $L^2(\mathbb{R}^{2n}_{p,q})$ onto $\mathcal{F}^2(\mathbb{R}^{2n}_{p,q})$.

Next, similarly to the Fourier transform, let us derive some commutation formulas for U.

Theorem 4 *One has the commutation formulas*

$$U \circ x = \left(q + ih\frac{\partial}{\partial p} \right) \circ U, \tag{23}$$

$$U \circ \left(-ih\frac{\partial}{\partial x} \right) = \left(-ih\frac{\partial}{\partial q} \right) \circ U \tag{24}$$

(as usual, the equality of two unbounded operators implies that their domains coincide).

Definition 2 Let $k \in \mathbb{Z}_+ = \{0, 1, 2, \ldots\}$. By $H^{1/h}_k(\mathbb{R}^n_x)$ we denote the space of functions $f(x)$ with finite norm

$$\|f\|^2_{k,n} = \int_{\mathbb{R}^n} \overline{f(x)} \left[\left(1 + x^2 - h^2\frac{\partial^2}{\partial x^2} \right)^k f(x) \right] dx,$$

$$x^2 = \sum x_j^2, \quad \frac{\partial^2}{\partial x^2} = \sum \frac{\partial^2}{\partial x_j^2}. \tag{25}$$

Obviously, $H_k^{1/h}(\mathbb{R}_x^n)$ is a Hilbert space and we have the filtration

$$L^2(\mathbb{R}^n) = H_0^{1/h}(\mathbb{R}^n) \supset H_1^{1/h}(\mathbb{R}^n) \supset \ldots \supset H_k^{1/h}(\mathbb{R}^n) \supset \ldots .$$

Similarly, we introduce the spaces $H_k^{1/h}(\mathbb{R}_{p,q}^{2n})$ equipped with the norms

$$\|\psi\|_{k,h}^2 = \int_{\mathbb{R}^{2n}} \overline{\psi(p,q)} \left[\left(1 + q^2 + p^2 - h^2 \frac{\partial^2}{\partial q^2} - h^2 \frac{\partial^2}{\partial p^2} \right)^k \psi(p,q) \right] dp\, dq. \tag{26}$$

Next, let us consider functions $f(x,h)$ depending on the parameter $h \in (0,1]$. We introduce the norm

$$\|f\|_k = \sup_{h \in (0,1]} \|f\|_{k,h} \tag{27}$$

and denote by $H_k(\mathbb{R}^n)$ the space of functions with finite norm (27). Furthermore, we consider the Fréchet space

$$H_\infty(\mathbb{R}^n) = \bigcap_{k=0}^{\infty} H_k(\mathbb{R}^n) \tag{28}$$

with the topology defined by the countable system of seminorms (27). The spaces $H_k(\mathbb{R}^{2n})$ and $H_\infty(\mathbb{R}^{2n})$ are defined similarly.

Finally, let $H_{-k}^{1/h}(\mathbb{R}^n)$ be the dual space of $H_k^{1/h}(\mathbb{R}^n)$ with respect to the L_2 inner product. The elements of $H_{-k}^{1/h}$ are naturally interpreted as distributions, and we have the embeddings

$$\supset H_{-k}^{1/h}(\mathbb{R}^n) \supset \ldots \supset H_{-1}^{1/h}(\mathbb{R}^n) \supset H_0^{1/h}(\mathbb{R}^n) = L_2(\mathbb{R}^n) \supset H_1^{1/h}(\mathbb{R}^n) \supset \ldots .$$

The definition of the spaces H_k extends to negative k, and we set

$$H_{-\infty}(\mathbb{R}^n) = \bigcup_{k=\infty}^{-\infty} H_k(\mathbb{R}^n); \tag{29}$$

a net $\{\psi_k\}$ is said to be convergent in $H_{-\infty}$ if it converges in some H_k; with this topology, $H_{-\infty}$ is the dual of H_∞.

Theorem 5 (i) *For any $k \in \mathbb{Z}$ the mappings U and U^* are continuous in the spaces*

$$U : H_k(\mathbb{R}_x^n) \to H_k(\mathbb{R}_{p,q}^{2n}), \quad U^* : H_k(\mathbb{R}_{p,q}^{2n}) \to H_k(\mathbb{R}_x^n) \tag{30}$$

(for negative k, we extend these mappings from L_2 by continuity).

(ii) *Let* $H_{-\infty,R}(\mathbb{R}^{2n}_{p,q})$ *be the subspace of elements* $\psi \in H_{-\infty}(\mathbb{R}^{2n}_{p,q})$ *such that* supp $\psi \subset B_R = \{(q,p) \in \mathbb{R}^{2n} | q^2 + p^2 \le R\}$. *Let* $H_{-\infty,\mathrm{comp}}(\mathbb{R}^{2n}_{p,q}) = \cup_R H_{-\infty,\mathrm{comp}}(\mathbb{R}^{2n}_{p,q})$. *Then* U^* *is continuous in the spaces*

$$U^* : H_{-\infty,\mathrm{comp}}(\mathbb{R}^{2n}_{p,q}) \to H_\infty(\mathbb{R}^n_x). \tag{31}$$

Consequently, the projection $P = UU^*$ *is continuous in the spaces*

$$P : H_{-\infty,\mathrm{comp}}(\mathbb{R}^{2n}_{p,q}) \to H_\infty(\mathbb{R}^{2n}_{p,q}) \cap \mathcal{F}_2(\mathbb{R}^{2n}_{p,q}). \tag{32}$$

3. Let us summarize the preceding in a somewhat different form. Let $M = \mathbb{R}^{2n}_{q,p}$ be the $2n$-dimensional space equipped with the standard Lebesgue measure $dp\,dq$. In $L^2(\mathbb{R}^n_x)$ consider the system of vectors

$$e_{(q,p)}(x) = 2^{-n/2}(\pi h)^{-3n/4} G_h(q,p;x) \tag{33}$$

The system (33) is *complete* in $L^2(\mathbb{R}^n_x)$ in the sense that

$$(f,f)_{L^2(\mathbb{R}^n_x)} = \int |(f, e_{(q,p)})|^2 dq\,dp. \tag{34}$$

This is just the Parseval identity (15); note that the transform $U[f]$ in these terms is given by

$$U[f](q,p) = (f, e_{(q,p)}) \tag{35}$$

and defines an isometric embedding

$$L^2(\mathbb{R}^n_x) \to L^2(\mathbb{R}^{2n}_{q,p}), \quad f \mapsto U[f].$$

We have

$$f = \int (f, e_{(q,p)}) e_{(q,p)} dq\,dp \tag{36}$$

(the integral is understood in the weak sense); this is just the inversion formula (18).

Furthermore, there is an orthogonal projection

$$P = UU^* : L^2(\mathbb{R}^{2n}_{q,p}) \to L^2(\mathbb{R}^n_x)$$

(identified with its image $\mathcal{F}^2(\mathbb{R}^{2n}_{q,p}) = U(L^2(\mathbb{R}^n_x))$), and the operator P can be extended to a wider set including distributions that belong to $H_{-\infty,R}(\mathbb{R}^{2n})$ for some R. In particular, this set includes the delta functions (more precisely, the functions

$$\varphi_{q_0,p_0} = h^{n/2}\delta(q - q_0)\delta(p - p_0), \tag{37}$$

where $\delta(y)$ is the Dirac delta function).

Thus, we are in the situation of the papers [5, 16], which permits us to consider operators with co- and contravariant symbols (or Wick and anti-Wick symbols); this will be used in the next section.

2. Quantization of observables

In this section we use the wave packet transform to study h^{-1}-pseudodifferential operators in the scale $\{H_k(\mathbb{R}_x^n)\}$; as a by-product, we obtain some more properties of wave packet transforms.

We use the symbol class $S^{\infty}(\mathbb{R}_{q,p}^{2n})$ consisting of smooth functions

$$H(q, p, h), \quad h \in [0, 1],$$

such that

$$\left| \frac{\partial^{|\alpha|+|\beta|+k} H(q,p)}{\partial q^{\alpha} \partial p^{\beta} \partial h^k} \right| \leq C_{\alpha\beta k}(1 + |q| + |p|)^m, \quad |\alpha|, |\beta| = 0, 1, 2, \ldots, \quad (38)$$

where m is independent of k, α and β (but depends on H). For the detailed definition of functions of operators, we refer the reader to [3] and the textbook [17].

1. The idea of quantization of observables, that is, of constructing the correspondence "symbols $\rightarrow$ operators", is to use the conjugation of the symbol by U. Since $U^{-1} \neq U^*$ (recall that $U^{-1} = U^*|_{\mathcal{F}_2(\mathbb{R}^{2n})}$), there are two different candidates:

$$H(q,p) \mapsto \hat{H} = U^{-1} \circ H(q,p) \circ U, \quad (39)$$

$$H(q,p) \mapsto \hat{H} = U^* \circ H(q,p) \circ U, \quad (40)$$

where $H(q,p)$ on the right-hand side stands for the multiplication by $H(q,p)$ in both cases. After a brief study, we see that (39) must be rejected, since the multiplication by $H(q,p)$ need not preserve the set of solutions to (19) unless $H(q,p)$ is an analytic function of $q - ip$, and hence the subsequent application of U^{-1} is undefined. So we shall use (40), but first let us note that although (39) is meaningless "as is," the idea itself is not so absurd. Namely, from the commutation relations (23), (24), which mean that U is an intertwining operator for the representations

$$\left(x, -ih\frac{\partial}{\partial x} \right) \quad \text{and} \quad \left(q + ih\frac{\partial}{\partial p}, -ih\frac{\partial}{\partial q} \right)$$

of the Heisenberg algebra, we can obviously derive the following theorem.

Theorem 6 *For any symbol $H(q,p) \in S^\infty(\mathbb{R}^{2n})$ one has*

$$U^{-1} H \left(\overset{2}{q + ih\frac{\partial}{\partial p}}, - \overset{1}{ih\frac{\partial}{\partial q}} \right) U = H \left(\overset{2}{x}, - \overset{1}{ih\frac{\partial}{\partial x}} \right). \qquad (41)$$

Remark 3 Note that the left-hand side of (41) is well defined. Indeed,

$$\left[h\frac{\partial}{\partial q} - ih\frac{\partial}{\partial p} - ip, q + ih\frac{\partial}{\partial p} \right] = \left[h\frac{\partial}{\partial q} - ih\frac{\partial}{\partial p} - ip, -ih\frac{\partial}{\partial q} \right] = 0,$$

whence it follows that the operator

$$H \left(\overset{2}{q + ih\frac{\partial}{\partial p}}, - \overset{1}{ih\frac{\partial}{\partial q}} \right)$$

preserves the set of solutions to (19).

Let us now leave this topic and return to formula (40).

Theorem 7 *The operator*

$$\hat{H} = U^* \circ H(q,p) \circ U \qquad (42)$$

is the operator with anti-Wick symbol $H(q,p)$.

Note that the operator with anti-Wick symbol $H(x,p)$ is a special case of the general construction of operators with $\mathcal{A}$-symbols [18] for

$$\mathcal{A} = \frac{1}{2} \left(\begin{array}{cc} -iE & E \\ E & -iE \end{array} \right). \qquad (43)$$

Theorem 8 *The operator $\hat{H}$ given by (42) has a $\left(\overset{2}{x}, - \overset{1}{ih\partial/\partial x} \right)$-symbol. More precisely,*

$$\hat{H} = \tilde{H} \left(\overset{2}{x}, - \overset{1}{ih\frac{\partial}{\partial x}}, h \right), \qquad (44)$$

where

$$\tilde{H}(q,p,h) = \exp\left\{ \frac{h}{4} \left(\frac{\partial^2}{\partial q^2} + \frac{\partial^2}{\partial p^2} \right) - \frac{ih}{2} \frac{\partial^2}{\partial q \, \partial p} \right\} H(q,p). \qquad (45)$$

Here

$$\frac{\partial^2}{\partial q \, \partial p} = \sum_{j=1}^{n} \frac{\partial^2}{\partial q_j \, \partial p_j}, \quad \frac{\partial^2}{\partial q^2} + \frac{\partial^2}{\partial p^2} = \sum_{j=1}^{n} \left(\frac{\partial^2}{\partial q_j^2} + \frac{\partial^2}{\partial p_j^2} \right); \qquad (46)$$

note that the exponential is well defined, since both operators in (46) are self-adjoint and nonpositive with respect to the L^2 inner product.

It is easy to obtain the expansion in powers of h of the symbol $\tilde{H}(q,p,h)$:

$$\tilde{H}(q,p,h) = H(q,p) + \frac{h}{4}\{H_{qq}(q,p) + H_{pp}(q,p) - 2iH_{qp}\} + \dots . \qquad (47)$$

We see that in the leading term $\tilde{H}$ coincides with H and that the supports of $\tilde{H}$ and H are the same if for some N we neglect functions that are $O(h^N)$.

The representation (42) combined with Theorem 8 permits one to prove boundedness theorems for pseudodifferential operators easily; however, we do not dwell on this topic.

Using (45) and the usual composition formula for pseudodifferential operators, we arrive at the following theorem.

Theorem 9 *Let $\hat{H}$ and $\hat{G}$ be the operators with anti-Wick symbols $H(q,p)$ and $G(q,p)$, respectively. Then the product $\hat{H} \circ \hat{G}$ has the form*

$$\hat{H} \circ \hat{G} = \hat{W} + O(h^\infty), \qquad (48)$$

where the operator W has the anti-Wick symbol $W(q,p)$ with the following asymptotic expansion as $h \to 0$:

$$W(q,p) \cong \sum_{|\alpha|=0}^{\infty} \frac{(2h)^\alpha}{\alpha!} \frac{\partial^\alpha H}{\partial z^\alpha} \frac{\partial^\alpha G}{\partial \bar{z}^\alpha}, \qquad (49)$$

where $z = q - ip$, $\bar{z} = q + ip$, and, accordingly,

$$\frac{\partial}{\partial z} = \frac{1}{2}\left(\frac{\partial}{\partial q} + i\frac{\partial}{\partial p}\right), \quad \frac{\partial}{\partial \bar{z}} = \frac{1}{2}\left(\frac{\partial}{\partial q} - i\frac{\partial}{\partial p}\right). \qquad (50)$$

2. Now we shall apply this definition of $1/h$-pseudodifferential operators to study the behavior under U of *fronts of oscillations*. Let us recall this notion, well-known in semi-classical theory.

We start from the definition of the support of oscillations.

Definition 3 *Let $\psi \in H_\infty(\mathbb{R}^n)$. We say that a point $x_0 \in \mathbb{R}^n$ belongs to the oscillation support of ψ,*

$$x_0 \in \operatorname{osc\,supp}\psi,$$

if for any function $\varphi(x) \in C_0^\infty(\mathbb{R}^n)$ independent of h and satisfying

$$\varphi\psi = O(h^\infty)$$

we have $\varphi(x_0) = 0$.

Definition 4 Let $\psi \in H_\infty(\mathbb{R}^n)$. We say that a point (q_0, p_0) of the phase space $\mathbb{R}_q^n \oplus \mathbb{R}_p^n$ belongs to the *oscillation front* of ψ,

$$(q_0, p_0) \in \mathrm{OF}(\psi),$$

if for any symbol $H(q, p) \in C_0^\infty(\mathbb{R}^{2n})$ independent of h and satisfying

$$H\left(\overset{2}{x}, -ih\partial/\partial x \overset{1}{} \right) \psi = O(h^\infty)$$

we have $H(q_0, p_0) = 0$.

The sets $\mathrm{OF}(\psi)$ and osc supp ψ satisfy properties closely resembling those of $\mathrm{WF}(\psi)$ and sing supp ψ (see [19]). Some of those properties are collected in the following theorem.

Theorem 10 *Let $\psi \in H_\infty(\mathbb{R}^n)$. Then*
(i) osc supp $\psi = \pi(\mathrm{OF}(\psi))$, *where* $\pi : \mathbb{R}_{q,p}^{2n} \to \mathbb{R}_x^n$, $(q, p) \mapsto x = q$, *is the natural projection.*
(ii) osc supp $\hat{H}\psi \subset$ osc supp ψ *and* $\mathrm{OF}(\hat{H}\psi) \subset \mathrm{OF}(\psi)$ *for any pseudodifferential operator* $\hat{H}$.
(iii) *If* $H(q, p) = 0$ *in a neighborhood of* (q_0, p_0), *then* $(q_0, p_0) \notin \mathrm{OF}(\hat{H}\psi)$.
(iv) *Let*

$$\psi(x, h) = e^{\frac{i}{h}S(x)} \varphi(x), \tag{51}$$

where $S(x)$ and $\varphi(x)$ are smooth functions and $\mathrm{Im}\, S \geq 0$. *Then*

$$\mathrm{OF}(\psi) = \overline{\left\{ (q, p) \,\big|\, \varphi(q) \neq 0, \mathrm{Im}\, S(q) = 0, \text{ and } p = \frac{\partial S(q)}{\partial q} \right\}}. \tag{52}$$

Corollary 1 *The oscillation front of the Gaussian wave packet $G_h(q, p; x)$, considered as a function of x, has the form*

$$\mathrm{OF}(G_h(q, p, \cdot)) = \{(q, p)\}, \tag{53}$$

that is, consists of the single point (q, p).

Theorem 11 *For any $\psi \in H(\mathbb{R}_x^n)$ one has*

$$\mathrm{OF}[\psi] = \text{osc supp}\,[U[\psi]]. \tag{54}$$

This assertion readily follows from the fact that in the leading term the application of a pseudodifferential operator amounts to the multiplication of the wave packet transform by the principal symbol.

3. Quantization of states

Informally, quantization of states is a procedure that assigns ψ-functions (or classes of ψ-functions) on the configuration space to "objects" in (or on) the phase space $\mathbb{R}^{2n}_{q,p}$. In a sense, the simplest quantization rule is provided by the Gaussian wave packets themselves: to each point $(p,q) \in \mathbb{R}^{2n}_{q,p}$ we assign the Gaussian wave packet

$$\psi(x,h) = G_h(p,q,x)$$

that has the oscillation front $\mathrm{OF}[\psi]$ consisting of that very point. If we intend to obtain ψ-functions with oscillation fronts that do not amount to a single point but are some more general closed subsets (say, manifolds) of the phase space, then one of the possible approaches is to integrate the Gaussian packets with respect to the parameters (p,q) with some density. Naively, this density would be supported on the desired oscillation front; however, we shall see that this is not always the case.

The integration can be interpreted in two ways: we apply either U^{-1} or U^* to the density. More precisely, we set either

$$\psi = U^*[f], \tag{55}$$

where $f \in H_{-\infty}(\mathbb{R}^{2n})$ and is compactly supported[4], or

$$\psi = U^{-1}[\tilde{f}], \tag{56}$$

where $\tilde{f} \in \mathcal{F}_2(\mathbb{R}^{2n}_{q,p})$. (Note that we can always pass from (55) to (56) by setting $\tilde{f} = Pf$, but each description has its own geometric and analytical advantages.)

1. First, we briefly discuss formula (55), which can be reduced to a construction well-known in the literature. Our exposition mainly follows [2].

Suppose that a submanifold $\Lambda \subset (\mathbb{R}^{2n}_{q,p})$ is given, and we intend to construct functions $\psi \in H_\infty(\mathbb{R}^n)$ with $\mathrm{OF}[\psi] \subset \Lambda$. To this end, we apply the transform U^* to functions of the form

$$f(x,p) = (\pi h)^{n/4} e^{\frac{i}{h}S} \varphi \delta_{(\Lambda,d\sigma)}, \tag{57}$$

where S and φ are smooth functions on Λ, φ is compactly supported, S is real-valued, and $\delta_{(\Lambda,d\sigma)}$ is the delta function on Λ corresponding to a smooth measure $d\sigma$:

$$\langle \delta_{(\Lambda,d\sigma)}, \chi \rangle = \int_\Lambda \chi \, d\sigma, \quad \chi \in C_0^\infty(\mathbb{R}^{2n}_{q,p}). \tag{58}$$

---

[4]The last requirement can be weakened of course.

We have introduced the factor $e^{\frac{i}{\hbar}S}$ in (57) for the following reason. Integration over Λ may cancel out the oscillations, and we shall choose S so as to exclude this possibility. On substituting (57) into (55) we obtain

$$\psi(x) \overset{\text{def}}{=} K_{(\Lambda, d\sigma)}\varphi = \left(\frac{1}{2\pi h}\right)^{n/2} \int_\Lambda e^{\frac{i}{\hbar}\Phi(x,\alpha)}\varphi(\alpha)\,d\sigma(\alpha), \qquad (59)$$

where

$$\Phi(x,\alpha) = S(\alpha) + (x - q(\alpha))p(\alpha) + \frac{i}{2}(x - q(\alpha))^2 \qquad (60)$$

and $\alpha \mapsto (q(\alpha), p(\alpha))$ is the embedding $\Lambda \subset \mathbb{R}^{2n}_{q,p}$. Let $H(q,p)$ be a compactly supported symbol. Then

$$H\left(\overset{2}{x}, -ih\overset{1}{\frac{\partial}{\partial x}}\right)\psi(x)$$

$$= \left(\frac{1}{2\pi h}\right)^{3n/2} \int_{\Lambda \times \mathbb{R}^{2n}_{q,p}} e^{\frac{i}{\hbar}\{p(x-p)+\Phi(q,\alpha)\}} H(x,p)\varphi(\alpha)d\sigma(\alpha)\,dq\,dp. \quad (61)$$

Obviously, the function (61) is $O(h^\infty)$ if the phase function

$$\psi(x, q, p, \alpha) = p(x - q) + (q - q(\alpha))p(\alpha) + S(\alpha) + \frac{i}{2}(q - q(\alpha))^2 \qquad (62)$$

has no stationary points on the support of the integrand.

The stationary point equations read

$$\begin{cases} \dfrac{\partial\psi}{\partial p} = x - q = 0 \\[2mm] \dfrac{\partial\psi}{\partial q} = p(\alpha) - p + i(q - q(\alpha)) = 0 \\[2mm] \dfrac{\partial\psi}{\partial\alpha} = (q - q(\alpha))\dfrac{\partial p(\alpha)}{\partial\alpha} - p(\alpha)\dfrac{\partial q(\alpha)}{\partial\alpha} + \dfrac{\partial S(\alpha)}{\partial\alpha} + i(q(\alpha) - q)\dfrac{\partial q(\alpha)}{\partial\alpha}, \end{cases} \qquad (63)$$

whence we obtain

$$\begin{cases} q(\alpha) = q = x, \quad p = p(\alpha), \\ dS(\alpha) = p(\alpha)\,dx(\alpha) \end{cases} \qquad (64)$$

From (61) we see that the validity of these equations for some point (x, p) is necessary and sufficient for this point to belong to $OF(\psi)$ (provided $\varphi(\alpha) \neq 0$).

If we require that $OF(\psi) = \Lambda$ (more precisely, $OF(\psi) = \text{supp}\,\varphi$), we must require that

$$p\,dx = dS \qquad (65)$$

on Λ, that is, Λ is a Lagrangian manifold. In this case, formula (59) defines the *Maslov canonical operator* [8] on Λ in the form[5] considered by Karasev [20], which itself is a paraphrase of the construction suggested by Cordoba and Fefferman [14] for Fourier integral operators.

2. Let us now study formula (56). In this case,

$$\tilde{f} = U[\psi], \qquad (66)$$

and so an appropriate method is to start from the desired function ψ and try to see what $\tilde{f}$ must be.

We are primarily interested in the *semiclassical wave functions* of the form

$$\psi(x) = e^{i/h\, S(x)} \varphi(x) \qquad (67)$$

or Fourier transforms of such functions.

Consider the wave packet transform of the function (67):

$$U[\psi](q,p) = 2^{-\frac{n}{2}} (\pi h)^{-\frac{3n}{4}} \int_{\mathbb{R}_x^n} \exp \frac{i}{h} \left\{ S(x) + (q-x)p + \frac{i}{2}(q-x)^2 \right\} \varphi(x)\, dx. \qquad (68)$$

The first obvious property of the function (68) is that it satisfies equations (19). Furthermore, we can obtain the asymptotic expansion of $U[\psi](q,p)$ in powers of h by using the version [9] of the stationary phase method with complex-valued phase function.

To this end, let us write out the equations of stationary points of the phase function

$$\Phi(x,q,p) = S(x) + (q-x)p + \frac{i}{2}(q-x)^2 \qquad (69)$$

of the integral (68). They read

$$\frac{\partial \Phi}{\partial x} \equiv \frac{\partial S}{\partial x} - p + i(x-q) = 0. \qquad (70)$$

We are interested in *real* stationary points, i.e., points at which the phase function (69) is real. Then we have

$$x = q, \quad p = \frac{\partial S}{\partial x}(q). \qquad (71)$$

Thus, the integral (68) has a real stationary point $x = x(q,p)$ if and only if

$$p = \frac{\partial S}{\partial x}(q), \qquad (72)$$

---

[5]For lack of space, our considerations are purely local and we do not even touch any issues pertaining to quantization conditions on Λ.

that is, the point (p, q) lies on the Lagrangian manifold Λ_S generated by S. In this case,

$$x(q, p) = q. \tag{73}$$

This stationary point is nondegenerate. Indeed,

$$\frac{\partial^2 \Phi}{\partial x^2} = \frac{\partial^2 S}{\partial x^2} + iE \tag{74}$$

is a nondegenerate matrix since $\partial^2 S/\partial x^2$ is real symmetric.[6] Applying the stationary phase method, we obtain the following result:

$$U[\psi](q, p) = O(h^\infty) \tag{75}$$

outside a neighborhood of Λ_S, whereas in the vicinity of Λ_S for any $N > 0$ we have the asymptotic expansion

$$u[\psi](q, p) = h^{-n/4} e^{\frac{i}{h} \Phi(q, p)} \sum_{k=0}^{N=1} h^k a_k(q, p) + O(h^N), \tag{76}$$

where $a_k(q, p)$ are smooth functions independent of h and

$$\Phi(q, p) = {}^\infty\Phi(x(q, p), q, p) \tag{77}$$

is the almost *analytic continuation* of $\Phi(x, q, p)$ to the almost-solution $x(q, p)$ of equation (70) (see the details in [9]). The phase function $\Phi(q, p)$ has the following properties:

$$\operatorname{Im} \Phi(q, p) \geq 0; \quad \operatorname{Im} \Phi(q, p) \geq 0 \Leftrightarrow (q, p) \in \Lambda_S. \tag{78}$$

Similarly, we can consider a semiclassical wave functions of the form

$$\psi(x) = \left(\frac{1}{2\pi h}\right)^{n/2} \int e^{\frac{i}{h}(\xi x + \tilde{S}(\xi))} \varphi(\xi)\, d\xi. \tag{79}$$

3. Thus, we arrive at considering the following class of functions f to be used in the formula $\psi = U^{-1}[f]$.

Definition 5 Let $\Phi(q, p)$ be a smooth function on $\mathbb{R}^{2n}_{q,p}$,

$$\Phi(q, p) = \Phi_1(q, p) + i\Phi_2(q, p),$$

---

[6]The experienced reader will see that being appropriately modified, this argument remains valid for a complex-valued phase function $S(x)$ with nonnegative imaginary part.

such that $\Phi_2(q,p) \geq 0$, and let Γ be the set of zeros of $\Phi_2(q,p)$. By $I(\Phi)$ we denote the class of functions $f(q,p,h)$, $(q,p) \in \mathbb{R}^{2n}$, $h \in (0,1]$, that satisfy the following conditions:

(a) $f \in H_\infty(\mathbb{R}^{2n}_{p,q}) \cap \mathcal{F}_2(\mathbb{R}^{2n}_{p,q})$;

(b) for any integer $N > 0$ one has the asymptotic expansion

$$f(q,p) = e^{\frac{i}{h}\Phi(x,p)} \sum_{k=0}^{N-1} h^k a_k(q,p) + O(h^N), \tag{80}$$

where $a_k(q,p)$, $k = 1,2,\ldots$, are smooth functions independent of h and rapidly decaying at infinity. Furthermore, we set

$$C_h^\infty(\Phi) = U^{-1}[I(\Phi)]. \tag{81}$$

Let us study the class $I(\Phi)$ in some detail.

Lemma 1 *Let $f \in I(\Phi)$ and let $a_k(q,p)$ be the corresponding functions occurring in (80). Then*

(a) osc-supp $f = \mathrm{WF}(U^{-1}[f]) = \overline{\bigcup_{k=0}^{\infty} \mathrm{supp}\, a_k} \cap \Gamma$.

(b) *The functions $\Phi(x,p)$ and $a_k(q,p)$, $k = 0,1,2,\ldots$, satisfy the following system of equations in the interior of the support of a_0:*

$$i\frac{\partial\Phi}{\partial q} + \frac{\partial\Phi}{\partial p} - ip = O(\Phi_2^\infty), \quad \frac{\partial a_k}{\partial q} - i\frac{\partial a_k}{\partial p} = O(\Phi_2^\infty). \tag{82}$$

At this stage, it might seem that our construction provides functions $\psi = U^{-1}[f]$ with arbitrary closed oscillation fronts Γ. But this is not the case, as shown by the following remarkable theorem.

Theorem 12 *Let $f \in I(\Phi)$ have the asymptotic expansion (80) and let $(q_0,p_0) \in \Gamma$. Suppose, furthermore, that $a_0(q_0,p_0) \neq 0$ and Γ is a submanifold in a neighborhood of (q_0,p_0). Then Γ is isotropic in a neighborhood of (q_0,p_0), that is,*

$$dp \wedge dq|_\Gamma = 0. \tag{83}$$

If $\dim \Gamma = n$, then Γ is Lagrangian and the elements of $C_h^\infty(\Phi)$ correspond to the canonical operator on Γ. If, however, $\dim \Gamma < n$, then elements of C_h^∞ correspond to the canonical operator on the isotropic manifold Γ with Lagrangian complex germ. In fact, the interpretation of elements of $C_h^\infty(\Phi)$ as functions represented by the canonical operator corresponding to a general complex germ (e.g., see [21] and references therein) remains valid in the case of general set Γ. However, here we do not touch this subject any more; the corresponding study will be carried out elsewhere.

4. Quantization of symplectic transforms

In this section we shall show that the quantization of some symplectic transform

$$g : T^*\mathbb{R}^n \to T^*\mathbb{R}^n \tag{84}$$

is essentially the conjugation with the help of the U-transform of canonical change of variables (84). More exactly, the following assertion is valid.

Theorem 13 *The operator*

$$T_g = U^* e^{\frac{i}{\hbar}S(q,p)} g^* U,$$

or, in another form,

$$f(x) \mapsto U^* \left\{ \left(\frac{1}{2i}\right)^{n/2} e^{\frac{i}{\hbar}S(q,p)} U f[g(q,p)] \right\}(x), \tag{85}$$

where the function $S(q,p)$ is determined by the relation

$$dS = p\,dq - g^*(\xi\,dy),$$

is the Fourier integral operator $T(g,1)$ [22] with symbol 1 corresponding to the symplectic transform (84).

Remark 4 Similarly, the operator

$$f \mapsto U^{-1} \left\{ \left(\frac{1}{2i}\right)^{n/2} e^{\frac{i}{\hbar}S(q,p)} \varphi(q,p) U f[g(q,p)] \right\}(x)$$

coincides with the Fourier integral operator $T(g,\varphi)$.

Proof of Theorem 13. Let the functions

$$y = y(q,p), \quad \xi = \xi(q,p)$$

determine the symplectic transform (84). We write down the operator (85) in integral form by using the definitions of the transforms U and U^*:

$$U^* \left[\left(\frac{1}{2i}\right)^{n/2} e^{\frac{i}{\hbar}S(x,p)} U f[g(y,q)] \right](x) = \frac{i^{n/2}}{(2\pi h)^{3n/2}} \int G_{(x',p')}(x)$$

$$\times e^{\frac{i}{\hbar}S(x',p')} \left\{ \int \overline{G_{(y,q)}(y')} f(y')\,dy' \right\} \Bigg|_{y=y(x',p'),\, q=q(x',p')} dx'\,dp'.$$

Using formula (12), one can rewrite the latter formula in the form

$$
\begin{aligned}
T(g,1)\,f \;&=\; \frac{(-i)^{n/2}}{(2\pi h)^{3n/2}} \int \exp\left\{ \frac{i}{h} \left[S(x',p') + p'(x - x') + \frac{i}{2}(x - x')^2 \right.\right. \\
&\quad -\; q(x',p')(y' - y(x',p')) + \frac{i}{2}(y' - y(x',p'))^2 \Big] \Big\} f(y')\,dy'\,dx'\,dp' \\
&=\; \left(-\frac{i}{2\pi h} \right)^{n/2} \int K(x,y')f(y')\,dy',
\end{aligned}
$$

where the kernel $K(x,y')$ is given by

$$
\begin{aligned}
K(x,y') \;&=\; \left(\frac{1}{2\pi h} \right)^n \int \exp\left\{ \frac{i}{h} \left[S(x',p') + p'(x - x') + \frac{i}{2}(x - x')^2 \right.\right. \\
&\quad -\; q(x',p')(y' - y(x',p')) + \frac{i}{2}(y' - y(x',p'))^2 \Big] \Big\}\, dx'\,dp'.
\end{aligned}
$$

The latter expression exactly coincides with the expression for the canonically represented function

$$
K(x,y') = K_{(\Lambda_g,d\sigma)}(1)
$$

on the Lagrangian manifold $\Lambda_g = \operatorname{graph} g$ with the measure given by $d\sigma = (dp \wedge dx)^{\wedge n}$, written in the coordinates (x',p') of the manifold Λ_g. This follows from the fact that the nonsingular action S on the Lagrangian manifold Λ_g is determined by the formula

$$
S = \int (p\,dq - \xi\,dy)|_{\Lambda_g} = \int p\,dq - g^*(\xi\,dy). \quad \square
$$

References

1. Nazaikinskii, V. and Sternin, B.: Wave Packet Transform in Symplectic Geometry and Asymptotic Quantization, Preprint MPI 96-91, Max-Planck Institut für Mathematik, Bonn, 1996.
2. Sternin, B. and Shatalov, V.: Fourier–Gauss Transform and Quantization, Preprint MPI 94-102, Max-Planck Institut für Mathematik, Bonn, 1994.
3. Maslov, V.P.: *Operational Methods*, Nauka, Moscow, 1973; English transl.: Mir, Moscow, 1976.
4. Karasev, M.V. and Maslov, V.P.: Asymptotic and geometric quantization, *Uspekhi Matem. Nauk* **39** (1984) no. 6, 115–173 (in Russian).
5. Berezin, F.A.: Wick and anti-Wick symbols of operators, *Matem. Sb.* **86** (1971) no. 4, 578–610.
6. Schrödinger, E.: Quantisierung als Eigenwertproblem, *Ann. Phys.* **79** (1926) 361; **79** 489; **80** 437; **81** 109.

7. Fock, V.A.: On canonical transformation in classical and quantum mechanics, *Vestnik Leningr. Gos. Univ.* **16** (1959) 67–71; English transl.: *Acta Phys. Acad. Sci. Hungar.* **27** (1969) 219–224.

8. Maslov, V.P.: *Perturbation Theory and Asymptotic Methods.* Izd-vo MGU, Moscow, 1965 (in Russian); French transl.: Théorie des *Perturbations et Méthodes Asymptiques*, Dunod, Paris, 1972.

9. Mishchenko, A., Shatalov, V. and Sternin, B.: *Lagrangian Manifolds and the Maslov Operator*, Springer-Verlag, Berlin–Heidelberg, 1990.

10. Weinstein, A.: The order and symbol of a distribution, *Trans. Amer. Math. Soc.* **241** (1977) 1–54.

11. Bargmann, V.: On a Hilbert space of analytic functions and an associated integral transform. Part I, *Comm. Pure Appl. Math.* **14** (1961) no. 3, 187–214.

12. Fock, V.A.: Konfigurationsräum und zweite Quantelung, *Z. Phys.* **75** (1932) 622–647.

13. Fock, V.A.: Zur Quantenelectrodynamik, *Sowiet Phys.* **6** (1934) 425.

14. Cordoba, A. and Fefferman, Ch.: Wave packets and Fourier integral operators, *Comm. in Partial Differential Equations* (1978) 979–1005.

15. Shubin, M.A.: *Pseudodifferential Operators and Spectral Theory*, Nauka, Moscow, 1978; English transl.: Springer-Verlag, Berlin-Heidelberg, 1985.

16. Berezin, F.A.: Covariant and contravariant symbols of operators, *Izv. Akad. Nauk SSSR* **36** (1972) no. 5, 1134–1167.

17. Nazaikinskii, V., Sternin, B. and Shatalov, V.: *Methods of Noncommutative Analysis. Theory and Applications*, Mathematical Studies. Walter de Gruyter Publishers, Berlin–New York, 1995.

18. Karasev, M.V. and Nazaikinskii, V.E.: On the quantization of rapidly oscillating symbols, *Matem. Sb.* **106** (1978) no. 2, 183–213, (in Russian).

19. Hörmander, L.: Fourier integral operators I, *Acta Math.* **127** (1971) 79–183.

20. Karasev, M.V.: Connections on Lagrangian submanifolds and certain problems of semiclassical approximation, *Zap. Nauch. Sem. LOMI* **172** (1989) 41–54, (in Russian).

21. Dubnov, V.L., Maslov, V.P. and Nazaikinskii, V.E.: The complex Lagrangian germ and the canonical operator, *Russian J. of Mathematical Physics* **3** (1995) no. 2, 141–190.

22. Nazaikinskii, V., Oshmyan, V., Sternin, B. and Shatalov, V.: Fourier integral operators and the canonical operator, *Usp. Mat. Nauk* **36** (1981) no. 2, 81 – 140; English transl.: *Russ. Math. Surv.* **36** (1981) no 2, 93 – 161.

ON QUANTUM METHODS IN THE CLASSICAL
THEORY OF REPRESENTATIONS

D.P. ZHELOBENKO
International Sophus Lie Center,
People's Friendship Russian University,
Independent University of Moscow,
ul. Kedrova 13-2-73, 117036, Moscow, Russia

Abstract. An elementary introduction to the theory of Drinfeld–Jimbo algebras is given, including the famous "crystallization method" of M. Kashiwara and its applications. The exposition is based on a study of the so-called Schubert filtration of the underlying Serre algebras. An application to the classical reduction problem is considered for the case of finite-dimensional representations of semisimple Lie algebras.

Mathematics Subject Classification (1991): 17B35, 22E46, 17B37, 81R50.

Key words: quantum groups, quantum modules, Drinfeld–Jimbo algebras, Lie algebras, enveloping algebras, Schubert filtration, reduction problem.

The term "quantization" in modern mathematics means a passage from a certain algebraic object to its parametric deformation, in the framework of a given type of an algebraic structure. First examples of such kind, connected with a deformation of power series, are known from the end of the preceding century. In that sense, quantum mathematics is older than quantum mechanics.

A new stimulus for the development of quantum mathematics arose in 1985, after the known construction of the theory of quantum groups [3]. Here the term "quantum" means, in the spirit of algebraic geometry, the passage from a pointwise description to the language of algebraic structures. Of great interest, in this context, is the quantization of certain classical objects associated with Lie groups and Lie algebras.

B. P. Komrakov et al. (eds.), Lie Groups and Lie Algebras, 71–84.
© 1998 *Kluwer Academic Publishers. Printed in the Netherlands.*

It is remarkable that the theory of quantum groups opens new possibilities in the classical theory of representations of Lie groups and Lie algebras. An important example in this direction is given by G. Lusztig [5] and M. Kashiwara [8], [9], for solving an old construction problem of so-called "canonical" bases for universal enveloping algebras $U(\mathfrak{g})$, where $\mathfrak{g}$ is a complex semisimple Lie algebra. An essential step of this construction is to replace $U(\mathfrak{g})$ by its quantization $U_q(\mathfrak{g})$ (the so-called Drinfeld–Jimbo algebra).

It is known also that the use of $U_q(\mathfrak{g})$ allows one to clarify the connection between representations of semisimple Lie algebras over the classical fields $(\mathbb{Q}, \mathbb{R}, \mathbb{C})$ and representations of the corresponding structures over finite fields, see, for example [2].

Below we present an elementary introduction to the theory of Drinfeld–Jimbo algebras and associated quantum structures. As a starting point, we consider the corresponding Serre algebras (Section 1). The study of the special "Schubert filtration" of Serre algebras (Section 3) allows us to give a new approach to the famous "crystallization method" of M. Kashiwara. In Section 5, an application of this method to the classical reduction problem is considered.

1. Serre algebras

1.1. Let Δ be a root system in Euclidean space E over the field $\mathbb{C}$, $\dim E = n$, with simple roots α_i $(i \in I)$, where $\operatorname{card} I = n$. Let Q be the lattice (additive subgroup) in E generated by the roots α_i $(i \in I)$. Recall [1] that $2(\alpha, \beta) \in \mathbb{Z}$ for any $\alpha, \beta \in \Delta$. The function

$$\varepsilon(\alpha, \beta) = q^{2(\alpha,\beta)}, \qquad (1.1)$$

where $0 \neq q \in \mathbb{C}$ (a fixed constant), is a bicharacter of the group Q. This means that any of the functions $\varepsilon(\alpha, \cdot)$, $\varepsilon(\cdot, \beta)$ is a character of the group Q. We shall consider q as independent variable and interpret (1.1) as a function $\varepsilon : Q \times Q \to \mathbb{Z}[q, q^{-1}]$.

1.2. Let A be an associative algebra with unity over the field $\mathbb{C}(q)$ of rational functions on q. Assume that A is graded by the group Q, i.e., we have

$$A = \bigoplus_{\alpha} A_\alpha, \qquad A_\alpha A_\beta \subset A_{\alpha+\beta}, \qquad (1.2)$$

for any $\alpha, \beta \in Q$. The function

$$\varepsilon(x, y) = \varepsilon(\alpha, \beta) \qquad \text{for } x \in A_\alpha,\ y \in A_\beta, \qquad (1.3)$$

associated with bicharacter (1.1), is called a *structure coefficient* of the graded algebra A. The expression

$$[x, y]_\varepsilon = xy - \varepsilon(x, y)yx, \tag{1.4}$$

defined for homogeneous elements $x, y \in A$, is uniquely extended to a bilinear operation $A \otimes A \to A$ and is called the ε-*commutator* of the elements $x, y \in A$.

An operator $a \in \text{End}A$ is called *homogeneous*, of degree $\gamma \in Q$, if $aA_\alpha \subset A_{\alpha+\gamma}$ for any $\alpha \in Q$. The linear hull $E(A)$ of all homogeneous operators is a Q-graded algebra (subalgebra of $\text{End}A$).

We shall identify an element $a \in A$ with the corresponding operator of left multiplication in the algebra A. Hence the algebra A is identified with a graded subalgebra of $E(A)$.

1.3. Let A be a free algebra over $\mathbb{C}(q)$ with generators x_i ($i \in I$). Setting $\deg x_i = \alpha_i$ ($i \in I$), we endow A with a Q-grading, together with the structure coefficient (1.4).

It is easy to see that the algebra A possesses a unique family of operators $\partial_i \in E(A)$ ($i \in I$) satisfying the following commutation relations:

$$[\partial_i, x_j]_\varepsilon = \delta_{ij} \qquad \text{for any } i, j \in I, \tag{1.5}$$

where δ_{ij} is the Kronecker symbol, with the complementary condition $\partial_i(1) = 0$ for any $i \in I$. It is easy to see also that the algebra A possesses a unique bilinear form $(\cdot, \cdot)$ satisfying the normalizing condition $(1, 1) = 1$ and the following contravariance conditions:

$$(x_i f, g) = (f, \partial_i g), \qquad (\partial_i f, g) = (f, x_i g), \tag{1.6}$$

for any $i \in I$ and any $f, g \in A$. Moreover, the form (1.6) is symmetric, i.e., $(f, g) = (g, f)$ for any $f, g \in A$.

We shall use the symbol $(\cdot, \cdot)$ also for the associated form in $A \otimes A$, namely

$$(a \otimes b, c \otimes d) = (a, c)(b, d), \tag{1.7}$$

for any $a, b, c, d \in A$. On the other hand, we shall consider $A \otimes A$ as an algebra with respect to the following "twisted" rule of multiplication:

$$(a \otimes b)(c \otimes d) = \varepsilon(b, c)^{-1}(ac \otimes bd). \tag{1.8}$$

The rule (1.8), given for homogeneous elements of A, is extended uniquely to arbitrary elements of the algebra A.

1.4. Proposition *Let* $\delta : A \to A \otimes A$ *be a homomorphism of algebras, defined on generators by the rule* $\delta(1) = 1 \otimes 1$, $\delta(x_i) = x_i \otimes 1 + 1 \otimes x_i$ *for any* $i \in I$. *Then we have:*

$$(\delta(a), x \otimes y) = (a, xy), \tag{1.9}$$

for any $a, x, y \in A$. *In other words, the map* δ *is conjugated to the multiplication map* $\mu : A \otimes A \to A$ *given by the formula* $\mu(x \otimes y) = xy$.

The proof is straightforward.

1.5. Corollary *The kernel* N *of the bilinear form* (1.6) *is an ideal of the algebra* A.

1.6. Proposition *The ideal* N *is generated by the following elements of the algebra* A:

$$y_{ij} = (ad_\varepsilon x_i)^{1-a_{ij}} x_j \qquad \text{for } i \neq j, \tag{1.10}$$

where $(ad_\varepsilon x)y = [x, y]_\varepsilon$ *and* a_{ij} *is the Cartan matrix associated with the system* Δ, *i.e.,*

$$a_{ij} = 2(\alpha_i, \alpha_j)/(\alpha_i, \alpha_i). \tag{1.11}$$

The proof may be found, for example, in [9].

1.7. Definition *The quotient algebra* A/N *is denoted by* $X = X_q(a)$ *and is called* quantum Serre algebra *associated with the Cartan matrix* (1.11).

The quantum Serre conditions $y_{ij} = 0$ $(i \neq j)$ are the defining relations for the algebra X. For a more detailed description of these conditions, we introduce the following (standard) quantum symbols [5]:

$$\begin{aligned}
q_i &= q^{(\alpha_i, \alpha_i)}, \qquad \theta_i = q_i - q_i^{-1}, \\
[n]_i &= \theta_i^{-1}(q_i^n - q_i^{-n}), \\
[n]_i! &= [1]_i \cdots [n]_i \qquad \text{for } n \in \mathbb{Z}_+, \\
x_i^{(n)} &= x_i^n/[n]_i! \text{ for } n \in \mathbb{Z}_+, \qquad x_i^{(n)} = 0 \quad \text{for } n < 0.
\end{aligned}$$

Then the condition $y_{ij} = 0$ $(i \neq j)$ may be written in the following way:

$$\sum_n (-1)^n x_i^{(b-n)} x_j x_i^{(n)} = 0, \tag{1.12}$$

where $b = 1 - a_{ij}$.

1.8. We endow the Serre algebra X with the corresponding Q-grading (deg $x_i = \alpha_i$ for any $i \in I$) and with the family of operators $\partial_i \in \text{End} X$ (for any $i \in I$).

It is clear also that the algebra A inherits the contravariant bilinear form (1.6). In accordance with the definition of N, this bilinear form is nondegenerate on the space X.

We denote by $D_\varepsilon(X)$ the subalgebra with unity of $E(X)$ generated by the operators x_i, ∂_i ($i \in I$). It is not difficult to give an abstract characterization of this algebra of (twisted) differential operators on the algebra X.

1.9. Let $x \mapsto \bar{x}$ be an automorphism of the algebra X defined on the generators by the rule $\bar{x}_i = x_i$, $\bar{q} = q^{-1}$. Let $\bar{a}\,\bar{x} = \overline{ax}$ for $a \in \mathrm{End}A$ be the corresponding automorphism of the algebra $\mathrm{End}A$. Note that the operators $\bar{\partial}_i$ ($i \in I$) satisfy the modified condition (1.5) with $\varepsilon \mapsto \bar{\varepsilon}$.

On the other hand, let $t_i \in \mathrm{Aut}X$ be defined by the following rule:

$$t_i(x_j) = \varepsilon(\alpha_i, \alpha_j)x_j, \tag{1.13}$$

for any $i, j \in I$. Then we have:

$$[t_i\partial_i, x_j] = \delta_{ij}t_i. \tag{1.14}$$

1.10. Changing $\mathbb{C}(q)$ to $\mathbb{C}[q, q^{-1}]$ and computing the specialization of $X_q(a)$ at the point $q = 1$, we obtain the classical Serre algebras $X(a)$ over $\mathbb{C}$ with defining relations (1.12), where $x^{(n)} = x^n/n!$.

It is well known (see, for example, [10]) that $X(a) \approx U(\mathfrak{n})$, where $\mathfrak{n}$ is the nilpotent radical of Borel subalgebra in the semisimple Lie algebra $\mathfrak{g} = \mathfrak{g}(a)$ associated with the Cartan matrix (1.11).

2. Drinfeld–Jimbo algebras

2.1. Definition Let $U_q(\mathfrak{g})$ be an associative algebra with unity over the field $\mathbb{C}(q)$ generated by elements e_i, f_i, $t_i^{\pm 1}$ ($i \in I$) and by defining conditions of the following form:

$$t_i t_i^{-1} = t_i^{-1} t_i = 1, \tag{2.1}$$

$$t_i e_j t_i^{-1} = q^{a_{ij}} e_j, \qquad t_i f_j t_i^{-1} = q^{-a_{ij}} f_j, \tag{2.2}$$

$$[e_i, f_j] = \theta_i^{-1}\delta_{ij}(t_i - t_i^{-1}), \tag{2.3}$$

$$y_{ij}(e) = y_{ij}(f) = 0 \qquad \text{for } i \neq j, \tag{2.4}$$

where y_{ij} are polynomials defined by the equality (1.10) (with the change $x_i \mapsto e_i, f_i$ in (2.4)).

Let $U_1(\mathfrak{g})$ be the specialization of the algebra $U_q(\mathfrak{g})$ at the point $q = 1$ (with preliminary passage from $\mathbb{C}(q)$ to $\mathbb{C}[q, q^{-1}]$), and let $U(\mathfrak{g})$ be the quotient algebra of $U_1(\mathfrak{g})$ with respect to the ideal generated by the elements $t_i - 1$ ($i \in I$). It is easy to verify that $U(\mathfrak{g})$ is the universal enveloping algebra of the complex semisimple Lie algebra $\mathfrak{g} = \mathfrak{g}(a)$ associated to the Cartan matrix (1.11).

The definition of $U_q(\mathfrak{g})$ presented here is a slight modification [5] of the original ones given by V. Drinfeld [3] and M. Jimbo [4]. Correspondingly, the algebra $U_q(\mathfrak{g})$ is called *Drinfeld–Jimbo algebra* associated with the Lie algebra $\mathfrak{g} = \mathfrak{g}(a)$.

2.2. The algebra $U_q(\mathfrak{g})$ possesses a Q-grading defined on generators by the rule

$$\deg e_i = -\deg f_i = \alpha_i, \qquad \deg t_i = 0.$$

The algebra $U_q(\mathfrak{g})$ possesses an automorphism $a \mapsto \bar{a}$ and an involution (antiautomorphism) $a \mapsto a'$ of the following form:

$$\begin{aligned}
\bar{e}_i &= e_i, & \bar{f}_i &= f_i, & \bar{t}_i &= t_i^{-1}, & \bar{q} &= q^{-1}, \\
e_i' &= f_i, & f_i' &= e_i, & t_i' &= t_i, & q' &= q.
\end{aligned}$$

Sometimes it is also convenient to consider the involution $x^+ = \bar{x}'$ $(x \in U_q(\mathfrak{g}))$.

Let H (respectively X, X^+) be the subalgebra with unity of $U_q(\mathfrak{g})$ generated by the elements $t_i^{\pm 1}$ (respectively, f_i, e_i). It is clear that the algebra H is commutative and $(X)^+ = X^+$. It is clear also that each of the algebras X, X^+ is isomorphic to the Serre algebra $X_q(a)$ for $\mathfrak{g} = \mathfrak{g}(a)$.

2.3. Proposition [7], [10] *The map $x \otimes y \otimes z \mapsto xyz$ defines the following isomorphism of vector spaces:*

$$U_q(\mathfrak{g}) = XHX^+ \approx X \otimes H \otimes X^+ \tag{2.5}$$

(*the triangular decomposition of $U_q(\mathfrak{g})$*).

Hence the algebra $U_q(\mathfrak{g})$ is a Cartan type contragradient algebra, in the sense of [11], [13].

2.4. Let $\mathcal{P}$ be the weight lattice in E, i.e., the set of all vectors $\lambda \in E$ with integer coordinates

$$\lambda_i = 2(\alpha_i, \lambda)/(\alpha_i, \alpha_i). \tag{2.6}$$

Recall that $\mathcal{P} = \sum_i \mathbb{Z}\omega_i$, where ω_i $(i \in I)$ are fundamental weights, i.e., we have $2(\alpha_i, \omega_j) = \delta_{ij}(\alpha_i, \alpha_i)$, for any $i, j \in I$. We set $\mathcal{P}_+ = \sum_i \mathbb{Z}_+\omega_i$.

A module V over the algebra $U_q(\mathfrak{g})$ is called *$\mathcal{P}$-graded* if it is graded by the weight subspaces

$$V_\mu = \{v \in V \mid t_i v = q^{\mu_i} v \qquad \text{for any } i \in I\}, \tag{2.7}$$

where $\mu \in \mathcal{P}$. Remark that in this case any H-submodule $N \subset V$ is $\mathcal{P}$-graded (by the subspaces $N_\mu = N \cap V_\mu$).

For example, let V^e be the *extremal subspace* of V with respect to the family $e = (e_i)_{i \in I}$, i.e.,

$$V^e = \{v \in V \mid e_i v = 0 \qquad \text{for any } i \in I\}. \tag{2.8}$$

Then V^e is $\mathcal{P}$-graded, for any $\mathcal{P}$-graded $U_q(\mathfrak{g})$-module V.

2.5. A module V is called *admissible* if it is locally nilpotent with respect to the family e, i.e., for any $v \in V$ there exists $n = n(v) \geq 1$ such that

$$e_{i_1} \cdots e_{i_n} v = 0 \qquad \text{for any } i_1, \cdots, i_n \in I. \tag{2.9}$$

A module V is called *integrable* if any of the operators e_i, f_i $(i \in I)$ is locally nilpotent, i.e., we have $e_i^n v = f_i^n v = 0$ for $n = n(v) \geq 1$.

Let $\mathcal{O}$ be the category of $\mathcal{P}$-graded admissible modules, and let $\mathcal{O}_{\mathrm{int}}$ be its subcategory consisting with all integrable modules. It is known [10], [13] that the category $\mathcal{O}_{\mathrm{int}}$ is semisimple, with the following simple objects:

$$V_q(\lambda) = U_q(\mathfrak{g})/N(\lambda) \qquad \text{for } \lambda \in \mathcal{P}_+, \tag{2.10}$$

where $N(\lambda)$ is a left ideal of the algebra $U_q(\mathfrak{g})$ generated by the elements

$$e_i, \qquad t_i - q^{\lambda_i}, \qquad f_i^{1+\lambda_i}, \tag{2.11}$$

for any $i \in I$.

It is remarkable that every module $V_q(\lambda)$ is finite-dimensional and is nothing but the quantization of the corresponding $U(\mathfrak{g})$-module $V(\lambda)$ with highest weight $\lambda \in \mathcal{P}_+$. In particular, we have

$$V_q(\lambda)^e = \mathbb{C}(q)u_\lambda, \tag{2.12}$$

where $u_\lambda = 1 + N(\lambda)$ is a unique, up to a normalization, highest vector of weight λ.

2.6. On the other hand, we may define an action of the algebra $U_q(\mathfrak{g})$ on its Serre components $X = X_q(a)$. Namely, we set

$$e_i = \theta_i^{-1}(t_i \partial_i - \bar{t}_i \bar{\partial}_i), \tag{2.13}$$

$$f_i = \text{left multiplication on } x_i \in X, \tag{2.14}$$

where $t_i \in \mathrm{Aut} X$ are defined in 1.9. Using (1.14), it is easy to verify that this definition is correct, i.e., $U_q(\mathfrak{g})$ acts on X by the operators (2.13), (2.14).

It is easy to see that X is a cyclic $U_q(\mathfrak{g})$-module generated by the one-dimensional extremal subspace $X^e = \mathbb{C}(q)1$.

2.7. The subalgebra

$$\mathcal{B}(X) = HX = XH \approx X \otimes H \tag{2.15}$$

is called a *Borel subalgebra* of $U_q(\mathfrak{g})$.

The action of the algebra $U_q(\mathfrak{g})$ in the space $X = X_q(a)$ (see 2.6) may be lifted to its Borel envelope $\mathcal{B} = \mathcal{B}(X)$. Namely, we set

$$e_i(xh) = e_i(x)h \qquad \text{for any } x \in X, h \in H, \tag{2.16}$$

and we identify t_i, f_i ($i \in I$) with the corresponding operators of left multiplication on $\mathcal{B}(X)$.

It is known [10] that this action of $U_q(\mathfrak{g})$ on $\mathcal{B}(X)$ is exact. In other words, the algebra $U_q(\mathfrak{g})$ is isomorphic to its image in $\mathrm{End}\mathcal{B}(X)$.

Hence the algebra $U_q(\mathfrak{g})$ may be realized as an algebra of (twisted) differential operators on the space $\mathcal{B}(X)$.

3. Schubert filtration

3.1. Let W be the Weyl group associated with the system Δ. Recall [1] that $W = \Gamma/\Gamma_0$, where Γ is the corresponding braid group with canonical generators r_i ($i \in I$) and Γ_0 is its normal subgroup generated by the elements r_i^2 ($i \in I$). Recall also that the group W acts on E and preserves the system Δ.

On the other hand, G. Lusztig [7] defines an action of the group Γ by the automorphisms of the algebra $U_q(\mathfrak{g})$. The image of the group Γ in $\mathrm{Aut}U_q(\mathfrak{g})$ is denoted by $W_q(\mathfrak{g})$ and is called *quantum Weyl group* associated to the algebra $\mathfrak{g} = \mathfrak{g}(a)$.

Recall that a decomposition $w = r_{i_1} \cdots r_{i_n}$ (in Γ or in W) is called "reduced" if it has the minimal possible length $n = l(w)$. On the other hand, for any $w \in W$ we set

$$\Delta_w = \{\alpha \in \Delta_+ \mid w^{-1}\alpha < 0\}, \tag{3.1}$$

where Δ_+ is the system of positive roots generated by the simple roots α_i ($i \in I$). Then we have $l(w) = \mathrm{card}\,\Delta_w$. The group W contains a unique element $w_{\max}$ such that $l(w_{\max}) = \mathrm{card}\,\Delta_+$.

3.2. The Lusztig construction allows one to define generalized root vectors in the algebra $U_q(\mathfrak{g})$.

Let us fix a reduced decomposition $w_{\max} = r_{i_1} \cdots r_{i_m}$ and set

$$w(0) = 1, \qquad w(n) = r_{i_1} \cdots r_{i_n} \qquad \text{for } 1 \le n \le m.$$

The sequence

$$\alpha(n) = w(n-1)\alpha_{i_n} \qquad \text{for } 1 \le n \le m \tag{3.2}$$

defines a so-called *normal ordering* [10] of the system Δ_+. Writing (3.2) briefly as $\alpha = w\alpha_i$, we set

$$e_\alpha = we_i, \qquad f_\alpha = wf_i, \tag{3.3}$$

for any $\alpha \in \Delta_+$.

It is easy to verify that $e_\alpha \in X^+$, $f_\alpha \in X$ for any $\alpha \in \Delta_+$. The elements (3.3) are called *generalized root vectors* of $U_q(\mathfrak{g})$ (associated with the chain (3.2)). Note that $e_\alpha = f_\alpha^+$ for any $\alpha \in \Delta_+$.

We set $f_\alpha^{(n)} = f_\alpha^n/[n]_\alpha!$, where $[n]_\alpha! = [n]_i!$ for $\alpha = w\alpha_i$. The monomials

$$f(k) = f_{\alpha(1)}^{(k_1)} \cdots f_{\alpha(m)}^{(k_m)}, \tag{3.4}$$

where $k = (k_1, \cdots, k_m) \in \mathbb{Z}_+^m$, constitute a basis F (the *Lusztig basis*) of the vector space $X = X_q(a)$.

3.3. Definition Introduce a partial ordering of the group W, setting $u \leq w$ for $\Delta_u \subset \Delta_w$, and define the following decreasing family of subalgebras in X:

$$X_w = X \cap w \cdot X, \tag{3.5}$$

where $w \in W$. The family (3.5) is called [12] the (decreasing) *Schubert filtration of the algebra* X.

We set also $X_w = w \cdot Y_{w'}$ for $w_{\max} = w \cdot w'$. Using Lusztig bases of the form (3.4), it is easy to verify the following assertions:

(A) Let F_w be the part of basis (3.4) given by the condition $k_1 = \cdots = k_n = 0$ for $w = w_n$. Then we have:

$$X_w = \mathbb{C}(q) F_w. \tag{3.6}$$

(B) Let G_w be the part of basis (3.4) given by the conditions $k_{n+1} = \cdots = k_m = 0$. Then we have:

$$Y_w = \mathbb{C}(q) G_w. \tag{3.7}$$

(C) For any $w \in W$, we have the following decomposition of the algebra X:

$$X = Y_w X_w \approx Y_w \otimes X_w. \tag{3.8}$$

3.4. Let us write the root vectors f_α as a polynomials $f_\alpha = p_\alpha(f)$ on the set $f = (f_i)_{i \in I}$, and let $p_\alpha(\partial)$ be the result of the formal replacement $f_i \mapsto \partial_i$, for any $i \in I$, in the polynomial f_α. It is shown in [12] (Theorem 4.5) that the differential operators $p_\alpha(\partial)$ do not depend on the choice of representations $f_\alpha = p_\alpha(f)$. Moreover, there exists a set of numbers $\lambda_\alpha \in \mathbb{C}[q, q^{-1}]$ such that the operators

$$\partial_\alpha = \lambda_\alpha^{-1} p_\alpha(\partial) \tag{3.9}$$

satisfy the following relations:

$$[\partial_\alpha, f_\beta]_\varepsilon = \delta_{\alpha\beta} \qquad \text{for } \alpha \leq \beta \tag{3.10}$$

(with respect to the ordering (3.2)).

3.5. Theorem [12] *The subspace X_w coincides with the set of all solutions in X of the following system of equations:*

$$\partial_\alpha x = 0 \qquad for\ any \quad \alpha \in \Delta_w. \tag{3.11}$$

In other words, let ∇_w be the family of operators ∂_α for $\alpha \in \Delta_w$; then we have

$$X_w = \ker \nabla_w. \tag{3.12}$$

3.6. Corollary *The Lusztig basis F is biorthogonal to the conjugate basis $\overline{F}$ with respect to the canonical bilinear form (1.6).*

3.7. Remark The normal orderings (3.2) possess the following abstract characterization [10]: any root of the form $\alpha_i = \alpha_j + \alpha_k$ lies between its components (i.e., we have $j < i < k$ or $k < i < j$).

From this it is clear that for any normal ordering τ there exists an inverse ordering τ^*. The root vectors e_α^*, f_α^* corresponding to τ^* differ only by normalizing factors of the corresponding vectors $\overline{e}_\alpha, \overline{f}_\alpha$ [12].

4. Crystallization

4.1. The passage from the classical objects $(U(\mathfrak{g}), V(\lambda))$ to the quantum $(U_q(\mathfrak{g}), V_q(\lambda))$ ones allows one to use the new degree of freedom (i.e., the quantum parameter q) to study classical objects from a new point of view. The main interest in this direction arises from the consideration the limiting process $q \to 0$ (or $q \to \infty$ after the automorphism $\overline{q} = q^{-1}$). This procedure is called *crystallization*, in the sense of M. Kashiwara [5], being the analog of an interpretation of the parameter q known in statistical physics as the temperature, where $q = 0$ is the absolute zero of temperature.

A detailed description of crystallization is based on certain nontrivial techniques, described by M. Kashiwara in [5]. Below we present a slight modification of the method of M. Kashiwara.

4.2. Let $(\cdot, \cdot)$ be the canonical bilinear form (1.6) for the algebra X. We set

$$L = \{x \in X \mid (x, x) \in A\}, \tag{4.1}$$

where A is the subalgebra of all functions from $\mathbb{C}(q)$ regular at the point $q = 0$.

To define an appropriate analog of (4.1) in the space $V_q(\lambda)$, we fix the unique bilinear form $(\cdot, \cdot)$ on the space $V_q(\lambda)$, satisfying the normalizing condition $(u_\lambda, u_\lambda) = 1$ and the following contravariance condition:

$$(ax, y) = (x, a^*y) \qquad for\ x, y \in V_q(\lambda), \tag{4.2}$$

where $a \mapsto a^*$ is the involution of $U_q(\mathfrak{g})$ given by the following rule:

$$e_i^* = q_i f_i \bar{t}_i, \qquad f_i^* = \bar{q}_i t_i e_i, \qquad t_i^* = t_i, \qquad (4.3)$$

for any $i \in I$. Then we set

$$L(\lambda) = \{x \in V_q(\lambda) \mid (x, x) \in A\} \qquad (4.4)$$

for any $\lambda \in \mathcal{P}_+$.

It is shown in [5] that the set L (respectively $L(\lambda)$) is a subspace of X (respectively $V_q(\lambda)$). Moreover, we have

$$X = \mathbb{C}(q) \otimes_A L, \qquad V_q(\lambda) = \mathbb{C}(q) \otimes_A L(\lambda), \qquad (4.5)$$

$$L(\lambda) = Lu_\lambda. \qquad (4.6)$$

The space L (respectively $L(\lambda)$) is called the *crystal lattice of X* (respectively $V_q(\lambda)$). It is clear that this lattice is $\mathcal{P}$-graded with respect to the given action of the commutative algebra H.

4.3. Let us consider the following transformations in the spaces X, $V_q(\lambda)$:

$$\tilde{f}_i(f_i^{(n)} x) = f_i^{(n+1)} x, \qquad \tilde{e}_i(f_i^{(n)} x) = f_i^{(n-1)} x, \qquad (4.7)$$

where $x \in \ker \partial_i$ in X, $x \in \ker e_i$ in $V_q(\lambda)$.

It is shown in [5] that the transformations (4.7) may be uniquely continued to linear maps of the corresponding space $X, V_q(\lambda)$. Moreover, this map preserves the corresponding crystal lattice $L, L_q(\lambda)$. We set

$$f_{i_1 \cdots i_n} = \tilde{f}_{i_1} \cdots \tilde{f}_{i_n} \cdot 1 \qquad \text{in } X, \qquad (4.8)$$

$$f_{i_1 \cdots i_n}(\lambda) = \tilde{f}_{i_1} \cdots \tilde{f}_{i_n} u_\lambda \qquad \text{in } V_q(\lambda), \qquad (4.9)$$

for any $i_1, \cdots, i_n \in I$, including the vectors $f_0 = 1$, $f_0(\lambda) = u_\lambda$. Then the set of vectors (4.8) (respectively (4.9)) constitutes a basis of the vector space X (respectively $V_q(\lambda)$) and of the corresponding crystal lattice L (respectively $L_q(\lambda)$) as a free A-module.

On the other hand, we may consider (4.7) as operators of the quotient spaces

$$K = L/qL, \qquad K(\lambda) = L(\lambda)/qL(\lambda). \qquad (4.10)$$

Then, identifying the cyclic vectors $1, u_\lambda$ with their images in (4.10) (respectively), we see that the set $\mathcal{B}$ (respectively $\mathcal{B}(\lambda)$) given by (4.8) (respectively (4.9)) is a basis of the vector space K (respectively $K(\lambda)$).

The set $\mathcal{B}$ (respectively $\mathcal{B}(\lambda)$) is called a *crystal basis* of the space K (respectively $K(\lambda)$).

4.4. Definition The family $L_w = L \cap X_w$ (respectively $L_w(\lambda) = L_w u_\lambda$) is called the *Schubert filtration of the lattice* L (respectively $L_w(\lambda)$). The corresponding family $K_w = L_w/qL_w$ (respectively $K_w(\lambda) = K_w u_\lambda$) is called the *Schubert filtration of the space* K (respectively $K(\lambda)$).

Using Lusztig bases (3.4), we may easily describe the module L_w (respectively $L_w(\lambda)$) as a free A-module with basis F_w (respectively $F_w(\lambda) = F_w u_\lambda$). Consequently, let $\mathcal{B}_w$ (respectively $\mathcal{B}_w(\lambda)$) be the canonical image of F_w (respectively $F_w(\lambda)$) in the space K (respectively $K(\lambda)$), then we see that $\mathcal{B}_w$ (respectively $\mathcal{B}_w(\lambda)$) is a basis of the vector space K_w (respectively $K_w(\lambda)$).

4.5. Using the quantum Weyl group $W_q(\mathfrak{g})$ (see 3.1), we may introduce a new system of generators of the algebra $U_q(\mathfrak{g})$, namely the elements e_α, f_α, $t_\alpha = wt_i$ (see 3.2) for a fixed element $w \in W$.

It is essential [14] that the crystal lattice $L(\lambda)$ does not depend on the choice of $w \in W$ (i.e., on the choice of the corresponding generators in $U_q(\mathfrak{g})$) provided the appropriate normalizing condition for $L(\lambda)$ is satisfied.

Moreover, we may continue the basic definition (4.7) to define the analogues $\hat{e}_\alpha$, $\hat{f}_\alpha$ of the operators $\bar{e}_i$, $\bar{f}_i$ (such that $\hat{e}_\alpha = \bar{e}_i$, $\hat{f}_\alpha = \bar{f}_i$ for $\alpha = \alpha_i$).

4.6. Theorem [14] *The space $K_w(\lambda)$, for any $\lambda \in \mathcal{P}_+$ and any $w \in W$, coincides with the set of all solutions of the following system of equations:*

$$\tilde{e}_\alpha x = 0 \qquad \text{for any} \qquad \alpha \in \Delta_w. \tag{4.11}$$

In other words, let $\tilde{E}_w(\lambda)$ be the set of all operators $\tilde{e}_\alpha \in \operatorname{End} K(\lambda)$ for any $\alpha \in \Delta_w$. Then we have:

$$K_w(\lambda) = \ker \tilde{E}_w(\lambda). \tag{4.12}$$

4.7. Remark The proof of Theorem 4.6 given in [14] consists in using Theorem 3.5 together with the asymptotic correspondence between the operators ∂_α, $\tilde{e}_\alpha$ for $\alpha \in \Delta_+$. This correspondence is described in detail by M. Kashiwara [5] for simple roots $\alpha = \alpha_i$.

Accordingly, Theorem 4.6 may be considered as the crystal version of Theorem 3.5 for the family of modules $V_q(\lambda)$, where $\lambda \in \mathcal{P}_+$.

5. The reduction problem

5.1. We present an application of Theorem 4.6 to the well-known reduction problem for classical and quantum modules.

The reduction problem $\mathfrak{g} \downarrow \mathfrak{g}_0$, where $\mathfrak{g}_0$ is a reductive subalgebra of $\mathfrak{g} = \mathfrak{g}(a)$, consists of the study of $\mathfrak{g}$-module $V(\lambda)$ considering as a $\mathfrak{g}_0$-module.

It is known that the $\mathfrak{g}_0$-module $V(\lambda)$ is semisimple, i.e.,

$$V(\lambda) = \bigoplus_\mu n(\lambda, \mu) V^0(\mu), \tag{5.1}$$

where $V^0(\mu)$ is a simple $\mathfrak{g}_0$-module of highest weight μ. An essential part of this reduction problem consists of the computation of the spectral multiplicities $n(\lambda, \mu)$ in (5.1).

There is no good analog of this formulation for quantum algebras $U_q(\mathfrak{g})$ in general, due to the difficulties in defining appropriate embeddings of the type $U_q(\mathfrak{g}_0) \subset U_q(\mathfrak{g})$. But this question not arise for the class of the so-called simple embeddings.

5.2. Let $\mathfrak{h}$ be a fixed Cartan subalgebra of $\mathfrak{g}$ with root vectors e_i, f_i $(i \in I)$. An embedding $\mathfrak{g}_0 \subset \mathfrak{g}$ is called *simple* if $\mathfrak{g}_0$ is generated by the subalgebra $\mathfrak{h}_0 = \mathfrak{h} \cap \mathfrak{g}_0$ and by the family of elements e_i, f_i $(i \in I_0)$, for a fixed subset $I_0 \subset I$.

It is clear, in this case there exists a natural embedding $U_q(\mathfrak{g}_0) \subset U_q(\mathfrak{g})$ (the symbol $U_q(\mathfrak{g}_0)$ is naturally extended to the case when $\mathfrak{g}_0$ is reductive). The problem of reduction (5.1) is completed by its quantum analog, namely

$$V_q(\lambda) = \bigoplus_\mu n_q(\lambda, \mu) V_q^0(\mu), \tag{5.2}$$

of type $U_q(\mathfrak{g}) \downarrow U_q(\mathfrak{g}_0)$.

It is not difficult to see from the general theory of [5] that the quantum multiplicities (5.2) coincide with a classical ones, namely $n(\lambda, \mu) = n_q(\lambda, \mu)$ for any pair λ, μ. Moreover, let $K(\lambda)$ be the crystal version of $V_q(\lambda)$ (see 4.3), and let $K(\lambda)^e$ be its extremal subspace with respect to $U_q(\mathfrak{g}_0)$ (i.e., the common kernel of $\tilde{e}_i$ for $i \in I_0$). Then we have:

$$n(\lambda, \mu) = n_q(\lambda, \mu) = \dim K(\lambda)_\mu^e, \tag{5.3}$$

where $K(\lambda)_\mu^e$ stands for the μ-homogeneous component of $K(\lambda)^e$ (with respect to induced weight gradation).

Hence the (classical or quantum) reduction problem $\mathfrak{g} \downarrow \mathfrak{g}_0$ may be essentially reduced to the computation of the weight multiplicities (5.3).

5.3. Let $\mathfrak{g}_0 \subset \mathfrak{g}$ be a simple embedding, and let $W_0 \subset W$ be the corresponding Weyl group of $\mathfrak{g}_0$. It is clear that W_0 is generated by the elements r_i for $i \in I_0$. Using Theorem 4.6, we find:

$$K(\lambda)^e = K_{w_0}(\lambda), \tag{5.4}$$

where $w_0 \in W_0$ is the maximal element of the Weyl group W_0. Hence we may characterize the extremal space $K(\lambda)^e$ as the span of the system

$\mathcal{B}_{w_0} \subset \mathcal{B}$ generated by the corresponding part $F_{w_0} \subset F$ of the Lusztig basis F.

The last result may be naturally formulated in terms of the induced bilinear form on the space $K(\lambda)$, namely

$$(x, y)_0 = (x, y) \mid_{q=0}, \tag{5.5}$$

for $x, y \in L(\lambda)$ (identified with its images in $K(\lambda)$).

5.4. Theorem *Let $\mathcal{B}(\lambda, \mu)$ be the set of all elements of the form $x = f(k)u_\lambda \in F_{w_0}u_\lambda$ having the weight μ and satisfying the condition $(x, x)_0 \neq 0$. Then we have*

$$n(\lambda, \mu) = \operatorname{card} \mathcal{B}(\lambda, \mu). \tag{5.6}$$

5.5. Remark It is known [5] (see also [10]) that $(x, x)_0 = 0, 1$ for any $x \in B(\lambda)$. Hence, the condition $(x, x)_0 \neq 0$ in the Theorem 5.4 may be also written as $(x, x)_0 = 1$.

5.6. It is interesting that an answer to the classical reduction problem (5.1) may be given in crystal terms (5.6).

There also exists another approach to this question, given in the recent work of P. Littelmann [6], in terms of certain combinatorial calculus for crystal bases.

References

1. Bourbaki, N.: *Groupes et algébras de Lie*, Ch. 4–6, Hermann, Paris, 1968.
2. Concini, C. and Kac, V.G.: Representations of quantum groups at roots of 1, *Progr. Math.* **92** (1990) 471–506.
3. Drinfeld, V.G.: Quantum groups, *Proc. Int. Congr. Math. Berkeley*, (1986) 798–820.
4. Jimbo, M.: A q-difference analogue of $U(\mathfrak{g})$ and Yang-Baxter equation, *Lett. Math. Phys.* **10** (1985) 63–69.
5. Kashiwara, M. Crystallizing the q-analogue of universal enveloping algebras, *Comm. Math. Phys.* (1990) 249–260.
6. Littelmann, P.: *Path and root operators in representation theory*, Preprint, 1994.
7. Lusztig, G.: On quantum groups, *J. of Algebra* **131** (1990) 466–475.
8. Lusztig, G.: Canonical bases arising from quantized enveloping algebras, *J. Amer. Math. Soc.* **3** (1990) 447–498.
9. Lusztig, G.: *Introduction to Quantum Groups*, Progr. in Math., Birkhäuser, 1994.
10. Zhelobenko, D.P.: *Representations of Reductive Lie Algebras*, Nauka, Moscow, 1994 (in Russian).
11. Zhelobenko, D.P.: Cartan type algebras, *Dokl. RAN* **339** (1994) no. 4, 137–140.
12. Zhelobenko, D.P.: Quantum boson algebra, Schubert filtration and Lusztig bases, *Izvestia RAN, Ser. Math.* **57** (1993) no. 6, 3–28.
13. Zhelobenko, D.P.: Constructive modules and the reductivity problem in the category $\mathcal{O}$, *Adv. in Math. Sci., Ser. 2*, **175** (1996) 207–224.
14. Zhelobenko, D.P.: Crystal bases and the reduction problem for classical and quantum modules, *Adv. in Math. Sci., Ser. 2*, **169** (1995) 183–202.

MULTIVALUED GROUPS, n-HOPF ALGEBRAS
AND n-RING HOMOMORPHISMS

V.M.BUCHSTABER
Department of Mathematics and Mechanics,
Moscow State University, 119899, Moscow, Russia

AND

E.G.REES
Department of Mathematics and Statistics,
University of Edinburgh,
King's Buildings, Edinburgh EH9 3JZ, Scotland

Abstract. This paper discusses the concept of an n-valued group. The theory that is developed here can be regarded as having more precise conditions than those in the much more developed theory of hypergroups; it is also more oriented towards algebraic and topological ideas. A number of methods for constructing examples are given and the classification of the algebraic 2-valued group structures on the Riemann sphere is presented. We introduce the (purely algebraic) notion of an n-Hopf algebra and show that the ring of functions on an n-valued group has an n-Hopf algebra structure. The crucial property of an n-Hopf algebra is that the diagonal is an n-ring homomorphism that has a weaker multiplicative property than that satisfied by a ring homomorphism. This concept arises in a number of other contexts, in particular as a trace. Convolution of two ring homomorphisms from an n-Hopf algebra to any commutative ring yields an n-ring homomorphism. There is a duality between symmetric products of topological spaces and the set of all n-ring homomorphisms from the ring of functions on the space to $\mathbb{C}$. The development of the properties of n-ring homomorphisms involves a number of combinatorial identities related closely to the classical properties of symmetric polynomials.

Mathematics Subject Classification (1991): Primary: 20N20, 16W30; Secondary: 05E05.

B. P. Komrakov et al. (eds.), Lie Groups and Lie Algebras, 85–107.

Key words: multivalued group, n-Hopf algebra, n-ring homomorphism, symmetric product, convolution, group algebra.

1. Introduction

There is a considerable literature on hypergroups and hypergroup algebras (for example, see the survey article [12]). We present a new theory of this type but with more precise conditions than previous versions. This allows us to introduce and study the concept of an n-Hopf algebra. The principal examples arise from n-valued groups in the same way that Hopf algebras arise from groups [7]. The key defining property is that the diagonal is what we call an n-ring homomorphism; this is a linear mapping with a multiplicative property weaker than that satisfied by a ring homomorphism. This property is satisfied by normalized traces and by the average of n ring homomorphisms. The major part of this paper is devoted to studying the properties of n-ring homomorphisms and this involves proving certain combinatorial results; since some of them may be of independent interest, they are collected together and proved in the final section.

2. Main definitions

If X is a space, let $(X)^n$ denote its n-fold symmetric product, i.e., $(X)^n = X/\Sigma_n$ where the symmetric group Σ_n acts by permuting the co-ordinates. An element of $(X)^n$ is called an n-subset of X or just an n-set; it is a subset with multiplicities of total cardinality n. An n-valued multiplication on X is a map

$$\mu : X \times X \to (X)^n.$$

which we sometimes write as

$$\mu(x, y) = x * y = [(x * y)_1, (x * y)_2, ..., (x * y)_n]$$

or

$$x * y = [z_1, z_2, ..., z_n].$$

    1. The multiplication μ is *associative* if the two n^2-sets $[x*(y*z)_1, x*(y*z)_2, ..., x*(y*z)_n]$ and $[(x*y)_1*z, (x*y)_2*z, ..., (x*y)_n*z]$ are equal for all $x, y, z \in X$.

    2. The element $e \in X$ is a *unit* if $e * x = x * e = [x, x, ...x]$ for all $x \in X$.

    3. When there is a unit, e, the multiplication $*$ has an *inverse* if there is a map inv : $X \to X$ such that $e \in \mathrm{inv}(x) * x$ and $e \in x * \mathrm{inv}(x)$ for all $x \in X$.

The map $\mu : X \times X \to (X)^n$ defines an *n-valued group structure* on X if it is associative, has a unit and an inverse.

An n-valued group X *acts on the set* Y if there is a mapping $\phi : X \times Y \to (Y)^n$, also denoted $x * y$ such that the two n^2-subsets $u * (v * y)$ and $(u * v) * y$ of Y are equal for all $u, v \in X$ and $y \in Y$ and also $e * y = [y, y, ..., y]$ for all $y \in Y$.

If A is an algebra, a *representation* of the n-valued group X in A is a mapping $\rho : X \to A$ such that $\rho(e) = 1$ and $\rho(x)\rho(y) = \sum \rho((x * y)_k)/n$ for all $x, y \in X$. An important special case is that where A is the endomorphism algebra of a vector space V.

3. Constructions

The principal examples are given by the following construction.

Construction 1. Let G be any group and A a group of automorphisms of G with $\#A = n$, then one can define an n-valued group structure on $X = G/A$ as follows:

Let $\pi : G \to G/A$ be the quotient map and define $\mu : X \times X \to (X)^n$ by

$$\mu(x, y) = \pi(\mu_o(\pi^{-1}(x), \pi^{-1}(y)))$$

where μ_o denotes the multiplication on G. The unit and inverses are the images of those in G.

An n-valued group of this type will be called an *n-coset group*.

An important special case is when $\#A = 2$, say $A = \{1, \alpha\}$, then the elements of X can be written as $\{g, g^\alpha\}$ and μ as

$$\{g, g^\alpha\} * \{h, h^\alpha\} = [\{gh, g^\alpha h^\alpha\}, \{gh^\alpha, g^\alpha h\}].$$

When G is a commutative group, we can consider the involution defined by the map $g \to g^{-1}$. In the special case of a cyclic group, it is easy to construct the multiplication table for G/A.

Construction 2. If G is a group and x is an indeterminate, let $\hat{G} = G \cup \{x\}$ then $\hat{G}$ has a 2-valued group structure by the rules

$$\mu(x, x) = e,$$
$$\mu(x, e) = \mu(e, x) = x,$$
$$\mu(x, g) = \mu(g, x) = g \text{ for all } g \in G \setminus \{e\},$$
$$\mu(g_1, g_2) = \begin{cases} g_1 g_2, & \text{if } g_1 g_2 \neq e, \\ \{e, x\}, & \text{if } g_1 g_2 = e \text{ and } g_1 \neq e. \end{cases}$$

Construction 3. Let $\mathbb{Z}_m$ denote the cyclic group of order m generated by the element x, we define a 2-valued "deformation" $\tilde{\mathbb{Z}}_m$ of $\mathbb{Z}_m$. Its elements can be identified with those of $\mathbb{Z}_m$ and the multiplication is given by (with $0 \le r, s < m$)

$$\mu(x^r, x^s) = \begin{cases} x^{r+s} & \text{for } r + s < m, \\ \{x^{r+s}, x^{r+s+1}\} & \text{for } m \le r + s. \end{cases}$$

It is shown in [7] that every 2-valued group with three elements (of which there are four) arises from using one of these constructions. There are eight 2-valued groups with four elements and they arise from the above constructions and two other general ones that we now describe.

Construction 4. Let G be any group, $m \ge 2$ and a, b be elements in the centre of G. We construct a 2-valued group X for which there is an exact sequence

$$1 \to G \to X \to \tilde{\mathbb{Z}}_m \to 0.$$

Each element of X is represented uniquely in the form gy^k with $g \in G$ and the multiplication is defined by

$$g_1 y^r * g_2 y^s = \begin{cases} g_1 g_2 y^{r+s} & \text{for } r + s < m, \\ [g_1 g_2 a y^{r+s-m}, g_1 g_2 b y^{r+s-m+1}] & \text{for } r + s \ge m. \end{cases}$$

The identity element of X is ey^0 where $e \in G$ is the identity. The inverse of the element gy^k is $g^{-1}a^{-1}y^{m-k}$. Associativity also holds and can be checked using a case by case analysis.

The map $X \to \tilde{\mathbb{Z}}_m$ defined by $gy^k \to y^k$ is a homomorphism to the "deformed" 2-valued cyclic group $\tilde{\mathbb{Z}}_m$.

Remark. There are one generator 2-valued group X which have $G = \mathbb{Z} \times \mathbb{Z}$ as a subgroup. As a specific example with $m = 2$, let $G = \mathbb{Z} \times \mathbb{Z}$ with generators a, b and $y^2 = [a, by]$, also the inverse of y is $a^{-1}y$ and $by * a^{-1}y = [b, a^{-1}b^2 y]$. Hence the subgroup generated by y (defined to be the set of elements that arise by successively multiplying all the elements obtained from y and their inverses) is the whole of the 2-valued group X.

Construction 5. Let G be a group and A an abelian group with multiplications μ_1, μ_2 respectively, we describe n-valued extensions X of A by G. Let $\xi : G \times G \to (A)^n$ be a map denoted

$$\xi(g_1, g_2) = [\xi_1(g_1, g_2), \xi_2(g_1, g_2), ..., \xi_n(g_1, g_2)]$$

such that

1. $\xi(g, e) = \xi(e, g) = [e, e, ..., e]$ for all $g \in G$

2. the following diagram commutes

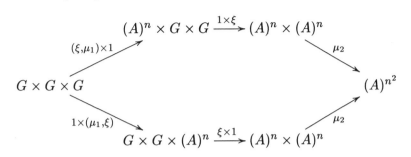

or, equivalently,

$$\xi(g_1, g_2) * \xi(g_1 g_2, g_3) = \xi(g_1, g_2 g_3) * \xi(g_2, g_3)$$

as n^2–sets.

If X denotes the product $G \times A$, the multiplication on X is given by

$$(g_1, a_1) * (g_2, a_2) = [z_1, z_2, ..., z_n],$$

where $z_k = (g_1 g_2, a_1 a_2 \xi_k(g_1, g_2))$; this gives an n-valued group structure on X.

As a specific example, let $G = A = \mathbb{Z}/2$; the only possibilities for ξ are $\xi(1, 1) = 0, 1$ and $[0, 1]$. The first two cases give the groups $\mathbb{Z}/2 \times \mathbb{Z}/2$ and $\mathbb{Z}/4$; the third case gives a 2-valued group with four elements.

In the case $G = \mathbb{Z}/2, A = \mathbb{Z}$ and $\xi(1, 1) = [n, m]$ with $n \neq m$ we obtain the multiplication

$$(s, a) * (t, b) = (s + t, a + b)$$

unless $s = t = 1$ and

$$(1, a) * (1, b) = [(0, a + b + n), (0, a + b + m)].$$

4. Algebraic 2-valued group structures on the Riemann sphere

In [7] we described several examples of multivalued group structures on Euclidean spaces and spheres. In particular we considered the correspondence $\mu : \mathbb{CP}^1 \times \mathbb{CP}^1 \to (\mathbb{CP}^1)^2$ defined by

$$(x_0 y_0 - k^2 x_1 y_1)^2 z_1^2 + (x_1 y_0 - x_0 y_1)^2 z_0^2$$
$$-2[(x_0 y_1 + x_1 y_0)(x_0 y_0 + k^2 x_1 y_1) - 4\delta x_0 y_0 x_1 y_1] z_1 z_0 = 0.$$

It defines a 2-valued group structure on $\mathbb{CP}^1$ if and only if $k(k^2 - \delta^2) \neq 0$. The group is commutative, the unit is $(0 : 1)$ and every element is its own inverse.

We can consider the more general situation of a homogeneous symmetric algebraic 2–2 correspondence

$$p_0(x_1, x_0; y_1, y_0)z_1^2 - p_1(x_1, x_0; y_1, y_0)z_1 z_0 + p_2(x_1, x_0; y_1, y_0)z_0^2 = 0,$$

where every variable has degree at most two. The space of all such has (projective) dimension 11. We show that the subspace of those which give 2-valued groups has dimension four.

Because of the action of the group $PGL(2, \mathbb{C})$ we can work in the affine chart $x_0 = 1$, so the correspondence becomes

$$p_0(x, y)z^2 - p_1(x, y)z + p_2(x, y) = 0.$$

Corollary 6.5 of [4] proves that, setting $\Theta = p_1/p_0$, two 2-valued groups defined in this way (but allowing the polynomials p to have any degree) coincide as 2-valued formal groups in a neighbourhood of the identity if and only if

$$\left.\frac{\partial \Theta}{\partial y}\right|_{y=0} = \left.\frac{\partial \Theta'}{\partial y}\right|_{y=0}.$$

We use this result to understand which correspondences give rise to 2-valued groups. Assume that 0 is the identity and write

$$p_k(x, y) = p_{k0}(x) + p_{k1}(x)y + p_{k2}(x)y^2;$$

then $p_0 = 1 + p_{01}(x)y + p_{02}(x)y^2$ with $p_{01}(0) = 0$ and $p_1 = 2x + p_{11}(x)y + p_{12}(x)y^2$ with $p_{11}(0) = 2$.

So

$$\left.\frac{\partial \Theta}{\partial y}\right|_{y=0} = p_{11}(x) - 2xp_{01}(x)$$

at $y = 0$. Hence, by the result quoted above from [4], the 2-valued group is determined by

$$\left.\frac{\partial \Theta}{\partial y}\right|_{y=0} = 2 + a_1 x + a_2 x^2 + a_3 x^3,$$

so there are three independent parameters.

One can rewrite these three parameters as k, δ appearing in the examples above and another coming from the group $PGL(2, \mathbb{C})$ of fractional linear transformations; one of the three parameters in $PGL(2, \mathbb{C})$ translates the identity and another acts on the k, δ parameters by homothety. Hence the moduli space of algebraic 2-valued group structures on $\mathbb{CP}^1$ has dimension one; these can all be realized by the coset construction on elliptic curves.

5. n-Hopf algebras

Definition 1. An n-*Hopf algebra* structure on a commutative algebra A (over a ring k that contains $1/n$) consists of

1. an *augmentation* $\epsilon : A \to k$,
2. an *antipode* $s : A \to A$,
3. a *diagonal* $\Delta : A \to A \otimes A$ making A into a coassociative coalgebra and
4. a map $P : A \to (A \otimes A)[t]$ which assigns to each $a \in A$ a monic *polynomial* of degree n

$$P_a(t) = t^n - \beta_1 t^{n-1} + \ldots + (-1)^n \beta_n$$

with $\beta_r = \beta_r(a) \in A \otimes A$.

These are related in the following way:

For each $a \in A$, introduce the series

$$\alpha_a(t) := \sum_{q \geq 0} \frac{a^q}{t^{q+1}} \in A[[t^{-1}]].$$

Axiom 1 (diagonal): The polynomial $P_a(t)$ is such that

$$\Delta \alpha_a(t) = \sum_{q \geq 0} \frac{\Delta(a^q)}{t^{q+1}} = \frac{1}{n} \frac{d}{dt} \ln(P_a(t)).$$

Axiom 2 (unit): If the unit of A is denoted by 1, then $P_1(t) = (t - (1 \otimes 1))^n$.

Axiom 3 (counit): The extension $i : (A \otimes A)[t] \to A[t]$ of $i : A \otimes A \to A$ defined by

$$i(a \otimes b) = \epsilon(b)a$$

satisfies

$$iP_a(t) = (t - a)^n.$$

Axiom 4 (antipode): The antipode is an algebra homomorphism $s : A \to A$ such that $\mu(1 \otimes s)P_a(\epsilon(a)) = 0$ and $\mu(s \otimes 1)P_a(\epsilon(a)) = 0$.

Remark. For $n = 1$ an n-Hopf algebra is precisely a Hopf algebra in the usual sense.

An *n-bialgebra* will satisfy all these axioms except for the existence of an antipode will not be assumed.

Theorem 1. [7] *If X is an n-valued group, then the ring $\mathbb{C}[X]$ of $\mathbb{C}$-valued functions on X and the cohomology algebra $H^*(X; \mathbb{C})$ are n-Hopf algebras.*

As a consequence, one can prove that the spaces $\mathbb{CP}^m$ for $m > 1$ do not admit the structure of a 2-valued group.

6. n-ring homomorphisms

This section investigates the general properties of maps such as the diagonal in an n-Hopf algebra.

Let A and B be algebras with B commutative.

Definition 2. A linear mapping $f : A \to B$ is called *algebraic of degree n at* $a \in A$ if there is a polynomial $p(a, t) = t^n - \gamma_1(a)t^{n-1} + \ldots + (-1)^n \gamma_n(a) \in B[t]$ such that

$$\sum_{q \geq 0} \frac{f(a^q)}{t^{q+1}} = \frac{1}{n} \frac{d}{dt} \ln p(a, t).$$

Such a map is called an *algebraic ring homomorphism* if it is algebraic at each $a \in A$.

If $f : A \to B$ is a linear map, define $D_f(n, m)$ of the form

$$\begin{pmatrix} f(a) & \frac{1}{n} & 0 & 0 & & \cdots & 0 \\ f(a^2) & f(a) & \frac{2}{n} & 0 & & \cdots & 0 \\ \vdots & \vdots & \vdots & & \ddots & & \vdots \\ \vdots & \vdots & \vdots & & \ddots & & \vdots \\ f(a^{n-1}) & f(a^{n-2}) & f(a^{n-3}) & \cdots & f(a) & \frac{(n-1)}{n} & 0 \\ f(a^n) & f(a^{n-1}) & f(a^{n-2}) & \cdots & f(a^2) & f(a) & 1 \\ f(a^{n+m}) & f(a^{n+m-1}) & f(a^{n+m-2}) & \cdots & f(a^{m+2}) & f(a^{m+1}) & f(a^m) \end{pmatrix}$$

Lemma 1. *The following conditions on a linear map $f : A \to B$ are equivalent*

1. f *is algebraic of degree n at $a \in A$,*
2. $D_f(n, m) = 0$ *for all $m \geq 1$.*

Proof. If f is algebraic, consider the polynomial p; choose a splitting extension of B and suppose the roots of p are $\gamma_1, \gamma_2, \ldots, \gamma_n$; then, for all $q \geq 0$,

$$f(a^q) = \frac{1}{n}(\gamma_1^q + \gamma_2^q + \ldots + \gamma_n^q).$$

By Proposition 7 (below), p has the same roots as

$$\det \begin{pmatrix} f(a) & \frac{1}{n} & 0 & 0 & & \cdots & 0 \\ f(a^2) & f(a) & \frac{2}{n} & 0 & & \cdots & 0 \\ \vdots & \vdots & \vdots & & \ddots & & \vdots \\ \vdots & \vdots & \vdots & & \ddots & & \vdots \\ f(a^{n-1}) & f(a^{n-2}) & f(a^{n-3}) & \cdots & f(a) & \frac{n-1}{n} & 0 \\ f(a^n) & f(a^{n-1}) & f(a^{n-2}) & \cdots & f(a^2) & f(a) & 1 \\ t^n & t^{n-1} & t^{n-2} & \cdots & t^2 & t & 1 \end{pmatrix}$$

Hence, putting $t = \gamma_i$ and multiplying the last row by γ_i^m we get the equation

$$\det \begin{pmatrix} f(a) & \frac{1}{n} & 0 & 0 & \cdots & 0 \\ f(a^2) & f(a) & \frac{2}{n} & 0 & & \cdots & 0 \\ \vdots & \vdots & \vdots & & \ddots & & \vdots \\ \vdots & \vdots & \vdots & & & \ddots & \vdots \\ f(a^{n-1}) & f(a^{n-2}) & f(a^{n-3}) & \cdots & f(a) & \frac{n-1}{n} & 0 \\ f(a^n) & f(a^{n-1}) & f(a^{n-2}) & \cdots & f(a^2) & f(a) & 1 \\ \gamma_i^{n+m} & \gamma_i^{n+m-1} & \gamma_i^{n+m-2} & \cdots & \gamma_i^{m+2} & \gamma_i^{m+1} & \gamma_i^m \end{pmatrix} = 0.$$

Adding these equations for $1 \le i \le n$ gives the result.

Conversely, consider the recurrence given by the determinantal equation $D_f(n, m) = 0$ it defines $f(a^q)$ for all q in terms of $f(a^q)$ for $1 \le q \le n$ and, by the general properties of linear recurrence relations, any solution is of the form $f(a^q) = c_1 \gamma_1^q + c_2 \gamma_2^q + \ldots + c_n \gamma_n^q$, where $\gamma_1, \gamma_2, \ldots, \gamma_n$ are the roots of the polynomial

$$\det \begin{pmatrix} f(a) & \frac{1}{n} & 0 & 0 & \cdots & 0 \\ f(a^2) & f(a) & \frac{2}{n} & 0 & & \cdots & 0 \\ \vdots & \vdots & \vdots & & \ddots & & \vdots \\ \vdots & \vdots & \vdots & & & \ddots & \vdots \\ f(a^{n-1}) & f(a^{n-2}) & f(a^{n-3}) & \cdots & f(a) & \frac{n-1}{n} & 0 \\ f(a^n) & f(a^{n-1}) & f(a^{n-2}) & \cdots & f(a^2) & f(a) & 1 \\ t^n & t^{n-1} & t^{n-2} & \cdots & t^2 & t & 1 \end{pmatrix}$$

But

$$f(a^q) = \frac{1}{n}(\gamma_1^q + \gamma_2^q + \ldots + \gamma_n^q)$$

for $1 \le q \le n$ and so $c_i = 1/n$ for each i. $\square$

We note that the proof shows that if f is algebraic of degree n, then $f(a^q)$ is determined for all $q \ge 1$ by those with $q \le n$.

Definition 3. A linear mapping $f : A \to B$ is called an *n-ring homomorphism* if $f(1) = 1$ and if it is an algebraic ring homomorphism and the degree n can be chosen to be independent of $a \in A$.

Examples.

1. A ring homomorphism f (in the usual sense) can be regarded as an n-ring homomorphism for any $n \ge 1$ by taking $p(a, t)$ to be $(t - f(a))^n$. More generally, for any $k \ge 1$, an n-ring homomorphism is an nk-ring homomorphism, using the polynomial $p(a, t)^k$.

2. If $f_1, f_2, ..., f_n : A \to B$ are ring homomorphisms then

$$(f_1 + f_2 + ... + f_n)/n : A \to B$$

is an n-ring homomorphism. The polynomial $p(a, t)$ is

$$(t - f(a_1))(t - f(a_2))...(t - f(a_n)).$$

More generally, if $\hat{B}$ is an extension of B and $f_k : A \to \hat{B}$ are ring homomorphisms such that the values of all the elementary symmetric functions in the variables $f_k(a)$ all lie in B, then $f = (f_1 + f_2 + ... + f_n)/n : A \to B$ is an n-ring homomorphism.

3. The diagonal of an n-Hopf algebra is an n-ring homomorphism.

4. If A is a finite extension of the ring B whose degree is n, (i.e., A is a free B module of rank n) then the trace map

$$\frac{1}{n}\mathrm{tr} : A \to B$$

is an n-ring homomorphism and the polynomial $p(a, t)$ is the characteristic polynomial of the element $a \in A$. Theorem 4 below gives an explicit formula (depending on n) that describes a multiplicative property of the trace. For small values of n, the formula can be checked easily. For example, the map $\mathrm{Re} : \mathbb{C} \to \mathbb{R}$ is a 2-ring homomorphism.

5. Let $\pi : X \to Y$ be a fiber bundle projection, with compact fibers F, then there is an associated transfer map

$$\pi_* : \{X, Z\} \to \{Y, Z\}$$

as defined by J. Becker and D. Gottlieb [1]. ($\{X, Z\}$ denotes the set of stable homotopy classes of maps $X \to Z$.)

In particular, there is a map

$$\pi_* : h^*(X) \to h^*(Y)$$

for any cohomology theory h^*. If h^* is multiplicative, π_* is not usually a ring homomorphism. However, in the case where $h^*(X)$ is flat over $h^*(Y)$ it is an n-ring homomorphism, for $n = |\chi(F)|$, where $\chi(F)$ denotes the Euler–Poincaré characteristic of F (see [8]).

Remark. The diagonal of a coalgebra structure on $k[[x_1, x_2, ..., x_m]]$ which has the form given in Example 2 for some n has investigated previously. The case $m = 1$ was considered in [2] and for $m > 1$ in [9] and [11]. The complete classification in the case $n = 2$ was obtained in [2] and [11].

Theorem 2 below calculates $p(a, t)$ in general; to prove it we need the following result:

Proposition 1. *Let $\dot{f} : A \to B$ be an n-ring homomorphism then $\gamma_1(a) = nf(a)$ and for $2 \le r \le n$,*

$$
\gamma_r(a) = \frac{n^r}{r!} \det
\begin{pmatrix}
f(a) & \frac{1}{n} & 0 & 0 & \cdots & 0 \\
f(a^2) & f(a) & \frac{2}{n} & 0 & \cdots & 0 \\
\vdots & \vdots & \vdots & \ddots & & \vdots \\
\vdots & \vdots & \vdots & & \ddots & \vdots \\
f(a^{r-1}) & f(a^{r-2}) & f(a^{r-3}) & \cdots & f(a) & \frac{r-1}{n} \\
f(a^r) & f(a^{r-1}) & f(a^{r-2}) & \cdots & f(a^2) & f(a)
\end{pmatrix}.
$$

Proof. The definition of an n-ring homomorphism and Lemma 6 below give the result immediately. $\square$

The following Theorem gives a direct description of the polynomial $p(a, t)$:

Theorem 2. *The polynomial $p(a, t)$ is of the form*

$$
\frac{(-n)^n}{n!} \det
\begin{pmatrix}
f(a) & \frac{1}{n} & 0 & 0 & \cdots & 0 \\
f(a^2) & f(a) & \frac{2}{n} & 0 & \cdots & 0 \\
\vdots & \vdots & \vdots & \ddots & & \vdots \\
\vdots & \vdots & \vdots & & \ddots & \vdots \\
f(a^{n-1}) & f(a^{n-2}) & f(a^{n-3}) & \cdots & f(a) & \frac{n-1}{n} & 0 \\
f(a^n) & f(a^{n-1}) & f(a^{n-2}) & \cdots & f(a^2) & f(a) & 1 \\
t^n & t^{n-1} & t^{n-2} & \cdots & t^2 & t & 1
\end{pmatrix}.
$$

Proof. The proof is a consequence of the combinatorial result given as Proposition 7. $\square$

The set of all n-ring homomorphisms from A to B will be denoted by $R^{(n)}(A, B)$ and $R^{(1)}(A, B)$ simply by $R(A, B)$.

If Y is a topological space and $\mathbb{C}[Y]$ is the algebra of continuous functions on Y, then the map

$$(Y)^n \to R^{(n)}(\mathbb{C}[Y], \mathbb{C})$$

defined by

$$[y_1, y_2, ..., y_n] \to \left(f \to \frac{1}{n} \sum f(y_k) \right)$$

is an embedding. Using the relationship that $R(\mathbb{C}[Y], \mathbb{C}) \cong Y$ one sees that this is a special case of the fact that $(R(A, B))^n \to R^{(n)}(A, B)$ is an embedding.

Theorem 3. *If $f : A \to B$ is an n-ring homomorphism and $g : B \to C$ is an m-ring homomorphism, then the composition $gf : A \to C$ is an nm-ring homomorphism.*

This is obvious from the definitions when either m or n is 1, but the proof in general will be given in another paper. The next result follows immediately from the definitions.

Proposition 2. 1. *A linear mapping $f : A \to B$ is an n-ring homomorphism if and only if the composition $fi : \mathbb{C}[\tau] \to A \to B$ is an n-ring homomorphism for each ring homomorphism $i : \mathbb{C}[\tau] \to A$.*

2. *A linear mapping $f : \mathbb{C}[a] \to B$ is an n-ring homomorphism if and only if it is the composition*

$$\mathbb{C}[a] \xrightarrow{i} B[t]/(p(a,t) = 0) \xrightarrow{\frac{1}{n}\mathrm{tr}} B,$$

where $i(a) = t$.

It is therefore important to describe the set $R^{(n)}(\mathbb{C}[\tau], B)$.

Proposition 3. *The mapping*

$$\rho : R^{(n)}(\mathbb{C}[\tau], B) \to R(\mathbb{C}[\tau_1, \tau_2, \ldots, \tau_n], B)$$

defined by $\rho(f)(\tau_k) = f(\tau^k)$ for $1 \le k \le n$ is a bijection.

The proof will follow from the first part of Theorem 4 below but the result will be used in the proof of the second part of that theorem.

Let M be the matrix $m_{r,s}$, where

$$
\begin{aligned}
m_{r,s} &= 0 && \text{for} \quad s > r+1, \\
m_{r,r+1} &= r/n, \\
m_{s+t,s} &= f(a_s a_{s+1} \ldots a_{s+t}) && \text{for} \quad t \ge 0,
\end{aligned}
$$

and M_σ, for $\sigma \in \Sigma_{n+1}$, be the matrix with entries $f(a_{\sigma(s)} a_{\sigma(s+1)} \cdots a_{\sigma(s+t)})$. So the matrix M is

$$
\begin{pmatrix}
F_{11} & \frac{1}{n} & 0 & 0 & \cdots & 0 \\
F_{21} & F_{22} & \frac{2}{n} & 0 & \cdots & 0 \\
\vdots & \vdots & \vdots & \ddots & & \vdots \\
F_{n-1,1} & F_{n-1,2} & F_{n-1,3} & \cdots & \frac{n-1}{n} & 0 \\
F_{n,1} & F_{n,2} & F_{n,3} & \cdots & F_{n,n} & 1 \\
F_{n+1,1} & F_{n+1,2} & F_{n+1,3} & \cdots & F_{n+1,n} & F_{n,n}
\end{pmatrix},
$$

where $F_{ij} = f(a_j a_{j+1} \ldots a_i)$, $j \le i$.

Lemma 2. *The mapping ψ_f defined by*

$$\psi_f(a_1, a_2, \ldots, a_{n+1}) = \sum_{\sigma \in \Sigma_{n+1}} \det M_\sigma$$

is multilinear.

Proof. It is enough to show that det(M) is multilinear. If a monomial term T in the expansion of det(M) involves $y = f(a_r a_{r+1} \ldots a_s)$ then it also involves the factor $(r/n)(r+1)/n \ldots (s-1)/n$ and hence does not contain any other factors involving entries from the rows and columns that include these entries. Hence each term T is a product of some consecutive terms on the upper diagonal together with the term at the opposite vertex of a right angled triangle. A specific example is a term such as $x_1 x_2 \ldots x_6 y$ (where each x_i is a rational with denominator n) consisting of entries of the matrix placed as follows:

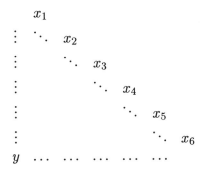

The multilinearity of each matrix follows. □

Theorem 4. *A linear mapping $f : A \to B$ is an n-ring homomorphism if and only if $\psi_f(a_1, a_2, \ldots, a_{n+1}) = 0$ for all $a_1, a_2, \ldots, a_{n+1} \in A$.*

From this theorem we can give an iterative procedure for calculating the value of $f(a_1 a_2 \ldots a_{n+1})$ in terms of the value of f on elements expressible as linear combinations of products of shorter length. In particular, an n-ring homomorphism is determined by its values on elements $a_1 a_2 \ldots a_j$, where $j \leq n$. In the case $n = 1$ this is the usual defining property of a ring homomorphism. For example, for low values of n, we have the following identities:

$$n = 1: \quad f(a_1 a_2) = f(a_1) f(a_2).$$

$$n = 2: \quad f(a_1 a_2 a_3) = f(a_1) f(a_2 a_3) + f(a_2) f(a_3 a_1)$$
$$+ f(a_3) f(a_1 a_2) - 2 f(a_1) f(a_2) f(a_3).$$

$$n = 3: \quad 2 f(a_1 a_2 a_3 a_4) = 2[f(a_1 a_2 a_3) f(a_4) + f(a_1 a_2 a_4) f(a_3)$$
$$+ f(a_1 a_3 a_4) f(a_2) + f(a_2 a_3 a_4) f(a_1)] + f(a_1 a_4) f(a_2 a_3)$$
$$+ f(a_1 a_2) f(a_3 a_4) + f(a_1 a_3) f(a_2 a_4)$$
$$- 3[f(a_1 a_2) f(a_3) f(a_4) + f(a_1 a_3) f(a_2) f(a_4)$$
$$+ f(a_1 a_4) f(a_2) f(a_3) + f(a_2 a_3) f(a_1) f(a_4)$$
$$+ f(a_2 a_4) f(a_1) f(a_3) + f(a_3 a_4) f(a_1) f(a_2)]$$
$$+ 9 f(a_1) f(a_2) f(a_3) f(a_4).$$

Proposition 4. *If $f : A \to B$ is a 2-ring homomorphism then*

1. $\det \begin{bmatrix} f(a_1b_1) - f(a_1)f(b_1) & f(a_1b_2) - f(a_1)f(b_2) \\ f(a_2b_1) - f(a_2)f(b_1) & f(a_2b_2) - f(a_2)f(b_2) \end{bmatrix} = 0.$

2. $(f(ab) - f(a)f(b))^2 = (f(a^2) - f(a)^2)(f(b^2) - f(b)^2).$

3. $f(a^2b^2) - f(a^2)f(b^2) = 4f(a)f(b)(f(ab) - f(a)f(b)).$

Proof. Expand $f(a_1b_1a_2b_2)$ in two different ways using the identity given above for a 2-ring homomorphism and equate the results. First, using $(a_1b_1).a_2.b_2$ we get

$$f(a_1b_1a_2b_2) = f(a_1b_1)f(a_2b_2) + f(a_2)f(b_2a_1b_1)$$
$$+f(b_2)f(a_1b_1a_2) - 2f(a_1b_1)f(a_2)f(b_2).$$

Then, using $a_1(b_1a_2)b_2$ we get

$$f(a_1b_1a_2b_2) = f(a_1)f(b_1a_2b_2) + f(b_1a_2)f(b_2a_1)$$
$$+f(b_2)f(a_1b_1a_2) - 2f(a_1)f(b_1a_2)f(b_2).$$

Now, equating these and expanding the terms with triple products, we get that

$$f(a_1b_1)f(a_2b_2) - 2f(a_1b_1)f(a_2)f(b_2)$$
$$+f(a_2)\{f(b_2)f(a_1b_1) + f(a_1)f(b_1b_2) + f(b_1)f(b_2a_1)$$
$$-2f(b_2)f(a_1)f(b_1)\}$$
$$+f(b_2)\{f(a_1)f(b_1a_2) + f(b_1)f(a_2a_1) + f(a_2)f(a_1b_1)$$
$$-2f(a_1)f(b_1)f(a_2)\}$$

equals to

$$f(a_1)\{f(b_1)f(a_2b_2) + f(a_2)f(b_2b_1) + f(b_2)f(b_1a_2) - 2f(b_1)f(a_2)f(b_2)\}$$
$$+f(b_1a_2)f(b_2a_1) - 2f(a_1)f(b_1a_2)f(b_2)$$
$$+f(b_2)\{f(a_1)f(b_1a_2) + f(b_1)f(a_2a_1) + f(a_2)f(a_1b_1)$$
$$-2f(a_1)f(b_1)f(a_2)\},$$

which is easily manipulated to get the first identity.

The second identity is an immediate consequence of the first.

We also have

$$f(a^2b^2) = f(a^2)f(b^2) + 2f(b)f(a^2b) - 2f(a^2)f(b)^2.$$

Substituting $f(a^2b) = 2f(a)f(ab) + f(a^2)f(b) - 2f(a)^2f(b)$ into this identity, one gets the third identity immediately. $\square$

Corollary 1. *If $f : A \to B$ is a 2-ring homomorphism and B has no nilpotent elements, then $\{a \mid f(a^2) = f(a)^2\}$ is a subalgebra of A on which f is a ring homomorphism.*

Indeed, it follows from the second identity that $f(ab) = f(a)f(b)$ if either $f(a^2) = f(a)^2$ or $f(b^2) = f(b)^2$.

Proof of Theorem 3. First assume that $f : A \to B$ is an n-ring homomorphism. By using the combinatorial Theorem 8 (below) with $t_i = \frac{i}{n}$ and $f_{i_1 i_2 \ldots i_\ell} = f(a_{i_1} a_{i_2} \ldots a_{i_\ell})$, it is enough to prove that $E_I = 0$ for each subset $I \subset \{1, 2, \ldots, n+1\}$. By applying Lemma 6 with $Z_s = f((a_{i_1} + a_{i_2} + \ldots + a_{i_\ell})^s) = F_I^{(s)}$ we obtain that $E_I = 0$ since $E_I = \dfrac{(n+1)!}{n^{n+1}} \mathcal{G}_{n+1}$ and $\mathcal{G}_{n+1} = 0$.
□

Proof of Proposition 3. The equation $\psi_f(a_1, a_2, \ldots, a_{n+1}) = 0$ has the form

$$\frac{(n+1)!n!}{n^n} f(a_1 a_2 \ldots a_{n+1}) = \overline{\psi}(f(a_{r_1} \ldots a_{r_k})),$$

where $\overline{\psi}$ is a function of a set of $2^{n+1} - 1$ variables. Putting $a_1 = a_2 = \ldots = a_n = a$ and $a_{n+1} = a^m$, we get $f(a^{n+m}) = \psi_{n+m}(f(a^r) : r < n+m)$, where ψ_{n+m} is a function of $n + m - 1$ variables.

Now consider $f_1, f_2 \in R^{(n)}(\mathbb{C}[\tau], B)$, then by the part of Theorem 3 that has been proved, we know that $\psi_{f_i}(\tau, \tau, \ldots, \tau, \tau^m) = 0$ for $i = 1, 2$. By induction, using the above formula one sees that $f_i(\tau^{n+m})$ is determined by $f_i(\tau^k)$ for $k < n + m$. But if $\rho(f_1) = \rho(f_2)$, then $f_1(\tau^k) = f_2(\tau^k)$ for $1 \leq k \leq n$ and so $f_1 = f_2$.

To show that ρ is onto, take $h \in R(\mathbb{C}[\tau_1, \tau_2, \ldots, \tau_n], B)$. Define $h(\tau, t)$ to be

$$\frac{(-n)^n}{n!} \det \begin{pmatrix} h(\tau_1) & \frac{1}{n} & 0 & 0 & \cdots & 0 \\ h(\tau_2) & h(\tau_1) & \frac{2}{n} & 0 & \cdots & 0 \\ \vdots & \vdots & \vdots & & \ddots & \vdots \\ \vdots & \vdots & \vdots & & \ddots & \vdots \\ h(\tau_{n-1}) & h(\tau_{n-2}) & h(\tau_{n-3}) & \cdots & h(\tau_1) & \frac{n-1}{n} & 0 \\ h(\tau_n) & h(\tau_{n-1}) & h(\tau_{n-2}) & \cdots & h(\tau_2) & h(\tau_1) & 1 \\ t^n & t^{n-1} & t^{n-2} & \cdots & t^2 & t & 1 \end{pmatrix}.$$

Now we work in a finite extension $B(h)$ of B in which $h(\tau, t)$ factors as

$$\prod_{i=1}^{n} (t - \beta_i).$$

The linear mapping $h_0 : \mathbb{C}[\tau] \to B$ is defined for $p(\tau) \in \mathbb{C}[\tau]$ by

$$h_0(p(\tau)) = \frac{1}{n} \sum_{\ell=1}^{n} p(\beta_\ell).$$

Then

$$\sum_{q \geq 0} \frac{h_0(p^q)}{t^{q+1}} = \frac{1}{n} \sum_{q \geq 0} \sum_{\ell=1}^{n} \frac{p(\beta_\ell)^q}{t^{q+1}} = \frac{1}{n} \sum_{\ell=1}^{n} \frac{1}{t - p(\beta_\ell)} = \frac{1}{n} \frac{d}{dt} \ln \prod_{\ell=1}^{n} (t - p(\beta_\ell))$$

which is of the required form since $\prod_{\ell=1}^{n} (t - p(\beta_\ell))$ is a monic polynomial of degree n whose coefficients are in B. These coefficients are the elementary symmetric functions in the variables $p(\beta_\ell)$ and so are polynomials in the elementary symmetric polynomials in $\beta_1, \beta_2, \ldots \beta_n$. Hence h_0 is an n-ring homomorphism. Taking the polynomial $p(\tau) = \tau$ in Theorem 3 we see that $h_0(\tau^\ell) = h(\tau_\ell)$ for $1 \leq \ell \leq n$, and so $\rho(h_0) = h$.

To prove the second part of Theorem 3, first note that by Proposition 2 it is enough to prove the result when A is taken to be a polynomial ring $\mathbb{C}[\tau]$. Let $f : \mathbb{C}[\tau] \to B$ satisfy $\psi_f(a_1, a_2, \ldots, a_{n+1}) = 0$ for all $a_i \in \mathbb{C}[\tau]$. Use Proposition 3 to find a unique n-ring homomorphism $h : \mathbb{C}[\tau] \to B$ such that $h(\tau^k) = f(\tau^k)$ for $1 \leq k \leq n$. By induction one has

$$\begin{aligned}
f(\tau^{n+m}) &= \psi_{n+m}(f(\tau^r) : r < n + m) \\
&= \psi_{n+m}(h(\tau^r) : r < n + m) \\
&= h(\tau^{n+m})
\end{aligned}$$

since, by the first part of the theorem,

$$\psi_h(\tau, \tau, \ldots, \tau, \tau^m) = 0.$$

$\square$

Proposition 5. *If $\mathcal{A} = \{\alpha_1, \alpha_2, \ldots\}$ is a multiplicative generating set for the algebra A, then a linear map $f : A \to B$ is an n-ring homomorphism if and only if the identity of Theorem 3 holds for all monomials in the elements of the set $\mathcal{A}$.*

Proof. By Lemma 5 each determinant in the formula is multilinear in each of the variables a_i and so the Proposition follows from Theorem 3. $\square$

We will now use this result to illustrate the close connection between n-ring homomorphisms and symmetric products.

Theorem 5. *The space of all 2-ring homomorphisms $\mathbb{C}[x_1, x_2, \ldots x_m] \to \mathbb{C}$ is naturally bijective with the symmetric square $(\mathbb{C}^m)^2$.*

Proof. It is convenient to identify $(\mathbb{C}^m)^2$ with $\mathbb{C}^m \times (\mathbb{C}^m / \pm 1)$ using the homeomorphism $[v_1, v_2] \to (v_1 + v_2, \pm(v_1 - v_2))$. Define

$$\text{ev} : (\mathbb{C}^m)^2 \to R^{(2)}(\mathbb{C}[x_1, x_2, ..., x_m], \mathbb{C})$$

and

$$\mu : R^{(2)}(\mathbb{C}[x_1, x_2, ..., x_m], \mathbb{C}) \to \mathbb{C}^m \times (\mathbb{C}^m / \pm 1)$$

by the following

$$\text{ev}[v_1, v_2](p(x_1, x_2, ..., x_m)) = \frac{1}{2}(p(v_1) + p(v_2))$$

We now construct the map μ. For $f \in R^{(2)}(\mathbb{C}[x_1, x_2, ..., x_m], \mathbb{C})$, let $u = (f(x_1), f(x_2), ..., f(x_m)) \in \mathbb{C}^m$ and M_f denote the symmetric $m \times m$ matrix with entries $f(x_i x_j) - f(x_i)f(x_j)$. By Proposition 4, the rank of M is at most one, so there is a vector $w \in \mathbb{C}^m$, unique up to sign such that $M = ww^T$. Then $\mu(f) = (u, w) \in \mathbb{C}^m \times (\mathbb{C}^m / \pm 1)$.

It is easily checked that ev and μ are mutually inverse. $\square$

Theorem 6. *Let* $f : \mathbb{C}[a_1, a_2, ...] \to B$ *be a linear mapping and define* $\phi(\omega) \in B$ *to be* $f(a^\omega)$ *where* $\omega = (\omega_1, \omega_2, ...)$ *lies in the positive part of the integer lattice. Then* f *is an* n-*ring homomorphism if and only if* ϕ *satisfies the functional equation of additive type in* $n + 1$ *variables* $\omega_1, \omega_2, ..., \omega_{n+1}$ *arising from the identity given in the statement of Theorem 3.*

These functional equations are

$$n = 1 : \quad \phi(\omega_1 + \omega_2) = \phi(\omega_1)\phi(\omega_2)$$
$$n = 2 : \quad \phi(\omega_1 + \omega_2 + \omega_3) = \phi(\omega_1)\phi(\omega_2 + \omega_3) + \phi(\omega_2)\phi(\omega_3 + \omega_1)$$
$$+ \phi(\omega_3)\phi(\omega_1 + \omega_2) - 2\phi(\omega_1)\phi(\omega_2)\phi(\omega_3)$$
$$n = 3 : \quad 2\phi(\omega_1 + \omega_2 + \omega_3 + \omega_4) =$$
$$2(\phi(\omega_1 + \omega_2 + \omega_3)\phi(\omega_4) + \phi(\omega_1 + \omega_2 + \omega_4)\phi(\omega_3)$$
$$+ \phi(\omega_1 + \omega_3 + \omega_4)\phi(\omega_2) + \phi(\omega_2 + \omega_3 + \omega_4)\phi(\omega_1))$$
$$+ (\phi(\omega_1 + \omega_2)\phi(\omega_3 + \omega_4) + \phi(\omega_1 + \omega_3)\phi(\omega_2 + \omega_4)$$
$$+ \phi(\omega_1 + \omega_4)\phi(\omega_2 + \omega_3)) - 3(\phi(\omega_1 + \omega_2)\phi(\omega_3)\phi(\omega_4)$$
$$+ \phi(\omega_1 + \omega_3)\phi(\omega_2)\phi(\omega_4) + \phi(\omega_1 + \omega_4)\phi(\omega_2)\phi(\omega_3)$$
$$+ \phi(\omega_2 + \omega_3)\phi(\omega_1)\phi(\omega_4) + \phi(\omega_2 + \omega_4)\phi(\omega_1)\phi(\omega_3)$$
$$+ \phi(\omega_3 + \omega_4)\phi(\omega_1)\phi(\omega_2)) + 9\phi(\omega_1)\phi(\omega_2)\phi(\omega_3)\phi(\omega_4).$$

When $B = \mathbb{C}$ these functional equations can be regarded as a discrete form of continuous functional equations. Any solution of these equations will give rise to an n-ring homomorphism

$$f : \mathbb{C}[a_1, a_2, ..., a_m] \to \mathbb{C}$$

for each m. For $n = 1$ the continuous equation is called the Cauchy functional equation whose only solution is given by the exponential function $\exp(\xi.\omega)$.

Proposition 6. *The general analytic solution of the functional equation*

$$\phi(z_1 + z_2 + z_3) = \phi(z_1)\phi(z_2 + z_3) + \phi(z_2)\phi(z_3 + z_1)$$

$$+\phi(z_3)\phi(z_1 + z_2) - 2\phi(z_1)\phi(z_2)\phi(z_3)$$

for $\phi : \mathbb{C}^m \to \mathbb{C}$ is $\phi(z) = \exp(\xi_1.z)\cosh(\xi_2.z)$.

Proof. It is easy to check that the stated functions are solutions. It remains to check that these are the only solutions. Note that these solutions are parametrized by $\mathbb{C}^m \times (\mathbb{C}^m / \pm 1)$.

Differentiating the given equation with respect to z_1 and z_2 and then setting $z_1 = z_2 = 0$ gives the following system of differential equations

$$\phi''(z) + 2\phi'(0)^2\phi(z) = 2\phi'(0)\phi'(z) + \phi(z)\phi''(0).$$

For a fixed vector $\mathbf{a} \in \mathbb{C}^m$, let $g(t)$ denote the function $\phi(t\mathbf{a})$ and then g satisfies the equation

$$g''(t) + 2g'(0)^2 g(t) = 2g'(0)g'(t) + g(t)g''(0).$$

This has the unique solution $g(t) = \exp(\alpha t)\cosh(\beta t)$ where $\alpha = g'(0) = d\phi_0(\mathbf{a})$ and $\beta^2 = g''(0) + g'(0)^2 = d^2\phi_0(\mathbf{a}, \mathbf{a}) + (d\phi_0(\mathbf{a}))^2$. By multiplying the solution by $\exp(-\xi_1.z)$ where ξ_1 is dual to $d\phi_0$, one can assume that $\phi'(0) = 0$ and by change of coordinates that the symmetric matrix $\phi''(0)$ is diagonal with diagonal entries consisting of 0's and 1's. If rank$\phi''(0) > 1$, then there are two coordinates x and y such that on the plane spanned by x and y the function ϕ satisfies the system

$$\frac{\partial^2 \phi}{\partial x^2} = \phi, \qquad \frac{\partial^2 \phi}{\partial x \partial y} = 0, \quad \text{and} \quad \frac{\partial^2 \phi}{\partial y^2} = \phi$$

Hence,

$$\phi = \frac{\partial^2(\phi)}{\partial y^2} = \frac{\partial^2}{\partial y^2}\left(\frac{\partial^2 \phi}{\partial x^2}\right) = \frac{\partial^2}{\partial x \partial y}\left(\frac{\partial^2 \phi}{\partial x \partial y}\right) = 0$$

and this is a contradiction. So, the rank of the matrix $\phi''(0) = d^2\phi_0$ is at most one. Hence $d^2\phi_0 = \xi_2\xi_2^T$ for some vector ξ_2. The general solution is therefore of the form stated. $\square$

7. Convolution

Given an algebra (B, μ, η) where μ is the multiplication and η is the unit, and a coalgebra (C, Δ, ϵ) where Δ is the diagonal and ϵ is the counit, we define a bilinear map

$$* : \hom(C, B) \otimes \hom(C, B) \to \hom(C, B)$$

called *convolution* and denoted by $f * g$; it is the composition

$$C \xrightarrow{\Delta} C \otimes C \xrightarrow{f \otimes g} B \otimes B \xrightarrow{\mu} B$$

The triple $(\hom(C, B), *, \eta\epsilon)$ is an algebra (see [10], Prop III.3.1). When B is commutative and C is a polynomial algebra in n variables, the set of all ring homomorphisms $R(C, B)$ is called the *set of B-points of n-dimensional affine space* ([10], page 8). When C is a Hopf algebra, then $R(C, B)$ is a group under the multiplication $*$ ([10], Prop III.3.7).

Theorem 7. *Let C be an n-bialgebra and B a commutative algebra; then the convolution $*$ induces a map*

$$R(C, B) \times R(C, B) \to R^{(n)}(C, B)$$

which is associative in the appropriate sense that the following diagram commutes:

$$
\begin{array}{ccc}
R(C, B) \times R(C, B) \times R(C, B) & \xrightarrow{\;*\times 1\;} & R(C, B) \times R^{(n)}(C, B) \\
{\scriptstyle 1 \times *} \downarrow & & \downarrow {\scriptstyle *} \\
R(C, B) \times R^{(n)}(C, B) & \xrightarrow{\quad * \quad} & R^{(n^2)}(C, B)
\end{array}
$$

Proof. We first prove that the image of $R(C, B) \times R(C, B)$ under convolution lies in $R^{(n)}(C, B)$. From the definition of an n-bialgebra, one has for each $c \in C$ elements $\beta_r(c) \in C \otimes C$ which are the coefficients of the polynomial $P_c(t)$. For $f, g \in R(C, B)$ define a polynomial of degree n with coefficients $\gamma_r(c) \in B$ given by $\mu(f \otimes g)\beta_r(c)$. Then,

$$\sum_{q \geq 0} \frac{(f * g)c^q}{t^{q+1}} = \mu(f \otimes g) \sum_{q \geq 0} \frac{\Delta c^q}{t^{q+1}} = \frac{1}{n} \frac{d}{dt} \ln \mu(f \otimes g) P_c(t)$$

and $\mu(f \otimes g)P_c(t)$ is a monic polynomial of degree n as required.

The fact that the diagram commutes follows from Theorem 3 about composition of n-ring homomorphisms. $\square$

Corollary 2. *Let $C=\mathbb{C}[x_1, x_2, ..., x_m]$ have a 2-Hopf algebra structure, then one has an induced 2-valued group structure on $\mathbb{C}^m$.*

Proof. We take $B = \mathbb{C}$ and $n = 2$ in the theorem and note that, by Theorem 4, $R^{(2)}(\mathbb{C}[x_1, x_2, ..., x_m], \mathbb{C})$ can be identified with the symmetric product $(\mathbb{C}^m)^2$. $\square$

8. Combinatorial results

For $I = \{i_1, i_2, \ldots, i_q\} \subset \{1, 2, \ldots, n+1\}$, define $F_I = f_{i_1} + f_{i_2} + \ldots + f_{i_q}$ and $F_I^{(k)}$ to be the kth symmetric tensor power of F_I, e.g. $F_I^{(2)} = \sum f_{ii} + 2 \sum_{i<j} f_{ij}$.

If $t_1, t_2, \ldots, t_n$ is a set of indeterminates, define

$$D = \det \begin{pmatrix} f_1 & t_1 & 0 & 0 & \cdots & 0 \\ f_{12} & f_2 & t_2 & 0 & \cdots & 0 \\ \vdots & \vdots & \ddots & \ddots & & \vdots \\ \vdots & \vdots & & \ddots & \ddots & \vdots \\ f_{12\ldots n} & f_{23\ldots n} & & \cdots & f_n & t_n \\ f_{12\ldots n+1} & f_{23\ldots n+1} & & \cdots & f_{nn+1} & f_{n+1} \end{pmatrix}$$

The entries of the matrix defining D are given by

$$\begin{aligned} d_{ij} &= 0 & \text{for} \quad j > i+1 \\ d_{ii+1} &= t_i \\ d_{ii} &= f_i \\ d_{ij} &= f_{jj+1\ldots i} & \text{for} \quad j \le i. \end{aligned}$$

For $\sigma \in \Sigma_{n+1}$, let D_σ denote the same determinant but with the f_i permuted by σ and the t_i unchanged.

Similarly, define

$$E_I = \det \begin{pmatrix} F_I & t_1 & 0 & 0 & \cdots & 0 \\ F_I^{(2)} & F_I & t_2 & 0 & \cdots & 0 \\ \vdots & \vdots & & \ddots & & \vdots \\ \vdots & \vdots & & & \ddots & \vdots \\ F_I^{(n)} & F_I^{(n-1)} & & \cdots & F_I & t_n \\ F_I^{(n+1)} & F_I^{(n)} & & \cdots & F_I^{(2)} & F_I \end{pmatrix}$$

The entries of the matrix defining E_I are given by

$$\begin{aligned} e_{ij} &= 0 & \text{for} \quad j > i+1 \\ e_{ii+1} &= t_i \\ e_{ii} &= F_I \\ e_{ij} &= F_I^{(i-j+1)} & \text{for} \quad j \le i. \end{aligned}$$

Theorem 8. *If $\#I$ is denoted by q, then*

$$\sum_{\sigma\in\Sigma_{n+1}} D_\sigma = \sum_{I\subset\{1,2,\ldots,n+1\}} (-1)^{n+1+q} E_I.$$

For $\epsilon = \{\epsilon_1, \epsilon_2, \ldots, \epsilon_n\} \in \{0,1\}^n$ let t^ϵ denote the monomial $t_1^{\epsilon_1} t_2^{\epsilon_2} \ldots t_n^{\epsilon_n}$ and define a partition π_ϵ of $\{1, 2, \ldots, n+1\}$ by $\pi_\epsilon = [\ell_1, \ell_2, \ldots]$ where

$$\{\ell_1, \ell_1 + \ell_2, \ldots, \ell_1 + \ell_2 \ldots + \ell_k, \ldots\} = \{i : \epsilon_i = 0\} \cup \{n+1\}.$$

For example, when all ϵ_i are zero, one obtains the partition $[1, 1, \ldots, 1]$.
For a partition $\pi = [\ell_1, \ell_2, \ldots]$ of $\{1, 2, \ldots, n+1\}$, define

$$F_{I,\pi} = F_I^{(\ell_1)} F_I^{(\ell_2)} \ldots = \prod F_I^{(\ell_k)}$$

Lemma 4. $E_I = \sum_\epsilon t^\epsilon (-1)^\epsilon F_{I,\pi_\epsilon}.$

Now let

$$F_\pi = f_{12\ldots\ell_1} f_{\ell_1+1\ell_1+2\ldots\ell_1+\ell_2} \cdots$$

Lemma 5. $D = \sum_\epsilon t^\epsilon (-1)^\epsilon F_{\pi_\epsilon}.$

There is a similar expression for D_σ.

Claim. *For every partition π,*

$$\sum_{I\subset\{1,2,\ldots,n+1\}} (-1)^{n+1+q} F_{I,\pi} = \sum_{\sigma\in\Sigma_{n+1}} F_{\sigma(\pi)}.$$

Clearly, the Claim implies Theorem 8.

Proof of Claim. We consider individual terms in the expansion of the left hand side. First consider one that does not contain all the $n+1$ different indices, say it involves only k indices. It appears $\binom{k}{r}$ times in $F_{I,\pi}$ where $r + q = n + 1$ and q is the cardinality of I. Hence the algebraic sum of all these terms is $(1 - 1)^k = 0$. Now consider a term that involves all the $n+1$ indices, it appears only in the expansion of the expression $F_{\{1,2,\ldots,n+1\}}$ and has coefficient $\pi!$ where $\pi!$ denotes the order of the normalizer of the group $\Sigma_\pi = \prod \Sigma_{\ell_i}$ in Σ_{n+1}. Let h_π denote the "smallest" symmetric polynomial containing the term F_π. Then the right hand side of the equation stated in the Claim is $((n+1)!/\pi!)h_\pi$ and the result is proved. $\square$

Lemma 6. *Let $G(t) = t^n - G_1 t^{n-1} + \ldots + (-1)^n G_n$ and write*

$$\frac{1}{n}\frac{d}{dt}\ln G(t) = \sum_{q\geq 0} \frac{Z_q}{t^{q+1}}.$$

Then, putting

$$
\mathcal{G}_r = \frac{n^r}{r!} \det
\begin{pmatrix}
Z_1 & \frac{1}{n} & 0 & 0 & \cdots & 0 \\
Z_2 & Z_1 & \frac{2}{n} & 0 & \cdots & 0 \\
\vdots & \vdots & \vdots & \ddots & & \vdots \\
\vdots & \vdots & \vdots & & \ddots & \vdots \\
Z_{r-1} & Z_{r-2} & Z_{r-3} & \cdots & Z_1 & \frac{r-1}{n} \\
Z_r & Z_{r-1} & Z_{r-2} & \cdots & Z_2 & Z_1
\end{pmatrix}
$$

one has that, $\mathcal{G}_r = G_r$ for all $1 \le r \le n$ and $\mathcal{G}_r = 0$ for $r > n$.

Proof. By adjoining the roots α_i of the polynomial $G(t)$ to k and setting the elementary symmetric polynomials $e_i = G_i$ and the power sums of these roots $p_i = nZ_i$, the formula follows from the standard Newton formulae (see [13] page 20)

$$
p_r - p_{r-1}e_1 + \ldots + (-1)^{r-1}p_1e_{r-1} + (-1)^r re_r = 0.
$$

□

Proposition 7. *Let p_r be the rth power sum of the variables $x_1, x_2, \ldots, x_n$ and let $d(t)$ denote the polynomial*

$$
\det
\begin{pmatrix}
p_1 & 1 & 0 & 0 & & \cdots & 0 \\
p_2 & p_1 & 2 & 0 & & \cdots & 0 \\
p_3 & p_2 & p_1 & 3 & \ddots & & \vdots \\
\vdots & \vdots & \vdots & & \ddots & & \vdots \\
p_n & p_{n-1} & p_{n-2} & \cdots & p_2 & p_1 & n \\
t^n & t^{n-1} & t^{n-2} & \cdots & t^2 & t & 1
\end{pmatrix}.
$$

Then, the roots of $d(t)$ are $x_1, x_2, \ldots, x_n$.

Proof. Let e_r denote the rth elementary symmetric function in the variables $x_1, x_2, \ldots, x_n$. Denote also the columns of the matrix above by $(c_1, c_2, \ldots, c_{n+1})$. Replace the first column of the matrix by the column vector

$$
c_1 - e_1c_2 + e_2c_3 - \ldots + (-1)^n e_n c_{n+1}.
$$

By the Newton formulae, this gives a matrix whose first column entries are all zero except for the last entry, and this is

$$
p(t) = t^n - e_1 t^{n-1} + e_2 t^{n-2} - \ldots + (-1)^n e_n.
$$

So $d(t) = (-1)^n n! p(t)$ and hence the roots of $d(t)$ are the same as those of $p(t)$, i.e., $\{x_1, x_2, \ldots, x_n\}$. □

Acknowledgments.

The authors are grateful to Drs. Toby Bailey and Antony Maciocia who calculated all the 2-valued groups of order ≤ 4 in 1993. The list they produced showed that there was a wide number of possibilities.

During the period when this work was carried out, V.M. Buchstaber was supported by a Royal Society Kapitza Fellowship in April 1993 and by an EPSRC Visiting Fellowship at the University of Edinburgh for a total of six months during 1994–96.

References

1. Becker, J.C. and Gottlieb, D.H.: The transfer map and fiber bundles, *Topology* **14** (1975) 1–12.
2. Buchstaber, V.M.: Two-valued formal groups. Some applications to cobordisms, *Uspekhi Mat. Nauk,* **26** (1971) no. 3, 195–196.
3. Buchstaber, V.M.: Characteristic classes in cobordism and topological applications of the theory of single-valued and two-valued formal groups, *Current problems in mathematics*, VINITI, Moscow, **10** (1978) 5–178; English translation: *J. Soviet Mathematics*, **11** (1979) no. 6.
4. Buchstaber, V.M.: Two-valued formal groups. Algebraic theory and applications to cobordism I, *Izv. Akad. Nauk SSSR, Ser. Mat.* **39** (1975) 1044–1064.
5. Buchstaber, V.M. and Gorbunov V.G.: Theory and applications of multivalued groups, Manchester University preprint, 1991.
6. Buchstaber, V.M. and Novikov, S.P.: Formal groups, power systems and Adams operators, *Mat. Sb. (N.S.)* **84** (1971) 81–118.
7. Buchstaber, V.M. and Rees, E.G.: Multivalued groups, their representations and Hopf algebras, University of Edinburgh preprint, 1996.
8. Dold, A.: The fixed point transfer of fibre-preserving maps, *Math. Zeit.* **148** (1976) 215–244.
9. Gurevich, D.I.: Multivalued Lie groups, *Uspekhi Mat. Nauk.* **39** (1984) 195–196.
10. Kassel, C.: *Quantum groups*, Springer–Verlag, New York, 1995.
11. Kholodov, A.N.: Multidimensional two-valued commutative formal groups, *Uspekhi Mat. Nauk.* **43** (1988) 213–214.
12. Litvinov, G.L.: Hypergroups and hypergroup algebras, *Current Problems in Math.*, VINITI, Moscow, **26** (1985) 57–106.
13. Macdonald, I.G.: *Symmetric functions and Hall polynomials*, Clarendon Press, Oxford, 1979.

HYPERGROUPS AND DIFFERENTIAL EQUATIONS

WILLIAM C. CONNETT AND ALAN L. SCHWARTZ
Department of Mathematics and Computer Science
University of Missouri–St. Louis, St. Louis, MO 63121, U.S.A.
E-mail: connett@@umslvma.bitnet, schwartz@@arch.umsl.edu

Abstract. The relationships between the characters of an algebra of measures supported on an interval H, and the eigenfunctions of a Sturm–Liouville problem on H are utilized to study the Jacobi polynomials and the spheroidal wave functions.

Mathematics Subject Classification (1991): Primary: 43A62, 34B24; Secondary: 42C10.

Key words: hypergroups, spheroidal wave functions, Sturm–Liouville problems.

1. What is a hypergroup?

A hypergroup is an algebra of measures supported on a set H, which imitates the structure of the algebra of measures on a group. The goal is to create an algebraic structure in the collection of measures sufficiently rich in detail to allow various theorems from the harmonic analysis of groups to be carried out in a setting where an algebraic structure is not available in the underlying space H. The goal is clear. What is not clear is what properties of the group measure algebra will be needed. Since different choices have been made by different authors, it is important to review which properties will be imitated.

1.1. THE GROUP MEASURE ALGEBRA

If $(G, \cdot)$ is a compact topological group, then the group operation can be used to define an algebraic structure on $M(G)$, the collection of regular

B. P. Komrakov et al. (eds.), Lie Groups and Lie Algebras, 109–115.
© *1998 Kluwer Academic Publishers. Printed in the Netherlands.*

bounded Borel measures supported on G, see [17]. For $x, y \in G$, define $\delta_x * \delta_y = \delta_{x \cdot y}$, and extend this product to all $\mu, \nu \in M(G)$ by

$$\langle \mu * \nu, f \rangle = \int_G \int_G \langle \delta_x * \delta_y, f \rangle \, d\mu(x) \, d\nu(y), \qquad f \in C(G).$$

This makes $M(G)$ into a Banach algebra. Some useful properties of $*$ are:

1. If μ and ν are probability measures, so is $\mu * \nu$.
2. If e is the group identity, then δ_e is a unit in the algebra. That is $\delta_e * \delta_x = \delta_{e \cdot x} = \delta_x$.
3. $\delta_x * \delta_y = \delta_e$ if and only if $x \cdot y = e$, or equivalently, $y = x^{-1}$.
4. Define μ^{-1} by its action on continuous functions:

$$\langle \mu^{-1}, f \rangle = \int_G f(x) \, d\mu^{-1}(x) = \int_G f(x^{-1}) \, d\mu(x).$$

Then $\mu^{-1}(E) = \mu\{x^{-1} | x \in E\}$, so

$$(\delta_x * \delta_y)^{-1} = (\delta_{x \cdot y})^{-1} = \delta_{y^{-1} \cdot x^{-1}} = \delta_y^{-1} * \delta_x^{-1},$$

and for all measures μ and ν

$$(\mu * \nu)^{-1} = \nu^{-1} * \mu^{-1}.$$

For a commutative compact group, a continuous character, χ, is a bounded continuous function from G to the complex numbers of modulus one, such that $\chi(x \cdot y) = \chi(x) \cdot \chi(y)$. There exists an invariant measure m, called Haar measure, such that

$$\delta_x * m = m * \delta_x = m, \qquad x \in G.$$

1.2. THE HYPERGROUP MEASURE ALGEBRA

Now to construct an algebra of measures that imitates the group measure algebra. We will focus on compact hypergroups. The same axioms with slight modifications are required for the locally compact case.

Definition Let H be a compact Hausdorff space. Denote the regular bounded Borel measures supported on H by $M(H)$. If $\|\cdot\|$ denotes the total variation norm, then $(M(H), \|\cdot\|)$ is a Banach space. Assume that there is a bilinear mapping, $*$, usually called convolution, from $M(H) \times M(H)$ into $M(H)$, and that with this mapping the above Banach space is a Banach algebra. $(H, *)$ is a *compact hypergroup*, if the following axioms are satisfied:

1. A convolution of probability measures is a probability measure.

2. There is an element $e \in H$ such that $\delta_e * \mu = \mu * \delta_e = \mu$ for every $\mu \in M(H)$.
3. There is a homeomorphic mapping $x \mapsto x^\vee$ of H onto itself such that $x^{\vee\vee} = x$ and $e \in \mathrm{supp}(\delta_x * \delta_y)$ if and only if $y = x^\vee$.
4. For $\mu, \nu \in M(H)$, $(\mu * \nu)^\vee = \nu^\vee * \mu^\vee$, where $\mu^\vee$ is defined by

$$\int_H f(s)\, d\mu^\vee(x) = \int_H f(x^\vee)\, d\mu(x).$$

5. The mapping $(\mu, \nu) \mapsto \mu * \nu$ is continuous from $M(H) \times M(H)$ into $M(H)$ with the weak-* topology.
6. The mapping $(x, y) \mapsto \mathrm{supp}(\delta_x * \delta_y)$ is continuous from $H \times H$ into the space of compact subsets of H as topologized in [16].

By analogy to the group case, $\chi \in C(H)$ is a *character*, if

$$\int_H \chi\, d(\delta_x * \delta_y) = \chi(x)\chi(y) \qquad (x, y \in H).$$

And m is a *Haar measure* for $(H, *)$, if

$$m * \delta_x = \delta_x * m = m \qquad (x \in H).$$

Notice that property (1) is the same for both algebras. For the hypergroup measure algebra, (2) replaces the idea that the group has an identity (since H need not have an algebraic operation). In (3), the involution $^\vee$ on H is a surrogate for the inverse on the group.

The founding articles on the subject are by Dunkl [11], Jewett [14] and Spector [18]. See [9] and [10], and the references cited there, for the more recent literature on hypergroups. Another good source for hypergroups is the proceedings of the Joint Summer Research Conference on Applications of Hypergroups and Related Measure Algebras, which appeared as volume 183 of *Contemporary Mathematics*. For Haar measure, see [11] for the compact case, [18] for the commutative case. The existence of Haar measure in the noncompact noncommutative case is still a major open question in the field.

2. Characters and eigenfunctions

In this section we discuss how a Sturm–Liouville problem on a compact interval H may give rise to a hypergroup on H, and conversely, how a hypergroup on a compact interval H can give rise to a Sturm–Liouville problem on H. Frequently the key to the solution of problems about the harmonic analysis of functions defined on H, is the realization that a given set of basis functions can be considered both as the characters of a hypergroup on H, and as the eigenfunctions associated with a Sturm–Liouville problem on H,

2.1. FROM DIFFERENTIAL EQUATION TO HYPERGROUP.

We first became aware of hypergroup structures in our effort to establish an adequate L^1 theory for expansions in terms of families of eigenfunctions of Sturm–Liouville problems that arose in mathematical physics [5] and [7]. Some examples of such families (in order of increasing generality) are:

(1) The ultraspherical polynomials (I. I. Hirshman, [13]).
(2) The Jacobi polynomials for certain values of (α, β) (G. Gasper, [12]).
(3) The eigenfunctions associated with the Sturm–Liouville problem

$$(\rho^2 w')' + \mu \rho^2 w = 0, \qquad (1)$$
$$w'(0) = 0 = w'(\pi)$$

for certain choices of ρ. (H. Chébli [2], Achour and Trimeche [1])

In each of these three cases, the route from the Sturm–Liouville problem to the hypergroup is the following:

(1) The eigenfunctions and the differential equation from a Sturm–Liouville problem on H can be used to define a hyperbolic Cauchy problem on $H \times H$. The solution of this problem with boundary data $f \in C(H)$ for each $x, y \in H$ can be used to show the existence of a measure $\sigma_{x,y}$, with the property that for each χ an eigenfunction of the Sturm–Liouville problem, we have the following product formula

$$\int_H \chi \, d\sigma_{x,y} = \chi(x)\chi(y).$$

See [7] for the details.
(2) The next step is to show positivity of the measures $\sigma_{x,y}$. It may be possible to solve the hyperbolic boundary problem explicitly using the Riemann integration technique, and the positivity may be read off from the explicit solution, [4]. If this fails, it still may be possible to use a hyperbolic maximum principle along with other facts about the eigenfunctions to establish the positivity, [7].
(3) Finally, define $\delta_x * \delta_y = \sigma_{x,y}$, and then extend the product $*$ from point masses to all measures using the formula

$$\langle \mu * \nu, f \rangle = \int_H \int_H \langle \delta_x * \delta_y, f \rangle \, d\mu(x) \, d\nu(y) \qquad f \in C(H).$$

2.2. FROM HYPERGROUP TO DIFFERENTIAL EQUATION.

If a hypergroup has some regularity, then the characters of the hypergroup on a compact interval H are equivalent by a change of variables to the eigenfunctions of a Sturm–Liouville problem 1, where ρ^2 is completely determined by the hypergroup [6]. The route to the Sturm–Liouville problem involves the following steps:

(1) Use the family of measures $\delta_x * \delta_y$ to define the *moments*

$$M_\mu(x,y) = \int_H (r-y)^\mu d(\delta_x * \delta_y)(r).$$

If $M_\mu(x,y) = O((x-e)^k)$ for all y and some positive integer k, then define k_μ = the largest k for which this is true, if there is any such, otherwise, let $k_\mu = \infty$. In a similar manner define $\widetilde{M}_\mu$ and $\widetilde{k}_\mu$ by reversing the roles of x and y on the right.

(2) Regularity on the hypergroup is imposed by assuming that at least one of $k_1, k_2, \widetilde{k}_1$, and $\widetilde{k}_2$ is finite.

(3) These moments and Taylor's formula can be used to obtain two representations of $\langle \delta_x * \delta_y, \chi \rangle$, from which it is possible to obtain a linear differential equation that the characters must satisfy.

These ideas were inspired by the work of Levitan [15].

3. An application to polynomials

If the characters of a hypergroup are polynomials, then the two routes described above can be used to prove

Theorem 1 *A necessary and sufficient condition for a family of polynomials orthogonal on the interval $I \subset \mathbb{R}$ to be the characters of a hypergroup on I, is that up to a linear change of variables, these polynomials must be the Jacobi polynomials with parameters (α, β) in the region described by Gasper in [12].*

The sufficiency was established by Gasper [12], the necessity is in [8] and [3].

4. An application to the spheroidal wave functions

Another application of the above circle of ideas is to the spheroidal wave functions which arise in a natural way when the reduced wave equation $(\Delta^2 + k^2)W = 0$ is solved by separation of variables, using as coordinates confocal families of hyperbolas and ellipses.

After the separation of variables we obtain two sets of ordinary differential equations depending on whether oblate of prolate coordinates are used.

Each set contains two differential equations one with solutions supported on a compact interval (the angular solutions) and one with solutions supported on an infinite interval (the radial solutions). After another change of variables these equations are transformed into four new equations that depend on the parameters (α, γ), $\alpha \geq -1/2$ and $\gamma \in \mathbb{R}$ with families of eigenfunctions $\mathcal{S}^{\alpha,\gamma} = \{\mathcal{S}_n^{\alpha,\gamma}(\theta)\}$ for the angular equations, and $\mathcal{S}^{\alpha,\gamma} = \{\mathcal{S}_\lambda^{\alpha,\gamma}(x)\}$ for the radial equations. The oblate coordinates yield $\gamma \geq 0$, the prolate correspond to $\gamma \leq 0$. When $\gamma = 0$, the angular functions correspond to ultraspherical polynomials and the radial functions correspond to ultraspherical functions. In all cases except the radial oblate, there is only one bounded eigenfunction associated with a particular eigenvalue.

The surprising result is that even though these functions cannot be expressed in a simple closed form, for example they cannot be described by hypergeometric functions, we are able to find an explicit formula for the measures $\sigma_{\theta,\phi} = \delta_\theta * \delta_\phi$ or $\sigma_{x,y} = \delta_x * \delta_y$ as in Subsection 2.1, and using these formulas we are able to determine which values of (α, γ) lead to hypergroups and which values lead to more general measure algebras. The following is typical of a number of results that can be found in [4].

Theorem 2 Let $\alpha \geq -1/2$, $\gamma \in \mathbb{R}$, $0 < \theta$, $\phi < \pi$ and n be a nonnegative integer.

(1) For $\alpha \geq -1/2$, the angular spheroidal wave functions satisfy a product formula with an absolutely continuous measure

$$d\sigma_{\theta,\phi}(\xi) = \mathcal{K}^{\alpha,\gamma}(\xi, \theta, \phi)\, w^\alpha(\xi)\, d\xi,$$

where

$$\mathcal{K}^{\alpha,\gamma}(\xi, \theta, \phi) = K^\alpha(\xi, \theta, \phi)\mathcal{J}_{\alpha-1/2}((\gamma\phi_0)^{1/2}),$$
$$\phi_0 = [\cos(\theta - \phi) - \cos(\xi)][\cos(\xi) - \cos(\theta + \phi)],$$
$$w^\alpha(\xi) = \sin^{2\alpha+1}(\xi),$$

$\mathcal{J}_{\alpha-1/2}$ is a modified Bessel function of the first kind of order $\alpha - 1/2$ and $j_{\alpha-1/2}$ represents its first positive zero. $K^\alpha(\xi, \theta, \phi)$ is the measure in the product formula for the ultraspherical polynomials of order α. For $\alpha = -1/2$ a similar formula holds, but the measure is not absolutely continuous.

(2) For $\alpha > -1/2$ and $-\infty < \gamma \leq j_{\alpha-1/2}^2$ or $\alpha = -1/2$ and $\gamma \leq 0$, the measure in the product formula is positive.

(3) If $\gamma \geq 0$ and $\alpha \geq 0$, then the total variation of $\sigma_{\theta,\phi}$ is bounded by one.

Acknowledgments

Both authors were supported by the National Science Foundation, Grant No. DMS 9404316, and a grant from the University of Missouri Research Board.

References

1. Achour, A. and Trimèche, K.: Opérateurs de translation généralisée associés á un opérateur différentiel singulier sur un intervalle borné, *C. R. Acad. Sci. Paris* **288** (1979) 399–402.
2. Chébli, H.: Sur la positivité des opérateurs de "translation généralisée" associés à un opérateur de Sturm–Liouville sur]0, ∞[, *C. R. Acad. Sci. Paris* **275** (1972) 601–604.
3. Connett, W.C., Markett, C. and Schwartz, A.L.: Jacobi polynomials and related hypergroup structures, *Probability Measures on Groups X*, Proceedings Oberwolfach 1990, New York (H. Heyer, ed.), Plenum, 1990, 1991, 45–81.
4. Connett, W.C., Markett, C. and Schwartz, A.L.: Product formulas and convolutions for angular and radial spheroidal wave functions, *Trans. Amer. Math. Soc.* **338** (1993) 695–710.
5. Connett, W.C. and Schwartz, A.L.: The harmonic machinery for eigenfunction expansions, In: *Harmonic analysis in Euclidean spaces* (G. Weiss and S. Wainger, eds.), Proceedings of Symposia in Pure Mathematics **35** Amer. Math. Soc., 1979, 429–434.
6. Connett, W.C. and Schwartz, A.L.: Analysis of a class of probability preserving measure algebras on a compact interval, *Trans. Amer. Math. Soc.* **320** (1990) 371–393.
7. Connett, W.C. and Schwartz, A.L.: Positive product formulas and hypergroups associated with singular Sturm–Liouville problems on a compact interval, *Colloq. Math.* **LX/LXI** (1990) 525–535.
8. Connett, W.C. and Schwartz, A.L.: Product formulas, hypergroups, and Jacobi polynomials, *Bull. Amer. Math. Soc.* **22** (1990) 91–96.
9. Connett, W.C. and Schwartz, A.L.: Fourier analysis off groups, In: *The Madison symposium on complex analysis* (A. Nagel and L. Stout, eds.), Amer. Math. Soc., Providence, R. I., *Contemporary Mathematics* **137** (1992) 169–176.
10. Connett, W.C. and Schwartz, A.L.: Continuous 2-variable polynomial hypergroups", In: *Applications of hypergroups and related measure algebras* (W. C. Connett, O. Gebuhrer and A. L. Schwartz, eds.), Amer. Math. Soc., Providence, R. I., *Contemporary Mathematics* **183** (1995) 89–109.
11. Dunkl, C.F.: The measure algebra of a locally compact hypergroup, *Trans. Amer. Math. Soc.* **179** (1973) 331–348.
12. Gasper, G.: Banach algebras for Jacobi series and positivity of a kernel, *Ann. of Math.* **95** (1972) 261–280.
13. Hirschman, I.I., Jr.: Sur les polynomes ultraspheriques, *C. R. Acad. Sci. Paris*, **242** (1956) 2212–2214.
14. Jewett, R.I.: Spaces with an abstract convolution of measures, *Adv. in Math.* **18** (1975) 1–101.
15. Levitan, B.M.: *Generalized Translation Operators*, Israel Program for Scientific Translations, Jerusalem, 1964.
16. Michael, E.: Topologies on spaces of subsets, *Trans. Amer. Math. Soc.* **71** (1951) 152–182.
17. Rudin, W.: *Fourier Analysis on Groups*, Interscience Publishers, 1962.
18. Spector, R.: Mesures invariantes sur les hypergroupes, *Trans. Amer. Math. Soc.*, **239** (1978) 147–165.

HAAR MEASURE
ON LOCALLY COMPACT HYPERGROUPS

In memory of all whose death during the 900 days of the siege of Leningrad contributed to the survival of civilization.

MARC-OLIVIER GEBUHRER
Institut de Recherche Mathématique Avancée,
Université Louis Pasteur et C.N.R.S.
7, Rue René Descartes, 67084 Strasbourg, Cedex, FRANCE.
E-mail: gebuhrer@math.u-strasbrg.fr

Abstract. The paper presents the basic ideas leading to the proof of the existence of a Haar measure on locally compact hypergroups. A new axiomatics is introduced, allowing for a natural theory.

Mathematics Subject Classification (1991): 43A62, 47H10, 28C10.

Key words: locally compact hypergroups, Haar measure, fixed point theorems.

Introduction and a historical survey

The aim of this paper is to present within a new and natural axiomatics the steps of the proof of the following theorem.

Theorem *Let X be a locally compact hypergroup. There exists a nontrivial positive Radon measure σ on X, which is left invariant by translations. Moreover, the measure σ is unique up to a positive multiplicative constant.*

The detailed arguments will appear elsewhere.

Before entering the core of the subject, it might be of interest to stress some historical steps pertaining to the representation theory of locally compact hypergroups.

The new reference, book [1], already provides some basic material but we will keep on a different point of view, insisting on representation theory of a large class of natural ∗-Banach algebras.

B. P. Komrakov et al. (eds.), Lie Groups and Lie Algebras, 117–131.
© 1998 *Kluwer Academic Publishers. Printed in the Netherlands.*

Moreover, as we will deal with the general case, within a different axiomatics (in fact a larger one, and a more natural one from our point of
view) than presented in [1], which aside from the very first pages, mainly
develops the *commutative* theory, this paper is largely self-contained and
can be read independently from other sources.

We start with the following situation.

We are given a locally compact Hausdorff topological space X furnished
with an involutory homeomorphism $x \mapsto \check{x}$ $(X \to X)$. It may occasionally
happen that this homeomorphism is the identity, in which case the theory
becomes commutative. We suppose that $M_b(X)$, the complex Banach space
of bounded Radon measures on X with its natural dual norm $\| \cdot \|_1$ is
endowed with an associative, involutive Banach algebra structure whose
product is denoted by $*$ and called thereafter *convolution*. The involution
referred to on $M_b(X)$ is the natural one induced by the involution $x \mapsto \check{x}$
on X.

We would like to be able to describe as far as possible the representation
theory of such $*$-Banach algebras, but such a program is irrealistic. We now
suppose the two following requirements to be fulfilled:

(M_1) *The convex subset of probability measures $M^1(X)$ is a semi-group in*
$M_b(X)$ *under the convolution:* $M^1(X) * M^1(X) \subset M^1(X)$.

(M_2) *The mapping $(\mu, \nu) \mapsto \mu * \nu$ from $M_b(X) \times M_b(X)$ into $M_b(X)$ is*
w^*-*continuous, when $M_b(X)$ is endowed with its dual w^*-topology*
$\sigma(M_b(X), C_0(X))$, *where $C_0(X)$ denotes the complex Banach space*
of continuous functions on X, vanishing at infinity, furnished with its
natural uniform convergence norm $\| \cdot \|_\infty$.

Then for $\mu \in M_b(X)$, $x \in X$, one may define:

$$\mu f(x) := \langle \check{\mu} * \delta_x, f \rangle, \ f_\mu(x) := \langle \delta_x * \check{\mu}, f \rangle,$$

for any $f \in C_{\mathcal{K}}(x)$ ($C_{\mathcal{K}}(x)$ being the space of complex compactly supported
continuous functions).

So far, we have just mimicked the group measure algebra $M_b(G)$ for
$X = G$, where G is a locally compact group, with the important difference
that usually there is no algebraic structure on X.

The question then arises naturally whether there exists a positive Radon
measure $\sigma \neq 0$ such that $\langle \sigma, {}_x f \rangle = \langle \sigma, f \rangle$ for any $f \in C_{\mathcal{K}}(X)$, $x \in X$, where
${}_x f$ is just ${}_{\delta_x} f$ in the previous notation.

The first historical attempt in that direction was made by C. Dunkl in
1973 [3]; he observed that for the case in which X is compact and the measure algebra is commutative, the answer is positive and the proof is based on
the Markov–Kakutani fixed-point theorem. However, the invariant measure
thus obtained may be highly degenerated and therefore virtually irrelevant

to the representation theory of the ∗-Banach algebra under investigation. Its support might be reduced to a single point; see [6].

Anyhow, the set of axioms (M_1), (M_2) is not sufficient to allow generalizations even if X is compact, or discrete: this is because the measure algebras considered so far embody the class of measure algebras of *monoids*; for those, an invariant measure is virtually absent in general (aside precisely from the group-case!). At that point, further axioms have to be introduced; there is some reason for such an investigation if one bears in mind the following quotation: "without an invariant measure, most of harmonic analysis comes to a standstill" [5]. That this strong statement is disputable taking into account on the one hand how the theory of semigroups develops (e.g. the book "Harmonic Analysis on semigroups" C. Berg, J.P.R. Christensen, P. Ressel (Springer Grad. Texts in Maths, Vol. 100)) or, on the other hand, signed hypergroups, hypercomplex systems, ..., is another problem where only the future will have a say.

Further attempts were made independently and almost simultaneously by R.I. Jewett and R. Spector ([7] and [10] respectively) in 1975 and 1978. Their axioms differ to a certain extent.

The first one is purely algebraic in nature and is common to both authors:

(U) There exists a point $e \in X$ such that δ_e is the unit of $M_b(X)$ and $e \in \mathrm{supp}(\delta_x * \delta_{\check{y}})$ iff $x = y$.

This last axiom coupled with the former ones lead to a positive and completely satisfactory answer for the invariant measure problem in the case X is discrete (see [1]). The second requirement is topological and more troublesome. There is some parenthood between the formulations both authors adopted. Let us mention each of them.

For R. Spector one has the following:

(S_7) For any compact subset K of X, any neighborhood V of K, there exists a neighborhood V of e such that
  i) $\mathrm{supp}(\mu) \subset K$, $S(\nu) \subset V$ together imply $\mathrm{supp}(\mu * \nu) \subset V$ and $\mathrm{supp}(\nu * \mu) \subset V$;
  ii) $\mathrm{supp}(\mu) \subset K$, $S(\nu) \subset V^c$ together imply:

$$\mathrm{supp}(\mu * \check{\nu}) \cap K = K \cap \mathrm{supp}(\check{\mu} * \nu) = \mathrm{supp}(\nu * \check{\mu}) \cap K = \mathrm{supp}(\check{\nu} * \mu) \cap K = \varnothing,$$

while R.I. Jewett makes use of the following topology on $\mathcal{K}(X)$, the space of nonempty compact subsets of X: For $A, B \subset X$, we define $\mathcal{K}_A(B) = \{\Gamma \in \mathcal{K}(X) \mid \Gamma \cap A \neq \varnothing, \Gamma \subset B\}$. Then $\mathcal{K}(X)$ is given the topology generated by the subbasis of all $\mathcal{K}_U(V)$, for U, V open subsets of X ([8]). Then, Jewett's requirement is ([1] p. 9):

(HG$_4$) The mapping $(x, y) \mapsto \mathrm{supp}(\delta_x * \delta_y)$ of $X \times X$ *into* $\mathcal{K}(X)$ is continuous, where, in the contrast with R. Spector, it is assumed that $\mathrm{supp}(\delta_x * \delta_y) \in \mathcal{K}(X)$ if $x, y \in X$.

Mimicking A. Weil's procedure, R.I. Jewett proved the following

Theorem *Every locally compact "convo"* (M$_1$, M$_2$, U, HG$_4$ *for short*) *possesses a positive nontrivial subinvariant measure* τ: $\langle \sigma, {}_x f \rangle \leq \langle \tau, f \rangle$ *for any* $f \in C_{\mathcal{K}}(X)$, $f \geq 0$. *Moreover, if X is compact, σ is in fact left invariant.*

Attacking for its own purpose the locally compact case, R. Spector discovered that *a long way was to be covered* in order to attain the goal of an invariant measure, once you get a subinvariant measure; assuming the hypergroup to be commutative, he proved the following

Theorem *Every locally compact commutative hypergroup* (M$_1$, M$_2$, U, S$_7$ *for short*) *possesses a positive nontrivial invariant measure.*

The philosophy behind R. Spector's beautiful result was the necessity of *new fixed point theorems* adapted to the actual situation, *embodying* many of the classical fixed point theorems of functional analysis, thus making the link with C. Dunkl's elementary step.

The general case remained unsettled and mysterious for some years: even Spector did not have a clear-cut conjecture about the existence of an invariant measure in general.

Now, regarding the naturality of axioms, if (U) does not raise any problem (but of course, it is a strong requirement since it works as a kind of a substitute for the missing inverse), certainly both conditions (S$_7$) and (HG$_4$) are of the type you might wish to avoid. Some authors adopted immediately Jewett's point of view, Spector's condition (S$_7$) looking apparently even more cumbersome, and intriguing, if more general, but it is to be mentioned that the checking on examples of condition (HG$_4$) is far from evident (see [11] for instance).

On the other hand, there was no clue to hope for the general situation in Jewett's paper [7], and, for its own sake, Spector's paper [10] made the argument about the existence of the Haar measure and its basic properties extremely intricate. In fact, as the following quotation indicates: "It is the possibility of constructing the L_1 group algebra *more than any other fact, which accounts for the importance* of the category of locally compact groups in functional analysis" ([5]), the problem about hypergroups was not only the Haar measure problem, but also the possibility of constructing a "good" L_1-theory.

This is why Spector proved in [10] the following basic

Theorem *If X is a locally compact hypergroup with a left Haar measure σ, then one has*

$$\int_X f \cdot {}_x g \, d\sigma(x) = \int_X {}_{\check{x}} f \cdot g \, d\sigma(x)$$

for $f, g \in C_\mathcal{K}(X)$.

A property which is trivial for locally compact groups, highly nontrivial for locally compact hypergroups. Probably, the intricacy of the arguments in Spector's paper explains the very misleading remark in [1]: "The (preceding) theorem is the basis of Spector's proof of the existence of the Haar measure for noncommutative hypergroups". Anyhow, after some (largely yet unpublished) previous work within some naive axiomatics designed to free the theory of any axioms of the type (HG_4) or (S_7), I decided to rebuild the whole theory from the beginning in order to understand better the role of the axioms in the Haar measure problem.

The first part of the story is described in detail in [6]; restricting the scope to the case in which X is compact, a new set of axioms appeared, and as suspected, the Haar measure problem was solved positively and in a very satisfactory way by using a *noncommutative and deep result, the Ryll–Nardzewski fixed point theorem.* Moreover, it appeared that, while the distention of the Haar measure in that case was almost algebraic, a topological condition appeared in order to build an L^1-theory; at least, this topological condition was a natural generalization (easy to check on examples!) of the topological group situation. In any case, it appears that the *compact* case leads to an extremely rigid theory as shown in [6]. The second part of the story will be told in the next sections.

1. Fundamentals on locally compact hypergroups

1.0. *General notation to be used in the sequel of the paper.* Let X be a locally compact Hausdorff topological space. The space X is endowed with a given involutory homeomorphism denoted by $x \mapsto \check{x}$ $(X \to X)$.

Functions on X:

$C_\mathcal{K}(X)$ will denote the complex vector space of compactly supported continuous functions on X endowed with its natural inductive topology $\mathcal{C}_{\text{ind}}$.

$C_0(X)$ will denote the complex vector space of continuous functions on X vanishing at infinity. It will always be regarded as a Banach space under the uniform convergence topology. The norm of an element f of $C_0(X)$ is denoted by $\|f\|_\infty = \sup_{x \in X} |f(x)|$.

$C_b(X)$ will denote the complex vector space of bounded continuous functions on X. It will always be regarded as a Banach space under the uniform convergence topology.

$C_\mathcal{K}^+(X), C_0^+(X), C_b^+(X)$ denote the respective convex cones of positive elements within the respective spaces.

Measures on X:

$M(X)$ is the complex space of Radon measures in X, dual space of $(C_\mathcal{K}(X), C_{\mathrm{ind}})$.

$M^1(X)$ is the convex set of probability measures on X.

$M_b(X)$ is the complex space of bounded Radon measures on X, dual space of the space $(C_0(x), \| \cdot \|_\infty)$.

The norm of an element μ of $M_b(X)$ is denoted by

$$\|\mu\|_1 = \sup_{\|f\|_\infty \leq 1} |\langle \mu, f \rangle|.$$

$M_\mathcal{K}(X)$ is the complex vector space of compactly supported Radon measures on X; it is a subspace of $M_b(X)$.

$M_\mathcal{K}^1(X)$ is just $M^1(X) \cap M_\mathcal{K}(X)$.

$M^+(X)$, $M_b^+(X)$, $M_\mathcal{K}^+(X)$ denote the respective convex cones of positive elements to within the respective spaces.

Supports:

The support of a function f (respectively, a measure μ) on X is denoted by $S(f)$ (respectively $S(\mu)$).

Involutions:

The main involution to be used is the classical one:

$$\tilde{f}(x) := \overline{f(\tilde{x})} \text{ for a function } f \text{ on } X \text{ and } \langle \tilde{\mu}, f \rangle := \langle \mu, \tilde{f} \rangle$$

(whenever the coupling makes sense) for a measure μ on X.

We also allow the use of another involution (mainly the restriction of the former one to real elements) by letting $\check{f}(x) := f(\check{x})$ for a function f on X and $\langle \check{\mu}, f \rangle := \langle \mu, \check{f} \rangle$ (whenever the coupling makes sense) for a measure μ on X.

Weak*-topologies:

In general, unless otherwise stated, the w^*-topologies to be used in the various spaces of measures are the canonical duality w^*-topologies, i.e., on $M_b(X)$ one considers $\tau(M_b(X), C_0(X))$, etc.

1.1. *Preliminary observations.* As shown in [6], almost no choice is allowed relative to axioms whenever the base space X is compact. The situation changes if X is assumed to be merely locally compact. In fact, in the first case, the space $C(X)$ of continuous functions on X is immediately seen to be translation invariant.

This property is going to pass over to the space $C_0(X)$, if X is locally compact. But regarding integration theory, and further, representation theory, the space $C_0(X)$ is not convenient at all.

We would like to have $C_{\mathcal{K}}(X)$ translation invariant. However to make the theory consistent, we can not just *decide* $C_{\mathcal{K}}(X)$ to be translation invariant: that property should be deduced from basic structural properties of our measure algebras. This leads to some different axioms than in the compact case. However we hope the reader will convince himself of their naturality and simplicity.

1.2. Proposition and Definition   *There exists one-to-one correspondence between:*

i) Banach algebra products on $M_b(X)$ relative to its usual norm such that:

(NBA$_1$) $M^1(X)$ *is a semigroup of $M_b(X)$,*

(NBA$_2$) $(\mu, \nu) \to \mu * \nu$ *is separately w^*-continuous, $M^1(X) \times M^1(X) \mapsto M^1(X)$ (where $*$ denotes the product in $M_b(X)$).*

ii) Mappings $\pi : X \times X \to M^1(X)$ which are separately w^*-continuous.

Moreover, the mapping π is jointly w^*-continuous on $X \times X$ and formula (RF) *holds: For any $f \in C_{\mathcal{K}}(X)$, $\mu, \nu \in M_b(X)$, we have*

$$\langle \mu * \nu, f \rangle = \int_X \int_X \langle \pi(x, y), f \rangle \, d\mu(x) \, d\nu(y). \tag{RF}$$

Banach algebra products on $M_b(X)$ arising from such a mapping π are said to give rise to *natural Banach algebras of bounded measures* in the locally compact Hausdorff topological space X (NBA for short).

1.3. Definition   Let $M_b(X)$ be an associative, involutive (relatively to the involution $\sim$) NBA on the space X. We shall say that $(M_b(X), *)$ is a *hypergroup on the locally compact space X* (for short, a *locally compact hypergroup*), provided the following requirements are fulfilled:

(S) $S(\pi(x, y))$ is a compact subset of X for each $x, y \in X$ and for each open neighborhood Γ of $S(\pi(x, y))$ there exist open neighborhoods $W(x), W(y)$ of x, y respectively such that, if $(s, t) \in W(x) \times W(y)$, one has the inclusion $S(\pi(s, t)) \subset \Gamma$.

(V_1) There exists a unique point $e \in X$ such that δ_e is the unit of $M_b(X)$ and such that $e \in S(\pi(x, \check{y})) \Leftrightarrow x = y$.

1.3.1. Remark   As we shall mention, the combination of (S) and (V_1) entails the property (V_2) of [6]. Therefore, our system is consistent, although more stringent than Spector's axiomatics regarding the control over $S(\pi(x, y))$. However (S) is far from requiring any of the above-mentioned axioms (S_7) or (HG$_4$). Notice that Jewett's axioms imply ours.

1.4. Notation   For two subsets A, B of X we denote

$$A * B := \bigcup_{a \in A, b \in B} S(\delta_a * \delta_b).$$

1.5. Proposition Let $(M_b(X), *)$ be a (locally compact) hypergroup on the space X. Then $M_{\mathcal{K}}(X)$ is a subalgebra of $M_b(X)$.

1.6. Definition Let $(M_b(X), *)$ be a (locally compact) hypergroup on the space X. For $f \in C_0(X), \mu \in M^1(X)$ (or $M_b(X)$) we define

$$\mu f(x) := \langle \check{\mu} * \delta_x, f \rangle, \qquad f_\mu(x) := \langle \delta_x * \check{\mu}, f \rangle.$$

The linear operators $R_\mu : f \mapsto \mu f$, $L_\mu : f \mapsto f_\mu$ are called the *translation operators associated with* μ on $C_0(X)$. If μ is a Dirac measure δ_y, we denote the corresponding operators by R_y, L_y and $R_y(f) := {}_y f, L_y(f) := f_y$.

From Proposition 1.5 we obtain

1.7. Corollary Let $(M_b(X), *)$ be a (locally compact) hypergroup on the space X. Then:

i) Property (V_2) holds true: For each $x, y \in X$, such that $x \neq y$, there exist open neighborhoods $W(x), W(\check{y})$ of $x, \check{y}$ respectively such that $e \notin \overline{W(x) * W(\check{y})}$.

ii) The vector space $C_{\mathcal{K}}(X)$ is invariant with respect to the translation operators $R_\mu, L_\mu, (\mu \in M_{\mathcal{K}}(X))$.

In proving some results of the next sections, the following lemma is important:

1.8. Lemma Let $(M_b(X), *)$ be a (locally compact) hypergroup on the space X. Then for each relatively compact, open subset Ω of X, each compact subset Γ of X, such that $\overline{\Omega} \cap \Gamma = \varnothing$, there exists an open neighborhood V of e such that $V \cap (\Omega * \check{\Gamma}) = \varnothing$ (where $\check{\Gamma}$ stands for the image of Γ under the involution).

1.9. Proposition (nondegeneracy of the translation operators)

i) For $\mu \in M_{\mathcal{K}}(X)$, the subspaces $R_\mu(C_{\mathcal{K}}(X))$ (respectively $L_\mu(C_{\mathcal{K}}(X))$) are dense within $C_{\mathcal{K}}(X)$.

ii) For $\mu \in M_b(X)$, the subspaces $R_\mu(C_0(X))$ (respectively $L_\mu(C_0(X))$) are dense within $C_0(X)$.

iii) For $\mu \in M_{\mathcal{K}}^+(X)$, the subspaces $R_\mu(C_{\mathcal{K}}^+(X))$ (respectively $L_\mu(C_{\mathcal{K}}^+(X))$) are dense within $C_{\mathcal{K}}^+(X)$.

iv) For $\mu \in M_b^+(X)$, the subspaces $R_\mu(C_0^+(X))$ (respectively $L_\mu(C_0^+(X))$) are dense within $C_0^+(X)$.

2. Construction of a job invariant measure

We sketch here some of the steps toward the construction of a left-subinvariant measure on X. R.I. Jewett proved in fact such a result within his axiomatics but, what is really needed, as R. Spector observed, is to construct a translation invariant *cap* of the convex cone of subinvariant measures (such a cone is well-capped as soon as X is τ-compact but we do not need a restrictive hypothesis for our purposes, nor the theory of well-capped cones [9]). Therefore, producing one subinvariant measure is far from sufficient to cover the rest of the road. It was not possible to merely adapt the arguments of either author.

2.1. *The Left-and-Right Scheme.* Let us first notice that, letting $f, g \in C_{\mathcal{K}}^+(X), f \not\equiv 0$, there exists a $\mu \in M_{\mathcal{K}}^+(X)$ such that $g \leq f_\mu$ on X.

We define $A_f(g) := \inf\{\|\mu\|, \mu \in M_{\mathcal{K}}^+(X), g \leq f_\mu\}$. The following inequalities are readily checked for $f, g, h \in C_{\mathcal{K}}^+(X), \alpha \in M_{\mathcal{K}}'(X)$ ($f \not\equiv 0$):

$$1) \quad A_f(h) \leq A_f(g) A_g(h) \text{ if } g \not\equiv 0$$

$$2) \quad A_f(g) \geq \frac{\|g\|_\infty}{\|f\|_\infty}$$

$$3) \quad A_f(f) = 1$$

$$4) \quad A_f(g + h) \leq A_f(g) + A_f(h)$$

$$5) \quad A_{\alpha f}(\alpha g) \leq A_f(g).$$

Two more inequalities are needed. They are embodied in the following lemmata, where the proof of the first one relies upon 1.9.iv)

2.1.1. Lemma *Let $\alpha \in M_{\mathcal{K}}'(X)$; then for each $x \in X$, there exists a $\mu_x \in M^1(X)$ (uniquely defined), such that $_\alpha f(x) = f_{\mu_x}(x)$ for every $f \in C_K(X)$.*

The trouble being that μ_x might not belong to $M_{\mathcal{K}}^+(X)$, we avoid the difficulty by easily proving the following

2.1.2. Lemma *For $f, g \in C_{\mathcal{K}}^+(X), f \not\equiv 0$, one has $A_f(g) = \inf\{\|\mu\|_1, \mu \in M^+(X) g \leq f_\mu\}$.*

Using these two lemmata, one may show further that the following inequality holds for $f \not\equiv 0, \alpha \in M_{\mathcal{K}}^1(X)$:

$$6) \quad A_f(\alpha f) \leq 1.$$

Now

$$7) \quad A_f(\alpha g) \leq A_f(\alpha f) \cdot A_{\alpha f}(\alpha g) \leq A_f(g),$$

so that,

$$8) \quad A_f(\alpha g) \leq A_f(g).$$

Thus we get the following fundamental sequence of inequalities (see [10], p. 160):

$$9) \quad A_f(\alpha g) \leq A_{\alpha f}(\alpha g) \leq A_f(g) \leq A_{\alpha f}(g).$$

2.1.3. Remark In Spector's scheme, where only left translations are involved, inequalities 6) and 8) are trivial while 5) depends on commutativity.

R. Spector just mentions that he believes in the general validity of 5). This was the main reason for our change.

From this point on, it is possible, up to technicalities, but without further trouble (1.8 is used on the way), to follow the lines of R. Spector's proof for the existence of a left-subinvariant measure denoted by τ_0. This particular left-invariant measure depends further on the choice of a fixed function $f_0 \in C_K^+(X)$ ($f_0 \not\equiv 0$) normalized by the condition $\tau_0(f_0) = 1$.

Now, in order to prepare for the very last part of the proof of the existence of a *left-invariant* measure, we mention some important further properties of τ_0, which are not difficult to prove.

2.2. Notation

(i) The filter of open neighborhoods of the neutral element e of the (locally compact) hypergroup up($M_b(X), *$) is denoted by $\mathcal{U}$. Any function $f \in C_K^+(X)$ such that $f \not\equiv 0, S(f) \subset U, U \in \mathcal{U}$, is denoted by f_U.

(ii) We shall also denote by V the linear span of the collection

$$\{\alpha f_U\}_{U \in \mathcal{U}, \alpha \in M_K'(X)}.$$

2.3. Theorem *Let* $U \in \mathcal{U}, f_U$ *a fixed element in* $C_K^+(X)$ *of the type described in 2.2 i); there exists a positive Radon measure* σ_0 *on* X, *such that:*

i) $\sigma_0(\alpha f) \leq \sigma_0(f)$ *for any* $f \in C_K^+(X), \alpha \in M_K'(X)$,

ii) $\delta(\sigma_0) = x$,

iii) $\sigma_0(f_U) = 1$.

Furthermore, σ_0 *is independent of the choice of* $f_U \not\equiv 0, U \in \mathcal{U}$, *such that* $\sigma_0(f_U) = 1$.

2.4. Theorem *The vector space* V *(see 2.2 (ii)) is dense in* $C_K(X)$.

(The actual proof of 2.4 is inspired by the properties of an approximate unit, using σ_0 on the way).

2.5. Notation For a fixed $f_U \in C_K^+(X)$ such that $\sigma_0(f_0) = 1$, we let $a(g) := 1/A_g(f_U), b(g) := A_{f_U}(g)$ if $g \in C_K^+(X); 0 < a(g) \leq b(g)$ if $g \not\equiv 0$; $a(g) = b(g) = 0$ if $g \equiv 0$.

2.6. Proposition ([10]) *Let* σ_0 *be the left-invariant measure obtained in 2.3. We let* $\tau_\alpha(\sigma_0)(f) = \sigma_0(\alpha f)/\sigma_0(\alpha f_U)$ *for* $\alpha \in M_K'(X), f \in C_K(X)$. *Then*

i) τ_α *is* w^*-*continuous on* $M^+(X)$;

ii) $a(g) \leq \tau_\alpha(\sigma_0)(g) \leq b(g)$ *for any* $\alpha \in M_K^1(X), g \in C_K^+(X)$;

iii) $\tau_\beta \circ \tau_\alpha(\sigma_0) = \tau_{\beta*\alpha}(\sigma_0)$ *for* $\alpha, \beta \in M_K^1(X)$.

iv) *The closed orbit of* σ_0 *under the action of the semigroup* $M_K^1(X)$ *is a compact convex subset of* $M^+(X)$, *not containing the measure 0*.

3. An extension of the Ryll–Nardzewski fixed point theorem

3.1. Definition Let E be a real vector space, C a convex subset of E. We shall say that a mapping $\tau : C \to C$ is *segment preserving*, if $\tau([x, y]) = [\tau x, \tau y]$, where $[\alpha, \beta]$ denotes the segment in C of extremities α, β.

3.2. Problem If $E = \mathbb{R}$, any injective continuous function on an interval is segment preserving. On the other hand, one can easily deduce a general framework for producing such functions from the example of the τ_α (2.6). J.P. Kahane asked (private communication) to describe as completely as possible the class of segment preserving continuous functions in a finite dimensional vector space.

3.3. Definition Let E be a locally convex topological vector space, Q a nonempty subset of E. If $\mathcal{S}$ is a family of maps of Q into Q, then $\mathcal{S}$ is said to be *distal*, if for two distinct points $x, y \in Q$, $0 \notin cl\{Tx - Ty\}_{T \in \mathcal{S}}$.

3.4. Definition Let E be a locally convex Hausdorff topological vector space over $\mathbb{R}$; if p is a seminorm on E, A is a subset of E, we define the p-*diameter* of A to be the number $p\text{-diam}(A) := \sup\{p(x - y) : x, y \in A\}$.

3.5. Lemma ([2]; 10.2 Lemma) *If E is a locally convex Hausdorff topological vector space over $\mathbb{R}$, $\mathbb{K}$ a nonempty separable weakly compact convex subset of E, and p a continuous seminorm on E, then for every $\varepsilon > 0$, there exists a closed convex subset Γ of K such that:*

   i) $\Gamma \neq K$
   ii) $p\text{-diam}(K \backslash \Gamma) \leq \varepsilon$.

3.6. Theorem *If E is a locally convex Hausdorff topological vector space over $\mathbb{R}$, Q a weakly compact convex subset of E, $\mathcal{S}$ a distal semigroup of weakly continuous segment-preserving maps of Q into Q, then there is a point $x_0 \in Q$, such that $T(x_0) = x_0$ for every $T \in \mathcal{S}$.*

Proof. We modify accordingly to our assumptions the steps of the proof of the Ryll–Nardzewski fixed point theorem for *affine maps* (Theorem 10.8 in [2]).

3.6.1. Lemma *If $\{T_1, \ldots, T_n\} \subset \mathcal{S}$, then there is an $x_0 \in Q$ such that $T_k x_0 = x_0$ for $1 \leq k \leq n$.*

Proof of 3.6.1. Let $T_0 = (T_1 + \ldots + T_n)/n$; then $T_0 : Q \to Q$, T_0 is w^*-continuous. By the Schauder–Tychonoff theorem ([4], Theorem 5, p. 456), T_0 has a fixed point $x_0 \in Q$. (One may observe here that Spector's fixed point theorem is available here since T_0 is segment preserving; however Spector's fixed point theorem ([10]) relies for its proof on the Schauder–Tychonoff theorem.) We shall show that $T_k(x_0) = x_0$ for $1 \leq k \leq n$. If $T_k(x_0) \neq x_0$ for some k, by renumbering the T_k, it can be assumed that there is an integer m for which $T_k(x_0) \neq x_0$ for $1 \leq k \leq m, T_k(x_0) =$

$x_0, m \leq k \leq n$. Let $T_0' = (T_1 + \ldots + T_m)/m$. Then

$$x_0 = T_0(x_0) = \frac{[T_1(x_0) + \ldots + T_m(x_0)]}{n} + \frac{(n-m)x_0}{n}.$$

Hence

$$T'_0(x_0) = \frac{[T_1(x_0) + \ldots + T_m(x_0)]}{m} = \frac{n}{m} \cdot \frac{1}{n}[T_1(x_0) + \ldots + T_m(x_0)] =$$

$$\frac{n}{m}\left[x_0 - \frac{n-m}{n}x_0\right] = x_0.$$

Thus, it may be assumed that $T_k(x_0) \neq x_0$ for all k, but $T_0(x_0) = x_0$. We now make this assumption.

Since $\mathcal{S}$ is distal, there exists an $\varepsilon > 0$, and a continuous seminorm p on E, such that for every $T \in \mathcal{S}$ and $1 \leq k \leq n$

$$p(T(T_K(x_0)) - T(x_0)) > \varepsilon.$$

Let $\mathcal{S}_1$ be the semigroup generated by $\{T_1, \ldots, T_n\}$. Then $\mathcal{S}_1 \subset \mathcal{S}$ and $\mathcal{S}_1 = \{T_{l_1} \ldots T_{l_m} \, m \geq 1, 1 \leq l_j \leq n\}$. Thus $\mathcal{S}_1$ is a countable subsemigroup of $\mathcal{S}$. Put $K = \overline{co}(T(x_0); T \in \mathcal{S}_1)$. K is now a weakly compact convex subset of Q and K is separable. By Lemma 3.5, there is a closed convex subset Γ of k such that $\Gamma \neq K$ and p-diam$(K \backslash \Gamma) \leq \varepsilon$. Since $\Gamma \neq K$, there is an S in $\mathcal{S}_1$ such that $S(x_0) \in K \backslash \Gamma$. Hence

$$S(x_0) = ST_0(x_0) = S\left(\frac{T_1(x_0) + \ldots + T_n(x_0)}{n}\right) \in K \backslash \Gamma.$$

Now S is segment preserving and therefore sends barycenters onto barycenters; in other words

$$S\left(\frac{T_1(x_0) + \ldots + T_n(x_0)}{n}\right) \in co(ST_1(x_0), \ldots, ST_n(x_0)).$$

Then, there must exist a k, $1 \leq k \leq n$, such that $ST_k(x_0) \in K \backslash \Gamma$. This implies that $p(S(T_k(x_0)) - S(x_0)) \leq p$-diam$(K \backslash \Gamma) \leq \varepsilon$ which contradicts the distality of $\mathcal{S}$. This establishes 3.6.1. □

Now, let $\mathcal{F}$ be the collection of all finite nonempty subsets of $\mathcal{S}$. If $F \in \mathcal{F}$, let $Q_F = \{x \in Q : T(x) = x \text{ for every } T \in F\}$. By 3.6.1, $Q_F \neq \varnothing$, Q_F is compact and $\{Q_F, F \in \mathcal{F}\}$ has the finite intersection property; so there is an $x_0 \in \cap\{Q_F, F \in \mathcal{F}\}$. Theorem 3.6 is proved. □

4. The Haar measure: the final step

Now we come back to the setting of Section 2. Let $(M_b(X), *)$ be a (locally compact) hypergroup on the space X. Let σ_0 be the left subinvariant measure on X introduced in 2.3. Recall that by defining

$$\tau_\alpha(\sigma_0)(f) = \frac{\sigma_0(\alpha f)}{\sigma_0(\alpha f_U)} \text{ for } \alpha \in M^1_K(x), \ f \in C_K(X).$$

We have obtained w^*-continuous mappings acting upon the convex w^*-compact orbit of σ_0 under the action of the semigroup $M^1_K(X)$; that the mappings τ_α are segment-preserving is proved in [10]; convexity of the orbit results from this.

Therefore, with Q_{σ_0} being the closed orbit generated by σ_0 in $M^+(X)$ under the semigroup $M^1_K(X)$ ($Q_{\sigma_0} \not\ni 0$) we have to check the distality of the semigroup $M^1_K(X)$. This amounts to check that if

$$0 \in \overline{\{\tau_\alpha(\sigma_1) - \tau_\alpha(\sigma_2)\}}_{\alpha \in M^1_K(X)}, \ \sigma_1, \sigma_2 \in Q_{\sigma_0},$$

then $\sigma_1 = \sigma_2$.

First observe that distality is clear at "finite distance": if $0 = \tau_\alpha(\sigma_1) - \tau_\alpha(\sigma_2)$ for some $\alpha \in M^1_K(X)$, then appeal to 1.9.i) shows immediately that $\sigma_1 = \sigma_2$ (this was the argument in [6].). The reader will realize easily that, what remains to check, is the following situation:

Let $\mathcal{F}$ be a filter on $M^1_K(X)$, such that $\lim_{\mathcal{F}} \alpha_i = 0 - w^*$, and

$$\lim_{\mathcal{F}} \tau_\alpha(\sigma_1) = \lim_{\mathcal{F}} \tau_\alpha(\sigma_2).$$

One has the following property:

4.1. Proposition *Let M be a monoid acting continuously on the compact space X. Then for any $x \in X$, there exists a minimal closed orbit Z_x containing x. That is:*

    i) $M \cdot Z_x \subset Z_x,$

    ii) $\overline{M \cdot z} = Z_x$ *for any* $z \in Z_x.$

4.2. Theorem *Every commutative (locally compact) hypergroup possesses a nontrivial Haar measure.*

Proof. Using 4.1, we notice that the situation boils down to the following: let $\mathcal{F}$ be a contracting filter on $M'_K(X)$ such that $\lim_{\mathcal{F}} \alpha_i = 0$ (there can be no other possibilities); then for some $\sigma_1, \sigma_2 \in Q_{\sigma_0}$, one has $\lim_{\mathcal{F}} \tau_\alpha(\sigma_1) = \lim_{\mathcal{F}} \tau_\alpha(\sigma_2)$ (contracting property). We denote m_0 this common limit; then one has:

$$m_0 = \lim_{\mathcal{F}} \tau_\alpha(\sigma_1) = \lim_{\mathcal{F}} \tau_\alpha(\sigma_2),$$

while

$$\lim_{\mathcal{U}_1} \tau_\gamma m_0 = \sigma_1, \ \lim_{\mathcal{U}_2} \tau_\beta m_0 = \sigma_2$$

for some filters $\mathcal{U}_1, \mathcal{U}_2$ on $M'_{\mathcal{K}}(X)$.

Now using commutativity, it is easy to check that $\sigma_1 = \sigma_2$. In other words, $M'_{\mathcal{K}}(X)$ is acting in a distal way on Q_{σ_0}; $Q_{\sigma_0} = \{\sigma_0\}$. Therefore, our fixed point theorem embodies Spector's result. $\square$

4.3. Lemma Let $\mathcal{F}$ be a filter in $M'_{\mathcal{K}}(X)$ such that $\lim_{\mathcal{F}} \alpha = 0$. Then $\lim_{\mathcal{F}} \tau_\alpha \sigma(_\beta f_U) = \sigma(_\beta f_U)$ for any $\sigma \in Q_{\sigma_0}$, any $\beta \in M'_{\mathcal{K}}(X)$ (f_U being the normalizing function).

This lemma grew out of discussions with H. Chebli during his stay in Strasbourg (Fall and Winter Semester 94/95). It is a particular pleasure to thank him for his patience and inspiration.

4.4. Theorem Every discrete hypergroup possesses a nontrivial Haar measure.

Proof. By 4.3, upon taking $f_U = \delta_e$, we shall have:

$$\lim_{\mathcal{F}} \tau_\alpha(\sigma_1)(_x\delta_e) = \lim_{\mathcal{F}} \tau_\alpha(\sigma_2)(_x\delta_e) \text{ whence } \sigma_1(\{x\}) = \sigma_2(\{x\})$$

for any $x \in X$ as $_x\delta_e = \gamma(x)\delta_x$, where $\gamma(x) > 0$ for every $x \in X$. Hence $\sigma_1 = \sigma_2$. The theorem is proved. $\square$

4.5. Theorem Every locally compact hypergroup possesses a nontrivial Haar measure.

Proof. If $\lim_{\mathcal{F}} \tau_\alpha \sigma_1(_x f_U) = \lim_{\mathcal{F}} \tau_\alpha \sigma_2(_x f_U)$, then $\sigma_1(_x f_U) = \sigma_2(_x f_U)$ for any $x \in X$. But we have seen (2.6 iv)) that Q_{σ_0} does not depend on f_U ($\sigma_0(f_U) = 1$); therefore for any $U \in \mathcal{U}$, any $x \in X$, one has $\sigma_1(_x f_U) = \sigma_2(_x f_U)$ by 2.4; the theorem is proved. $\square$

4.6. *Final remarks.* Needless to say, with some technicalities, one may show that a good L^1 theory goes with the preceding result (4.5). The missing proofs will be published in a forthcoming paper.

There is no need to give a list of "remaining" open problems: virtually *any* question that has been raised for groups makes sense on hypergroups. Actually the only known example of a noncommutative hypergroup is the double class $x = K \backslash G / K$, where G is a locally compact group, K *any* *compact* subgroup of G. Other examples are mandatory and are waiting to be exhibited.

Acknowledgments

The paper elaborates on a lecture given by the author in Moscow (August 1994) at the International Sophus Lie Center. It is a great pleasure to have the opportunity to thank Prof. B. Komrakov and Prof. G. Litvinov for their wonderful hospitality.

References

1. Bloom, W.R. and Heyer, H.: *Harmonic Analysis of Probability Measure on Hypergroups*, De Gruyter Studies in Mathematics, **20**, Berlin, 1995.
2. Conway, J.B.: *A Course in Functional Analysis* (Second Edition), Springer Verlag Graduate Texts in Mathematics, **96**, Paris, 1990.
3. Dunkl, C.: The measure algebra of a locally compact hypergroup, *Trans. Amer. Math. Soc.* **179** (1973) 331–347.
4. Dunford, N. and Schwartz, J.: *Linear Operators, Part I: General Theory*, Jonh Wiley and Sons; Wiley Classical Library, 1957.
5. Fell, J.M.G. and Doran R.S.: *Representations of *-Algebras, Locally Compact Groups, and Banach *-Algebraic Bundles*, vol. 1. Academic Press (125) 1988.
6. Gebuhrer, M.O.: Bounded Measures Algebras: a fixed-point approach, *Contemporary Mathematics, Amer. Math. Soc., Providence RI* **183** (1995) 171–190.
7. Jewett, R.I.: Spaces with an abstract convolution of measures, *Adv. in Maths.* **18** (1975) no. 1, 1–101.
8. Michael, E.A.: Topologies on spaces of subsets, *Trans. Amer. Math. Soc.* **71** (1955) 152–182.
9. Phelps, R.R.: *Lectures on Choquet's theorem*, Van Nostrand Mathematical Studies, 1966.
10. Spector, R.: Mesures invariants sur le hypergroupes commutatifs, *Trans. Amer. Math. Soc.* **239** (1978) 147–165.
11. Zeuner, H.: One-dimensional hypergroups, *Adv. in Maths.* **76** (1989) no. 1, 1–18.

LAGUERRE HYPERGROUP AND LIMIT THEOREM

M.M. NESSIBI AND M. SIFI
Department de Mathematics,
Faculty of Sciences of Tunis, 1060 Tunis, TUNISIA.

Abstract. We provide $K = [0, +\infty[\times \mathbb{R}$ with a generalized convolution product $\star$ related to Laguerre functions. We prove that $(K, \star)$ is a hypergroup in the sense of Jewett called Laguerre hypergroup and we prove a central limit theorem on K.

Mathematics Subject Classification (1991): Primary 60F05, 60J15; Secondary 33C25, 43A62.

Key words: convolution, hypergroup, central limit theorem.

1. Introduction

This paper deals with a central limit theorem on $K = [0, +\infty[\times \mathbb{R}$, provided with a convolution product $\star$ related to Laguerre functions and generalizing the convolution product of radial functions on the $(2n + 1)$-dimensional Heisenberg group $\mathbb{H}^n$.

If we write $\mathbb{H}^n = \mathbb{C}^n \times \mathbb{R}$ with product $(z, t).(z', t') = (z + z', t + t' - \text{Im}(z/z'))$, where $(z/z') = \sum_{i=1}^{n} z_i \bar{z}'_i$ is the usual positive definite Hermitian form on $\mathbb{C}^n$, then the unitary group $\mathcal{U}(\mathbb{C}^n)$ acts on $\mathbb{H}^n$ via

$$u.(z, t) = (u.z, t), \quad u \in \mathcal{U}(\mathbb{C}^n);$$

hence $\mathcal{U}(\mathbb{C}^n)$ acts on functions F on $\mathbb{H}^n$ via

$$F^u(z, t) = F(u.(z, t)) = F(u.z, t).$$

Let $L^1(\mathbb{H}^n)$ be the space of integrable functions on $\mathbb{H}^n$. It is well known that the set $L^1_{\text{rad}}(\mathbb{H}^n) = \{F \in L^1(\mathbb{H}^n); \ F^u = F, \ \forall \ u \in \mathcal{U}(\mathbb{C}^n)\}$ is a commu-

B. P. Komrakov et al. (eds.), Lie Groups and Lie Algebras, 133–145.
© 1998 *Kluwer Academic Publishers. Printed in the Netherlands.*

tative subalgebra of $L^1(\mathbb{H}^n)$ and a radial function on $\mathbb{H}^n$ depends only on $(\|z\|, t)$, where $\|z\|^2 = \sum_{i=1}^{n} |z_i|^2$ is the usual norm on $\mathbb{C}^n$. (See [5], pp. 76–78).

Let F and G be in $L^1_{\mathrm{rad}}(\mathbb{H}^n)$ such that $F(z, t) = f(\|z\|, t)$ and $G(z, t) = g(\|z\|, t)$; then the ordinary convolution product of F and G on $\mathbb{H}^n$ can be written as follows

$$(F \star G)(z, t) = 2\pi^{n-1} \int_K T^{(n-1)}_{(\|z\|, t)} f(y, s) g(y, -s) dm_{n-1}(y, s),$$

where dm_{n-1} is the positive measure defined on K by

$$dm_{n-1}(y, s) = \frac{1}{\pi(n-1)!} y^{2n+1} dy ds,$$

and $T^{(n-1)}_{(x,t)}$, $(x, t) \in K$, are the generalized translation operators defined on K for real $\alpha \geq 0$, by
   1. if $\alpha = 0$, then

$$T^{(\alpha)}_{(x,t)} f(y, s) = \frac{1}{2\pi} \int_0^{2\pi} f(\sqrt{x^2 + y^2 + 2xy\cos(\theta)}, s + t + xy\sin(\theta)) d\theta;$$

   2. if $\alpha > 0$, then

$$T^{(\alpha)}_{(x,t)} f(y, s) =$$
$$\frac{\alpha}{\pi} \int_0^{2\pi} [\int_0^1 f(a(x, y, r, \theta), s + t + xyr\sin(\theta)) r(1 - r^2)^{\alpha-1} dr] d\theta,$$

where

$$a(x, y, r, \theta) = \sqrt{x^2 + y^2 + 2xyr\cos(\theta)}.$$

Hence $L^1_{\mathrm{rad}}(\mathbb{H}^n)$ can be identified with the commutative algebra $L^1_{n-1}(K)$ of integrable functions on K with respect to the measure dm_{n-1} and with convolution product defined by

$$(f \star g)(x, t) = \int_K T^{(n-1)}_{(x,t)} f(y, s) g(y, -s) dm_{n-1}(y, s), \quad (x, t) \in K.$$

We consider the system of partial differential operators

$$\begin{cases} D_1 = \dfrac{\partial}{\partial t}, \\ D_2 = \dfrac{\partial^2}{\partial x^2} + \dfrac{2\alpha + 1}{x} \dfrac{\partial}{\partial x} + x^2 \dfrac{\partial^2}{\partial t^2}; \quad (x, t) \in]0, +\infty[\times \mathbb{R}, \end{cases}$$

where α is a nonnegative number.

For $\alpha = n - 1$, $n \in \mathbb{N} \setminus \{0\}$, the operator D_2 is the radial part of the sublaplacian on the Heisenberg Lie group $\mathbb{H}^n$.

For all $(\lambda, m) \in \mathbb{R} \times \mathbb{N}$, we denote by $\varphi_{\lambda,m}$ the function defined on K by

$$\varphi_{\lambda,m}(x, t) = e^{i\lambda t} \mathcal{L}_m^\alpha(|\lambda| x^2),$$

where $\mathcal{L}_m^\alpha$ is the Laguerre function defined by $\mathcal{L}_m^\alpha(x) = e^{\frac{-x}{2}} L_m^{(\alpha)}(x) / L_m^{(\alpha)}(0)$, $L_m^{(\alpha)}$ being the Laguerre polynomial of degree m and order α (See [4]).

For all $(\lambda, m) \in \mathbb{R} \times \mathbb{N}$, the function $\varphi_{\lambda,m}$ is the unique solution of the system

$$\begin{cases} D_1 u(x, t) = i\lambda u(x, t), \\ D_2 u(x, t) = -4 |\lambda| (m + \dfrac{\alpha + 1}{2}) u(x, t); \\ u(0, 0) = 1, \quad \dfrac{\partial u}{\partial x}(0, t) = 0 \quad \text{for all } t \in \mathbb{R}. \end{cases}$$

For $\alpha = n - 1$, $n \in \mathbb{N} \setminus \{0\}$, the functions $(z, t) \to \varphi_{\lambda,m}(\|z\|, t)$ are zonal spherical functions of the Gelfand pair $(G, \mathcal{U}(\mathbb{C}^n))$, where $G = \mathcal{U}(\mathbb{C}^n) \ltimes \mathbb{H}^n$ is the semi-direct product of $\mathcal{U}(\mathbb{C}^n)$ by $\mathbb{H}^n$. (See [5] p. 78).

For all $(\lambda, m) \in \mathbb{R} \times \mathbb{N}$, the function $\varphi_{\lambda,m}$ satisfies the following product formula

$$\varphi_{\lambda,m}(x, t) \varphi_{\lambda,m}(y, s) = T_{(x,t)}^{(\alpha)} \varphi_{\lambda,m}(y, s), \quad \text{for all } (x, t) \text{ and } (y, s) \text{ in } K.$$

Using the generalized translation operators $T_{(x,t)}^{(\alpha)}$, $(x,t) \in K$, we define a generalized convolution product $\star$ on K by

$$(\delta_{(x,t)} \star \delta_{(y,s)})(f) = T_{(x,t)}^{(\alpha)} f(y, s),$$

where $\delta_{(x,t)}$ is the Dirac measure at (x, t).

In this work, we prove that K provided with this convolution product is a hypergroup and we establish a central limit theorem on K.

The content of this work is as follows: In the second section we define generalized translation operators, convolution product on K and we prove that K provided with this convolution is a hypergroup in the sense of Jewett ([3], [7]). In the third section we study a harmonic analysis on K. In the fourth section we define the notion of dispersion and Gaussian distributions on K and we prove a central limit theorem on K. Analogous results have been proved by K. Trimèche ([10], [11]), N. Ben Salem and M.N. Lazhari ([2]), N. Ben Salem ([1]), and M. Sifi ([9]).

2. Hypergroup structure on K.

Notations. We denote by

1. $C_*(K)$ the space of continuous functions on $\mathbb{R}^2$, even with respect to the first variable;
2. $M_b(K)$ the space of bounded Radon measures on K;
3. $L_\alpha^p(K)$, $p \in [1, +\infty]$, the space of functions $f : K \to \mathbb{C}$ measurable and such that

$$\|f\|_{\alpha,p} = [\int_K |f(x,t)|^p dm_\alpha(x,t)]^{\frac{1}{p}} < +\infty, \text{ if } p \in [1, +\infty[\,,$$

$$\|f\|_{\alpha,\infty} = \text{ess} \sup_{(x,t)\in K} |f(x,t)| < +\infty \,,$$

where $dm_\alpha(x,t)$ is the positive measure on K defined by

$$dm_\alpha(x,t) = \frac{x^{2\alpha+1}}{\pi\Gamma(\alpha+1)} dx\,dt.$$

Definition 1 Let f be in $C_*(K)$. For all (x,t) and (y,s) in K, we put
- for $\alpha = 0$

$$T_{(x,t)}^{(\alpha)} f(y,s) = \frac{1}{2\pi} \int_0^{2\pi} f(\sqrt{x^2+y^2+2xy\cos(\theta)}, s+t+xy\sin(\theta))d\theta$$

- for $\alpha > 0$

$$T_{(x,t)}^{(\alpha)} f(y,s) =$$
$$\frac{\alpha}{\pi} \int_0^{2\pi} [\int_0^1 f(a(x,y,r,\theta), s+t+xyr\sin(\theta))r(1-r^2)^{\alpha-1}dr]d\theta,$$

where

$$a(x,y,r,\theta) = \sqrt{x^2+y^2+2xyr\cos(\theta)}.$$

The operators $T_{(x,t)}^{(\alpha)}$, $(x,t) \in K$, are called *generalized translation operators* on K.

Proposition 1 1. *Let $\alpha > 0$. For all (x,t) and (y,s) in $]0,+\infty[\times\mathbb{R}$, we put*
- *for $(z,v) \in S_\alpha((x,t),(y,s))$*

$$W_\alpha((x,t),(y,s),(z,v)) =$$
$$\frac{\alpha}{\pi} \cdot \frac{1}{(xyz)^{2\alpha}} [x^2y^2 - (\frac{z^2-(x^2+y^2)}{2})^2 - (v-(s+t))^2]^{(\alpha-1)},$$

- *for $(z,v) \notin S_\alpha((x,t),(y,s))$*

$$W_\alpha((x,t),(y,s),(z,v)) = 0,$$

where

$$S_\alpha((x,t),(y,s)) = \{(z,v) \in K_1; \ (\frac{z^2 - (x^2 + y^2)}{2})^2 + (v - (s+t))^2 \leq x^2 y^2\};$$

here

$$K_1 = \{(z,v) \in K_1; \ z \neq 0\}.$$

Then for all $f \in C_*(K)$, *we have*

$$T^{(\alpha)}_{(x,t)} f(y,s) = \int_K f(z,v) W_\alpha((x,t),(y,s),(z,v)) z^{2\alpha+1} dz dv.$$

2. *Let* $\alpha = 0$. *For all* (x,t) *and* (y,s) *in* $]0, +\infty[\times \mathbb{R}$, *we put*

$$S_0((x,t),(y,s)) = \{(z,v) \in K_1; \ (\frac{z^2 - (x^2 + y^2)}{2})^2 + (v - (s+t))^2 = x^2 y^2\}.$$

Then for all $f \in C_*(K)$, *we have*

$$T^{(0)}_{(x,t)} f(y,s) = \int_{S_0((x,t),(y,s))} f(z,v) dW_{0,(x,t),(y,s)}(z,v),$$

where $dW_{0,(x,t),(y,s)}$ *is the canonical measure on* $S_0((x,t),(y,s))$ *normalized to have total measure equal to one.*

Proposition 2 *Let* $\alpha > 0$. *For all* (x,t) *and* (y,s) *in* $]0, +\infty[\times \mathbb{R}$, *we have*
1. *For all* $(z,v) \in]0, +\infty[\times \mathbb{R}$,

    (a) $W_\alpha((x,t),(y,s),(z,v)) = W_\alpha((y,s),(x,t),(z,v))$,
    (b) $W_\alpha((x,t),(y,s),(z,v)) = W_\alpha((x,t),(z,-v),(y,s))$,
    (c) $W_\alpha((x,t),(y,s),(z,v)) \geq 0$,

2. $\int_K W_\alpha((x,t),(y,s),(z,v)) z^{2\alpha+1} dz dv = 1$.

Proposition 3 1. *Let* f *be in* $C_*(K)$. *Then for all* $(x,t) \in K$, *the function* $T^{(\alpha)}_{(x,t)} f$ *belongs to* $C_*(K)$.
    2. *Let* f *be in* $L^p_\alpha(K)$, $p \in [1, +\infty]$. *Then for all* $(x,t) \in K$, *the function* $T^{(\alpha)}_{(x,t)} f$ *belongs to* $L^p_\alpha(K)$ *and we have*

$$\|T^{(\alpha)}_{(x,t)} f\|_{\alpha,p} \leq \|f\|_{\alpha,p}.$$

Proposition 4 1. *Let f be in $C_*(K)$. For all (x,t) and (y,s) in K, we have*

$$\text{(a) } T^{(\alpha)}_{(x,t)}[T^{(\alpha)}_{(y,s)}f] = T^{(\alpha)}_{(y,s)}[T^{(\alpha)}_{(x,t)}f],$$

$$\text{(b) } T^{(\alpha)}_{(x,t)}f(y,s) = T^{(\alpha)}_{(y,s)}f(x,t),$$

$$\text{(c) } T^{(\alpha)}_{(0,0)}f(y,s) = f(y,s).$$

2. *For all $(\lambda,m) \in \mathbb{R} \times \mathbb{N}$, we have the following product formula:*

$$T^{(\alpha)}_{(x,t)}\varphi_{\lambda,m}(y,s) = \varphi_{\lambda,m}(x,t)\varphi_{\lambda,m}(y,s), (x,t) \in K, \ (y,s) \in K.$$

Definition 2 1. For all $(x,t), (y,s) \in K$ and $f \in C_*(K)$ we put

$$(\delta_{(x,t)} \star \delta_{(y,s)})(f) = T^{(\alpha)}_{(x,t)}f(y,s),$$

where $\delta_{(x,t)}$ is the Dirac measure at (x,t). The product $\star$ is called *generalized convolution product* on K.

2. The *convolution product* $\star$ on $M_b(K)$ is defined by

$$(\nu_1 \star \nu_2)(f) = \int_K \int_K T^{(\alpha)}_{(x,t)}f(y,s)d\nu_1(x,t)d\nu_2(y,s).$$

Proposition 5 *For all (x,t) and (y,s) in K, we have*
  1. $\delta_{(x,t)} \star \delta_{(y,s)} = \delta_{(y,s)} \star \delta_{(x,t)}$,
  2. $\delta_{(x,t)} \star \delta_{(0,s)} = \delta_{(x,s+t)}$,
  3. $\text{supp}(\delta_{(x,t)} \star \delta_{(y,s)}) = S_\alpha((x,t),(y,s))$, *here S_α is the set defined in Proposition 1,*
  4. $(0,0) \in \text{supp}(\delta_{(x,t)} \star \delta_{(y,s)}) \Leftrightarrow x = y$ *and $s = -t$.*

Theorem 1 $(K,\star,i)$ *is a hypergroup in the sense of Jewett, where $i : K \to K$ is the involution defined by*

$$i(x,t) = (x,-t).$$

Proof. From Proposition 5 it is easy to verify all the conditions of Jewett (See [7], p 12 and 17, See also [3], pp. 9–10), hence $(K,\star,i)$ is a hypergroup. $\square$

3. Harmonic analysis on K.

3.1. PROPERTIES OF THE FUNCTIONS $\varphi_{\lambda,m}$.

We consider in this subsection the functions $\varphi_{\lambda,m}$ defined in the Introduction.

Notations. We denote by

1. $\mathbb{R}^* = \mathbb{R} \setminus \{0\}$.
2. $F(\mathbb{R}^* \times \mathbb{N})$ the space of functions defined on $\mathbb{R}^* \times \mathbb{N}$.
3. Δ_+ and Δ_- the operators defined on $F(\mathbb{R}^* \times \mathbb{N})$ by

$$\Delta_+ g(\lambda, m) = g(\lambda, m+1) - g(\lambda, m).$$

$$\Delta_- g(\lambda, m) = \begin{cases} g(\lambda, m) - g(\lambda, m-1), & \text{if } m \geq 1, \\ g(\lambda, 0) & , \text{ if } m = 0. \end{cases}$$

4. Λ_1 and Λ_2 the operators defined on $F(\mathbb{R}^* \times \mathbb{N})$ by

$$\Lambda_1 g(\lambda, m) = \frac{1}{|\lambda|}(m\Delta_+\Delta_- g(\lambda, m) + (\alpha + 1)\Delta_+ g(\lambda, m))$$

$$\Lambda_2 g(\lambda, m) = \frac{1}{2\lambda}((\alpha + m + 1)\Delta_+ g(\lambda, m) + m\Delta_- g(\lambda, m)).$$

Proposition 6 (see [8]) *For all $(\lambda, m) \in \mathbb{R} \times \mathbb{N}$ and $(x, t) \in K$, we have*
1. $\displaystyle\sup_{(x,t)\in K} |\varphi_{\lambda,m}(x, t)| = 1$.
2. $-x^2 \varphi_{\lambda,m}(x, t) = \Lambda_1 \varphi_{\lambda,m}(x, t)$.
3. $it\varphi_{\lambda,m}(x, t) = \Lambda_2 \varphi_{\lambda,m}(x, t) + \dfrac{\partial \varphi_{\lambda,m}}{\partial \lambda}(x, t)$.

Proposition 7 *For all $(\lambda, m) \in \mathbb{R} \times \mathbb{N}$ we have*

$$\varphi_{\lambda,m}(x, t) = 1 + i\lambda t - |\lambda|(m + \frac{\alpha + 1}{2})\frac{x^2}{\alpha + 1} + \varepsilon_{\lambda,m}(x, t),$$

where

$$|\varepsilon_{\lambda,m}(x, t)| \leq \frac{\lambda^2 t^2}{2} + \frac{\lambda^2}{\alpha + 1}(m + \frac{\alpha + 1}{2})x^2|t| + \frac{|\lambda|^3|t|^3}{6} + \frac{|\lambda|^3}{2(\alpha + 1)} \times$$
$$(m + \frac{\alpha + 1}{2})x^2 t^2 + \frac{\lambda^2}{2}(\frac{1}{4} + \frac{m}{\alpha + 1} + \frac{m(m - 1)}{(\alpha + 1)(\alpha + 2)})x^4.$$

3.2. GENERALIZED FOURIER TRANSFORM ON K.

Notations. We denote by
1. $\mathcal{S}_*(K)$ the space of functions $f : \mathbb{R}^2 \to \mathbb{C}$, even with respect to the first variable, $\mathcal{C}^\infty$ on $\mathbb{R}^2$ and rapidly decreasing together with all their derivatives.
2. $L_\alpha^p(\mathbb{R} \times \mathbb{N})$, $p \in [1, +\infty[$, the space of functions defined on $\mathbb{R} \times \mathbb{N}$ measurable and such that

$$\|f\|_{L_\alpha^p} = [\int_{\mathbb{R}\times\mathbb{N}} |f(\lambda, m)|^p d\gamma_\alpha(\lambda, m)]^{\frac{1}{p}} < \infty,$$

where $d\gamma_\alpha$ is the positive measure on $\mathbb{R} \times \mathbb{N}$, defined by

$$\int_{\mathbb{R}\times\mathbb{N}} g(\lambda,m)d\gamma_\alpha(\lambda,m) = \sum_{m=0}^{+\infty} L_m^{(\alpha)}(0) \int_{\mathbb{R}} g(\lambda,m)|\lambda|^{\alpha+1}d\lambda.$$

3. $\mathcal{S}(\mathbb{R} \times \mathbb{N})$ the space of functions $g : \mathbb{R} \times \mathbb{N} \to \mathbb{C}$ satisfying

(a) $\forall m,p,q,r,s \in \mathbb{N}$, the function

$$\lambda \to \lambda^p (|\lambda|(m + \frac{\alpha+1}{2}))^q \Lambda_1^r (\Lambda_2 + \frac{\partial}{\partial\lambda})^s g(\lambda,m)$$

is continuous and bounded on $\mathbb{R}$, C^∞ on $\mathbb{R}^*$ such that the right and the left derivatives exist at 0.

(b) $\forall k,p,q \in \mathbb{N}$ the quantity

$$N_{k,p,q}(g) = \sup_{(\lambda,m)\in\mathbb{R}^*\times\mathbb{N}} \left\{ (1 + \lambda^2(1+m^2))^k |\Lambda_1^p (\Lambda_2 + \frac{\partial}{\partial\lambda})^q g(\lambda,m)| \right\}$$

is finite.

The norms $N_{k,p,q}$, $(k,p,q) \in \mathbb{N}^3$, define a Frechét space topology on the space $\mathcal{S}(\mathbb{R} \times \mathbb{N})$.

Definition 3 The *generalized Fourier transform* of a function f in $L_\alpha^1(K)$ is defined by

$$\mathcal{F}(f)(\lambda,m) = \int_K f(x,t)\varphi_{-\lambda,m}(x,t)dm_\alpha(x,t).$$

Proposition 8 1. *The space $\mathcal{S}(\mathbb{R} \times \mathbb{N})$ is dense in $L_\alpha^2(\mathbb{R} \times \mathbb{N})$.*

2. *The generalized Fourier transform $\mathcal{F}$ is an isomorphism between $\mathcal{S}_*(K)$ and $\mathcal{S}(\mathbb{R} \times \mathbb{N})$.*

Theorem 2 (Inversion Theorem) *Let f be in $L_\alpha^1(K)$ such that $\mathcal{F}(f)$ belongs to $L_\alpha^1(\mathbb{R} \times \mathbb{N})$, then we have the following inversion formula:*

$$f(x,t) = \int_{\mathbb{R}\times\mathbb{N}} \mathcal{F}(f)(\lambda,m)\varphi_{\lambda,m}(x,t)d\gamma_\alpha(\lambda,m), \quad \text{a.e. on } K.$$

Theorem 3 1. *For all $f \in \mathcal{S}_*(K)$, we have the following Plancherel formula*

$$\|f\|_{\alpha,2}^2 = \|\mathcal{F}(f)\|_{L_\alpha^2}^2.$$

2. *The generalized Fourier transform $\mathcal{F}$ can be extended to an isometric isomorphism from $L_\alpha^2(K)$ onto $L_\alpha^2(\mathbb{R} \times \mathbb{N})$.*

Definition 4 The *Fourier transform of a measure ν* in $M_b(K)$ is defined by

$$\mathcal{F}(\nu)(\lambda,m) = \int_K \varphi_{-\lambda,m}(x,t)d\nu(x,t), \quad (\lambda,m) \in \mathbb{R} \times \mathbb{N}.$$

Theorem 4 1. *For every ν in $M_b(K)$, the function $\mathcal{F}(\nu)$ is bounded on $\mathbb{R} \times \mathbb{N}$.*

2. *For all ν_1 and ν_2 in $M_b(K)$, we have*

$$\mathcal{F}(\nu_1 \star \nu_2) = \mathcal{F}(\nu_1)\mathcal{F}(\nu_2).$$

Theorem 5 (Levy's continuity theorem, see [6]) *Let $(\sigma_n)_{n\in\mathbb{N}}$ be a sequence of probability measures on K such that for all $(\lambda, m) \in \mathbb{R} \times \mathbb{N}$ we have*

$$\lim_{n\to+\infty} \mathcal{F}(\sigma_n)(\lambda, m) = f(\lambda, m),$$

where f is defined on $\mathbb{R} \times \mathbb{N}$.

Then the sequence $(\sigma_n)_{n\in\mathbb{N}}$ converges vaguely to a nonnegative measure σ on K with mass not larger than 1 and satisfying $\mathcal{F}(\sigma)(\lambda, m) = f(\lambda, m)$, for all $m \in \mathbb{N}$ and for λ almost everywhere in $\mathbb{R}$ with respect to the Lebesgue measure. Furthermore, if σ is a probability measure, then $(\sigma_n)_{n\in\mathbb{N}}$ converges weakly to σ.

4. Central limit theorem on K.

4.1. GAUSSIAN DISTRIBUTIONS

Definition 5 A continuous function $\Psi : \mathbb{R} \times \mathbb{N} \to \mathbb{C}$ is called *positive definite* if for all $h \in \mathcal{S}(\mathbb{R} \times \mathbb{N})$ and $(x, t) \in K$, the following property holds:

$$\int_{\mathbb{R}\times\mathbb{N}} h(\lambda, m)\varphi_{\lambda,m}(x, t)d\gamma_\alpha(\lambda, m) \geq 0 \Longrightarrow$$

$$\int_{\mathbb{R}\times\mathbb{N}} h(\lambda, m)\varphi_{\lambda,m}(x, t)\Psi(\lambda, m)d\gamma_\alpha(\lambda, m) \geq 0.$$

Proposition 9 *For all $r > 0$, the function*

$$(\lambda, m) \to \exp(-4r|\lambda|(m + \frac{\alpha+1}{2}),$$

is positive definite.

Proof. From Propositions 4 and 7, we deduce that for all $r, y \in]0, +\infty[$, the function $(\lambda, m) \to \varphi_{\lambda,m}(\sqrt{r}y, 0)$ is positive definite. It follows that the function

$$(\lambda, m) \to \exp[4(\alpha + 1)(\frac{\varphi_{\lambda,m}(\sqrt{r}y, 0) - 1}{y^2})]$$

is positive definite. Next we tend y to 0. The result follows. $\square$

Let Ψ be in $\mathcal{S}_*(K)$, nonnegative and satisfying

1. $\int_K \Psi(x,t)dm_\alpha(x,t) = 1$.
2. $\mathrm{Supp}(\Psi) \subset [-1,1] \times [-1,1]$.

For all $\varepsilon > 0$, we put

$$\Psi_\varepsilon(x,t) = \frac{1}{\varepsilon^{2\alpha+3}}\Psi(\frac{x}{\varepsilon}, \frac{t}{\varepsilon}), \quad \text{for all } (x,t) \in K.$$

Lemma 1 1. *For all $\varepsilon > 0$, the function Ψ_ε belongs to $\mathcal{S}_*(K)$ and we have*

(a) $\mathrm{Supp}(\Psi_\varepsilon) \subset [-\varepsilon, \varepsilon] \times [-\varepsilon, \varepsilon]$.

(b) $\Psi_\varepsilon(x,t) > 0$, *for all* $(x,t) \in K$.

(c) $\displaystyle\int_K \Psi_\varepsilon(x,t)dm_\alpha(x,t) = 1$.

2. *For all* $(\lambda, m) \in \mathbb{R} \times \mathbb{N}$, *we have*

(a) $|\mathcal{F}(\Psi_\varepsilon)(\lambda, m)| \leq 1, \forall \varepsilon > 0$.

(b) $\displaystyle\lim_{\varepsilon \to 0} \mathcal{F}(\Psi_\varepsilon)(\lambda, m) = 1$.

Theorem 6 *The function $\alpha_r : K \to \mathbb{C}$, defined by*

$$\alpha_r(x,t) = \int_{\mathbb{R} \times \mathbb{N}} \exp(-4r|\lambda|(m + \frac{\alpha+1}{2}))\varphi_{\lambda,m}(x,t)d\gamma_\alpha(\lambda, m),$$

is nonnegative.

Proof. For all $\varepsilon > 0$, the function $\mathcal{F}(\Psi_\varepsilon)$ belongs to $\mathcal{S}_*(\mathbb{R} \times \mathbb{N})$. Using Theorem 2 we obtain

$$\Psi_\varepsilon(x,t) = \int_{\mathbb{R} \times \mathbb{N}} \mathcal{F}(\Psi_\varepsilon)(\lambda, m)\varphi_{\lambda,m}(x,t)d\gamma_\alpha(x,t),$$

for all $(x,t) \in K$. Or Ψ_ε is nonnegative, then from Proposition 6, we deduce that for all $r > 0$ and $(x,t) \in K$, we have

$$\int_{\mathbb{R} \times \mathbb{N}} \mathcal{F}(\Psi_\varepsilon)(\lambda, m) \exp(-4r|\lambda|(m + \frac{\alpha+1}{2}))\varphi_{\lambda,m}(x,t)d\gamma(\lambda, m) \geq 0.$$

On the other hand, from Lemma 1 the last integral tends to $\alpha_r(x,t)$, when ε tends to 0. This proves the theorem. $\square$

Definition 6 The measures $(\alpha_r)_{r \geq 0}$ defined on K by

$$\alpha_r = \begin{cases} \alpha_r(x,t).dm_\alpha(x,t), & \text{if } r > 0, \\ \delta_{(0,0)} & , \quad \text{if } r = 0, \end{cases}$$

are called *Gaussian distributions*.

4.2. DISPERSION

Notation We denote by $M_*^2(K)$ the set of probability measures ν on K satisfying

1. $\int_K (t^2 + x^2) d\nu(x, t) < +\infty$,
2. $\int_K t d\nu(x, t) = 0$.

Definition 7 Let ν be a measure in $M_*^2(K)$. The *dispersion* of ν is defined by

$$V(\nu) = \int_K \frac{x^2}{4(\alpha + 1)} d\nu(x, t).$$

Proposition 10 1. *For all ν_1 and ν_2 in $M_*^2(K)$, we have*

$$V(\nu_1 \star \nu_2) = V(\nu_1) + V(\nu_2).$$

2. *For all $r \geq 0$, we have $V(\alpha_r) = r$.*

Proof. The property 1) follows from Definitions 2 and 7. From Proposition 6-b), we have for all $\nu \in M_*^2(K)$

$$V(\nu) = -\frac{1}{4(\alpha + 1)} \lim_{\lambda \to 0} \Lambda_1 \mathcal{F}(\nu)(\lambda, m), \quad \text{for all } m \in \mathbb{N}.$$

The property 2) follows from this relation. □

4.3. CENTRAL LIMIT THEOREM.

Let $(\nu_{n,j})_{1 \leq j \leq k_n}$, $n \in \mathbb{N}$, be a sequence of measures in $M_*^2(K)$. We put

$$\nu_n = \nu_{n,1} \star \nu_{n,2} \star \cdots \star \nu_{n,k_n}.$$

Theorem 7 (Central Limit Theorem) *We suppose that the measures $(\nu_{n,j})_{1 \leq j \leq k_n}$, and $(\nu_n)_{n \in \mathbb{N}}$, satisfy*

1. $\lim\limits_{n \to +\infty} \sum\limits_{j=1}^{k_n} \int_K (x^2 + |t|)(x^2 + |t| + t^2) d\nu_{n,j}(x, t) = 0$,

2. $\lim\limits_{n \to +\infty} \sup\limits_{1 \leq j \leq k_n} \int_K x^2 d\nu_{n,j}(x, t) = 0$,

3. $\lim\limits_{n \to +\infty} V(\nu_n) = r$.

Then the sequence $(\nu_n)_{n \in \mathbb{N}}$ converges weakly to the Gaussian distribution α_r, $r \geq 0$.

Proof. From Proposition 7, we have

$$1 - \mathcal{F}(\nu_{n,j})(\lambda, m) = 4|\lambda|(m + \frac{\alpha + 1}{2})V(\nu_{n,j}) + \int_K \epsilon_{\lambda,m}(x, t) d\nu_{n,j}(x, t).$$

From assumptions 1) and 2), we deduce

$$\lim_{n \to +\infty} \sup_{1 \leq j \leq k_n} |1 - \mathcal{F}(\nu_{n,j})(\lambda, m)| = 0.$$

It follows that for all $(\lambda, m) \in \mathbb{R} \times \mathbb{N}$ and for n sufficiently large we have

$$\sup_{1 \le j \le k_n} |1 - \mathcal{F}(\nu_{n,j})(\lambda, m)| \le \frac{1}{2}.$$

Using the principal branch of the logarithm we obtain

$$\text{Log}(\mathcal{F}(\nu_n)(\lambda, m)) = \sum_{j=1}^{k_n} \text{Log}(1 - d_{n,j}(\lambda, m)) = -\sum_{j=1}^{k_n} \sum_{p=1}^{+\infty} \frac{(d_{n,j}(\lambda, m))^p}{p},$$

where

$$d_{n,j}(\lambda, m) = 1 - \mathcal{F}(\nu_{n,j})(\lambda, m).$$

From assumptions 1), 2) and 3), we deduce that for $\alpha \ge 0$

$$\lim_{n \to +\infty} \sum_{j=1}^{k_n} d_{n,j}(\lambda, m)) = 4r|\lambda|(m + \frac{\alpha + 1}{2}),$$

and

$$\lim_{n \to +\infty} \sum_{j=1}^{k_n} \sum_{p=2}^{+\infty} \frac{(d_{n,j}(\lambda, m))^p}{p} = 0.$$

Therefore, for all $(\lambda, m) \in \mathbb{R} \times \mathbb{N}$, we have

$$\lim_{n \to +\infty} \text{Log}(\mathcal{F}(\nu_n)(\lambda, m)) = \exp\left(-4r|\lambda|(m + \frac{\alpha + 1}{2})\right).$$

From Theorem 5, we deduce that the sequence $(\nu_n)_{n \in \mathbb{N}}$ converges weakly to $\alpha_r, r \ge 0$. $\square$

Acknowledgments.

We are grateful to Professor K. Trimeche for helpful comments and remarks.

References

1. Ben Salem, N.: Convolution semi-groups and central limit theorem associated with dual convolution structure, *J. of Theoretical Probability* **7** (1994) 417–436.
2. Ben Salem, N. and Lazhari, M.N.: Limit theorems for some hypergroup structures on $\mathbb{R}^n \times [0, +\infty[$, *Contemporary Mathematics* **183** (1995) 1–13.
3. Bloom, W. and Heyer, H.: Harmonic Analysis of Probability Measures on Hypergroups, In: Bauer, H., Kazdan, J.L. and Achuder, E. (eds.) *De Gruyter Studies in Mathematics* **20**, De Gruyter, Berlin – New York, 1994.
4. Erdely, A., Magnus, W., Oberhettinger, F. and Tricomi, F.G.: *Higher transcendental functions, II*, MacGraw-Hill, New York, 1994.
5. Faraut, J. and Harzallah, K.: *Deux cours d'Analyse Harmonique*, École d'été d'-Analyse Harmonique de Tunis, Birkhauser, 1984.

6. Gallardo, L. and Geburher, O.: Lois de probabilité infiniment divisibles sur les hypergroups commutatifs, discrets, dénombrables, *Lecture Notes in Math.* **1064** (1984) Springer, Berlin–Heildelberg–New York.

7. Jewett, R.I.: Spaces with an abstract convolution of measures, *Advances in Math.* **18** (1975) 1–101.

8. Nessibi, M.M., Sifi, M. and Trimeche, K.: Continuous wavelet transform and continuous multiscale analysis on Laguerre hypergroup, *C.R. Math. Rep. Acad. Sci. Canada* (to appear).

9. Sifi, M.: Central limit theorem and infinitely divisible probabilities associated with partial differential operators, *J. of Theoretical Probability* **8** (1995) no. 3, 475–499.

10. Trimeche, K.: Probabilités indefiniment divisible et théorème de la limite centrale pour une convolution généralisée sur la demi-droite, *C.R. Acad. Sci. Paris, Sér. A* **286** (1978) 399–402.

11. Trimeche, K.: Opérateurs de permutation et théorème de la limite centrale associée à des opérateurs aux dérivées partielles, In: Heyer, H. (ed.) *Probability measures on groups*, Plenum Press.

THE REPRESENTATION OF THE REPRODUCING KERNEL
IN ORTHOGONAL POLYNOMIALS
ON SEVERAL INTERVALS

BORIS P. OSILENKER
Moscow State University of Civil Engineering,
Jaroslavskoe Shosse 26, 129337 Moscow, Russia

Abstract. For orthogonal polynomials with asymptotically N-periodic recurrence coefficients a representation of the trilinear reproducing kernel is obtained. This result is used to study weighted estimates for generalized translation operators and polynomial hypergroups.

Mathematics Subject Classification (1991): 42C05, 42C15, 43A62.

Key words: trilinear kernel, asymptotically N-periodic recurrence coefficients, reproducing kernel, generalized translation operators, polynomial hypergroups, convolution structure.

1. Introduction

Suppose μ is a positive Borel measure on a compact set on the real line. Then there is a unique sequence of polynomials

$$p_n(x) = k_n x^n + \dots \quad k_n > 0 \quad (n \in \mathbb{Z}_+),$$

such that

$$\int p_m(x)p_n(x)d\mu(x) = 1.$$

These orthonormal polynomials satisfy á three-term recurrence relation

$$xp_n(x) = a_{n+1}p_{n+1}(x) + b_n p_n(x) + a_n p_{n-1}(x), \ n \in \mathbb{Z}_+, \quad (1.1)$$
$$p_{-1}(x) = 0, \ p_0(x) = 1.$$

B. P. Komrakov et al. (eds.), Lie Groups and Lie Algebras, 147–162.
© 1998 *Kluwer Academic Publishers. Printed in the Netherlands.*

where $a_{n+1} = k_n/k_{n+1} > 0$ and $b_n \in \mathbb{R}$. If the support of μ is compact, then the recurrence coefficients a_n and b_n are bounded. Conversely, by Favard's Theorem [8], [12], if $p_n(x)$ are given by the recurrence formula (1.1) with $a_n > 0$ and $b_n \in \mathbb{R}$, then there exists a positive Borel measure μ such that $p_n(x)$, $n \in \mathbb{Z}_+$, is an orthonormal polynomial system with respect to the measure μ. If a_n and b_n are bounded, then the measure μ is unique and the support of μ is compact. In this paper we obtain a representation of the trilinear kernel in orthogonal polynomials that plays an important role in some problems of mathematical physics and in harmonic analysis (hypergroups associated with orthogonal polynomials, generalized translation operators, generalized product formula, computation of the infinite sums associated with orthogonal polynomials and so on); for details see [5], [9], [10], [13], [14], [22], [30]–[34], [37], [38], [43].

If the Jacobi matrix

$$
J=\begin{pmatrix}
b_o & a_1 & 0 & 0 & 0 & \cdots \\
a_1 & b_1 & a_2 & 0 & 0 & \cdots \\
0 & a_2 & b_2 & a_3 & 0 & \cdots \\
\vdots & \vdots & \vdots & \vdots & \vdots & \ddots
\end{pmatrix}
\tag{1.2}
$$

belongs to the Nevai class [28], i.e.,

$$
\lim_{n\to\infty} a_n = \frac{1}{2}, \quad \lim_{n\to\infty} b_n = 0,
\tag{1.3}
$$

the representation of the trilinear kernel associated with the orthonormal polynomial system $p_n(x)$, $n \in \mathbb{Z}_+$, was given in [30]–[34].

We will consider orthogonal polynomials on several intervals. Instead of the condition (1.3), we assume convergence modulo N, where N is a positive integer. This means that we will consider asymptotically periodic recurrence coefficients. Orthogonal polynomials with asymptotically periodic recurrence coefficients have been studied by many mathematicians [1]–[4], [6], [7], [15]–[20], [23], [24], [27], [35], [36], [40], [42].

These polynomials have a measure μ supported on at most N disjoint intervals and in addition μ may have a denumerable number of jumps which can only accumulate on these intervals. The special case when the intervals touch each other leads to sieved orthogonal polynomials [2], [7], [20]. Besides being of interest in its own right [3], [15]–[19], orthogonal polynomials on several intervals is used in numerical analysis [25], [39] and in quantum chemistry [44], [45].

2. The class AP_N

Assume that we are given two periodic sequences $a_{n+1}^o > 0$ and $b_n^o, n \in \mathbb{Z}_+$, such that

$$a_{n+N}^o = a_n^o, \quad n = 1, 2, \dots, \\ b_{n+N}^o = b_n^o, \quad n = 0, 1, 2, \dots \tag{2.1}$$

(here $N \geq 1$ is the period), and that the recurrence coefficients a_{n+1} and $b_n, n \in \mathbb{Z}_+$, satisfy

$$\lim_{n \to \infty} [|a_n - a_n^o| + |b_n - b_n^o|] = 0. \tag{2.2}$$

We will say that the orthogonal polynomials $p_n(x), n \in \mathbb{Z}_+$, have asymptotically periodic recurrence coefficients, or, the associated Jacobi matrix (1.2) has asymptotically periodic elements, i.e., $J \in AP_N$. We denote the orthonormal polynomials with periodic recurrence coefficients a_{n+1}^o and b_n^o by $q_n(x)$. Then

$$xq_n(x) = a_{n+1}^o q_{n+1}(x) + b_n^o q_n(x) + a_n^o q_{n-1}(x), \quad n \in \mathbb{Z}_+, \\ q_{-1}(x) = 0, \quad q_o(x) = 1. \tag{2.3}$$

The associated polynomials of order $k, k \geq 0$, can be introduced by the shifted recurrence formula

$$xq_n^{(k)}(x) = a_{n+k+1}^o q_{n+1}^{(k)}(x) + b_{n+k}^o q_n^{(k)}(x) + a_{n+k}^o q_{n-1}^{(k)}(x), \quad n \in \mathbb{Z}_+,$$

with initial conditions $q_{-1}^{(k)}(x) = 0$, $q_0^{(k)}(x) = 1$. Define $\omega^N(x) = \rho(T(x))$, where

$$T(x) = \frac{1}{2} \left\{ q_N(x) - \frac{a_N^o}{a_{N+1}^o} q_{N-2}^{(1)}(x) \right\} \tag{2.4}$$

and $\rho(x) = x + \sqrt{x^2 - 1}$ with the analytic square root and satisfying $|\rho(x)| > 1$ in $\mathbb{C} \setminus [-1, 1]$. On $[-1, 1]$ we define $\rho(x) = \rho(x + i0+)$, which gives $\rho(\cos \theta) = e^{i\theta}$. In particular for $x \in [-1, 1]$ we have $\Re[\rho(x)] = x$ and $\Im[\rho(x)] = \sqrt{1 - x^2}$, this square root being positive for every $x \in [-1, 1]$. Define

$$E = \{x, |\omega^N(x)| = 1\};$$

then E consists of N intervals, where $-1 \leq T(x) \leq 1$, and between every two consecutive intervals of E there is exactly one zero of each of the polynomials $T'(x)$ and $q_{N-1}^{(j)}(x), j \geq 0$, [15]. On E we have $\Re \left[\omega^N(x) \right] = T(x)$ and $\Im \left[\omega^N(x) \right] = \text{sign}\,[T'(x)]\,\sqrt{1 - T^2(x)}$, where the latter square root is always positive on E. The set E corresponds to the essential spectrum of the orthogonal polynomials $p_n(x)$ and the measure μ for these orthogonal polynomials has support $G = E \bigcup E^*$, where E^* is a denumerable set for which the accumulation points are on E. The set $\{\omega^{2N}(x) = 1\}$ consists of the endpoints of the intervals (it is possible that some of the intervals touch at a point where $\omega^{2N}(x) = 1$). The orthonormality of the polynomials $q_n(x)$ is given by

$$\frac{1}{\pi a_N^o} \int_E q_m(x) q_n(x) \frac{\sqrt{1 - T^2(x)}}{|q_{N-1}(x)|} \, dx$$

$$+ \frac{2}{a_N^o} \sum q_m(x_i) q_n(x_i) \frac{\sqrt{T^2(x_i) - 1}}{q'_{N-1}(x_i)} = \delta_{m,n},$$

where x_i are those zeros of $q_{N-1}(x)$ for which $q_N(x_i) \neq \omega^N(x_i)$, see [15].

Examples

1. *The sieved Legendre polynomials.* Let $w_N(x)$ be the sieved Legendre weight on $[-1, 1]$

$$w_N(x) = \frac{1}{2} |U_{N-1}(x)| \ (-1 \leq x \leq 1),$$

where $U_{N-1}(x)$ is the Chebyshev polynomial of the second kind of degree $N - 1$. The orthogonal polynomials corresponding to this weight are a special case of sieved ultraspherical polynomials [2], and of sieved symmetric Pollaczek polynomials [20]. The recurrence coefficients (see [17]) are

$$a_{nN+j} = \frac{1}{2}, \ j = 2, 3, \ldots, N - 1,$$

$$a_{nN}^2 = \frac{1}{2} \cdot \frac{n}{2n + 1}, \ n \in \mathbb{Z}_+,$$

$$a_{nN+1}^2 = \frac{1}{2} \cdot \frac{n + 1}{2n + 1}, \ b_n = 0.$$

2. *The nonsymmetric sieved Pollaczek polynomials* [7] (see also [29]). Given $k \in \mathbb{N}$ and $a, b, c, \lambda \in \mathbb{R}$, we define the k-sieved 4-parameter Pollaczek polynomials as the characteristic polynomials associated with the Jacobi matrix $J = J(k, a, b, c, \lambda)$, where

$$a_n = \sqrt{\frac{A_{n-1} D_n}{B_{n-1} B_n}}, \ b_n = \frac{C_n}{B_n}, \ n \in \mathbb{N}.$$

Here

$$A_n = \frac{n}{k} + c + 2\lambda, \ B_n = \frac{2n}{k} + 2a + 2c + 2\lambda,$$

$$C_n = -2b, \ D_n = \frac{n}{k} + c + 2\lambda - 1$$

for $n \equiv 0 \pmod{k}$ and

$$A_n = 1, \quad B_n = 2, \quad C_n = 0, \quad D_n = 1$$

otherwise. Naturally, the Pollaczek polynomials are orthogonal with respect to a positive measure if and only if all parameters are chosen in such a way that all subdiagonals in J are positive.

3. Denote $E_\xi = [-1, -\xi] \cup [\xi, 1]$, $0 \le \xi < 1$. a) Let

$$w_0(x) = \begin{cases} \sqrt{\dfrac{x+\xi}{x-\xi}} \cdot \sqrt{\dfrac{1-x}{1+x}}, & x \in E_\xi, \\ \\ 0, & \text{otherwise.} \end{cases}$$

In the paper [6] the following polynomials were introduced:

$$\hat{\phi}_n(x; \xi) = \hat{k}_n x^n + \ldots, \ \hat{k}_n > 0, \ n \in \mathbb{Z}_+, \ x \in E_\xi;$$

they are orthonormal with respect to the weight $w_0(x)$ on E_ξ and satisfy the following three-term recurrence relation

$$a_{n+1}\hat{\phi}_{n+1}(x; \xi) - (x + b_n)\hat{\phi}_n(x; \xi) + a_n\hat{\phi}_{n-1}(x; \xi) = 0$$

with

$$a_n = \frac{1 + (-1)^n \xi}{2}, \ n = 0, 1, \ldots, \quad b_0 = -\frac{1-\xi}{2}$$

and $b_n = 0$ for $n = 1, 2, \ldots$.

b) Denote

$$w^{(p,q)}(x) = \begin{cases} \left(\dfrac{2}{1-\xi^2}\right)^{p+q} (x+\alpha)(x^2 - \xi^2)^p(1-x^2)^q, & -1 \le x \le -\xi, \\ \\ -\left(\dfrac{2}{1-\xi^2}\right)^{p+q} (x+\alpha)(x^2 - \xi^2)^p(1-x^2)^q, & \xi \le x \le 1, \\ \\ 0, & x \notin E_\xi, \end{cases}$$

where $p > -1$, $q > -1$, $-\xi \le \alpha \le \xi$. Let $\{\phi_n^{(p,q)}(x)\}$ $(n \in \mathbb{Z}_+)$ be a polynomial system orthogonal on E_ξ with respect to the weight $w^{(p,q)}(x)$ with the leading coefficient is equal to 1. In [4] G. Barkov showed that

$$\phi_{2n}^{(p,q)}(x; \xi) = \left(\frac{1-\xi^2}{2}\right)^{n-p-q} I_n^{(p,q)}\left(\frac{2x^2 - \xi^2 - 1}{1-\xi^2}\right),$$

$$\phi_{2n+1}^{(p,q)}(x; \xi) = \frac{\phi_{2n+2}^{(p,q)}(x) - m_{2n}\phi_{2n}^{(p,q)}(x)}{x+\alpha}, \quad m_{2n} = \frac{\phi_{2n+2}^{(p,q)}(\alpha)}{\phi_{2n}(\alpha)},$$

where $I_n^{(p,q)}(x)$ are the Jacobi polynomials (with leading coefficients equal to 1) orthogonal with respect to the weight $(1+x)^p(1-x)^q$, $-1 \le x \le 1$. The polynomials $\phi_{2n}^{(p,q)}(x;\xi)$ satisfy the following differential equation [4]

$$x(x^2-1)(x^2-\xi^2)y'' + [2(p+1)x^2(x^2-1) + 2(q+1)(x^2-\xi^2)x$$
$$-(x^2-1)(x^2-\xi^2)]y' = 4n(n+p+q+1)x^3y$$

and the polynomials $\phi_{2n+1}^{(p,q)}(x)$ satisfy the more complicated differential equation [4]. The corresponding orthonormal polynomials $\hat{\phi}_n^{(p,q)}(x)$ can be represented in the form

$$\hat{\phi}_{2n}^{(p,q)}(x) = (1-\xi^2)^{-\frac{2n+1}{2}} 2^{-\frac{p+q}{2}} \Lambda_{2n}^{p,q} \phi_{2n}^{(p,q)}(x),$$
$$\hat{\phi}_{2n+1}^{(p,q)}(x) = (1-\xi^2)^{-\frac{2n+1}{2}} 2^{-\frac{p+q}{2}} \Lambda_{2n+1}^{p,q} \phi_{2n+1}^{(p,q)}(x),$$

where

$$\Lambda_{2n}^{(p,q)} = \sqrt{\frac{n!\Gamma(n+p+1)\Gamma(n+q+1)\Gamma(n+p+q+1)}{\Gamma(2n+p+q+1)\Gamma(2n+p+q+2)}},$$

$$\Lambda_{2n+1}^{(p,q)} = \sqrt{\frac{n!\Gamma(n+p+1)\Gamma(n+q+1)\Gamma(n+p+q+1)}{\Gamma(2n+p+q+1)\Gamma(2n+p+q+2)}} [-m_{2n}]^{-\frac{1}{2}}.$$

These polynomials satisfy the following recurrence relation

$$x\hat{\phi}_n^{(p,q)}(x) = a_{n+1}^{(p,q)} \hat{\phi}_{n+1}^{(p,q)}(x) + b_n^{(p,q)} \hat{\phi}_n^{(p,q)}(x) + a_n^{(p,q)} \hat{\phi}_{n-1}^{(p,q)}(x)$$

with

$$a_{2n+1}^{(p,q)} = \sqrt{-m_{2n}},$$

$$a_{2n+2}^{(p,q)} = \frac{1-\xi^2}{\sqrt{-m_{2n}}} \sqrt{\frac{(n+1)(n+p+1)(n+q+1)(n+p+q+1)}{(2n+p+q+1)(2n+p+q+2)^2(2n+p+q+3)}},$$

$$b_n^{(p,q)} = (-1)^n \alpha.$$

It is not difficult to see that the corresponding Jacobi matrix belongs to the class AP_2.

In the case $\alpha = \xi, \alpha = -\xi$, i.e.,

$$w(x) = \left(\frac{2}{1-\xi^2}\right)^{p+q} |x \pm \xi|(x^2-\xi^2)^p(1-x^2)^q$$

one obtains

$$a_{2n+1}^{(p,q)} = \sqrt{1-\xi^2}\sqrt{\frac{(n+p+1)(n+p+q+1)}{(2n+p+q+1)(2n+p+q+2)}},$$

$$a_{2n+2}^{(p,q)} = \sqrt{1-\xi^2}\sqrt{\frac{(n+1)(n+q+1)}{(2n+p+q+2)(2n+p+q+3)}},$$

$$b_n^{(p,q)} = (-1)^n\xi.$$

In particular, if $\alpha = \xi = 0$, i.e., $w(x) = |x|^{2p+1}(1-x^2)^q$, one gets the generalized Chebyshev polynomials, for which ([24], [28])

$$a_n^{(p,q)} = \frac{1}{2} + (-1)^n\frac{C_{p,q}}{n} + O(\frac{1}{n^2}), \quad b_n = 0,$$

where $C_{p,q} > 0$ is a constant independent of $n \in \mathbb{Z}_+$.

3. Representation of the kernels

Define

$$\lambda_{n,k} = \begin{cases} a_n & \text{for } k = n-1, \\ b_n & \text{for } k = n, \\ a_{n+1} & \text{for } k = n+1, \end{cases} \qquad \lambda_{n,k}^o = \begin{cases} a_n^o & \text{for } k = n-1, \\ b_n^o & \text{for } k = n, \\ a_{n+1}^o & \text{for } k = n+1, \end{cases}$$

where a_n, b_n, a_n^o, b_n^o are the recurrence coefficients (see (1.1), (2.3)). Introduce the notation

$$N_m = \{(k_1,\ldots,k_m) \mid k_i = 0,\pm 1\},$$
$$N_m^r = \{(k_1,\ldots,k_m) \in N_m \mid k_1 + \ldots + k_m = r\}.$$

Lemma 1 ([28], p.45) *Let m be a nonnegative integer and $n > m$. Then*

$$x^m p_n(x) = \sum_{N_m}\lambda_{n,n+k_1}\lambda_{n+k_1,n+k_1+k_2}\cdots$$

$$\lambda_{n+k_1+\ldots+k_{m-1},n+k_1+\ldots+k_{m-1}+k_m}p_{n+k_1+\ldots+k_m}(x).$$

Denote

$$\alpha_{n+s}^{(s,m)} = \delta\sum_{N_m^s}\lambda_{n,n+k_1}\lambda_{n+k_1,n+k_1+k_2}\cdots\lambda_{n+k_1+\ldots+k_{m-1},n+k_1+\ldots+k_{m-1}+k_m},$$

where $\delta = 1$ for $1 \le s \le m$ and $\delta = 1/2$ for $s = 0$, and

$$\alpha_n^{(s,m)} = \sum_{N_m^{-s}}\lambda_{n,n+k_1}\lambda_{n+k_1,n+k_1+k_2}\cdots\lambda_{n+k_1+\ldots+k_{m-1},n+k_1+\ldots+k_{m-1}+k_m}$$

for $1 \leq s \leq m$. In a similar way,

$$\overset{\circ}{\alpha}_{n+s}^{(s,m)} = \delta \sum_{N_m^{-s}} \overset{\circ}{\lambda}_{n,n+k_1} \overset{\circ}{\lambda}_{n+k_1,n+k_1+k_2} \cdots \overset{\circ}{\lambda}_{n+k_1+\ldots+k_{m-1},n+k_1+\ldots+k_m},$$

where $1 \leq s \leq m$ and δ is defined as above, and

$$\overset{\circ}{\alpha}_n^{(s,m)} = \sum_{N_m^{-s}} \overset{\circ}{\lambda}_{n,n+k_1} \cdots \overset{\circ}{\lambda}_{n+k_1+\ldots+k_{m-1},n+k_1+\ldots+k_m}, \quad 1 \leq s \leq m.$$

Lemma 2 *Let m be a nonnegative integer $(n > m)$ and $J \in AP_N$. Then*

$$x^m p_n(x) = \sum_{j=0}^{m} \alpha_{n+j}^{(j,m)} p_{n+j}(x) + \sum_{j=0}^{m} \alpha_n^{(j,m)} p_{n-j}(x), \quad m = 0, 1, \ldots, N, \quad (3.1)$$

and

$$x^m q_n(x) = \sum_{j=0}^{m} \overset{\circ}{\alpha}_{n+j}^{(j,m)} q_{n+j}(x) + \sum_{j=0}^{m} \overset{\circ}{\alpha}_n^{(j,m)} q_{n-j}(x), \quad m = 0, 1, \ldots, N. \quad (3.2)$$

Proof. We will prove formula (3.1); relation (3.2) can be obtained in a similar way. It follows from Lemma 1 that the following equation is valid

$$x^m p_n(x) = \sum_{j=0}^{m} \alpha_{n+s}^{(n+j)} p_{n+j}(x) + \sum_{j=0}^{m} \gamma_{n-j}^{(j,m)} p_{n-s}(x),$$

where

$$\alpha_{n+j}^{(j,m)} = \int x^m p_n(x) p_{n+j}(x) d\mu \qquad (3.3)$$

and

$$\gamma_{n-j}^{(j,m)} = \int x^m p_n(x) p_{n-j}(x) d\mu. \qquad (3.4)$$

Equation (3.3) yields $\alpha_l^{(j,m)} = \int x^m p_{l-j}(x) p_l(x) d\mu$ which coincides with (3.4) for $l = n$. $\square$

Define

$$\pi_N(x) := \left(\prod_{k=1}^{N} a_k^o \right) 2T(x), \qquad (3.5)$$

where the function $T(x)$ is defined by (2.4). Then by (2.3) and (2.4) one obtains

$$\pi_N(x) = \sum_{k=0}^{N} \gamma_k^{(N)} x^k, \quad \gamma_N^{(N)} = 1. \qquad (3.6)$$

We introduce

$$\begin{cases} d_{n+j}^{(j,N)} & = \sum_{k=j}^{N} \gamma_k^{(N)} [\alpha_{n+j}^{(j,N)} - \overset{o}{\alpha}_{n+j}^{(j,N)}], \quad j = 0, 1, \ldots, N-1, \\ d_n^{(j,N)} & = \sum_{k=j}^{N} \gamma_k^{(N)} [\alpha_n^{(j,N)} - \overset{o}{\alpha}_n^{(j,N)}], \quad j = 0, 1, \ldots, N-1, \\ d_{n+N}^{(N,N)} & = \prod_{k=1}^{N} a_{n+k}, \qquad\qquad d_n^{(N,N)} = \prod_{k=1}^{N} a_{n-k+1}. \end{cases} \quad (3.7)$$

Lemma 3 *Let N be a nonnegative integer and $n \geq N$. If $J \in AP_N$, then*

$$\pi_N(x) p_n(x) = \sum_{j=0}^{N} d_{n+j}^{(j,N)} p_{n+j}(x) + \sum_{j=0}^{N} d_n^{(j,N)} p_{n-j}(x), \quad n \geq N, \qquad (3.8)$$

where the following relations

$$\lim_{n\to\infty} d_{n+j}^{(j,N)} = \lim_{n\to\infty} d_n^{(j,N)} = 0, \quad j = 0, 1, \ldots, N-1, \qquad (3.9)$$

$$\lim_{n\to\infty} d_{n+N}^{(N,N)} = \lim_{n\to\infty} d_n^{(N,N)} = \prod_{k=1}^{N} a_k^o \qquad (3.10)$$

hold.

   Proof. For an arbitrary sequence $\{u_n(x)\}$ we consider (see (3.2))

$$\alpha^m u_n(x) = \sum_{j=0}^{m} \overset{o}{\alpha}_{n+j}^{(j,m)} u_{n+j}(x) + \sum_{j=0}^{m} \overset{o}{\alpha}_n^{(n)} u_{n-j}(x) \qquad (3.11)$$

with $u_0 = 1$ and $m = 0, 1, \ldots, N$, $m \leq n$. The following relation is known (see [15], [27]):

$$2T(x) u_n(x) = u_{n+N}(x) + u_{n-N}(x), \quad n \geq N.$$

So, by (3.5), (3.6), (3.11) and the last equation one obtains

$$\begin{aligned} \pi_N(x) u_n(x) &= \sum_{k=0}^{N} \gamma_k^{(N)} x^k u_n(x) \\ &= \sum_{k=0}^{n} \gamma_k^{(N)} \left(\sum_{j=0}^{k} \overset{o}{\alpha}_{n+j}^{(j,k)} u_{n+j}(x) + \sum_{j=0}^{k} \overset{o}{\alpha}_j^{(j,k)} u_{n-j}(x) \right) \qquad (3.12) \\ &= u_{n+j}(x) \prod_{k=1}^{N} a_k^o + u_{n-j}(x) \prod_{k=1}^{N} a_k^o \end{aligned}$$

On the other hand, by (3.1), (3.5)–(3.7) and (3.12) for $n \geq N$ we have

$$\pi_N(x)p_n(x) =$$

$$= \sum_{k=0}^{N} \gamma_k^{(N)} x^k p_n(x) = \sum_{k=0}^{N} \gamma_k^{(N)} \left(\sum_{j=0}^{k} \alpha_n^{(j,k)} p_{n+j}(x) + \sum_{j=0}^{k} \alpha_n^{(j,k)} p_{n-j}(x) \right)$$

$$= \sum_{k=0}^{N} \gamma_k^{(N)} \left(\sum_{j=0}^{k} (\alpha_{n+j}^{(j,k)} - \overset{o}{\alpha}_{n+j}^{(j,k)}) p_{n+j}(x) + \sum_{j=0}^{k} (\alpha_n^{(j,k)} - \overset{o}{\alpha}_n^{(j,k)}) p_{n-j}(x) \right)$$

$$+ \sum_{k=0}^{N} \left(\sum_{j=0}^{k} \overset{o}{\alpha}_n^{(j,k)} p_{n-j}(x) + \sum_{j=0}^{k} \overset{o}{\alpha}_n^{(j,k)} p_{n-j}(x) \right) = \sum_{j=0}^{N-1} d_{n+j}^{(j,N)} p_{n+j}(x)$$

$$+ \sum_{j=0}^{N-1} d_n^{(j,N)} + \left(\prod_{k=1}^{N} a_k^o \right) p_{n+N}(x) + \left(\prod_{k=1}^{N} a_k^o \right) p_{n-N}(x)$$

$$+ \gamma_N^{(N)} \left(\alpha_{n+N}^{(N,N)} - \overset{o}{\alpha}_{n+N}^{(N,N)} \right) p_{n+N}(x) + \gamma_N^{(N)} \left(\alpha_n^{(N,N)} - \overset{o}{\alpha}_n^{(N,N)} \right) p_{n-N}(x)$$

which coincides with (3.8) because by (2.1), (3.6) and Lemma 1 one gets

$$\gamma_N^{(N)} \left[\alpha_{n+N}^{(N,N)} - \overset{o}{\alpha}_{n+N}^{(N,N)} \right] + \prod_{k=1}^{N} a_k^o$$

$$= \prod_{k=1}^{N} a_{k+n} - \prod_{k=1}^{N} a_{k+n}^o + \prod_{k=1}^{N} a_k^o = \prod_{k=1}^{N} a_{k+n}$$

and

$$\gamma_N^{(N)} \left[\alpha_{n-N}^{(N,N)} - \overset{o}{\alpha}_{n-N}^{(N,N)} \right] + \prod_{k=1}^{N} a_k^o$$

$$= \prod_{k=1}^{N} a_{n-k+1} - \prod_{k=1}^{N} a_{n-k+1}^o + \prod_{k=1}^{N} a_k^o = \prod_{k=1}^{N} a_{n-k+1}.$$

Relations (3.9) and (3.10) follow from (2.2), (3.7). Lemma 3 is proved. $\square$

Corollary 1 *Let $J \in AP_N$; then for $n \geq (N-1)$ we have*

$$\pi_N(x)p_n(x) = d_{n+N}p_{n+N}(x) + d_n p_{n-N}(x) + \omega_n^{(n-1)}(x), \qquad (3.13)$$

where

$$d_n = d_n^{(N,N)}, \quad \lim_{n \to \infty} d_n = \prod_{k=1}^{N} a_k^o, \qquad (3.14)$$

$$\omega_n^{(N-1)}(x) = \sum_{j=0}^{N-1} d_{n+j}^{(j,N)} p_{n+j}(x) + \sum_{j=0}^{N-1} d_n^{(j,N)} p_{n-j}(x) \qquad (3.15)$$

with (3.9).

Corollary 2 Let $J \in AP_N$, then for $n \geq (2N-1)$ we have

$$\pi_N^2(x) p_n(x) = d_{n+N} d_{n+2n} p_{n+2N}(x) + (d_n^2 + d_{n+N}^2) p_n(x)$$
$$+ d_{n-N} d_n p_{n-2N}(x) + \Omega_n^{(N-1)}(x) \qquad (3.16)$$

with

$$\Omega_n^{(N-1)}(x) = d_{n+N} \omega_{n+N}^{(N-1)}(x) + d_n \omega_{n-N}^{(N-1)}(x)$$
$$+ \sum_{j=0}^{N-1} d_{n+j}^{(j,N)} \left(d_{n+N+j} p_{n+N+j}(x) + d_{n+j} p_{n+j-N} + \omega_{n+j}^{(N-1)}(x) \right)$$
$$+ \sum_{j=0}^{N-1} d_n^{(j,N)} \left(d_{n+N-j} p_{n-N-j}(x) + d_{n-j} p_{n-N-j}(x) + \omega_{n-j}^{(N-1)}(x) \right),$$
$$\qquad (3.17)$$

where we use definitions (3.14) and (3.15).

Corollary 2 follows from Corollary 1 by straightforward calculations. Let $\{p_n(x)\}$ $(n \in \mathbb{Z}_+)$ be the orthogonal polynomials associated with Jacobi matrix (1.2) which belongs to the class AP_N. We consider the trilinear kernel

$$D_n^N(x, y, z) = \sum_{n=2N-1}^{n} p_k(x) p_k(y) p_k(z),$$

which possesses the following reproducing property: for every polynomial

$$P_n(x) = \sum_{k=0}^{n} c_k p_k(x), \ n \in \mathbb{Z}_+, n \geq 2N-1,$$

the relation

$$\int P_n(x) D_n^N(x, y, z) \, d\mu = \sum_{k=2N-1}^{n} c_k p_k(y) p_k(z)$$

holds.

The main result of this paper is the following

Theorem 1 Let $J \in AP_N$ and define

$$\Delta_N(x, y, z) = 4d^2 + \frac{1}{d} \pi_N(x) \pi_N(y) \pi_N(z) - \pi_N^2(x) - \pi_N^2(y) - \pi_N^2(z) \quad (3.18)$$

with

$$d = \prod_{k=1}^{N} a_k^o$$

For all x, y, z and $n \in \mathbb{Z}_+, n \geq 2N - 1$, the following representation is valid

$$\Delta_N(x, y, z) D_n^N(x, y, z) = A_n^N(x, y, z) + B_n^N(x, y, z) + H_n^N(x, y, z), \quad (3.19)$$

where

$$A_n^N(x, y, z) = \frac{1}{d} \left(\sum_{n+1}^{n+N} d_k^3 p_k(x) p_k(y) p_k(z) - \sum_{2N-1}^{3N-2} d_k^3 p_k(x) p_k(y) p_k(x) \right.$$

$$+ \sum_{N-1}^{2N-2} \left[d_{k+N}^3 p_k(x) p_k(z) + d_{k+N} d_{k+2N}^2 [p_{k+2N}(x) p_{k+2N}(y) p_k(z) \right.$$

$$+ p_{k+2N}(x) p_k(y) p_{k+2N}(z) + p_k(x) p_{k+2N}(y) p_{k+2N}(z)] + d_{k+N}^2 d_{k+2N} \cdot$$

$$\cdot [p_{k+2N}(x) p_k(y) p_k(z) + p_k(x) p_{k+2N}(y) p_k(z) + p_k(x) p_k(y) p_{k+2N}(z)]]$$

$$- \sum_{n-N+1}^{n} \left[d_{k+N}^3 p_k(x) p_k(y) p_k(z) + d_{k+2N}^2(x) d_{k+N} \cdot \right.$$

$$\cdot [p_{k+2N}(x) p_{k+2N}(y) p_k(z) + p_{k+2N}(x) p_k(y) p_{k+2N}(z)$$

$$+ p_k(x) p_{k+2N}(y) p_{k+2N}(z)] + d_{k+N}^2 d_{k+2N} [p_{k+2N}(x) p_k(y) p_k(z)$$

$$\left. \left. + p_k(x) p_{k+2N}(y) p_k(z) + p_k(x) p_k(y) p_{k+2N}(z)] \right] \right)$$

$$- \sum_{0}^{2(N-1)} d_{k+N} d_{k+2N} [p_{k+2N}(x) p_{k+2N}(y) p_k(z) + p_{k+2N}(x) p_k(y) p_{k+2N}(z)$$

$$+ p_k(x) p_{k+2N}(y) p_{k+2N}(z)] + \sum_{n-2N+1}^{n} d_{k+N} d_{k+2N} [p_{k+2N}(x) p_{k+2N}(y) p_k(z)$$

$$+ p_{k+2N}(x) p_k(y) p_{k+2N}(z) + p_k(x) p_{k+2N}(y) p_{k+2N}(z)]$$

(all sums here are taken over k),

$$B_n^N(x, y, z) = \frac{1}{d} \left(\sum_{k=2N-1}^{n} \left[(d_k - d)(d_k^2 - 2d_k d - 2d^2) p_k(x) p_k(y) p_k(z) \right. \right.$$

$$+ (d_{k+N} - d)(d_{k+N}^2 - 2d_{k+N} d - 2d^2) p_k(x) p_k(y) p_k(z)$$

$$+ d_{k+N} d_{k+2N} (d_{k+2N} - d)[p_{k+2N}(x) p_{k+2N}(y) p_k(z)$$

$$+ p_{k+2N}(x) p_k(y) p_{k+2N}(z) + p_k(x) p_{k+2N}(y) p_{k+2N}(z)]$$

$$+d_{k+N}d_{k+2N}(d_{k+N}-d)[p_{k+2N}(x)p_k(y)p_k(z)$$

$$+p_k(x)p_{k+2N}(y)p_k(z)+p_k(x)p_k(y)p_{k+2N}(z)]]\Bigg)$$

and $H_n^N(x,y,z)=\sum_{k=2N-1}^n h_k(x,y,z)$, where

$$h_k(x,y,z)=\frac{1}{d}\{d_{k+N}^2 p_{k+N}(x)p_{k+N}(y)+d_k d_{k+N}p_{k+N}(x)p_{k-N}(y)$$

$$+d_k d_{k-N}p_{k-N}(x)p_{k+N}(y)+d_k^2 p_{k-N}(x)p_{k-N}(y)$$

$$+[d_{k+N}p_{k+N}(x)+d_k p_{k-N}(x)+\omega_k^{(N-1)}(x)]\omega_k^{(N-1)}(y)$$

$$+\omega_k^{(N-1)}(x)[d_{k+N}p_{k+N}(y)+d_k p_{k-N}(y)]\}\omega_k^{(N-1)}(z)$$

$$+\frac{1}{d}\{[d_{k+N}p_{k+N}(x)+d_k p_{k-N}(x)+\omega_k^{(N-1)}(x)]\}\omega_k^{(N-1)}(y)$$

$$+\omega_k^{(N-1)}(x)[d_{k+N}p_{k+N}(y)+d_k p_{k-N}(y)][d_{k+N}p_{k+N}(z)+d_k p_{k-N}(z)]$$

$$-\Omega_k^{(N-1)}(x)p_k(y)p_k(z)-p_k(x)\Omega_k^{(N-1)}(y)p_k(z)-p_k(x)p_k(y)\Omega_k^{(N-1)}(z).$$

Here $\omega_k^{(N-1)}(x)$ and $\Omega_k^{(N-1)}(x)$ are defined by (3.15) and (3.17).

Proof. In view of (3.13)–(3.18) one obtains

$$\Delta_N(x,y,z)p_k(x)p_k(y)p_k(z)=4d^2 p_k(x)p_k(y)p_k(z)$$

$$-\pi_N(x)^2 p_k(x)p_k(y)p_k(z)-p_k(x)\pi_N(y)^2 p_k(y)p_k(z)$$

$$-p_k(x)p_k(y)\pi_N(z)^2 p_k(z)+\frac{1}{d}\{d_{k+N}^2 p_{k+N}(x)p_{k+N}(y)$$

$$+d_k d_{k+N}p_{k+N}(x)p_{k-N}(y)+d_k d_{k+N}p_{k-N}(x)p_{k+N}(y)$$

$$+d_k^2 p_{k-N}(x)p_{k-N}(y)+[d_{k+N}p_k+N(x)+d_k p_{k-N}(x)$$

$$+\omega_k^{(N-1)}(x)]\omega_k^{(N-1)}(y)+\omega_k^{(N-1)}(x)][d_{k+N}p_{k+N}(y)+d_k p_{k-N}(y)]\}$$

$$\times[d_{k+N}p_{k+N}(z)+d_k p_{k-N}(z)+\omega_k^{(N-1)}(z)],$$

so that

$$\Delta_N(x,y,z)\sum_{2N-1}^n p_k(x)p_k(y)p_k(z)=4d^2\sum_{2N-1}^n p_k(x)p_k(y)p_k(z)$$

$$+\frac{1}{d}\Big(\sum_{2N-1}^n[d_{k+N}^3 p_k(x)p_k(y)p_k(z)+d_k d_{k+N}^2 p_{k+N}(x)p_{k+N}(y)p_{k-N}(z)$$

$$+d_k d_{k+N}^2 p_{k+N}(x)p_{k-N}(y)p_{k+N}(z)+d_k^2 d_{k+N}p_{k+N}(x)p_{k-N}(y)p_{k-N}(z)$$

$$+d_k d_{k+N}^2 p_{k-N}(x)p_{k+N}(y)p_{k+N}(z)+d_k^2 d_{k+N}p_{k-N}(x)p_{k+N}(y)p_{k-N}(z)$$

$$+d_k^2 d_{k+N}p_{k-N}(x)p_{k-N}(y)p_{k+N}(z)+d_k^3 p_{k-N}(x)p_{k-N}(y)p_{k-N}(z)]\Big)$$

$$-\sum_{2N-1}^{n}[d_{k+N}d_{k+2N}p_{k+2n}(x)p_k(y)p_k(z) + 3(d_k{}^2 + d_{k+N}{}^2) \cdot$$

$$\cdot p_k(x)p_k(y)p_k(z) + d_{k-N}d_kp_{k-2N}(x)p_k(y)p_k(z)$$

$$+d_{k+N}d_{k+2N}p_k(x)p_{k+2N}(y)p_k(z) + d_{k-N}d_kp_k(x)p_{k-2N}(y)p_k(z)$$

$$+d_{k+N}d_{k+2N}p_k(x)p_k(y)p_{k+2N}(z) + d_{k-N}d_kp_k(x)p_k(y)p_{k-2N}(z)]$$

$$+H_n(x,y,z),$$

where $H_n(x,y,z)$ is defined as above and all sums are taken over k. For the proof of formula (3.19), we consider the following "principal" terms (the other terms are treated in a similar manner):

$$\overset{(1)}{\sum} := 4d^2 \sum_{2N-1}^{n} p_k(x)p_k(y)p_k(z) +$$

$$\frac{1}{d}(\sum_{2N-1}^{n}[d_{k+N}^3 p_{k+N}(x)p_{k+N}(y)p_{k+N}(z) + d_k^3 p_{k-N}(x)p_{k-N}(y)p_{k-N}(z)])$$

$$-3\sum_{2N-1}^{n}(d_k^2 + d_{k+N}^2)p_k(x)p_k(y)p_k(z),$$

$$\overset{(2)}{\sum} := \frac{1}{d}\sum_{2N-1}^{n} d_k d_{k+N}{}^2 p_{k+N}(x)p_{k+N}(y)p_{k-N}(z)$$

$$-\sum_{2N-1}^{n} d_{k-N}d_k p_k(x)p_k(y)p_{k-2N}(z),$$

$$\overset{(3)}{\sum} := \frac{1}{d}\sum_{2N-1}^{n} d_k{}^2 d_{k+N}p_{k+N}(x)p_{k-N}(y)p_{k-N}(z)$$

$$-\sum_{2N-1}^{n} d_{k+N}d_{k+2N}p_{k+2N}(x)p_k(y)p_k(z).$$

As in [31], using Abel's summation by parts and the initial conditions (1.1) one gets the assertion of theorem. □

Remarks. 1. The particular case $N = 1$ of the Theorem was obtained in [30]–[34].

2. A similar representation is valid for the reproducing kernels composed of associated orthogonal polynomials (case $N = 1$ see in [32]), and for another systems (see, for example, [21], [26]).

Acknowledgments

This work was supported by the State Committee of the Russian Federation for Higher Education (grant no. 94-1.2-1.3) and by the International Scientific Foundation (grant no. MB 6300).

References

1. Akhiezer, N.I.: Orthogonal polynomials on several intervals, *Dokl. Akad. Nauk SSSR* **134** (1960) 9–12 (in Russian). English transl.: *Soviet Math. Dokl.* (1960) **1** 989–992.

2. Al-Salam, W., Allaway, W. and Askey, R.: Sieved ultraspherical polynomials, *Trans. Amer. Math. Soc.* **284** (1984) 39–55.

3. Aptekarev, A.I.: Asymptotic properties of polynomials orthogonal on a system of contours and periodic motions of Toda lattice, *Math. Sb.* **125 (167)** (1984) 231–258 (in Russian). English transl. *Math. USSR–Sb.* **53** (1986) 233–260.

4. Barkov, G.I.: On some systems of polynomials orthogonal on two symmetric intervals, *Izvestia Vyssch. Uchebn. Zaved., Mathem.* (1960) no. 4, 3–16 (in Russian).

5. Berezanskyi, Y.M. and Kalyuzhnyi, A.A.: *Harmonic Analysis in Hypercomplex Systems*, Naukova Dumka, Kiev, 1995 (in Russian).

6. Brjechka, V. F.: On a certain class of polynomials orthogonal on two finite symmetric intervals, *Zap. nauchno-issled. inst. of mat.and mechanics of Kharkov. Gosud Univ. and Kharkov. Math. Obsch.* **17** (1940) 75–98 (in Russian).

7. Charris, J. and Ismail M.E.H.: On sieved orthogonal polynomials: random walk polynomials, *Canad. J. Math.* **38** (1986) 397–415.

8. Chihara, T.: *On Introduction to Orthogonal Polynomials*, Gordon and Breach, New York, 1976.

9. Connett, W. C. and Schwartz, A. L.: The Theory of Ultraspherical Multipliers, *Memoirs of Amer. Math. Society* **183** (1977).

10. Connett, W. C. and Schwartz, A. L.: Product formula, hypergroups and Jacobi polynomials, *Bull. Amer. Math. Soc.* **22** (1990) 91–97.

11. Fischer, B. and Golub, G. H.: On generating polynomials which orthogonal over several intervals, *Math. Computation* **56** (1991) 711–730.

12. Freud, G.: *Orthogonal Polynomials*, Akad. Kiado, Budapest – Pergamon Press, Oxford, 1971.

13. Gasper, G.: Positivity and the convolution structure for Jacobi polynomials, *Ann. Math.* **93** (1971) 112–118.

14. Gasper, G.: Banach algebra for Jacobi series and positivity of kernel, *Ann. Math.* **95** (1972) 261–280.

15. Geronimo, J. S. and van Assche, W.: Orthogonal polynomials with asymptotically periodic recurrence coefficients, *J. Approx. Theory* **46** (1986) 251–283.

16. Geronimo, J. S. and van Assche, W.: Orthogonal polynomials on several intervals via a polynomial mapping, *Trans. Amer. Math. Soc.* **308** (1988) 559–579.

17. Geronimo, J. S. and van Assche, W.: Approximating the weight function for orthogonal polynomials on several intervals, *J. Approx. Theory* **65** (1991) 341–371.

18. Geronimus, Ya. L.: On the character of the solutions of the moment problem in the case of a limit-periodic associated fraction, *Bull. Acad. Sci USSR, Sect. Math.* **5** (1941) 203–210 (in Russian).

19. Geronimus, Ya. L.: On some finite difference equations and corresponding systems of orthogonal polynomials, *Zap. Mat. Otd. Fiz.-Math. Fak. Kharkov Gosud. Univ. i Kharkov Mat. Obsch.* **25** (1957) 87–100 (in Russian).

20. Ismail, M.E.H.: On sieved orthogonal polynomials, I: Symmetric Pollaczek polynomials, *SIAM J. Math. Anal.* **16** (1985) 1093–1113.

21. Koch, P.E.: On extension of the theory of orthogonal polynomials and Gaussian

quadrature to trigonometric and hyperbolic polynomials, *J. Approx. Theory,* **43** (1985) 1093–1113.

22. Koorwinder, T.H.: Jacobi polynomials, II. An analytic proof of the product formula, *SIAM J. Math. Anal.* **5** (1974) 125–137.

23. Laine, T.P.: The product formula and convolution structure for the generalized Chebyshev polynomials, *SIAM J. Math. Anal.* **11** (1980) 133–147.

24. Laschenov, K.V.: On a certain class of orthogonal polynomials, *Uchebn. Zap. Leningr. Gosud. Pedagog. Inst.* **89** (1953) 169–189 (in Russian).

25. Lebedev, V.I.: Iterative methods for solving operator equations with a spectrum contained in several intervals, *Zh. Vychisl. Mat. i Mat. Fiz.* **9** 1247–1252. English trans.: *USSR Comput. Math. and Math. Phys.* **9** (1969) 17–24.

26. Marcellan, F. and van Assche, W.: Relative asymptotics for orthogonal polynomials with a Sobolev inner product, *J. Approx. Theory* **72** (1993) 193–209.

27. Niman, P.B.: To the theory of periodic and limit-periodic Jacobi matrix, *Dokl. Akad. Nauk SSSR* **143** (1962) 277–279 (in Russian).

28. Nevai, P.: Orthogonal Polynomials, *Memoirs of Amer. Math. Soc.* **213** (1979).

29. Nevai, P.: Orthogonal polynomials, recurrences, Jacobi matrices, and measures. In: Gonchar, A.A. and Saff, E.B. (eds.) *Progress in Approximation Theory*, Springer–Verlag, New York–Berlin–Heidelberg, 1992, 79–104.

30. Osilenker, B.P.: Generalized translation operator and convolution structure for the orthogonal polynomials, *Dokl. Akad. Nauk SSSR* **298** 1072–1076. English transl. in: *Soviet Math. Dokl.* **.137** (1988).

31. Osilenker, B.P.: The representation of the trilinear kernel in general orthogonal polynomials and some applications, *J. Approx. Theory* **67** (1991) 93–114.

32. Osilenker, B.P.: Generalized product formula for orthogonal polynomials, *Contemporary Math.* **183** (1995) 269–285.

33. Osilenker, B.P.: Quasi-potential functions associated with orthogonal polynomials, *Dokl. Akad. Nauk* **342** (1995) no. 5, 589–591. English transl. in: *Dokl. Math.* **51** (1995) 394–396.

34. Osilenker, B.P.: *Orthogonal Polynomial Series*, Birkhäuser (in preparation).

35. Peherstorfer, F.: Orthogonal and Chebyshev polynomials on two intervals, *Acta Math. Hung.* **55** (1990) no. 3–4, 245–278.

36. Peherstorfer, F.: On Bernstein–Szegö orthogonal polynomials on several intervals, II. Orthogonal polynomials with periodic recurrence coefficients, *J. Approx. Theory* **64** (1991) 123–161.

37. Rahman, M.: A product formula for the continuous q-Jacobi polynomials, *J. Math. Anal. Appl.* **118** (1986) 309–322.

38. Rahman, M. and Shah, M.J.: An infinite series with products of Jacobi polynomials and Jacobi functions of the second kind, *SIAM J. Math. Anal.* **16** (1985) 859–875.

39. Saad, Y.: Iterative solution of indefinite symmetric linear systems by methods using orthogonal polynomials over two distinct intervals, *SIAM J. Numer. Anal.* **20** (1983) 784–811.

40. Stieltjes, T.: Recherches sur les fraction continuous, *Ann. Fac. Sci. Touluse* **8** (1894) J1–J122; **9** (1895) A1–A47.

41. Van Haeringen, H.: A class a Gegenbauer functions: twenty four sums in closed form, *J. Math. Phys.* **27** (1986) 938–952.

42. Van Assche, W.: Asymptotics for Orthogonal Polynomials, *Lect. Notes Math.* Springer–Verlag, Berlin **1265** (1987).

43. Vilenkin, N.Ya.: *Special Functions and the Theory of Group Representation*, Nauka, Moscow 1965 (in Russian).

44. Wheeler, J.C.: Modified moments and Gaussian quadrature, *Rocky Mountain J. Math.* **4** (1974) 287–296.

45. Wheeler, J.C.: Modified moments and continued fraction coefficients for the diatomic linear chain, *J. Chem. Phys.* **80** (1984) 472–476.

HOMOLOGY INVARIANTS OF

HOMOGENEOUS COMPLEX MANIFOLDS

BRUCE GILLIGAN

Department of Mathematics and Statistics
University of Regina, Regina, Canada S4S 0A2.
E-mail: gilligan@max.cc.uregina.ca

Abstract. A connected Lie group has an Iwasawa decomposition $G = K \times \mathbb{R}^{d_G}$, where K is a maximal compact subgroup of G. It is well-known that a complex Lie group G is compact, i.e., $d_G = 0$, if and only if G is a torus. One also has the following

Theorem *A connected complex Lie group G with $d_G \leq 2$ is abelian.*

One would like to have an invariant d similar to the above for studying the structure of noncompact homogeneous complex manifolds in settings where d is small. (For our present purposes a complex manifold X is homogeneous if there is a transitive action of a connected Lie group G acting on X by means of holomorphic automorphisms. Such an X can be written in a Klein form $X = G/H$, where H is a closed subgroup of G and G/H has a left G-invariant complex structure.) For a manifold X we define d_X to be the codimension of the top nonvanishing homology group of X with coefficients in $\mathbb{Z}_2$.

Now suppose G is a connected complex Lie group and H is a closed complex subgroup of H such that $d_{G/H} = 2$ and G/H has nonconstant holomorphic functions. Then there is the following classification result which is joint work with D.N. Akhiezer: Let $\pi : G/H \to G/J$ be the holomorphic reduction of G/H and set $F := J/H$ and $Y := G/J$. Then either F is connected with $d_F = 1$ and Y is an affine cone with its vertex removed, or F is compact and connected and $d_Y = 2$, where Y is $\mathbb{C}$, the affine quadric Q_2, $\mathbb{P}_2 - Q$ (with Q a quadric curve) or a homogeneous holomorphic $\mathbb{C}^*$-bundle over an affine cone minus its vertex which is itself an algebraic principal bundle or which admits a two-to-one covering that is. The basic ideas of the proof will be given. The most interesting situation occurs in the case of discrete isotropy.

B. P. Komrakov et al. (eds.), Lie Groups and Lie Algebras, 163–179.

We will also discuss some results in settings where $d_{G/H} = 1$. One can show that if X is any connected manifold with more than one end (in the sense of H. Freudenthal/L. Zippin), then $d_X = 1$. The Moebius band shows that the converse is false. However, one does have the following

Theorem *Suppose $X = G/H$ is a homogeneous complex manifold such that one of the following conditions holds:*

(i) *X has nonconstant holomorphic functions,*
(ii) *X is a Kähler manifold,*
(iii) *H has a finite number of connected components.*

Then $d_{G/H} = 1$ if and only if X has two ends.

All such G/H have been classified. The proofs involve standard fibration methods.

Mathematics Subject Classification (1991): 32M10.

Key words: Lie groups, solv-manifolds, homology invariants.

1. Introduction

One approach to understanding complex analysis on homogeneous complex manifolds is to study compact submanifolds, together with their relationship with the complex structure. An interesting example of this approach is furnished by a theorem of Matsushima–Morimoto in the case of complex Lie groups. Any connected Lie group G has an Iwasawa decomposition which *topologically* says that $G = K \times \mathbb{R}^d$, where K is a maximal compact subgroup of G, see [22]. Matsushima and Morimoto [25] showed that a connected complex Lie group G is Stein if and only if K is totally real, i.e., if and only if $\mathfrak{k} \cap i\mathfrak{k} = (0)$, where $\mathfrak{k}$ is the Lie algebra of K.

Efforts to extend this idea to homogeneous spaces are complicated by a couple of factors. First, if $X = G/H$ is a homogeneous complex manifold, where H has a finite number of connected components, then there exists a fibration $G/H \xrightarrow{\mathbb{R}^d} K/L$, where K (respectively L) is a maximal compact subgroup of G (respectively of H, contained in K), see [27] and [23]. But not every homogeneous space admits a deformation retract onto a compact submanifold. And it is possible for compact submanifolds to arise because of discrete isotropy, e.g., $\mathbb{C}^* = \mathbb{C}/\mathbb{Z}$. Note that this example depends on the Klein form chosen.

This note presents a survey on some recent results where one uses the codimension d of the top nonvanishing homology group with coefficients in $\mathbb{Z}_2$ in order to detect compact subvarieties. For any manifold X we define the invariant d_X, investigate its relationship to ends (in the sense of Freuden-thal/Zippin [7]) for certain homogeneous complex manifolds, and determine the structure of homogeneous complex manifolds G/H with $d_{G/H} = 2$ under the assumption $\mathcal{O}(G/H) \neq \mathbb{C}$. When G and H are complex groups, this is joint work with D.N. Akhiezer [3]. For the case of real groups we show that essentially the only "new" building block which can occur is the unit disk as the base of the holomorphic reduction of G/H, see [14]. It is then easy to see that a homogeneous complex manifold G/H with $d_{G/H} = 2$ has nonconstant *bounded* holomorphic functions if and only if $G/H = F \times \Delta$, where F is a connected compact homogeneous complex manifold and Δ denotes the unit disk.

A more detailed survey on the structure of homogeneous complex manifolds under various topological assumptions on the spaces in question can be found in the monograph [13].

This work was partially supported by NSERC Grant A3494 and SFB 237 of the Deutsche Forschungsgemeinschaft.

2. The invariant d_X

We now define the invariant d_X and discuss some of its properties for homogeneous complex manifolds, i.e., when $X = G/H$ is a homogeneous complex manifold of a Lie group G.

Definition. Suppose X is a connected manifold with $\dim_{\mathbb{R}} X = n$. Define

$$d_X := \min\{\ r \mid H_{n-r}(X, \mathbb{Z}_2) \neq 0\ \}.$$

Note that $d_X = 0$ exactly if the manifold X is compact. Otherwise, d_X is positive.

In passing we mention that d_X is dual to an invariant introduced by H. Abels; see [1]. For any locally compact topological space X and nonzero abelian group k define

$$n_c(X; k) := \inf\{\ i \mid H_c^i(X; k) \neq 0\ \}.$$

Abels calls this number the *noncompact dimension* of X with respect to k. In [1] the following properties of this invariant are noted, where we use the numbering from that article.

2.2 If X is compact, then $n_c(X; k) = 0$ for every $k \neq 0$. Conversely, if X is connected and $n_c(X; k) = 0$ for some abelian group $k \neq 0$, then X is compact.

2.7 There is a long exact sequence

$$\ldots \to H_c^i(X; k) \to H^i(X; k) \to H_\infty^i(X; k) \to H_c^{i+1}(X; k) \to \ldots$$

Thus $e(X) = \dim H_\infty^0(X; k)$ for k a field.

2.8 If X is connected, then $e(X) > 1 \Longrightarrow n_c(X; k) = 1$.

2.9 If the natural map $H_c^*(X; k) \to H^*(X; k)$ is the zero map (one such situation is described below), then one has the short exact sequence

$$0 \to H^i(X; k) \to H_\infty^i(X; k) \to H_c^{i+1}(X; k) \to 0.$$

Thus $e(X) = 1 + \dim_k H_c^1(X; k)$.

Remark S.T. Yau [34] proved that if M is a noncompact manifold admitting an infinite group which acts freely and properly discontinuously and if $\dim H_c^i(M) < \infty$, then the map $H_c^i(M) \to H^i(M)$ is the zero map. As noted under 2.9 above, this implies $H_c^1(X) = H_\infty^0(X)/H^0(X)$ and thus $n_c(X; k) = 1$ implies $e(X) > 1$; see also 2.7 above.

Our basic approach will be to assume that $d_{G/H}$ is small, where G/H is a homogeneous complex manifold, and use this assumption to draw conclusions about the structure of G/H. In order to do this we need to know how $d_{G/H}$ behaves with respect to fibrations. Using a spectral sequence argument, we noted the following in [3].

Lemma 1 1) Let $X \xrightarrow{F} B$ be an orientable fiber bundle (e.g., if $\pi_1(B) = 0$); then $d_X = d_F + d_B$.

2) If $X \xrightarrow{F} B$ is a fiber bundle and B has the homotopy type of a CW-complex of dimension q, then $d_X \geq d_F + (\dim B - q)$. Further, if B is homotopy equivalent to a compact manifold, then $d_X \geq d_F + d_B$.

3) Suppose X and Y are connected manifolds, $\pi : X \to Y$ is an unramified covering, $M \subset Y$ is a compact submanifold and Y is retractable onto M. Then $d_X = d_Y + d_{\pi^{-1}(M)}$. In particular, if π is infinite, then $d_X > d_Y$.

3. Spaces G/H having more than one end

3.1. THE CONDITIONS $D_X = 1$ AND $E(X) > 1$

For any connected manifold X the number of its ends $e(X)$, in the sense of Freudenthal/Zippin, e.g., see [7], is defined and so is the invariant d_X. How are they related? For a connected manifold $e(X) > 1$ implies $d_X = 1$; a proof of this follows from 2.8 in the paper of Abels cited above and is also given below. In many different settings involving homogeneous spaces, the condition $d_{G/H} = 1$ holds if and only if G/H has exactly two ends. Some settings where these two conditions are equivalent are presented here, see Theorem 2. However, these two conditions are not equivalent, e.g., the Moebius band M satisfies $d_M = 1$ and $e(M) = 1$. Note that the Moebius band is homogeneous, but is not a complex manifold. It is possible to give an example of a complex manifold X which is covered two-to-one by $\mathbb{C}^* \times T$, where T is an elliptic curve and which satisfies $d_X = 1$ and $e(X) = 1$. This yields a complex manifold, but one which is not homogeneous. However, we do not know whether $e(X) > 1$ is equivalent to $d_X = 1$ for arbitrary homogeneous complex manifolds. Just as a matter of speculation, there might be some Zariski dense discrete subgroup Γ of a connected complex semisimple Lie group S with $d_{S/\Gamma} = 1$ and $e(S/\Gamma) = 1$, and where $\mathcal{O}(S/\Gamma) = \mathbb{C}$, S/Γ has no analytic hypersurfaces, etc.

Lemma 2 *Let X be a connected n-dimensional manifold. If $e(X) > 1$, then $d_X = 1$.*

Proof.[1] (See also [1, p. 529].) By a result of H-C. Wang, see [32, p. 307] we have

$$e(X) - 1 = \dim(\ker H_c^1(X;k) \to H^1(X;k))$$

for any field k. If k is of characteristic two or X is orientable, then, by Poincaré duality

$$\ker(H_c^1(X;k) \to H^1(X;k)) \cong \ker(H_{n-1}(X;k) \to H_{n-1}^{cl}(X;k)),$$

where H_n^{cl} refers to homology with closed supports. In particular, this gives

$$e(X) > 1 \Longrightarrow H_c^1(X;\mathbb{Z}_2) \neq 0 \Longleftrightarrow H_{n-1}(X;\mathbb{Z}_2) \neq 0.$$

Since X is not compact, $H_n(X;\mathbb{Z}_2) = 0$, and so $d_X = 1$. □

We now address the converse for homogeneous complex manifolds. Before discussing the structure of spaces with $e(X) > 1$, we first note the following results from [16]. The proof of this theorem is somewhat involved and we just mention the main idea here. Suppose there is a holomorphic action of a (real) Lie group G on a holomorphic principal bundle $P \to X$ which is compatible (in a certain sense) with the action of the structure group on the bundle. If the complexification $G^{\mathbb{C}}$ of G acts on X, then $G^{\mathbb{C}}$ acts on P. This is then applied to the $\mathfrak{g}$-anticanonical fibration and the rest of the proof uses ideas similar to those presented in section five of this paper. The interested reader is referred to [16].

Theorem 1 *Suppose $X = G/H$ is a homogeneous complex manifold with $d_X = 1$. Then the complexification of G acts holomorphically and transitively on X.*

Corollary 1 *Suppose $X = G/H$ is a homogeneous complex manifold and assume X has more than one end. Then the complexification $G^{\mathbb{C}}$ of G acts holomorphically and transitively on X.*

---

[1]This proof was communicated to us by A. Borel, whom we thank.

Proof. This follows from the Theorem, since any connected manifold X with more than one end satisfies $d_X = 1$, see Lemma 2. $\square$

The conditions given in the next result, which appears in [12], may seem rather unrelated. But these conditions all play more or less the same role; they all eliminate S/Γ which might satisfy $d_{S/\Gamma} = 1$ and have any (positive) number of ends!

Theorem 2 *Suppose $X = G/H$ is a homogeneous complex manifold. Assume that one of the following conditions hold:*

1) *X is Kähler.*
2) *H is a closed subgroup of G having a finite number of connected components.*
3) *$\mathcal{O}(X) \neq \mathbb{C}$.*

Then $e(X) = 2$ if and only if $d_X = 1$.

There is a description in each of these three settings of those homogeneous complex manifolds which have two ends: see [18], [17], [8], and [20].

3.2. THE STRUCTURE OF G/H WITH $E(G/H) > 2$

In [17] we showed that for every integer $k > 2$ there exists a discrete subgroup Γ_k of $SL(2, \mathbb{C})$ such that $e(SL(2, \mathbb{C})/\Gamma_k) = k$. We now present a result from [16] concerning the structure of homogeneous complex manifolds X with $e(X) > 2$. All of these are basically of the form S/Γ, where Γ is a Zariski dense discrete subgroup of a semisimple complex Lie group S. This is an improvement on a result which appeared in [11], since the following tells us explicitly how to find the intermediate subgroups I and P.

Theorem 3 *Assume X is a connected homogeneous complex manifold with Klein form G/H, where G is a connected complex Lie group and H is a closed complex subgroup, and assume $e(X) > 2$. Then*

a) *H° is normal in P, where P is the smallest parabolic subgroup of G containing H. Let L denote the connected complex Lie group P/H°.*
b) *The R_L-orbits are compact in P/H and*
c) *$\Gamma := \pi_L(H/H^\circ)$ is a Zariski dense discrete subgroup of $S_L := L/R_L$. Moreover, $e(G/H) = e(S_L/\Gamma)$.*

Thus there is a closed complex subgroup I of P containing H (the stabilizer in P of the typical R_L-orbit) such that $I/H = R_L/R_L \cap (H/H^\circ)$ and $P/I = S_L/\Gamma$ and hence one has the fibrations

$$G/H \xrightarrow{I/H} G/I \xrightarrow{S_L/\Gamma} G/P,$$

where G/P is a homogeneous projective rational manifold and I/H is a compact complex solv-manifold.

The proof uses a very simple form of the following structure theorem for homogeneous spaces of the form G/H, where H° is solvable. We intend to return to this topic at some time in the future, see [15].

Theorem 4 *Suppose $X = G/H$ is a connected homogeneous complex manifold, where G is a complex Lie group which is not solvable and which has a finite number of connected components and H is a closed complex subgroup of G with H° solvable. Then there exist closed complex subgroups $I \subset J$ of G containing H such that I° is solvable, $I^\circ \supset R_G$, and I/H is connected, and J has a finite number of connected components with G/J having a Klein form S_G/A, where A is an algebraic subgroup of the semisimple complex Lie group $S_G := G/R_G$. Moreover, unless $I = J$, then J/I has a Klein form $\hat{S}/\hat{H}$, where $\hat{H}$ is a Zariski dense discrete subgroup of a semisimple complex Lie group $\hat{S}$.*

4. Homogeneous complex spaces G/H with $d_{G/H} = 2$

We now consider the structure of complex homogeneous spaces G/H which satisfy $d_{G/H} = 2$. In this section we discuss such spaces where G is a complex Lie group and H is a closed complex subgroup. The case of complex manifolds homogeneous under the holomorphic action of a real Lie group is treated in the next section. The main focus of our attention will be on situations when the homogeneous space G/H has a nontrivial function algebra, but we consider first the case of a complex Lie group.

4.1. COMPLEX LIE GROUPS

Recall that a compact connected complex Lie group (i.e., $d = 0$) is a torus and thus is abelian. The case $d_G = 1$ corresponds to the Lie group G having two ends, as one easily sees from its Iwasawa decomposition. Complex Lie groups with two ends were considered in [17] and there it was shown that such G are abelian. Note that if $d_G \geq 3$, then G need no longer be abelian, as this condition is satisfied by $SL(2, \mathbb{C})$ and also by the two-dimensional affine group $\text{Aut}(\mathbb{C})$ which is biholomorphic to $\mathbb{C}^* \times \mathbb{C}$ and has a solvable group structure. Thus the observation from [10] that a connected complex Lie groups G with $d_G \leq 2$ is abelian is the best possible. One should also note that this is a statement about complex Lie groups, because an even-dimensional compact Lie group with a left invariant complex structure (but no right invariant complex structure) is not necessarily abelian.

Theorem 5 *Suppose G is a connected complex Lie group with $d_G = 2$. Then G is abelian and G is one of the following:*

a) $\mathbb{C} \times T$, *where T is a torus,*

b) $(\mathbb{C}^*)^2 \times T$, *where T is again a torus,*

c) $\mathbb{C}^* \times$ (*a Cousin group*), *the latter fibering as a $\mathbb{C}^*$-bundle over a torus*

d) *a Cousin group which fibers as a $(\mathbb{C}^*)^2$-bundle over a torus.*

Idea of proof. For a connected complex Lie group G its holomorphic reduction is given by a fibration $G \to G/G_0$, where G_0 is a connected complex central subgroup of G, see [26], which is, in fact, a Cousin group; i.e., $\mathcal{O}(G_0) = \mathbb{C}$. Since the total space, base and fiber of this fibration are Lie groups, it follows directly from properties of the Iwasawa decomposition [22] (i.e., simply count the dimensions of the various maximal compact subgroups) that one has $d_G = d_{G_0} + d_{G/G_0}$. Hence $d_{G/G_0} \le 2$. Now G/G_0 is a holomorphically separable complex Lie group which is thus Stein by the theorem of Matsushima–Morimoto [25]. From this it follows that G/G_0 is trivial, $\mathbb{C}^*, \mathbb{C}$ or $\mathbb{C}^* \times \mathbb{C}^*$. In the first case $G = G_0$ is clearly abelian. In the other cases one has to analyze the possibilities. In the second and fourth cases one tool that helps is the observation that if $G = \tilde{K}$ and H is a closed connected normal solvable complex subgroup of G such that $G/H = (\mathbb{C}^*)^k$, then G is abelian. In the third case one shows directly that a torus bundle over $\mathbb{C}$ splits as a product, see [10]. □

4.2. HOLOMORPHICALLY SEPARABLE SOLV-MANIFOLDS

Holomorphically separable complex solv-manifolds X with $d_X \le 2$ are described in [10].

Theorem 6 *Suppose G is a connected complex solvable Lie group and H is a closed complex subgroup such that G/H is holomorphically separable and $d_{G/H} = 1$ or 2. Then G/H is biholomorphic to $\mathbb{C}, \mathbb{C}^*, \mathbb{C}^* \times \mathbb{C}^*$, or the complex Klein bottle.*

Sketch of proof. Since the homology of a Stein manifold vanishes above the middle (real) dimension, if a connected Stein manifold X with $n :=$ $\dim_{\mathbb{C}} X$ fibers as a vector bundle of rank d over a compact manifold, then $n \le d$. (This idea was the basis of [10, Lemma 2].) Now a holomorphically separable complex solv-manifold is Stein [21] and fibers as a vector bundle over a compact solv-manifold. Hence $n \le d \le 2$ and thus $n = 1$ or 2. Now a check of the various possibilities in the classification of homogeneous Riemann surfaces and homogeneous complex surfaces, see [19], shows that X is biholomorphic to $\mathbb{C}, \mathbb{C}^*, \mathbb{C}^* \times \mathbb{C}^*$, or the complex Klein bottle. □

One can use the result of Loeb [24], together with the results of Winkelmann [33], in order to classify holomorphically separable complex solv-manifolds with $d = 3$. This is substantially more difficult than the cases discussed above and will not be dealt with here at all.

One can also analyze the situation where G is a connected complex solvable Lie group and H is a closed complex subgroup with G/H Kähler and $d_{G/H} = 1$ or 2. This is because of the fact that in this setting the fiber J/H of the holomorphic reduction $G/H \to G/J$ is a Cousin group and its base G/J is Stein (see [30]). Using the inequality $d_{G/H} \geq d_{J/H} + d_{G/J}$ one can easily see what the various possibilities are for this fiber and base.

4.3. HOLOMORPHICALLY SEPARABLE G/Γ

We consider next what happens if G/Γ is holomorphically separable with Γ a discrete subgroup of the connected complex Lie group G. The above observation turns out to be one ingredient which is needed for this. One should consult [3], where a somewhat stronger result is proved in detail.

Theorem 7 *Suppose G is a connected complex Lie group and Γ is a discrete subgroup of G such that G/Γ is holomorphically separable and satisfies $d_{G/\Gamma} = 1$ or 2. Then G is solvable and G/Γ is biholomorphic to $\mathbb{C}, \mathbb{C}^*, \mathbb{C}^* \times \mathbb{C}^*$, or the complex Klein bottle. In particular, $\dim_{\mathbb{C}} G = 1$ or 2.*

Indication of proof. The case when G is solvable is handled by the result outlined above. The case when G is semisimple is easily eliminated, because Γ is algebraic, see [5], and so finite and thus $d_{G/\Gamma} \geq 3$. The proof for mixed groups is more involved. Using Theorem 2 in [8] one can prove the existence of a closed complex subgroup J of G containing Γ and the radical of G. By induction $\dim_{\mathbb{C}} J \leq 2$ and thus the radical R of G has complex dimension at most two. Now one has to analyze two possibilities. If the adjoint action of a maximal semisimple subgroup S of G on the radical R is nontrivial, the complete reducibility of the S-action implies R is abelian. One then shows that the union of all conjugates of S contains a Zariski open set which is the pull-back by the projection of the complement of a proper Zariski closed subset of S. If $G = S \times R$ is a direct product, a theorem of Tits [31] implies that the Zariski closure of the projection of Γ to S is an algebraic subgroup of S whose connected component is solvable. In other words, in each of these two cases one essentially shows that Γ is contained in a proper algebraic subgroup B of G. Again by induction B is solvable and $\dim B \leq 2$. Then using the fact that B is solvable one obtains an inequality of the form $d_{G/\Gamma} \geq d_{G/B}$. Thus $d_{G/B} \leq 2$. If $d_{G/B} = 0$, then B is parabolic in G. Since $\dim B \leq 2$, this contradicts the assumption that G is mixed. If $d_{G/B} = 1$, by using Akhiezer's theorem one finds a parabolic subgroup P of G and shows that $\dim P \leq 2$. This also contradicts the assumption that G is mixed. If $d_{G/B} = 2$, then B/Γ is compact and thus B is discrete. But then $d_G \leq 2$ and so G is abelian, by the theorem above, which is also a contradiction. Hence G has to be solvable.

There are several technical points which we have omitted here. For the precise proof, see Proposition 3 in [3] and its corollaries. □

4.4. THE CASE $\mathcal{O}(G/H)$ NONTRIVIAL

Complex homogeneous spaces $X := G/H$ with $d_X = 2$ and $\mathcal{O}(X) \neq \mathbb{C}$ were described in a joint work with D.N. Akhiezer [3]. We recall next the main result proved there. Note that the situation for $d_X = 1$ and $\mathcal{O}(X) \neq \mathbb{C}$ was described earlier. Because of the reduction to the algebraic category, one should note the connection between the spaces appearing in part b) in the next result and those given by the following construction.

Homogeneous spaces of algebraic groups with $d = 2$ were described in [2]. We note the following method of seeing (geometrically) what these spaces are. Suppose G is a connected linear group and H is an algebraic subgroup of G such that $d_{G/H} = 2$. Now there exists a parabolic subgroup P of G containing H such that P/H is affine. Since $\pi_1(G/P) = 0$, it follows that $d_{P/H} = d_{G/H} = 2$. But P/H is Stein and thus $\dim P/H \leq d_{P/H} = 2$. The list of Stein manifolds with $d = 2$ and which are a quotient of algebraic groups and have dimension one or two is well-known: $\mathbb{C}$, the affine quadric Q_2, $\mathbb{P}_2 - Q$, where Q is a quadric curve and $\mathbb{C}^* \times \mathbb{C}^*$.

Theorem 8 *Suppose G is a connected complex Lie group and H is a closed complex subgroup such that $X := G/H$ satisfies $\mathcal{O}(X) \neq \mathbb{C}$ and $d_X = 2$. Let $X \xrightarrow{F} Y$ be the holomorphic reduction of X. Then one of the following two cases occurs:*

a) *The fiber F of the holomorphic reduction of X is connected with $d_F = 1$ and its base Y is an affine cone with its vertex removed.*

b) *The fiber F of the holomorphic reduction of X is compact and connected and its base Y is one of the following:*

   1) *The complex line $\mathbb{C}$;*
   2) *The affine quadric Q_2;*
   3) *$\mathbb{P}_2 - Q$, where Q is a quadric curve;*
   4) *A homogeneous holomorphic $\mathbb{C}^*$-bundle over an affine cone with its vertex removed which is either itself an algebraic principal $\mathbb{C}^*$-bundle or is covered two-to-one by such.*

Idea of proof. Let $G/H \to G/J$ be the holomorphic reduction of G/H and set $\tilde{J} := H \cdot J^\circ$, $N := N_G(J^\circ)$ and $\tilde{N} := \tilde{J} \cdot N^\circ$. By making use of the fact that $G/\tilde{N}G'$ is an abelian Lie group, the $R_{G'}$-orbits in $G'/G' \cap \tilde{N}$ are closed and $S \cap NR_{G'}$ is an algebraic subgroup of S along with repeated use of the fibration lemma, one can show (e.g., see Lemma 9 in [3]) that

$$d_{G/H} \geq d_{\tilde{J}/H} + d_{\tilde{N}/\tilde{J}} + d_{R_{G'}/R_{G'} \cap \tilde{N}} + d_{S/S \cap \tilde{N}R_{G'}} + d_{G/\tilde{N}G'}.$$

The sum $\Sigma = d_{\tilde{N}/\tilde{J}} + d_{R_{G'}/R_{G'} \cap \tilde{N}} + d_{S/S \cap \tilde{N}R_{G'}} + d_{G/\tilde{N}G'}$ of the last four terms on the right hand side of the above inequality is not equal to zero. Otherwise, G/J would be compact and thus a point, contradicting our assumption that $\mathcal{O}(G/H) \neq \mathbb{C}$. Therefore, if $d_X = 2$ and $\mathcal{O}(X) \neq \mathbb{C}$, it follows that $\Sigma = 1$ or 2. We look at these two cases separately.

If $\Sigma = 1$, then one can use this together with the fact that $G/\tilde{J}$ satisfies the maximal rank condition to prove that $G/\tilde{J}$ has two ends and thus is an affine cone minus its vertex. And from the general theory of fiber bundles one can easily see that it is not possible for $\tilde{J}/H$ to be compact and $\Sigma = 1$. For, in this case G/H would be homeomorphic to the product of some compact manifold with $\mathbb{R}$. But such a product would have $d_X = 1$, a contradiction!

The remaining case occurs when $\tilde{J}/H$ is compact and $\Sigma = 2$. If the base G/N of the normalizer fibration $G/J \to G/N$ is compact, then G/N is a projective rational manifold and so it follows from the fibration lemma that $d_{N/\tilde{J}} = 2$. Thus $N/\tilde{J}$ is biholomorphic to $\mathbb{C}, \mathbb{C}^* \times \mathbb{C}^*$ or the complex Klein bottle. In each of these settings it is relatively straightforward to show that either there is a transitive holomorphic action of $S \times \mathbb{C}^*$ on $G/\tilde{J}$ or on a two-to-one covering $G/\hat{J}$ of $G/\tilde{J}$, where S is a maximal semisimple subgroup of G, or else G/N is a point. In this last situation $G/\tilde{J}$ is itself $\mathbb{C}, \mathbb{C}^* \times \mathbb{C}^*$ or the complex Klein bottle, as the case may be.

The situation when G/N is not compact is more involved. One first proves that if the connected fiber $\tilde{N}/\tilde{J}$ of the normalizer fibration $\tilde{Y} := G/\tilde{J} \to G/\tilde{N}$ is biholomorphic to $\mathbb{C}^*$ and its base $G/\tilde{N}$ has two ends with $\dim G/\tilde{N} > 1$, then a group of the form $S \times \mathbb{C}^*$ is transitive on $\tilde{Y}$ or a two-to-one covering of $\tilde{Y}$. If G' is not transitive on G/N, then we have the fibration $G/\tilde{J} \to G/\tilde{N}G' = \mathbb{C}^k \times (\mathbb{C}^*)^l$ and thus $d_{G/\tilde{N}G'} = 2k + l > 0$, by assumption. Recall

$$\Sigma = d_{\tilde{N}/\tilde{J}} + d_{R_{G'}/R_{G'} \cap \tilde{N}} + d_{S/S \cap \tilde{N} R_{G'}} + d_{G/\tilde{N}G'}.$$

The sum of the first three terms on the right hand side of this equation cannot equal zero. For, if it were, then the corresponding spaces would be compact, and thus points, since $\tilde{Y}$ satisfies the maximal rank condition. It is easy to see that this would imply $G' \subset J^\circ$ and thus $N = G$, contrary to the assumption that G' is not transitive on G/N. It follows that this sum equals one. Hence there exists a homogeneous fibration $\tilde{N}G'/\tilde{J} \xrightarrow{\mathbb{C}^*} \tilde{N}G'/\tilde{N}$, where $\tilde{N}G'/\tilde{N} = G'/G' \cap \tilde{N}$ is a homogeneous projective rational manifold. As well, it is clear that $k = 0$ and $l = 1$. Since one has the fibration $G/\tilde{N} \to G/\tilde{N}G' = \mathbb{C}^*$ and its fiber is compact, it follows that $G/\tilde{N}$ has two ends. If $\dim G/\tilde{N} > 1$, the result follows from above. Otherwise, $\tilde{Y}$ is a homogeneous surface which, under our assumptions, is biholomorphic to either $\mathbb{C}^* \times \mathbb{C}^*$ or the complex Klein bottle.

Now assume G' is transitive on G/N. We claim first that the covering

$$G/\tilde{N} = G'/G' \cap \tilde{N} \to G'/G' \cap N = G/N$$

is finite in this setting. Since $R_{G'}$ is a unipotent group, its orbits in G/N and in $G/\tilde{N}$ are both biholomorphic to $\mathbb{C}^k$, for some k, i.e., $R_{G'}/R_{G'} \cap \tilde{N} = \mathbb{C}^k = R_{G'}/R_{G'} \cap N$. Also the covering $S/S \cap \tilde{N}R_{G'} \to S/S \cap NR_{G'}$ is finite, since each of these spaces is a quotient of algebraic groups and the two isotropy subgroups have the same connected component of the identity. As a consequence, the covering $G/\tilde{N} \to G/N$ is also finite.

Since G' is transitive on $G/\tilde{N}$, one has $d_{G/\tilde{N}G'} = 0$. Hence

$$2 = \Sigma = d_{\tilde{N}/\tilde{J}} + d_{R_{G'}/R_{G'} \cap \tilde{N}} + d_{S/S \cap \tilde{N}R_{G'}}. \tag{1}$$

Since $G/\tilde{N}$ is not compact, by assumption,

$$\delta := d_{R_{G'}/R_{G'} \cap \tilde{N}} + d_{S/S \cap \tilde{N}R_{G'}} \neq 0.$$

From (1) it follows that $\delta = 1$ or 2.

Now suppose $\delta = 1$. Then $d_{\tilde{N}/\tilde{J}} = 1$ and from the discrete case one concludes that $\tilde{N}/\tilde{J} = \mathbb{C}^*$, as a manifold. As well, $d_{R_{G'}/R_{G'} \cap \tilde{N}} + d_{S/S \cap NR_{G'}} = 1$. Since $R_{G'}$ is a unipotent group, its orbits are biholomorphic to $\mathbb{C}^k$ and thus $d_{R_{G'}/R_{G'} \cap \tilde{N}} = 2k$. Hence $k = 0$ and $R_{G'}$ acts trivially on $G/\tilde{N}$. Thus S is transitive on $G/\tilde{N}$ and $d_{G/\tilde{N}} = d_{S/S \cap \tilde{N}} = 1$. Since $S/S \cap \tilde{N}$ is the quotient of algebraic groups, $G/\tilde{N}$ has two ends. Note that because of the fact that S is transitive on $G/\tilde{N}$, it follows that $\dim G/\tilde{N} > 1$. (A semisimple group S cannot act transitively on $\mathbb{C}^*$.) The result is now a consequence of the above.

If $\delta = 2$, then $d_{\tilde{N}/\tilde{J}} = 0$, i.e., $\tilde{N}/\tilde{J}$ is compact. Because $\tilde{Y}$ satisfies the maximal rank condition, $\tilde{N} = \tilde{J}$. As noted above, the covering $G/\tilde{N} \to G/N$ is finite. But $N \supset J \supset \tilde{J}$ and thus the covering $G/J \to G/N$ is also finite. If follows that G/N is holomorphically separable and because $G/N = G'/G' \cap N$ is the quotient of algebraic groups, the result now follows from the classification in the case of algebraic groups.

For details see [3]. $\quad\square$

5. Real groups

The purpose of this section is to present the classification of homogeneous complex manifolds $X = G/H$, where G is a connected (real) Lie group, H is a closed subgroup and G/H has a G-invariant complex structure, when $d_X = 2$ and $\mathcal{O}(X) \neq \mathbb{C}$. We show that either a complex Lie group is acting transitively on the base of the holomorphic reduction of X or else the base of this holomorphic reduction is biholomorphic to the unit disk. The idea of the proof is to assume that the former does not occur and then show that one has the unit disk. The methods used are similar to those in [3], along with a result from [16], see Corollary 1.

We set up some notation. Suppose G/H is a homogeneous complex manifold (with real or complex groups) and assume $\mathcal{O}(G/H) \neq \mathbb{C}$. Let $G/H \to G/I$ be the holomorphic reduction of G/H and set $\tilde{I} = H \cdot I^\circ$. Also let $N := N_G(I^\circ)$, $\tilde{N} := \tilde{I} \cdot N^\circ$ and $\eta : \tilde{N} \to \tilde{N}/I^\circ$ be the canonical map. Put $\Gamma := \eta(\tilde{I}) = \tilde{I}/I^\circ$. Now suppose Q is any closed subgroup of $\tilde{N}$ which contains $\tilde{I}$ and which has the property that $L := \eta(\tilde{Q})$ is a complex Lie group, where $\tilde{Q} := \tilde{I} \cdot Q^\circ$. (We apply this when $G/I \to G/Q$ is the $\mathfrak{g}$-anticanonical fibration of G/I; the subgroup Q has the required property, see [20], i.e., $\tilde{Q}/I^\circ$ is a complex Lie group.) Now we have the following two fibrations

$$G/H \to G/\tilde{I} \overset{L/\Gamma}{\to} G/\tilde{Q},$$

where both of the fibers $\tilde{I}/H$ and L/Γ are connected complex manifolds. The analogue of Proposition 3 in [3] holds in this setting.

Lemma 3 *Suppose* $d_{G/H} \leq 2$ *and* $\mathcal{O}(G/H) \neq \mathbb{C}$. *Then with the above set-up, it follows that the group* L° *is solvable and* L/Γ *is biholomorphic to* $\mathbb{C}^*, \mathbb{C}, \mathbb{C}^* \times \mathbb{C}^*$ *or the complex Klein bottle. In particular,* $\dim_{\mathbb{C}} L \leq 2$.

5.1. A FIBRATION LEMMA

As above, $G/H \to G/I$ is the holomorphic reduction of G/H. Let $G/I \to G/J$ be the $\mathfrak{g}$-anticanonical fibration of G/I. Set $\tilde{J} := I \cdot J^\circ$. The base G/J of this fibration is an open orbit in the orbit of the complexification $G^\mathbb{C}$ of G, i.e., $G/J \hookrightarrow G^\mathbb{C}/J^\mathbb{C}$, see [20]. We will study the structure of G/J and its covering space $G/\tilde{J}$ by considering fibrations which are induced by fibrations of the complex orbit $G^\mathbb{C}/J^\mathbb{C}$. Let $S^\mathbb{C}$ be a maximal semisimple subgroup of $G^\mathbb{C}$. Then $S^\mathbb{C}$ is also a maximal semisimple subgroup of $(G^\mathbb{C})'$. The group $(G^\mathbb{C})'$ acts as an algebraic group on projective space, see Chevalley's theorem [6], and so it has closed orbits in $G^\mathbb{C}/J^\mathbb{C}$. The base of the fibration given by the orbits of the commutator subgroup is $G/U = G^\mathbb{C}/J^\mathbb{C} \cdot (G^\mathbb{C})'$, see diagram (2), where $U := G \cap J^\mathbb{C} \cdot (G^\mathbb{C})'$.

$$
\begin{array}{ccccccc}
G/H & \longrightarrow & G/\tilde{I} & \longrightarrow & G/J & \longrightarrow & G^\mathbb{C}/J^\mathbb{C} \\
 & & & & \downarrow & & \downarrow \ (G^\mathbb{C})'/(G^\mathbb{C})' \cap J^\mathbb{C} \quad (2) \\
 & & & & G/U & = & G^\mathbb{C}/J^\mathbb{C} \cdot (G^\mathbb{C})'
\end{array}
$$

Further, for the $(G^{\mathbb{C}})'$-orbits one has the following decompositions, which are explained below.

$$U/J \longrightarrow (G^{\mathbb{C}})'/(G^{\mathbb{C}})' \cap J^{\mathbb{C}} = J^{\mathbb{C}} \cdot (G^{\mathbb{C}})'/J^{\mathbb{C}}$$

$$L/J \downarrow \qquad\qquad \downarrow R_{(G^{\mathbb{C}})'}/R_{(G^{\mathbb{C}})'} \cap J^{\mathbb{C}}$$

$$U/L = S/S \cap L \longrightarrow S^{\mathbb{C}}/S^{\mathbb{C}} \cap J^{\mathbb{C}} R_{(G^{\mathbb{C}})'} = J^{\mathbb{C}} \cdot (G^{\mathbb{C}})'/J^{\mathbb{C}} \cdot R_{(G^{\mathbb{C}})'} \qquad (3)$$

$$M/M \cap L \downarrow \qquad\qquad \downarrow M/M \cap J^{\mathbb{C}} R_{(G^{\mathbb{C}})'}$$

$$S/M \cdot (S \cap L) \longrightarrow S^{\mathbb{C}}/M \cdot (S^{\mathbb{C}} \cap J^{\mathbb{C}} R_{(G^{\mathbb{C}})'})$$

Since $(G^{\mathbb{C}})'$ acts algebraically on $G^{\mathbb{C}}/J^{\mathbb{C}}$, the orbits of its radical $R_{(G^{\mathbb{C}})'}$ are closed in $(G^{\mathbb{C}})'/(G^{\mathbb{C}})' \cap J^{\mathbb{C}}$, and thus also in $G^{\mathbb{C}}/J^{\mathbb{C}}$, and one has the top part of diagram (3), where $L := U \cap J^{\mathbb{C}} R_{(G^{\mathbb{C}})'} = G \cap J^{\mathbb{C}} R_{(G^{\mathbb{C}})'}$. Let $\psi : G^{\mathbb{C}} \to S^{\mathbb{C}}$ be the natural quotient map. Set $S := \psi(U)$ and let $\mathfrak{s}$ denote the Lie algebra of S. Finally, let M be the complex normal subgroup in $S^{\mathbb{C}}$ corresponding to the ideal $\mathfrak{s} \cap i\mathfrak{s}$ in the Lie algebra $\mathfrak{s}^{\mathbb{C}}$ of $S^{\mathbb{C}}$. Note that $M \subset S$. The group M has closed orbits, because it is a product of some of the simple factors of $S^{\mathbb{C}}$. This gives rise to the lower part of diagram (3).

Lemma 4 *With the above notation assume further that $\tilde{J}/\tilde{I}$ has the homotopy type of a CW-complex of dimension q. Then*

$$d_{G/H} \geq d_{\tilde{I}/H} + (\dim \tilde{J}/\tilde{I} - q) + d_{L/\tilde{J}} + d_{M/M \cap L} + d_{S/M \cdot (S \cap L)} + d_{G/U}. \quad (4)$$

Moreover, if $d_{G/H} \leq 2$ and $\mathcal{O}(G/H) \neq \mathbb{C}$, then we can replace $\dim \tilde{J}/\tilde{I} - q$ by $d_{\tilde{J}/\tilde{I}}$.

The proof is a consequence of applying the fibrations described above, along with part 2) of Lemma 1, and is based on the proof of Lemma 9 in [3]. For details see [14].

5.2. AN INEQUALITY FOR SOLV-MANIFOLDS

We also consider what happens when G is solvable. Recall that any solv-manifold X admits a fibration $X \to M_X$ (see Auslander–Tolimieri [4] and Mostow [28]), where M_X is a compact solv-manifold and the fiber is a vector space of real dimension d_X. We now relate this to the complex structure in the case when the function algebra of X satisfies the maximal rank condition. In [25] Matsushima–Morimoto showed that a holomorphically separable complex Lie group G is Stein. As a consequence, $d_G \geq \dim_{\mathbb{C}} G$. Also Huckleberry and Oeljeklaus [21] proved that if G/H is holomorphically separable, where G is a complex solvable Lie group, then G/H is Stein and so $d_{G/H} \geq \dim_{\mathbb{C}} G/H$. But a holomorphically separable solv-manifold of a real Lie group need not be Stein, e.g., $\mathbb{C}^2 - \mathbb{R}^2$ is not Stein. However, the inequality $d_{G/H} \geq \dim_{\mathbb{C}} G/H$ does hold for solv-manifolds G/H of real Lie groups provided G/H satisfies the maximal rank condition.

Proposition 1 *Suppose G is a connected solvable Lie group, H is a closed subgroup such that $X := G/H$ has a left G-invariant complex structure and with this structure X satisfies the maximal rank condition. Then $d_X \geq \dim_{\mathbb{C}} X$.*

The proof is by induction on the dimension of the space G/H and can be found in [14].

The following is a technical observation about the space $L/\tilde{J}$ which is used to prove the classification result.

Lemma 5 *The space $L/\tilde{J}$ satisfies*

$$d_{L/\tilde{J}} \geq \dim_{\mathbb{C}} L/\tilde{J} + 1. \tag{5}$$

As a consequence,

a) *it is not possible for $d_{L/\tilde{J}}$ to equal 1, and*

b) *if $d_{L/\tilde{J}} = 2$, U/L is compact, and there is no complex Lie group acting transitively on $U/\tilde{J}$, then $L/\tilde{J}$ is biholomorphic to the unit disk, and*

c) *moreover, if $U/\tilde{J}$ satisfies the maximal rank condition, then $L = U$ and thus $U/\tilde{J}$ is biholomorphic to the unit disk.*

The proof of this can also be found in [14].

5.3. THE CLASSIFICATION

Note that the case $d_{G/H} = 1$ and $\mathcal{O}(G/H) \neq \mathbb{C}$ was handled previously, see Theorem 2. It follows easily from that result that $e(G/H) = 2$ and thus the holomorphic reduction of G/H has compact connected fiber and its base is an affine cone with its vertex removed.

Theorem 9 *Suppose G is a connected Lie group and H is a closed subgroup such that $X := G/H$ has an invariant complex structure, satisfies $\mathcal{O}(X) \neq \mathbb{C}$ and $d_X = 2$. Let $Y := G/I$ be the base of the holomorphic reduction of X and $F := I/H$ be its fiber. Then one of the following two cases occurs:*

a) *The fiber F is connected and satisfies $d_F = 1$ and the base Y is an affine cone minus its vertex.*

b) *The fiber F is compact and connected and $d_Y = 2$; moreover, either a complex Lie group is transitive on Y and thus Y is one of the spaces described in Theorem 8 or else Y is biholomorphic to the unit disk Δ. In this latter case, X is biholomorphic to $F \times \Delta$.*

The proof of this result is rather complicated. One uses Lemma 4 in much the same way that its analogue is used in the proof of Theorem 8. A result from [16] about the lifting of holomorphic actions in principal bundles allows one to restrict one's attention to the setting where Y does not admit the transitive action of a complex Lie group. The interested reader can find the details in [14].

References

1. Abels, H.: Some topological aspects of proper group actions; noncompact dimension of groups, *J. London Math. Soc.* **25** (1982) no. 2, 525–538.
2. Akhiezer, D.N.: Dense orbits with two ends, *Izv. AN SSSR, Ser. matem.* **41** (1977) 308–324.
3. Akhiezer, D.N. and Gilligan, B.: On complex homogeneous spaces with top homology in codimension two, *Canad. J. Math.* **46** (1994) 897–919.
4. Auslander, L. and Tolimieri, R.: Splitting theorems and the structure of solvmanifolds, *Ann. of Math.* **92** (1970) no. 2, 164–173.
5. Barth, W. and Otte, M.: Invariante holomorphe Funktionen auf reduktiven Liegruppen, *Math. Ann.* **201** (1973) 97–112.
6. Chevalley, C.: *Théorie des groupes de Lie II: Groupes algébriques*, Hermann, Paris, 1951.
7. Freudenthal, H.: Über die Enden topologischer Räume und Gruppen, *Math. Z.*, **33** (1931) 692–713.
8. Gilligan, B.: Ends of homogeneous complex manifolds having nonconstant holomorphic functions, *Arch. Math.* **37** (1981) 544–555.
9. Gilligan, B.: On the ends of complex manifolds homogeneous under a Lie group, *Proc. Sympos. Pure Math.* **52** (1991) Part 2, 217–224.
10. Gilligan, B.: On a topological invariant of complex Lie groups and solv-manifolds, *C.R. Math. Rep. Acad. Sci. Canada* **14** (1992) 109–114.
11. Gilligan, B.: On homogeneous complex manifolds having more than two ends, *C.R. Math. Rep. Acad. Sci. Canada* **15** (1993) 29–34.
12. Gilligan, B.: Comparing two topological invariants for homogeneous complex manifolds, *C.R. Math. Rep. Acad. Sci. Canada* **16** (1994) 155–160.
13. Gilligan, B.: *Structure of complex homogeneous spaces with respect to topological invariants*, Schriftenreihe, Heft Nr. 218, Forschungsschwerpunkt Komplexe Mannigfaltigkeiten, Bochum, 1994.
14. Gilligan, B.: Complex homogeneous spaces of real groups with top homology in codimension two, *Ann. Global Anal. Geom.* **13** (1995) 303–314.
15. Gilligan, B.: On closed radical orbits in certain homogeneous complex manifolds, *Bull. Austr. Math. Soc.* **54** (1996), to appear.
16. Gilligan, B. and Heinzner, P.: Globalization of holomorphic actions on principal bundles, preprint, 1995.
17. Gilligan, B. and Huckleberry, A.T.: Complex homogeneous manifolds with two ends, *Michigan Math. J.*, **28** (1981) 183–198.
18. Gilligan, B., Oeljeklaus, K. and Richthofer, W.: Homogeneous complex manifolds with more than one end, *Canad. J. Math.* **41** (1989) 163–177.
19. Huckleberry, A.T. and Livorni, L.: A classification of homogeneous surfaces, *Canad. J. Math.* **33** (1981) 1097–1110.
20. Huckleberry, A.T. and Oeljeklaus, E.: Classification Theorems for Almost Homogeneous Spaces, *Publ. Inst. E. Cartan*, **9**, Nancy, 1984.
21. Huckleberry, A.T. and Oeljeklaus, E.: On holomorphically separable complex solv-manifolds, *Ann. Inst. Fourier* **36** (1986) 57–65.

22. Iwasawa, K.: On some types of topological groups, *Ann. of Math.* **50** (1949) 507–558.
23. Karpelevich, F.I.: On a fibration of homogeneous spaces, *Uspekhi Matem. Nauk* **11** (1956) no. 3 (69), 131–138 (in Russian).
24. Loeb, J-J.: Action d'une forme réelle d'un groupe de Lie complexe sur les fonctions plurisousharmoniques, *Ann. Inst. Fourier* **35** (1985) 49–87.
25. Matsushima, Y. and Morimoto, A.: Sur certains espaces fibrés holomorphes sur une variété de Stein, *Bull. Soc. Math. France* **88** (1960) 137–155.
26. Morimoto, A.: Noncompact complex Lie groups without nonconstant holomorphic functions, *Proc. Conf. in Complex Analysis*, Minneapolis (1964) 256–272.
27. Mostow, G.D.: On covariant fiberings of Klein spaces, I, II, *Amer. J. Math.* **77** (1995) 247–278; **84** (1962) 466–474.
28. Mostow, G.D.: Some applications of representative functions to solv-manifolds, *Amer. J. Math.* **93** (1971) 11–32.
29. Oeljeklaus, K. and Richthofer, W.: Homogeneous complex surfaces, *Math. Ann.* **268** (1984) 273–292.
30. Oeljeklaus, K. and Richthofer, W.: On the structure of complex solvmanifolds, *J. Diff. Geom.* **27** (1988) 399–421.
31. Tits, J. Free subgroups in linear groups, *J. of Algebra* **20** (1972) no. 2, 250–270.
32. Wang, H-C.: One-dimensional cohomology group of locally compact metrically homogeneous spaces, *Duke Math. J.* **19** (1952) 303–310.
33. Winkelmann, J.: The classification of three-dimensional homogeneous complex manifolds, dissertation, Bochum, 1987.
34. Yau, S.T.: Remarks on the group of isometries of a Riemannian manifold, *Topology* **16** (1977) 239–247.

MICROMODULES

BORIS KOMRAKOV AND ALEXEI TCHURYUMOV
International Sophus Lie Center,
P.O. Box 70, Minsk, 220123, Belarus

Abstract. The modules described in this paper are a natural generalization of microweights in the sense of Bourbaki [1], and are encountered quite often as, for example, odd components of simple Lie superalgebras, isotropic representations of symmetric spaces and homogeneous spaces of semisimple Lie groups with stationary subgroups of maximal rank, etc. They also appear in isotropically irreducible homogeneous spaces and in inclusions of simple irreducible subalgebras of simple classical Lie algebras.

Mathematics subject classification (1991): 17B10.

Key words: semisimple Lie algebras, irreducible representations, simple weights, microweights.

Classification of micromodules

Definition. Let $\mathfrak{g}$ be a reductive complex Lie algebra and V a semisimple finite-dimensional $\mathfrak{g}$-module. The module V is called a *micromodule*, if all its nonzero weights are simple, that is, if they all have multiplicity 1. We say that a micromodule is *strict* (respectively *nonstrict*), if all its weights are simple (respectively if its zero weight has multiplicity > 1).

The next theorem reduces the study of simple micromodules over reductive Lie algebras to the study of simple micromodules over commutative and simple Lie algebras.

Proposition *Let $(\mathfrak{g}, \mathfrak{h})$ be a split reductive Lie algebra, and V a simple $\mathfrak{g}$-module. Suppose $\mathfrak{g} = \oplus_{i=0}^{r}\mathfrak{g}_i$ is a decomposition of the Lie algebra $\mathfrak{g}$ into a direct sum of its center $\mathfrak{g}_0$ and simple ideals $\mathfrak{g}_i$ $(1 \leqslant i \leqslant r)$, $h = \oplus_{i=0}^{r}\mathfrak{h}_i$ is the corresponding decomposition of the Cartan subalgebra, $R = R(\mathfrak{g}, \mathfrak{h})$*

B. P. Komrakov et al. (eds.), Lie Groups and Lie Algebras, 181–198.
© 1998 *Kluwer Academic Publishers. Printed in the Netherlands.*

is a root system of $(\mathfrak{g}, \mathfrak{h})$, B is a basis of the system R, ω is the highest weight of the $\mathfrak{g}$-module V with respect to the basis B, and ω_i $(1 \leqslant i \leqslant r)$ is the restriction of ω to the subalgebra $\mathfrak{h}_i$. Further, let V_0 denote the simple $\mathfrak{g}_0$-module corresponding to the weight ω_0, and V_i $(1 \leqslant i \leqslant r)$ the simple $\mathfrak{g}_i$-module whose highest weight with respect to the basis $B_i = B \cap R(\mathfrak{g}_i, \mathfrak{h}_i)$ is ω_i. We can regard the $\mathfrak{g}_i$-module V_i $(0 \leqslant i \leqslant r)$ as a simple $\mathfrak{g}$-module by extending its action to $\mathfrak{g}_j$ $(j \neq i)$ in a trivial way.

(a) The $\mathfrak{g}$-module V is isomorphic to the tensor product of the $\mathfrak{g}$-modules V_i $(0 \leqslant i \leqslant r)$.
(b) In order that the $\mathfrak{g}$-module V be a strict micromodule, it is necessary and sufficient that all $\mathfrak{g}_i$-modules V_i be strict micromodules.
(c) In order that the $\mathfrak{g}$-module V be a nonstrict micromodule, it is necessary and sufficient that exactly one $\mathfrak{g}_{i_0}$-module V_{i_0} $(0 \leqslant i_0 \leqslant r)$ be a nonstrict micromodule, while all others be one-dimensional.

Proof. (a) Let e_0 be a nonzero element of V_0, and e_i $(0 \leqslant i \leqslant r)$ a primitive element of V_i whose weight is equal to ω_i. It is clear that the tensor $e = \otimes e_i$ is a primitive element of the $\mathfrak{g}$-module $\otimes_{i=0}^r V_i$, and its weight is equal to $\omega = \sum \omega_i$. Let $(e_{k_i}^i)_{k_i \in K_i}$ $(0 \leqslant i \leqslant r)$ be a weight basis of V_i. It is easy to show that every tensor $\otimes_{i=0}^r e_{k_i}^i$ lies in the simple submodule of the $\mathfrak{g}$-module $\otimes_{i=0}^r V_i$ that contains e. Therefore $\otimes_{i=0}^r V_i$ is a simple $\mathfrak{g}$-module, w being its highest weight.

(b)–(c) The proof is obvious. □

Theorem 1 *A nontrivial simple micromodule over a simple Lie algebra is determined by one of the following diagrams:*

(1) $\overset{t}{\underset{\alpha_1 \ \alpha_2}{\circ\!\!-\!\!\circ\ \ldots\!-\!\circ}}\ \underset{\alpha_l}{}$ $(l \geqslant 1,\ t \geqslant 1)$

(2) $\underset{\alpha_1}{\circ}\!-\ \ldots\!-\!\circ\!\!-\!\!\overset{t}{\underset{\alpha_l}{\circ}}\ \underset{\alpha_{l-1}}{}$ $(l \geqslant 1,\ t \geqslant 1)$

$\left.\begin{array}{l}\\ \\ \\ \end{array}\right\}$ dim $V^0 = 1$, *if* $l+1$ *divides* t, dim $V^0 = 0$, *if* $l+1$ *does not divide* t;

(3) $\underset{\alpha_1}{\circ}\!-\ \ldots\!-\!\underset{\alpha_{t-1}}{\circ}\!\!-\!\!\overset{1}{\underset{\alpha_t}{\circ}}\!\!-\!\!\underset{\alpha_{t+1}}{\circ}\ \ldots\!-\!\underset{\alpha_l}{\circ}$ $(1 < t < l)$ dim $V^0 = 0$;

(4) $\overset{1}{\underset{\alpha_1}{\circ}}\!\!-\!\!\underset{\alpha_2}{\circ}\ \ldots\!-\!\circ\!\!-\!\!\overset{1}{\underset{\alpha_l}{\circ}}\ \underset{\alpha_{l-1}}{}$ $(l \geqslant 2)$ dim $V^0 = l$;

(5) $\underset{\alpha_1}{\circ}\!\!-\!\!\overset{2}{\underset{\alpha_2}{\circ}}\!\!-\!\!\underset{\alpha_3}{\circ}$ dim $V^0 = 2$;

(6) $\overset{1}{\underset{\alpha_1}{\circ}}\!\!-\!\!\underset{\alpha_2}{\circ}\ \ldots\!-\!\circ\!\!\Rightarrow\!\!\circ\ \underset{\alpha_{l-1}\ \alpha_l}{}$ $(l \geqslant 2)$ dim $V^0 = 1$;

(7) $\underset{\alpha_1}{\circ}\!\!-\!\!\underset{\alpha_2}{\circ}\ \ldots\!-\!\circ\!\!\Rightarrow\!\!\overset{1}{\underset{\alpha_l}{\circ}}\ \underset{\alpha_{l-1}}{}$ $(l \geqslant 2)$ dim $V^0 = 0$;

(8)　$\underset{\alpha_1\ \ \alpha_2\ \ \alpha_3\ \ \ \ \ \ \alpha_{l-1}\ \alpha_l}{\circ\overset{1}{-\!\!-}\circ-\!\!-\circ-\ \ldots\ -\circ\!\!\Rightarrow\!\!\circ}$　$(l \geqslant 3)$　　　　$\dim V^0 = l;$

(9)　$\underset{\alpha_1\ \ \alpha_2\ \ \ \ \ \alpha_{l-1}\ \alpha_l}{\overset{2}{\circ}-\!\!-\circ-\ \ldots\ -\circ\!\!\Rightarrow\!\!\circ}$　$(l \geqslant 2)$　　　　$\dim V^0 = l;$

(10)　$\underset{\alpha_1\ \ \alpha_2\ \ \ \ \ \alpha_{l-1}\ \alpha_l}{\overset{1}{\circ}-\!\!-\circ-\ \ldots\ -\circ\!\!\Leftarrow\!\!\circ}$　$(l \geqslant 3)$　　　　$\dim V^0 = 0;$

(11)　$\underset{\alpha_1\ \ \alpha_2\ \ \alpha_3}{\circ-\!\!-\circ\!\!\overset{1}{\Leftarrow}\!\!\circ}$　　　　　$\dim V^0 = 0;$

(12)　$\underset{\alpha_1\ \ \alpha_2\ \ \alpha_3\ \ \ \ \ \alpha_{l-1}\ \alpha_l}{\circ\overset{1}{-\!\!-}\circ-\!\!-\circ-\ \ldots\ -\circ\!\!\Leftarrow\!\!\circ}$　$(l \geqslant 3)$　　　　$\dim V^0 = l - 1;$

(13)　$\underset{\alpha_1\ \ \alpha_2\ \ \ \ \ \alpha_{l-1}\ \alpha_l}{\overset{2}{\circ}-\!\!-\circ-\ \ldots\ -\circ\!\!\Leftarrow\!\!\circ}$　$(l \geqslant 2)$　　　　$\dim V^0 = l;$

(14)　$\underset{\alpha_1\ \ \alpha_2\ \ \alpha_3\ \ \alpha_4}{\circ-\!\!-\circ-\!\!-\circ\!\!\overset{1}{\Leftarrow}\!\!\circ}$　　　　　$\dim V^0 = 2;$

(15)　$\underset{\alpha_1\ \ \alpha_2}{\overset{1}{\circ}-\!\!-\circ-\ \ldots\ -\circ}\!\!\Big\langle{\!\!\!\overset{\circ}{}\atop\!\!\!\underset{\alpha_l}{\circ}}^{\alpha_{l-1}}$　$(l \geqslant 4)$　　　　$\dim V^0 = 0;$

(16)　$\underset{\alpha_1}{\circ}-\ \ldots\ -\circ\!\!\Big\langle{\!\!\!\overset{\overset{1}{\circ}}{}\atop\!\!\!\underset{\alpha_l}{\circ}}^{\alpha_{l-1}}$　$(l \geqslant 4)$　　　　$\dim V^0 = 0;$

(17)　$\underset{\alpha_1}{\circ}-\ \ldots\ -\circ\!\!\Big\langle{\!\!\!\overset{\circ}{}\atop\!\!\!\underset{\alpha_l}{\circ}}\!\!\underset{1}{}\,\alpha_{l-1}$　$(l \geqslant 4)$　　　　$\dim V^0 = 0;$

(18)　$\underset{\alpha_1\ \ \alpha_2\ \ \alpha_3}{\circ\overset{1}{-\!\!-}\circ-\!\!-\circ-\ \ldots\ -\circ}\!\!\Big\langle{\!\!\!\overset{\circ}{}\atop\!\!\!\underset{\alpha_l}{\circ}}^{\alpha_{l-1}}$　$(l \geqslant 4)$　　　　$\dim V^0 = l;$

(19)　$\underset{\alpha_1\ \ \alpha_2}{\overset{2}{\circ}-\!\!-\circ-\ \ldots\ -\circ}\!\!\Big\langle{\!\!\!\overset{\circ}{}\atop\!\!\!\underset{\alpha_l}{\circ}}^{\alpha_{l-1}}$　$(l \geqslant 4)$　　　　$\dim V^0 = l - 1;$

(20)　$\overset{1}{\circ}-\!\!-\circ-\!\!-\underset{|}{\circ}-\!\!-\circ-\!\!-\circ$ (with a node below the middle)　　　　$\dim V^0 = 0;$

(21)　$\circ-\!\!-\circ-\!\!-\underset{|}{\circ}-\!\!-\circ-\!\!-\overset{1}{\circ}$ (with a node below the middle)　　　　$\dim V^0 = 0;$

(22)　$\circ-\!\!-\circ-\!\!-\underset{|_1}{\circ}-\!\!-\circ-\!\!-\circ$ (with a node below the middle)　　　　$\dim V^0 = 6;$

(23)　$\circ-\!\!-\circ-\!\!-\underset{|}{\circ}-\!\!-\circ-\!\!-\circ-\!\!-\overset{1}{\circ}$ (with a node below the middle)　　　　$\dim V^0 = 0;$

(24)　$\overset{1}{\circ}-\!\!-\circ-\!\!-\underset{|}{\circ}-\!\!-\circ-\!\!-\circ-\!\!-\circ$ (with a node below the middle)　　　　$\dim V^0 = 7;$

(25)　$\circ-\!\!-\circ-\!\!-\underset{|}{\circ}-\!\!-\circ-\!\!-\circ-\!\!-\circ-\!\!-\overset{1}{\circ}$ (with a node below the middle)　　　　$\dim V^0 = 8;$

(26) $\underset{\alpha_1 \quad \alpha_2 \quad \alpha_3 \quad \alpha_4}{\overset{1}{\circ}\!\!-\!\!\circ\!\!\Rightarrow\!\!\circ\!\!-\!\!\circ}$ $\dim V^0 = 4;$

(27) $\underset{\alpha_1 \quad \alpha_2 \quad \alpha_3 \quad \alpha_4}{\circ\!\!-\!\!\circ\!\!\Rightarrow\!\!\circ\!\!-\!\!\overset{1}{\circ}}$ $\dim V^0 = 2;$

(28) $\underset{\alpha_1 \quad \alpha_2}{\overset{1}{\circ}\!\!\Lleftarrow\!\!\circ}$ $\dim V^0 = 1;$

(29) $\underset{\alpha_1 \quad \alpha_2}{\circ\!\!\Lleftarrow\!\!\overset{1}{\circ}}$ $\dim V^0 = 2.$

Proof. Let us prove that the diagrams listed above indeed determine micromodules. The diagrams 4, 8, 13, 18, 22, 24, 25, 26, 29 correspond to the adjoint representations of the corresponding Lie algebras. Hence they determine micromodules.

It is known that the modules corresponding to the diagrams 1, 2, 3, 6, 7, 10, 15, 16, 17, 20, 21, 23 are strict micromodules (see [1], Chapter VIII, §7, no. 3; §13, no. 1 and Exercise 5).

It is easy to see that the Lie algebras corresponding to the diagrams 5, 9, 12, 14, 19, 27 can be regarded as the subalgebras of fixed points for second-order automorphisms s of some Lie algebras $\mathfrak{g}$, and the diagrams themselves determine the $\mathfrak{g}^1(s)$-modules $\mathfrak{g}^{-1}(s)$ in this case.

The results of these observations appear in Table 1.

With the root system $R(\mathfrak{g}, \mathfrak{h}^s)$, we can canonically associate a set of numerical labels; namely, to the black vertex we assign the number $1/2$, whereas the white ones are assigned a zero.

Attached to the diagram of the $\mathfrak{g}^1(s)$-module $\mathfrak{g}^{-1}(s)$ is the multiplicity of the zero weight, namely $\dim V^0$. The other weights of this module are simple, since the corresponding weight subspaces are root subspaces of $\mathfrak{g}$ with respect to $\mathfrak{h}^s$.

It remains to prove our assertion for diagrams 11 and 28.

(11) Let $\tilde{\mathfrak{g}}$ be the simple Lie algebra with the diagram

$$\underset{\alpha_4 \quad \alpha_3 \quad \alpha_2 \quad \alpha_1}{\circ\!\!-\!\!\circ\!\!\Leftarrow\!\!\circ\!\!-\!\!\circ}.$$

Consider the subalgebra $\mathfrak{g}$ of $\tilde{\mathfrak{g}}$ generated by the vectors $X_{\pm\alpha_2}$, $X_{\pm\alpha_3}$, and $X_{\pm\alpha_4}$. It is clear that the Lie algebra $\mathfrak{g}$ is simple and has the following diagram:

$$\underset{\alpha_4 \quad \alpha_3 \quad \alpha_2}{\circ\!\!-\!\!\circ\!\!\Leftarrow\!\!\circ}.$$

Consider the $\mathfrak{g}$-module $\tilde{\mathfrak{g}}$, that is, the restriction of the adjoined representation. It is clear that the vector $X_{-\alpha_1}$ is primitive, and the $\mathfrak{g}$-submodule generated by $X_{-\alpha_1}$ has the following diagram:

$$\circ\!\!-\!\!\circ\!\!\overset{1}{\Leftarrow}\!\!\circ.$$

Table 1.

Diagram of $\mathfrak{g}$ and s	Diagram of $R(\mathfrak{g}, \mathfrak{h}^s)$ with respect to h^s	Diagram of the $\mathfrak{g}^1(s)$-module $\mathfrak{g}^{-1}(s)$
(diagram)	*(diagram)*	*(diagram)* $\dim V^0 = 2$
(diagram) $\alpha_1 \quad \alpha_l \; \alpha_{l+1} \quad \alpha_{2l}$	*(diagram)*	*(diagram)* $\dim V^0 = l$
(diagram) $\alpha_1 \quad \alpha_{l-1} \; \alpha_l \; \alpha_{l+1} \quad \alpha_{2l-1}$	*(diagram)*	*(diagram)* $\dim V^0 = l-1$
(diagram)	*(diagram)*	*(diagram)* $\dim V^0 = 2$
(diagram) $\alpha_1 \quad \alpha_{l-1} \; \alpha_l \; \alpha_{l+1} \quad \alpha_{2l-1}$	*(diagram)*	*(diagram)* $\dim V^0 = l-1$
(diagram)	*(diagram)*	*(diagram)* $\dim V^0 = 2$

It is easy to see that the weight subspaces of this submodule coincide with the root subspaces of $\tilde{\mathfrak{g}}$ (corresponding to some negative roots of $\tilde{\mathfrak{g}}$). This proves that all its weights are simple.

(28) Let us construct the $\mathfrak{g}$-module V corresponding to the diagram

$$\underset{\alpha_1 \quad \alpha_2}{\overset{1}{\circ}\!\Lleftarrow\!\overset{}{\circ}} .$$

Let $\mathfrak{s}_i$ $(i = 1, 2)$ denote the simple subalgebra of $\mathfrak{g}$ generated by the vectors $X_{\pm\alpha_i}$, and let e_1 be a primitive element of the $\mathfrak{g}$-module V. By definition, we have

$$\begin{aligned} H_{\alpha_1} e_1 &= e_1, & X_{\alpha_1} e_1 &= 0, \\ H_{\alpha_2} e_1 &= 0, & X_{\alpha_2} e_1 &= 0, \end{aligned}$$

so e_1 is a primitive element of the $\mathfrak{s}_1$-module V (respectively, of the $\mathfrak{s}_2$-module V) whose weight is equal to 1 (respectively, 0). Therefore (see [1], Chapter VIII, §1, no. 2), $X_{-\alpha_1} e_1 \neq 0$, while $X_{-\alpha_1}^2 e_1 = 0$. Let $e_2 = X_{-\alpha_1} e_1$. Then we have

$$\begin{aligned} H_{\alpha_1} e_2 &= (1 - 2)e_2 = -e_2, & X_{\alpha_1} e_2 &= X_{\alpha_1} X_{-\alpha_1} e_1 = -H_{\alpha_1} e_1 = -e_1, \\ H_{\alpha_2} e_2 &= (0 + 1)e_2 = e_2, & X_{\alpha_2} e_2 &= X_{\alpha_2} X_{-\alpha_1} e_1 = X_{-\alpha_1} X_{\alpha_2} e_1 = 0, \end{aligned}$$

so e_2 is a primitive element of the $\mathfrak{s}_2$-module V whose weight is equal to 1. Therefore $X_{-\alpha_1} e_2 = X^2_{-\alpha_1} e_2 = 0$, $X_{-\alpha_2} e_2 \neq 0$, while $X^2_{-\alpha_2} e_2 = 0$.

We set $e_3 = X_{-\alpha_2} e_2 = X_{-\alpha_2} X_{-\alpha_1} e_1$. Then we have

$$
\begin{aligned}
H_{\alpha_1} e_3 &= (-1 + 3)e_3 = 2e_3, \\
&\quad X_{\alpha_1} e_3 = X_{\alpha_1} X_{-\alpha_2} e_2 = X_{-\alpha_2} X_{\alpha_1} e_2 = -X_{\alpha_2} e_1 = 0, \\
H_{\alpha_2} e_3 &= (1 - 2)e_3 = -e_3, \\
&\quad X_{\alpha_2} e_3 = X_{\alpha_2} X_{-\alpha_2} e_2 = -H_{\alpha_2} e_2 + X_{-\alpha_2} X_{\alpha_2} e_2 \\
&\quad = -e_2 + 0 = -e_2,
\end{aligned}
$$

so e_3 is a primitive element of the $\mathfrak{s}_1$-module V whose weight is equal to 2. Therefore $X_{-\alpha_1} e_3 \neq 0$, $X^2_{-\alpha_1} e_3 \neq 0$, while $X^3_{-\alpha_1} e_3 = 0$, $X_{-\alpha_2} e_3 = X^2_{-\alpha_2} e_2 = 0$.

Let

$$
e_4 = X_{-\alpha_1} e_3 = X_{-\alpha_1} X_{-\alpha_2} X_{-\alpha_1} e_1, \quad e_5 = X^2_{-\alpha_1} e_3 = X^2_{-\alpha_1} X_{-\alpha_2} X_{-\alpha_1} e_1.
$$

We have

$$
\begin{aligned}
H_{\alpha_1} e_4 &= (2 - 2)e_4 = 0, \\
&\quad X_{\alpha_1} e_4 = X_{\alpha_1} X_{-\alpha_1} e_3 = -H_{\alpha_1} e_3 + X_{-\alpha_1} X_{\alpha_1} e_3 \\
&\quad = -2e_3 + 0 = -2e_3, \\
H_{\alpha_2} e_4 &= (-1 + 1)e_4 = 0, \\
&\quad X_{\alpha_2} e_4 = X_{\alpha_2} X_{-\alpha_1} e_3 = X_{-\alpha_1} X_{\alpha_2} e_3 = -X_{-\alpha_1} e_2 = 0, \\
H_{\alpha_1} e_5 &= (0 - 2)e_5 = -2e_5, \\
&\quad X_{\alpha_1} e_5 = X_{\alpha_1} X_{-\alpha_1} e_4 = -H_{\alpha_1} e_4 + X_{-\alpha_1} X_{\alpha_1} e_4 \\
&\quad = (0 - 2)e_4 = -2e_4, \\
H_{\alpha_2} e_5 &= (0 + 1)e_5 = e_5, \\
&\quad X_{\alpha_2} e_5 = X_{\alpha_2} X_{-\alpha_1} e_4 = X_{-\alpha_1} X_{\alpha_2} e_4 = 0,
\end{aligned}
$$

so e_4 and e_5 are primitive elements of the $\mathfrak{s}_2$-module V whose weights are equal to 0 and 1, respectively. Therefore, $X_{-\alpha_1} e_4 = e_5$, $X_{-\alpha_2} e_4 = 0$, and $X_{-\alpha_1} e_5 = X^3_{-\alpha_1} e_3 = 0$, $X_{-\alpha_2} e_5 \neq 0$, while $X^2_{-\alpha_2} e_5 = 0$.

We set $e_6 = X_{-\alpha_2} e_5 = X_{-\alpha_2} X^2_{-\alpha_1} X_{-\alpha_2} X_{-\alpha_1} e_1$. Then we have

$$
\begin{aligned}
H_{\alpha_1} e_6 &= (-2 + 3)e_6 = e_6, \\
&\quad X_{\alpha_1} e_6 = X_{\alpha_1} X_{-\alpha_2} e_5 = X_{-\alpha_2} X_{\alpha_1} e_5 = 2X_{\alpha_2} e_4 = 0, \\
H_{\alpha_2} e_6 &= (1 - 2)e_6 = -e_6, \\
&\quad X_{\alpha_2} e_6 = X_{\alpha_2} X_{-\alpha_2} e_5 = -H_{\alpha_2} e_5 + X_{-\alpha_2} X_{\alpha_2} e_5 \\
&\quad = -e_5 + 0 = -e_5,
\end{aligned}
$$

so e_6 is a primitive element of the $\mathfrak{s}_1$-module V whose weight is equal to 1. Therefore $X_{-\alpha_1}e_6 \neq 0$, while $X^2_{-\alpha_1}e_6 = 0$, $X_{-\alpha_2}e_6 = X^2_{-\alpha_2}e_5 = 0$.

Let $e_7 = X_{-\alpha_1}e_6 = X_{-\alpha_1}X_{-\alpha_2}X_{-\alpha_1}X_{-\alpha_2}X_{-\alpha_1}e_1$. We have

$$
\begin{aligned}
H_{\alpha_1}e_7 &= (1-2)e_7 = -e_7, \\
X_{\alpha_1}e_7 &= X_{\alpha_2}X_{-\alpha_1}e_6 = -H_{\alpha_1}e_6 + X_{-\alpha_1}X_{\alpha_1}e_6 \\
&= -e_6 + 0 = -e_6, \\
H_{\alpha_2}e_7 &= (-1+1)e_7 = 0, \\
X_{\alpha_2}e_7 &= X_{\alpha_2}X_{-\alpha_1}e_6 = X_{-\alpha_1}(X_{\alpha_2}e_6) = -X_{-\alpha_1}e_5 = 0,
\end{aligned}
$$

so e_7 is a primitive element of the $\mathfrak{s}_2$-module V whose weight is equal to 0. Therefore $X_{-\alpha_1}e_7 = X^2_{-\alpha_1}e_7 = 0$, $X_{-\alpha_2}e_7 = 0$.

We have thus shown that $\dim V = 7$ and that V is a strict micromodule.

We shall now prove that there are no other strict micromodules apart from those listed in the Theorem.

Lemma 1 *A strict micromodule has exactly one nonzero label.*

Proof. Let $(\mathfrak{g}, \mathfrak{h})$ be a split simple Lie algebra, B a basis of the system $R(\mathfrak{g}, \mathfrak{h})$, and V a simple micromodule over $\mathfrak{g}$ whose highest weight is λ.

Suppose that there exists a chain $\alpha_1, \ldots, \alpha_n$ of roots in B such that $\lambda(H_{\alpha_1}) \neq 0$, $\lambda(H_{\alpha_i}) = 0$ $(2 \leqslant i \leqslant n-1)$, $\lambda(H_{\alpha_n}) \neq 0$ and such that $\alpha_i(H_{\alpha_j}) \neq 0 \iff |i - j| \leqslant 1$.

Let e be a primitive element of the $\mathfrak{g}$-module V. Let $\alpha_{ij} = \alpha_i(H_{\alpha_j})$, $\alpha_i = \lambda(H_{\alpha_i})$ $(1 \leqslant i, j \leqslant n)$;

$$
\begin{aligned}
e^k &= X_{-\alpha_k}\ldots X_{-\alpha_1}e \quad (1 \leqslant k \leqslant n), \quad e^0 = e; \\
e_k &= X_{-\alpha_{n+1-k}}\ldots X_{-\alpha_n}e \quad (1 \leqslant k \leqslant n), \quad e_0 = e;
\end{aligned}
$$

then

$$
\begin{aligned}
X_{\alpha_1}e^1 &= -\lambda_1 e^0, \\
X_{\alpha_k}e^k &= \alpha_{k-1,k}e^{k-1} \quad (2 \leqslant k \leqslant n-1), \\
X_{\alpha_n}e^n &= (\alpha_{n-1,n} - \lambda_n)e^{n-1}.
\end{aligned}
$$

Indeed, we have

$$
X_{\alpha_k}e^k = X_{\alpha_k}X_{-\alpha_k}e^{k-1} = -H_{\alpha_k}e^{k-1} + X_{-\alpha_k}X_{\alpha_k}e^{k-1} \quad (1 \leqslant k \leqslant n).
$$

Now $H_{\alpha_1}e^0 = \lambda_1 e^0$ and

$$
H_{\alpha_k}e^{k-1} = (\lambda_k - \alpha_{1k} - \ldots - \alpha_{k-1,k})e^{k+1} \quad (2 \leqslant k \leqslant n).
$$

Therefore $H_{\alpha_k}e^{k-1} = -\alpha_{k-1,k}$ $(2 \leqslant k \leqslant n-1)$ and $H_{\alpha_n}e^{n-1} = (\lambda_n - \alpha_{n-1,n})e^{n-1}$. In addition $X_{\alpha_1}e^0 = X_{\alpha_1}e = 0$ and

$$
X_{\alpha_k}e^{k-1} = X_{\alpha_k}X_{-\alpha_{k-1}}\ldots X_{-\alpha_1}e = X_{-\alpha_{k-1}}\ldots X_{-\alpha_1}X_{\alpha_k}e = 0,
$$

where $2 \leqslant k \leqslant n$. It follows that

$$X_{\alpha_1} \ldots X_{\alpha_n} e^n = (-1)^n \lambda_1 (-\alpha_{12}) \ldots (-\alpha_{n-2,n-1})(\lambda_n - \alpha_{n-1,n}) e \neq 0 \quad (1)$$

(since $-\alpha_{i,i+1} > 0$ for $1 \leqslant i \leqslant n-1$).

Since $X_{\alpha_k} X_{-\alpha_k} e = -H_{\alpha_k} e + X_{-\alpha_k} X_{\alpha_k} e = -\lambda_k e$ $(1 \leqslant k \leqslant n)$, we have

$$X_{\alpha_n} \ldots X_{\alpha_1} e^n = (-1)^n \lambda_1 \ldots \lambda_n e. \quad (2)$$

If we set $\beta_i = \alpha_{n+1-i}$, $\beta_{ij} = \beta_i(H_{\beta_j})$ and $\lambda^i = \lambda(\beta_i)$ $(1 \leqslant i, j \leqslant n)$, we obtain $\lambda^i = \lambda_{n+1-i}$, $e_k = X_{-\beta_k} \ldots X_{-\beta_1} e$ and $\beta_{ij} = \alpha_{n+1-i,n+1-j}$ $(1 \leqslant i, j, k \leqslant n)$.

It can be similarly proved that

$$X_{\beta_1} \ldots X_{\beta_n} e_n = (-1)^n \lambda^1 (-\beta_{12}) \ldots (-\beta_{n-2,n-1})(\lambda^n - \beta_{n-1,n}) e \neq 0,$$
$$X_{\beta_n} \ldots X_{\beta_1} e_n = (-1)^n \lambda^1 \ldots \lambda^n e.$$

In the former notation, we have

$$X_{\alpha_n} \ldots X_{\alpha_1} e = (-1)^n \lambda_n (-\alpha_{n,n-1}) \ldots (-\alpha_{3,2})(\lambda_1 - \alpha_{2,1}) e; \quad (3)$$
$$X_{\alpha_1} \ldots X_{\alpha_1} e = (-1)^n \lambda_n \ldots \lambda_1 e. \quad (4)$$

The vectors e_n and e^n are, as relations (1) and (3) show, nonzero and have the same weight, namely $\lambda - \alpha_1 - \ldots - \alpha_n$. Therefore $e = \varphi e$ for a certain $\varphi \in \mathbb{C}^*$.

It follows from (1) and (4) that

$$(-1)^n \lambda_n \ldots \lambda_1 = \varphi(-1)^n \lambda_1 (-\alpha_{12}) \ldots (-\alpha_{n-2,n-1})(\lambda_n - \alpha_{n-1,n}).$$

Relations (3) and (2) yield

$$(-1)^n \lambda_n (-\alpha_{n,n-1}) \ldots (-\alpha_{32})(\lambda_1 - \alpha_{21}) = \varphi(-1)^n \lambda_1 \ldots \lambda_n.$$

If $n > 2$, then one of the sides in each of these equations is zero (since, $\lambda_2 = 0$), while the other is nonzero.

If $n = 2$, then we have

$$\lambda_2 \lambda_1 = \varphi \lambda_1 (\lambda_2 - \alpha_{12}), \quad \lambda_2(\lambda - \alpha_{21}) = \varphi \lambda_1 \lambda_2,$$

so that $\lambda_1 \lambda_2 = (\lambda_1 - \alpha_{21})(\lambda_2 - \alpha_{12})$. However, $0 < \lambda_1 < \lambda_1 - \lambda_2$ and $0 < \lambda_2 < \lambda_2 - \alpha_{12}$, and hence $\lambda_1 \lambda_2 < (\lambda_1 - \alpha_{21})(\lambda_2 - \alpha_{12})$. This contradiction proves that there is no chain $\alpha_1, \ldots, \alpha_n$ with the properties specified above. $\square$

Lemma 2 *The modules corresponding to the diagrams*

$$\underset{t}{\circ\!\!\!\gg\!\!\!\circ} \qquad (t > 1);$$

$$\overset{t}{\circ\!\!\!\gg\!\!\!\circ} \qquad (t > 1);$$

$$\overset{t}{\circ\!\!\!-\!\!\!\circ\!\!\!-\!\!\!\circ} \qquad (t > 1);$$

$$\overset{t}{\circ\!\!\!\lll\!\!\!\circ} \qquad (t > 1);$$

$$\underset{t}{\circ\!\!\!\lll\!\!\!\circ} \qquad (t \geqslant 1)$$

are not strict micromodules.

Proof. At first, we construct several upper levels of the modules V corresponding to the diagrams:

$$\overset{1}{\circ\!\!\!\gg\!\!\!\circ},$$

$$\overset{1}{\circ\!\!\!\gg\!\!\!\circ},$$

$$\overset{1}{\circ\!\!\!-\!\!\!\circ\!\!\!-\!\!\!\circ},$$

$$\overset{1}{\circ\!\!\!\lll\!\!\!\circ} \quad \text{(it has already been constructed)},$$

$$\overset{1}{\circ\!\!\!\lll\!\!\!\circ},$$

and then we consider the submodule of the module $S^t(V)$ generated by the element e_1^t (where e_1 is a primitive element of the module V).

a) Let us construct the module V corresponding to the diagram $\overset{t}{\circ\!\!\!\lll\!\!\!\circ}$ $(t > 1)$. Its weight is equal to $t\bar{\omega}_1$. Using the computation for the diagram $\underset{\alpha_1 \;\; \alpha_2}{\overset{1}{\circ\!\!\!\lll\!\!\!\circ}}$, we obtain:

$$X_{-\alpha_1} e_1^t = t e_1^{t-1} e_2,$$
$$X_{-\alpha_1}^2 e_1^t = t(t-1) e_1^{t-2} e_2^2,$$
$$X_{-\alpha_2} X_{-\alpha_1} e_1^t = t e_1^{t-1} e_3,$$
$$X_{-\alpha_2} X_{-\alpha_1}^2 e_1^t = 2t(t-1) e_1^{t-2} e_2 e_3,$$
$$X_{-\alpha_1} X_{-\alpha_2} X_{-\alpha_1} e_1^t = t(t-1) e_1^{t-2} e_2 e_3 + t e_1^{t-1} e_4.$$

If $t > 1$, the elements $2t(t-1) e_1^{t-2} e_2 e_3$ and $t(t-1) e_1^{t-2} e_2 e_3 + t e_1^{t-1} e_4$ are not proportional, but have the same weight, which is equal to $t\bar{\omega}_1 - 2\alpha_1 - \alpha_2 \neq 0$.

b) Let us construct several weight vectors of the $\mathfrak{g}$-module V corresponding to the diagram $\underset{\alpha_1 \;\; \alpha_2}{\overset{1}{\circ\!\!\!\lll\!\!\!\circ}}$.

Denote by $\mathfrak{s}_i$ $(i = 1, 2)$ the simple subalgebra of $\mathfrak{g}$ generated by the vectors $X_{\pm\alpha_i}$, and let e_1 be a primitive element of the $\mathfrak{g}$-module V.

By definition, we have

$$H_{\alpha_1} e_1 = 0, \qquad X_{\alpha_1} e_1 = 0,$$
$$H_{\alpha_2} e_1 = e_1, \qquad X_{\alpha_2} e_1 = 0,$$

so e_1 is a primitive element of weight 1 (respectively, 0) in the $\mathfrak{s}_2$-module V (respectively, in the $\mathfrak{s}_1$-module V). Therefore $X_{-\alpha_1}e_1 = 0$, $X_{-\alpha_2}e_1 \neq 0$, while $X^2_{-\alpha_2}e_1 = 0$.

If $e_2 = X_{-\alpha_2}e_1$, we have

$$
\begin{aligned}
H_{\alpha_1}e_2 &= (0+3)e_2 = 3e_2, \\
&\quad X_{\alpha_1}e_2 = X_{\alpha_1}X_{-\alpha_2}e_1 = X_{-\alpha_2}X_{\alpha_1}e_1 = 0, \\
H_{\alpha_2}e_2 &= (1-2)e_2 = -e_2, \\
&\quad X_{\alpha_2}e_2 = X_{\alpha_2}X_{-\alpha_2}e_1 = -H_{\alpha_2}e_1 + X_{-\alpha_2}X_{\alpha_2}e_1 = -e_1,
\end{aligned}
$$

so that e_2 is a primitive element of weight 3 in the $\mathfrak{s}_1$-module V. Therefore $X_{-\alpha_1}e_2 \neq 0$, $X^2_{-\alpha_1}e_2 \neq 0$, $X^3_{-\alpha_1}e_2 \neq 0$, while $X^4_{-\alpha_1}e_2 = 0$, and $X_{-\alpha_2}e_2 = X^2_{-\alpha_2}e_2 = 0$.

Let

$$
\begin{aligned}
e_3 &= X_{-\alpha_1}e_2 = X_{-\alpha_1}X_{-\alpha_2}e_1, \\
e_4 &= X_{-\alpha_1}e_3 = X^2_{-\alpha_1}X_{-\alpha_2}e_1, \\
e_5 &= X_{-\alpha_1}e_4 = X^3_{-\alpha_1}X_{-\alpha_2}e_1.
\end{aligned}
$$

Then we have

$$
\begin{aligned}
H_{\alpha_1}e_3 &= (3-2)e_3 = e_3, \\
&\quad X_{\alpha_1}e_3 = X_{\alpha_1}X_{-\alpha_1}e_2 = -H_{\alpha_1}e_2 + X_{-\alpha_1}X_{\alpha_1}e_2 \\
&\quad = -3e_2 + 0 = -3e_2, \\
H_{\alpha_2}e_3 &= (-1+1)e_3 = 0, \\
&\quad X_{\alpha_2}e_3 = X_{\alpha_1}X_{-\alpha_2}e_2 = -X_{\alpha_1}e_1 = 0,
\end{aligned}
$$

so that e_3 is a primitive element of weight 0 in the $\mathfrak{s}_2$-module V. Therefore $X_{-\alpha_1}e_3 = e_4$ and $X_{-\alpha_2}e_3 = 0$.

Consider now the $\mathfrak{g}$-module corresponding to the diagram $\underset{\alpha_1 \quad \alpha_2}{\circ\!\!\overset{t}{\Longleftarrow}\!\!\circ}$ $(t > 1)$. We have

$$
\begin{aligned}
X_{-\alpha_2}e_1^t &= te_1^{t-1}e_2, \\
X_{-\alpha_1}X_{-\alpha_2}e_1^t &= te_1^{t-1}e_3, \\
X^2_{-\alpha_1}X_{-\alpha_2}e_1^t &= te_1^{t-1}e_4, \\
X_{-\alpha_2}X_{-\alpha_1}X_{-\alpha_2}e_1^t &= t(t-1)e_1^{t-2}e_2e_3, \\
X_{-\alpha_2}X^2_{-\alpha_1}X_{-\alpha_2}e_1^t &= t(t-1)e_1^{t-2}e_2e_4, \\
X_{-\alpha_1}X_{-\alpha_2}X_{-\alpha_1}X_{-\alpha_2}e_1^t &= t(t-1)e_1^{t-2}e_3 + t(t-1)e_1^{t-2}e_2e_4.
\end{aligned}
$$

If $t > 1$, then the elements $t(t-1)e_1^{t-2}e_2e_4$ and $t(t-1)e_1^{t-2}e_3 + t(t-1)e_1^{t-2}e_2e_4$ are not proportional, but have the same weight, which is equal to $t\bar\omega_2 - 2\alpha_1 - 2\alpha_2 \neq 0$.

c) Let us construct the $\mathfrak{g}$-module V corresponding to the diagram $\underset{\alpha_1 \quad \alpha_2}{\circ\!\!\overset{1}{\Longleftarrow}\!\!\circ}$.

Denote by $\mathfrak{s}_i$ $(i = 1, 2)$ the simple subalgebra of $\mathfrak{g}$ generated by the vectors $X_{\pm\alpha_i}$, and let e_1 be a primitive element of the $\mathfrak{g}$-module V.

By definition, we have

$$H_{\alpha_1}e_1 = e_1, \qquad X_{\alpha_1}e_1 = 0,$$
$$H_{\alpha_2}e_1 = 0, \qquad X_{\alpha_2}e_1 = 0,$$

so that e_1 is a primitive element of weight 1 (respectively, 0) in the $\mathfrak{s}_1$-module V (respectively, in the $\mathfrak{s}_2$-module V). Therefore $X_{-\alpha_1}e_1 \neq 0$, while $X^2_{-\alpha_1}e_1 = 0$, $X_{-\alpha_2}e_1 \neq 0$. Let $e_2 = X_{-\alpha_1}e_1$. Then we have

$$
\begin{aligned}
H_{\alpha_1}e_2 &= (1 - 2)e_2 = -e_2, \\
& X_{\alpha_1}e_2 = X_{\alpha_1}X_{-\alpha_1}e_1 = -H_{\alpha_1}e_1 + X_{-\alpha_1}X_{\alpha_1}e_1 = -e_1, \\
H_{\alpha_2}e_2 &= (0 + 1)e_2 = e_2, \\
& X_{\alpha_2}e_2 = X_{-\alpha_1}X_{\alpha_2}e_1 = 0,
\end{aligned}
$$

so that e_2 is a primitive element of weight 1 in the $\mathfrak{s}_2$-module V. Therefore $X_{-\alpha_1}e_2 = X^2_{-\alpha_1}e_2 = 0$, $X_{-\alpha_2}e_2 \neq 0$, while $X^2_{-\alpha_2}e_2 = 0$.

Let $e_3 = X_{-\alpha_2}e_2 = X_{-\alpha_2}X_{-\alpha_1}e_1$. Then we have

$$
\begin{aligned}
H_{\alpha_1}e_3 &= (-1 + 2)e_3 = e_3, \\
& X_{\alpha_1}e_3 = X_{-\alpha_2}X_{\alpha_1}e_2 = -X_{\alpha_2}e_1 = 0, \\
H_{\alpha_2}e_3 &= (1 - 2)e_3 = -e_3, \\
& X_{\alpha_2}e_3 = X_{\alpha_2}X_{-\alpha_2}e_2 = -H_{\alpha_2}e_2 + X_{-\alpha_2}X_{\alpha_2}e_2 \\
& = -e_2 + 0 = -e_2,
\end{aligned}
$$

so that e_3 is a primitive element of weight 1 in the $\mathfrak{s}_1$-module V. Therefore $X_{-\alpha_1}e_3 \neq 0$, while $X^2_{-\alpha_1}e_3 = 0$, $X_{-\alpha_2}e_3 = X^2_{-\alpha_2}e_2 = 0$.

Let $e_4 = X_{-\alpha_1}e_3 = X_{-\alpha_1}X_{-\alpha_2}X_{-\alpha_1}e_1$. We have

$$
\begin{aligned}
H_{\alpha_1}e_4 &= (1 - 2)e_4 = -e_4, \\
& X_{\alpha_1}e_4 = X_{\alpha_1}X_{-\alpha_1}e_3 = -H_{\alpha_1}e_3 + X_{-\alpha_1}X_{\alpha_1}e_3 \\
& = -e_3 + 0 = -e_3, \\
H_{\alpha_2}e_4 &= (-1 + 1)e_4 = 0, \\
& X_{\alpha_2}e_4 = X_{-\alpha_1}X_{\alpha_2}e_3 = -X_{-\alpha_1}e_2 = 0,
\end{aligned}
$$

so that e_4 is a primitive element of weight 0 in the $\mathfrak{s}_2$-module V. Therefore $X_{-\alpha_1}e_4 = X^2_{-\alpha_1}e_4 = 0$ and $X_{-\alpha_2}e_4 \neq 0$.

Let us consider the $\mathfrak{g}$-module corresponding to the diagram $\overset{1}{\underset{\alpha_1}{\circ}}\!\!\Longleftarrow\!\!\underset{\alpha_2}{\circ}$ $(t >$ 1). We have

$$X_{-\alpha_1}e_1^t = te_1^{t-1}e_2,$$
$$X_{-\alpha_1}^2 e_1^t = t(t-1)e_1^{t-2}e_2^2,$$
$$X_{-\alpha_2}X_{-\alpha_1}e_1^t = te_1^{t-1}e_3,$$
$$X_{-\alpha_2}X_{-\alpha_1}^2 e_1^t = 2t(t-1)e_1^{t-2}e_2e_3,$$
$$X_{-\alpha_1}X_{-\alpha_2}X_{-\alpha_1}e_1^t = t(t-1)e_1^{t-2}e_2e_3 + te_1^{t-1}e_4.$$

If $t > 1$, then the elements $2t(t-1)e_1^{t-2}e_2e_3$ and $t(t-1)e_1^{t-2}e_2e_3+te_1^{t-1}e_4$ are not proportional, but have the same weight, which is equal to $t\bar{\omega}_1 - 2\alpha_1 - \alpha_2$ (it equals 0 if $t = 2$, and is nonzero whenever $t > 2$).

d) Let us construct the $\mathfrak{g}$-module V corresponding to the diagram $\overset{1}{\underset{\alpha_1}{\circ}}\!\!\Longrightarrow\!\!\underset{\alpha_2}{\circ}$.
Denote by $\mathfrak{s}_i$ $(i = 1, 2)$ the simple subalgebra of $\mathfrak{g}$ generated by the vectors $X_{\pm\alpha_i}$, and let e_1 be a primitive element of the $\mathfrak{g}$-module V.

By definition,

$$H_{\alpha_1}e_1 = e_1, \qquad X_{\alpha_1}e_1 = 0,$$
$$H_{\alpha_2}e_1 = 0, \qquad X_{\alpha_2}e_1 = 0,$$

so that e_1 is a primitive element of weight 1 (respectively, 0) in the $\mathfrak{s}_1$-module V (respectively, in the $\mathfrak{s}_2$-module V). Therefore $X_{-\alpha_1}e_1 \neq 0$, while $X_{-\alpha_1}^2 e_1 = 0$, $X_{-\alpha_2}e_1 = 0$. Let $e_2 = X_{-\alpha_1}e_1$. We have

$$
\begin{aligned}
H_{\alpha_1}e_2 &= (1-2)e_2 = -e_2, \\
&X_{\alpha_1}e_2 = X_{\alpha_1}X_{-\alpha_1}e_1 = -H_{\alpha_1}e_1 + X_{-\alpha_1}X_{\alpha_1}e_1 = -e_1, \\
H_{\alpha_2}e_2 &= (0+2)e_2 = 2e_2, \\
&X_{\alpha_2}e_2 = X_{-\alpha_1}X_{\alpha_2}e_1 = 0,
\end{aligned}
$$

so that e_2 is a primitive element of weight 2 in the $\mathfrak{s}_2$-module V. Therefore $X_{-\alpha_1}e_2 = X_{-\alpha_1}^2 e_1 = 0$, $X_{-\alpha_2}e_2 \neq 0$, $X_{-\alpha_2}^2 e_2 \neq 0$, while $X_{-\alpha_2}^3 e_2 = 0$.
Let $e_3 = X_{-\alpha_2}e_2 = X_{-\alpha_2}X_{-\alpha_1}e_1$. We have

$$
\begin{aligned}
H_{\alpha_1}e_3 &= (-1+1)e_3 = 0, \\
&X_{\alpha_1}e_3 = X_{-\alpha_2}X_{\alpha_1}e_2 = -X_{\alpha_2}e_1 = 0, \\
H_{\alpha_2}e_3 &= (2-2)e_3 = 0, \\
&X_{\alpha_2}e_3 = X_{\alpha_2}X_{-\alpha_2}e_2 = -H_{\alpha_2}e_2 + X_{-\alpha_2}X_{\alpha_2}e_2 = -2e_2,
\end{aligned}
$$

so that e_3 is a primitive element of weight 0 in the $\mathfrak{s}_1$-module V. Therefore $X_{-\alpha_1}e_3 = 0$.

NEW BOOK

Let $e_4 = X_{-\alpha_2} e_3 = X^2_{-\alpha_2} X_{-\alpha_1} e_1$. We have

$$
\begin{aligned}
H_{\alpha_1} e_4 &= (0+1)e_4 = e_4, \\
&\quad X_{\alpha_1} e_4 = X_{-\alpha_2} X_{\alpha_1} e_3 = -X_{\alpha_2} e_2 = 0, \\
H_{\alpha_2} e_4 &= (0-2)e_4 = -2e_4, \\
&\quad X_{\alpha_2} e_4 = X_{\alpha_2} X_{-\alpha_2} e_3 = -H_{\alpha_2} e_3 + X_{-\alpha_2} X_{\alpha_2} e_3 = -2e_3,
\end{aligned}
$$

so that e_4 is a primitive element of weight 1 in the $\mathfrak{s}_1$-module V. Therefore $X_{-\alpha_1} e_4 \neq 0$, $X^2_{-\alpha_1} e_4 = 0$, $X_{-\alpha_2} e_4 = X^3_{-\alpha_2} e_2 = 0$.

Let $e_5 = X_{-\alpha_1} e_4 = X_{-\alpha_1} X^2_{-\alpha_2} X_{-\alpha_1} e_1$; then

$$
\begin{aligned}
H_{\alpha_1} e_5 &= (1-2)e_5 = -e_5, \\
&\quad X_{\alpha_1} e_5 = X_{\alpha_1} X_{-\alpha_1} e_4 = -H_{\alpha_1} e_4 + X_{-\alpha_1} X_{\alpha_1} e_4 = -e_4, \\
H_{\alpha_2} e_5 &= (-2+2)e_5 = 0, \\
&\quad X_{\alpha_2} e_5 = X_{-\alpha_1} X_{\alpha_2} e_4 = -2X_{-\alpha_1} e_3 = 0,
\end{aligned}
$$

so that e_5 is a primitive element of weight 0 in the $\mathfrak{s}_2$-module V. Therefore $X_{-\alpha_1} e_5 = X^2_{-\alpha_1} e_4 = 0$, $X_{-\alpha_2} e_5 = 0$.

Now consider the $\mathfrak{g}$-module corresponding to the diagram $\underset{\alpha_1 \;\; \alpha_2}{\circ\!\!\Rrightarrow\!\!\circ}$ $(t > 1)$. We have

$$
\begin{aligned}
X_{-\alpha_1} e_1^t &= t e_1^{t-1} e_2, \\
X^2_{-\alpha_1} e_1^t &= t(t-1)e_1^{t-2} e_2^2, \\
X_{-\alpha_2} X_{-\alpha_1} e_1^t &= t e_1^{t-1} e_3, \\
X_{-\alpha_2} X^2_{-\alpha_1} e_1^t &= 2t(t-1)e_1^{t-2} e_2 e_3, \\
X^2_{-\alpha_2} X_{-\alpha_1} e_1^t &= t e_1^{t-1} e_4, \\
X^2_{-\alpha_2} X^2_{-\alpha_1} e_1^t &= 2t(t-1)e_1^{t-2} e_3^2 + 2t(t-1)e_1^{t-2} e_2 e_4, \\
X_{-\alpha_1} X^2_{-\alpha_2} X_{-\alpha_1} e_1^t &= t(t-1)e_1^{t-2} e_2 e_4 + t e_1^{t-1} e_5.
\end{aligned}
$$

If $t > 1$, the elements $2t(t-1)e_1^{t-2} e_3^2 + 2t(t-1)e_1^{t-2} e_2 e_4$ and $t(t-1)e_1^{t-2} e_2 e_4 + t e_1^{t-1} e_5$ are not proportional and have the same weight equal to $t\bar{\omega}_1 - 2\alpha_1 - 2\alpha_2$ (it equals 0 if $t = 2$, and is nonzero whenever $t > 2$).

e) Let us construct the $\mathfrak{g}$-module V corresponding to the diagram

$$
\underset{\alpha_1 \quad \alpha_2 \quad \alpha_3}{\circ\!\!-\!\!\overset{1}{\circ}\!\!-\!\!\circ}.
$$

Denote by $\mathfrak{s}_i$ $(i = 1, 2, 3)$ the simple subalgebra of the Lie algebra $\mathfrak{g}$ generated by the vectors $X_{\pm\alpha_i}$, and let e_1 be a primitive element of the $\mathfrak{g}$-module V.

By definition, we have

$$
\begin{array}{ll}
H_{\alpha_1} e_1 = 0, & X_{\alpha_1} e_1 = 0, \\
H_{\alpha_2} e_1 = e_1, & X_{\alpha_2} e_1 = 0, \\
H_{\alpha_3} e_1 = 0, & X_{\alpha_3} e_1 = 0,
\end{array}
$$

so that e_1 is a primitive element of the $\mathfrak{s}_i$-module V $(i = 1, 2, 3)$ whose weight is equal to λ_i, where $\lambda_1 = 0$, $\lambda_2 = 1$, $\lambda_3 = 0$. Therefore $X_{-\alpha_1} e_1 = 0$, $X_{-\alpha_2} e_1 \neq 0$, while $X^2_{-\alpha_2} e_1 = 0$, $X_{-\alpha_3} e_1 = 0$.

Let $e_2 = X_{-\alpha_2} e_1$. We have

$$
\begin{aligned}
H_{\alpha_1} e_2 &= (0 + 1)e_2 = e_2, & X_{\alpha_1} e_2 &= X_{-\alpha_2} X_{\alpha_1} e_1 = 0, \\
H_{\alpha_2} e_2 &= (1 - 2)e_2 = -e_2, & X_{\alpha_2} e_2 &= -H_{\alpha_2} e_1 + X_{-\alpha_2} X_{\alpha_2} e_1 = -e_1, \\
H_{\alpha_3} e_2 &= (0 + 1)e_2 = e_2, & X_{\alpha_3} e_2 &= X_{-\alpha_2} X_{\alpha_1} e_1 = 0,
\end{aligned}
$$

so that e_2 is a primitive element of weight 1 in the $\mathfrak{s}_1$-module V and $\mathfrak{s}_3$-module V. Therefore $X_{-\alpha_1} e_2 \neq 0$, while $X^2_{-\alpha_1} e_2 = 0$, $X_{-\alpha_2} e_2 = X^2_{-\alpha_2} e_1 = 0$, $X_{-\alpha_3} e_2 \neq 0$, while $X^2_{-\alpha_3} e_2 = 0$.

Let $e_3 = X_{-\alpha_1} e_2 = X_{-\alpha_1} X_{-\alpha_2} e_1$ and $e_4 = X_{-\alpha_3} e_2 = X_{-\alpha_3} X_{-\alpha_2} e_1$. We have

$$
\begin{aligned}
H_{\alpha_1} e_3 &= (1 - 2)e_3 = -e_3, \\
& X_{\alpha_1} e_3 = X_{\alpha_1} X_{-\alpha_1} e_2 = -H_{\alpha_1} e_2 + X_{-\alpha_1} X_{\alpha_1} e_2 = -e_2, \\
H_{\alpha_2} e_3 &= (-1 + 1)e_3 = 0, \\
& X_{\alpha_2} e_3 = X_{-\alpha_1} X_{\alpha_2} e_2 = -X_{\alpha_1} e_1 = 0, \\
H_{\alpha_3} e_3 &= (1 - 0)e_3 = e_3, \\
& X_{\alpha_3} e_3 = X_{-\alpha_1} X_{\alpha_3} e_2 = 0,
\end{aligned}
$$

so that e_3 is a primitive element of weight 0 (respectively, 1) in the $\mathfrak{s}_2$-module V (respectively, in the $\mathfrak{s}_3$-module V). Thus we have $X_{-\alpha_1} e_3 = X^2_{-\alpha_1} e_2 = 0$, $X_{-\alpha_2} e_3 = 0$, $X_{-\alpha_3} e_3 \neq 0$, while $X^2_{-\alpha_3} e_3 = 0$.

Let $e_5 = X_{-\alpha_3} e_3 = X_{-\alpha_3} X_{-\alpha_1} X_{-\alpha_2} e_1$. We have

$$
\begin{aligned}
H_{\alpha_1} e_4 &= (1 - 0)e_4 = e_4, \\
& X_{\alpha_1} e_4 = X_{-\alpha_3} X_{\alpha_1} e_2 = 0, \\
H_{\alpha_2} e_4 &= (-1 + 1)e_4 = 0, \\
& X_{\alpha_2} e_4 = X_{-\alpha_3} X_{\alpha_2} e_2 = X_{-\alpha_3} e_1 = 0, \\
H_{\alpha_3} e_4 &= (1 - 2)e_4 = -e_4, \\
& X_{\alpha_3} e_4 = X_{\alpha_3} X_{-\alpha_3} e_2 = -H_{\alpha_3} e_2 + X_{-\alpha_3} X_{\alpha_3} e_2 = -e_2,
\end{aligned}
$$

so that e_4 is a primitive element of weight 0 (respectively, 1) in the $\mathfrak{s}_2$-module V (respectively, in the $\mathfrak{s}_1$-module V). Therefore $X_{-\alpha_1} e_4 \neq 0$, while $X^2_{-\alpha_1} e_4 = 0$, $X_{-\alpha_2} e_4 = 0$, $X_{-\alpha_3} e_4 = X^2_{-\alpha_3} e_2 = 0$.

It is clear that $X_{-\alpha_1} e_4 = X_{-\alpha_1} X_{-\alpha_3} e_2 = X_{-\alpha_2} X_{-\alpha_1} e_2 = X_{-\alpha_3} e_3 = e_5$. We have

$$
H_{\alpha_1} e_5 = (1 - 2)e_5 = -e_5,
$$

$$X_{\alpha_1} e_5 = X_{-\alpha_3} X_{\alpha_1} e_3 = X_{-\alpha_3} X_{\alpha_2} e_4 = -e_4,$$

$$H_{\alpha_2} e_5 = (0+1)e_5 = e_5,$$

$$X_{\alpha_2} e_5 = X_{-\alpha_1} X_{\alpha_2} e_4 = 0,$$

$$H_{\alpha_3} e_5 = (-1-0)e_5 = -e_5,$$

$$X_{\alpha_3} e_5 = X_{-\alpha_1} X_{\alpha_3} e_4 = -X_{-\alpha_1} e_2 = -e_3,$$

so that e_5 is a primitive element of weight 0 in the $\mathfrak{s}_2$-module V. Therefore $X_{-\alpha_1} e_5 = X^2_{-\alpha_1} e_4 = 0$, $X_{-\alpha_2} e_5 \neq 0$ while $X^2_{-\alpha_2} e_5 = 0$, $X_{-\alpha_3} e_5 = X^2_{-\alpha_3} e_3 = 0$.

Let $e_6 = X_{-\alpha_2} e_5 = X_{-\alpha_2} X_{-\alpha_3} X_{-\alpha_1} X_{-\alpha_2} e_1$. We have

$$H_{\alpha_1} e_6 = (-1+1)e_6 = 0,$$

$$X_{\alpha_1} e_6 = X_{-\alpha_2} X_{\alpha_1} e_5 = -X_{-\alpha_2} e_4 = 0,$$

$$H_{\alpha_2} e_6 = (1-2)e_6 = -e_6,$$

$$X_{\alpha_2} e_6 = X_{\alpha_2} X_{-\alpha_2} e_5 = -H_{\alpha_2} e_5 + X_{-\alpha_2} X_{\alpha_2} e_5 = -e_5,$$

$$H_{\alpha_3} e_6 = (-1+1)e_6 = 0,$$

$$X_{\alpha_3} e_6 = X_{-\alpha_2} X_{\alpha_3} e_5 = -X_{-\alpha_2} e_3 = 0,$$

so that e_6 is a primitive element of weight 0 in the $\mathfrak{s}_i$-module V ($i = 1, 3$). Therefore $X_{-\alpha_1} e_6 = 0$, $X_{-\alpha_2} e_6 = X^2_{-\alpha_2} e_6 = 0$, $X_{-\alpha_3} e_6 = 0$.

Consider now the $\mathfrak{g}$-module corresponding to the diagram $\underset{\alpha_1 \quad \alpha_2 \quad \alpha_3}{\circ\!\!-\!\!\overset{t}{\circ}\!\!-\!\!\circ}$ ($t > 1$). We have

$$X_{-\alpha_2} e_1^t = t e_1^{t-1} e_2,$$

$$X_{-\alpha_1} X_{-\alpha_2} e_1^t = t e_1^{t-1} e_3,$$

$$X_{-\alpha_2} X_{-\alpha_1} X_{-\alpha_2} e_1^t = t(t-1) e_1^{t-2} e_2 e_3,$$

$$X_{-\alpha_3} X_{-\alpha_1} X_{-\alpha_2} e_1^t = t e_1^{t-1} e_5,$$

$$X_{-\alpha_3} X_{-\alpha_2} X_{-\alpha_1} X_{-\alpha_2} e_1^t = t(t-1) e_1^{t-2} e_4 e_3 + t(t-1) e_1^{t-2} e_2 e_5,$$

$$X_{-\alpha_2} X_{-\alpha_3} X_{-\alpha_1} X_{-\alpha_2} e_1^t = t(t-1) e_1^{t-2} e_2 e_5 + t e_1^{t-1} e_6.$$

If $t > 1$, then the elements $t(t-1)e_1^{t-2} e_3 e_4 + t(t-1)e_1^{t-2} e_2 e_5$ and $t(t-1)e_1^{t-2} e_2 e_5 + t e_1^{t-1} e_6$ are not proportional and have the same weight $t\bar\omega_2 - \alpha_1 - 2\alpha_2 - \alpha_3$ (it equals 0 if $t = 2$, and is nonzero whenever $t > 2$). $\quad\square$

Lemma 3 *None of the following diagrams can be contained as a subdiagram in a diagram of a strict simple micromodule:*

$$\overset{t}{\circ\!\!-\!\!\circ} \; \ldots \; \!-\!\!\circ\!\!\Rrightarrow\!\!\circ \quad (t > 1),$$

$$\overset{t}{\circ\!\!-\!\!\circ} \; \ldots \; \!-\!\!\circ\!\!\Lleftarrow\!\!\circ \quad (t > 1),$$

$$\overset{t}{\circ\!\!-\!\!\circ} \; \ldots \; \!-\!\!\circ\!\!\!<\!\!\!{\overset{\circ}{\underset{\circ}{}}} \quad (t > 1),$$

$$\overset{t}{\circ\!\!-\!\!\circ\!\!-\!\!\circ}.$$

The same is true for diagrams 4, 5, 8, 9, 12, 13, 14, 18, 19, 22, 24, 25, 26, 27.

Proof. a) Suppose that the diagram of the module V contains a sub-diagram that does not determine a strict micromodule, and that e_1 is a primitive element of the module V. Restricting the action from the Lie algebra to the subalgebra given by the subdiagram, we obtain a module with this diagram and with the same primitive element e_1. In this module there exist nonproportional elements with the same weight. These elements have the same weight in the module V as well.

b) We shall prove by induction that the diagram $\overset{t}{\circ}\!\!-\!\!\circ \ldots -\!\!\circ\!\!\Rrightarrow\!\!\circ$ $(t > 1)$ does not determines a strict micromodule.

By Lemma 2, this is already true for the diagram $\overset{t}{\circ}\!\!\Rrightarrow\!\!\circ$. Assume that this is true for B_l $(l \geqslant 2)$. We wish to prove this assertion for B_{l+1}. Suppose e_1 is a primitive element of the module V with the diagram

$$\overset{t}{\underset{\alpha_1}{\circ}}\!\!-\!\!\underset{\alpha_2}{\circ}\; \ldots -\!\!\circ\!\!\underset{\alpha_l}{\Rrightarrow}\!\!\underset{\alpha_{l+1}}{\circ}.$$

Then $H_{\alpha_1}e_1 = te_1$, $X_{\alpha_1}e_1 = 0$, and hence e_1 is a primitive element of the $\mathfrak{s}_{\alpha_1}$-module V whose weight equals $t > 1$. Therefore $e_2 = X^t_{-\alpha_1}e_1 \neq 0$. We have

$$X_{\alpha_i}e_2 = 0 \quad (2 \leqslant i \leqslant l_1),$$
$$H_{\alpha_2}e_2 = -t\alpha_1(H_{\alpha_2})e_2 = te_2,$$
$$H_{\alpha_i}e_2 = 0 \quad (3 \leqslant i \leqslant l_1),$$

so that e_2 is a primitive element of the module with the diagram

$$\overset{t}{\underset{\alpha_2}{\circ}}\!\!-\!\!\underset{\alpha_3}{\circ}\; \ldots -\!\!\circ\!\!\underset{\alpha_l}{\Rrightarrow}\!\!\underset{\alpha_{l+1}}{\circ}.$$

(This module is not a strict micromodule.)

c) For the diagrams

$$\overset{t}{\circ}\!\!-\!\!\circ \ldots -\!\!\circ\!\!\Lleftarrow\!\!\circ \quad (t > 1) \quad \text{and} \quad \overset{t}{\circ}\!\!-\!\!\circ \ldots -\!\!\circ\!\!\!<\!\!\begin{smallmatrix}\circ\\[-2pt]\circ\end{smallmatrix} \quad (t > 1)$$

the proof is similar. $\square$

It follows from Lemmas 1–3 that all possible diagrams of strict micromodules appear in the statement of the theorem.

We shall finally prove that the statement of the theorem contains the diagrams of all possible nonstrict micromodules.

Lemma 4 *All proper subdiagrams of a nonstrict micromodule are diagrams of some strict micromodules.*

Proof. Let V be a $\mathfrak{g}$-module with a primitive element e of weight ω, and let $\mathfrak{a}$ be the subalgebra of $\mathfrak{g}$ corresponding to a proper subdiagram of the diagram of $\mathfrak{g}$.

Let E be the $\mathfrak{a}$-submodule generated by e. Assume that E is not a strict micromodule. It is clear that there exist nonproportional elements of E that have the same weight and have the form

$$X_{-\alpha_{i_1}} \dots X_{-\alpha_{i_k}} e \quad \text{and} \quad X_{-\alpha_{j_1}} \dots X_{-\alpha_{j_k}} e,$$

where $\alpha_{i_1}, \dots, \alpha_{i_k}, \alpha_{j_1}, \dots, \alpha_{j_k}$ are the vertices of the subdiagram. It is easy to see that these vectors have the same weight in the $\mathfrak{g}$-module V. Let α be a vertex of the diagram which is not joined with at least one of the vertices of the subdiagram. Then

$$H_\alpha(X_{-\alpha_{i_1}} \dots X_{-\alpha_{i_k}} e) = \left(\omega(H_\alpha) - \sum_{t=1}^{k} \alpha_{i_t}(H_\alpha)\right)(X_{-\alpha_{i_1}} \dots X_{-\alpha_{i_k}} e),$$

and moreover

$$\omega(H_\alpha) - \sum_{t-1}^{k} \alpha_{i_t}(H_\alpha) > 0,$$

so that the weight of the vectors specified above is nonzero. Thus V is not a micromodule. □

Lemma 5 *The modules corresponding to the diagrams*

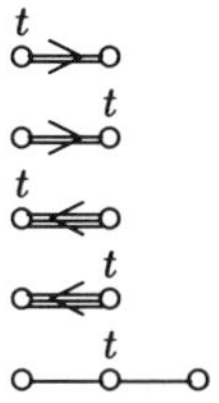

$(t > 2)$;
$(t > 2)$;
$(t > 1)$;
$(t > 1)$;
$(t > 2)$

are not micromodules.

Proof. See the proof of Lemma 2. □

Lemma 6 *The modules corresponding to the diagrams*

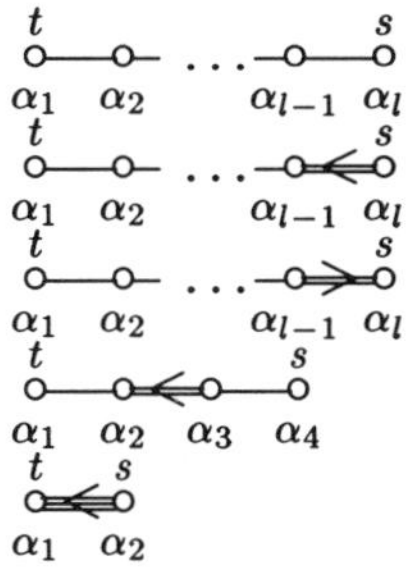

$(t > 1, s \geqslant 1)$;
$(t \geqslant 1, s \geqslant 1)$;
$(t \geqslant 1, s \geqslant 1)$;
$(t \geqslant 1, s \geqslant 1)$;
$(t \geqslant 1, s \geqslant 1)$

are not micromodules.

Proof. We shall use the notation introduced in the proof of Lemma 1. We have already proved that the multiplicity of the weight $\lambda - \alpha_1 - \ldots - \alpha_l$ is greater than 1.

For $\overset{t}{\underset{\alpha_1}{\circ}}\!\!-\!\!\overset{}{\underset{\alpha_2}{\circ}}\ \ldots-\overset{}{\underset{\alpha_{l-1}}{\circ}}\!\!-\!\!\overset{s}{\underset{\alpha_l}{\circ}}$ $(t > 1, s \geqslant 1)$, we have

$$(\lambda - \alpha_1 - \alpha_2 - \ldots - \alpha_l)(H_{\alpha_1}) = t - 2 + 1 - \ldots - 0 - 0 = t - 1 \neq 0.$$

For $\overset{t}{\underset{\alpha_1}{\circ}}\!\!-\!\!\overset{}{\underset{\alpha_2}{\circ}}\ \ldots-\overset{}{\underset{\alpha_{l-1}}{\circ}}\!\!\Leftarrow\!\!\overset{s}{\underset{\alpha_l}{\circ}}$ $(t \geqslant 1, s \geqslant 1)$, we have

$$(\lambda - \alpha_1 - \ldots - \alpha_{l-2} - \alpha_{l-1} - \alpha_l)(H_{\alpha_{l-1}}) = 0 - 0 - \ldots + 1 - 2 + 2 \neq 0.$$

For $\overset{t}{\underset{\alpha_1}{\circ}}\!\!-\!\!\overset{}{\underset{\alpha_2}{\circ}}\ \ldots-\overset{}{\underset{\alpha_{l-1}}{\circ}}\!\!\Rightarrow\!\!\overset{s}{\underset{\alpha_l}{\circ}}$ $(t \geqslant 1, s \geqslant 1)$, we have

$$(\lambda - \alpha_1 - \ldots - \alpha_{l-1} - \alpha_l)(H_{\alpha_l}) = s - 0 - \ldots + 2 - 2 \neq 0.$$

For $\overset{t}{\underset{\alpha_1}{\circ}}\!\!-\!\!\overset{}{\underset{\alpha_2}{\circ}}\!\!\Leftarrow\!\!\overset{}{\underset{\alpha_3}{\circ}}\!\!-\!\!\overset{s}{\underset{\alpha_4}{\circ}}$ $(t \geqslant 1, s \geqslant 1)$, we have

$$(\lambda - \alpha_1 - \alpha_2 - \alpha_3 - \alpha_4)(H_{\alpha_2}) = 0 + 1 - 2 + 2 - 0 \neq 0.$$

For $\overset{t}{\underset{\alpha_1}{\circ}}\!\!\Lleftarrow\!\!\overset{s}{\underset{\alpha_2}{\circ}}$ $(t \geqslant 1, s \geqslant 1)$, we have

$$(\lambda - \alpha_1 - \alpha_2)(H_{\alpha_2}) = 5 + 3 - 2 \neq 0.$$

Thus the weight $\lambda - \alpha_1 - \ldots - \alpha_l$ is nonzero for every module in question. □

It follows from Lemmas 4–6 that the diagrams of all nonstrict micromodules appear in the statement of the theorem.

The value of $\dim V^0$ for nonstrict micromodules is computed in accordance with Proposition 5 ([1], Chapter VIII, §7, no. 2; Chapter VI, Tables).

This completes the proof of the theorem.   □

References

1. N. Bourbaki. *Groupes et algébres de Lie.* Paris, Hermann, Ch. I (2nd ed.), 1971; Ch. II, III, 1972; Ch. IV–VI, 1968; Ch. VII, VIII, 1974.

A SPECTRAL SEQUENCE
FOR THE TANGENT SHEAF COHOMOLOGY
OF A SUPERMANIFOLD

A.L. ONISHCHIK

Yaroslavl University, 150 000 Yaroslavl, Russia

Abstract. We construct a spectral sequence permitting to compute the tangent sheaf cohomology of a complex analytic supermanifold in terms of the tangent sheaf cohomology of the associated split supermanifold. The first coboundary operators of this sequence are determined, and the example of the superquadric in the projective superplane is studied.

Mathematics subject classification (1991): 58A50, 17B70.

Key words: supermanifold, tangent sheaf, cohomology, spectral sequence, superquadric.

1. Introduction

We consider complex analytic supermanifolds in the sense of Berezin–Leites (see [4, 6]). An important problem is to calculate the cohomology of the tangent sheaf $\mathcal{T} = \mathcal{D}er\,\mathcal{O}$ of a supermanifold $(M, \mathcal{O})$. If $(M, \mathcal{O})$ is split, i.e., is determined by a holomorphic vector bundle over the complex manifold M, then $\mathcal{T}$ is a locally free analytic sheaf on M, and its cohomology can be calculated in many cases, using the well elaborated tools of complex analytic geometry. In the nonsplit case, these methods cannot be applied directly, but one can use the associated split supermanifold $(M, \mathcal{O}_{\mathrm{gr}})$. Our goal is to construct a spectral sequence (E_r, d_r) of differential bigraded Lie superalgebras such that $E_2 = H^*(M, \mathcal{T}_{\mathrm{gr}})$, where $\mathcal{T}_{\mathrm{gr}} = \mathcal{D}er\,\mathcal{O}_{\mathrm{gr}}$, and E_∞ is associated with a filtration in $H^*(M, \mathcal{T})$. The first nonzero operators d_r are determined, and some examples are studied.

B. P. Komrakov et al. (eds.), Lie Groups and Lie Algebras, 199–215.
© 1998 *Kluwer Academic Publishers. Printed in the Netherlands.*

This work was supported in part by the International Science Foundation (Grant RO4000). The author expresses his deep gratitude to the Sonderforschungsbereich 288 for hospitality at Humboldt University, Berlin (June – July 1994), where the paper was written.

2. Preliminaries

We consider here complex analytic supermanifolds in the sense of Berezin–Leites (see [4, 6]). The supermanifold having a complex manifold M as its reduction and a sheaf $\mathcal{O}$ on M as its structure sheaf, will be denoted by $(M, \mathcal{O})$. As in [6, 7], we say that a supermanifold is *split* if it is isomorphic to $(M, \bigwedge_{\mathcal{F}} \mathcal{E})$, where $M = (M, \mathcal{F})$ is an ordinary complex analytic manifold and $\mathcal{E}$ is a locally free analytic sheaf on M. The structure sheaf $\mathcal{O} = \bigwedge_{\mathcal{F}} \mathcal{E}$ admits in this case the $\mathbb{Z}$-grading by the subsheaves $\mathcal{O}_p = \bigwedge_{\mathcal{F}}^p \mathcal{E}$, $p \geq 0$. Note that the $\mathbb{Z}_2$-grading of $\mathcal{O}$ is compatible with this $\mathbb{Z}$-grading, in the sense that

$$\mathcal{O}_{\bar{0}} = \bigoplus_{p \geq 0} \mathcal{O}_{2p}, \quad \mathcal{O}_{\bar{1}} = \bigoplus_{p \geq 0} \mathcal{O}_{2p+1}. \tag{1}$$

It is well known that with an arbitrary supermanifold $(M, \mathcal{O})$ a split supermanifold $(M, \mathcal{O}_{\mathrm{gr}})$, having the same reduction M and the same dimension as $(M, \mathcal{O})$, can be associated. It is defined in the following way. Let $\mathcal{J}$ denote the subsheaf of ideals in $\mathcal{O}$, generated by the odd elements. Consider the filtration

$$\mathcal{O} = \mathcal{J}^0 \supset \mathcal{J}^1 \supset \mathcal{J}^2 \supset \dots \tag{2}$$

of $\mathcal{O}$ by the powers $\mathcal{J}^p$ of $\mathcal{J}$, and denote by $\mathrm{gr}\,\mathcal{O}$ the associated $\mathbb{Z}$-graded sheaf of superalgebras. Thus,

$$\mathrm{gr}\,\mathcal{O} = \bigoplus_{p \geq 0} \mathrm{gr}_p \mathcal{O}, \quad \text{where } \mathrm{gr}_p \mathcal{O} = \mathcal{J}^p / \mathcal{J}^{p+1}.$$

One checks that $\mathrm{gr}\,\mathcal{O} \simeq \bigwedge_{\mathcal{F}} \mathcal{E}$, where $\mathcal{F} = \mathrm{gr}_0 \mathcal{O}$, $\mathcal{E} = \mathrm{gr}_1 \mathcal{O}$.

A sheaf $\mathcal{S}$ on a supermanifold $(M, \mathcal{O})$ will be called *analytic* if $\mathcal{S}$ is a sheaf of $\mathcal{O}$-modules. An important example is the *tangent sheaf* $\mathcal{T} = \mathcal{D}er\,\mathcal{O}$ of $(M, \mathcal{O})$, i.e., the sheaf of derivations of $\mathcal{O}$ over $\mathbb{C}$ in the $\mathbb{Z}_2$-graded sense. This is a locally free analytic sheaf of rank $n|m = \dim(M, \mathcal{O})$. It has a natural structure of the sheaf of complex Lie superalgebras, the bracket being defined by

$$[u, v] = uv + (-1)^{p(u)p(v)+1} vu. \tag{3}$$

We consider the following filtration of $\mathcal{T}$:

$$\mathcal{T} = \mathcal{T}_{(-1)} \supset \mathcal{T}_{(0)} \supset \dots \supset \mathcal{T}_{(m)} \supset \mathcal{T}_{(m+1)} = 0, \tag{4}$$

where

$$\mathcal{T}_{(p)} = \{u \in \mathcal{T} \mid u(\mathcal{O}) \subset \mathcal{J}^p, \ u(\mathcal{J}) \subset \mathcal{J}^{p+1}\}, \ p \geq 0.$$

One sees easily that $\mathcal{T}_{(p)}$ are analytic subsheaves of $\mathcal{T}$ such that

$$\mathcal{J}\mathcal{T}_{(p)} \subset \mathcal{T}_{(p+1)}, \ p \geq -1. \tag{5}$$

On the other hand,

$$[\mathcal{T}_{(p)}, \mathcal{T}_{(q)}] \subset \mathcal{T}_{(p+q)},$$

and so we have a filtered sheaf of Lie superalgebras.

Now we make some remarks concerning the case of a split supermanifold. If $(M, \mathcal{O})$ is split, then $\mathcal{T}$ is a $\mathbb{Z}$-graded analytic sheaf, the grading being given by

$$\mathcal{T} = \bigoplus_{p \geq -1} \mathcal{T}_p, \tag{6}$$

where

$$\mathcal{T}_p = \mathcal{D}er_p \mathcal{O} = \{u \in \mathcal{T} \mid u(\mathcal{O}_q) \subset \mathcal{O}_{q+p} \text{ for all } q \in \mathbb{Z}\}.$$

As well as for the structure sheaf, the $\mathbb{Z}_2$-grading is compatible with this $\mathbb{Z}$-grading, i.e.,

$$\mathcal{T}_{\bar{0}} = \bigoplus_{p \geq 0} \mathcal{T}_{2p}, \ \mathcal{T}_{\bar{1}} = \bigoplus_{p \geq 0} \mathcal{T}_{2p-1}.$$

Under the bracket (3) and the grading (6), the sheaf $\mathcal{T}$ becomes a $\mathbb{Z}$-graded sheaf of Lie superalgebras.

One also verifies that the filtration (4) of $\mathcal{T}$ coincides with that associated with the grading (6), so that

$$\mathcal{T}_{(p)} = \bigoplus_{r \geq p} \mathcal{T}_r. \tag{7}$$

Since in the split case the inclusion $\mathcal{F} \subset \mathcal{O}$ is valid, any analytic sheaf on $(M, \mathcal{O})$ is an analytic sheaf on the complex manifold M as well. In particular, $\mathcal{T}$ turns out to be a locally free $\mathbb{Z}$-graded analytic sheaf of a finite rank on M.

Now we return to the general case. The filtration (4) of the tangent sheaf $\mathcal{T}$ of a supermanifold $(M, \mathcal{O})$ gives rise to the associated graded analytic sheaf

$$\text{gr} \, \mathcal{T} = \bigoplus_{p \geq -1} \text{gr}_p \mathcal{T},$$

where

$$\text{gr}_p \mathcal{T} = \mathcal{T}_{(p)} / \mathcal{T}_{(p+1)}.$$

This sheaf is actually a graded analytic sheaf on $(M, \mathrm{gr}\,\mathcal{O})$. In fact, due to (5), we have

$$\mathcal{J}^p \mathcal{T}_{(q)} \subset \mathcal{T}_{(p+q)}, \; \mathcal{J}^{p+1}\mathcal{T}_{(q+1)} \subset \mathcal{T}_{(p+q+2)} \subset \mathcal{T}_{(p+q+1)},$$

which gives rise to a multiplication $\mathrm{gr}_p\mathcal{O} \times \mathrm{gr}_q\mathcal{T} \to \mathrm{gr}_{p+q}\mathcal{T}$. We want to fix the following simple fact.

Proposition 1 *For any supermanifold $(M, \mathcal{O})$, the sheaf $\mathrm{gr}\,\mathcal{T}$ is isomorphic (as a $\mathbb{Z}$-graded analytic sheaf or a $\mathbb{Z}$-graded sheaf of complex Lie superalgebras on $(M, \mathrm{gr}\,\mathcal{O})$) to the tangent sheaf $\mathcal{D}er(\mathrm{gr}\,\mathcal{O})$.*

Proof. For any $u \in (\mathcal{T}_{(p)})_x$, $x \in M$, we have $u(\mathcal{J}_x^q) \subset \mathcal{J}_x^{q+p}$, $u(\mathcal{J}_x^{q+1}) \subset \mathcal{J}_x^{q+p+1}$, and thus u defines a $\mathbb{C}$-linear mapping $\hat{u} : (\mathrm{gr}\,\mathcal{O})_x \to (\mathrm{gr}\,\mathcal{O})_x$ which clearly is a derivation of degree p. Now, $\hat{u} = 0$ if and only if $u(\mathcal{J}_x^q) \subset \mathcal{J}_x^{q+p+1}$ for all q, i.e., if $u \in (\mathcal{T}_{(p+1)})_x$. One verifies easily that the correspondence $\sigma_p : u + (\mathcal{T}_{(p+1)})_x \mapsto \hat{u}$ gives the desired isomorphism. □

Let $(M, \mathcal{O})$ be a supermanifold. In the study of $(M, \mathcal{O})$, the cohomology groups $H^q(M, \mathcal{T})$ of its tangent sheaf are of great importance. For example, $H^0(M, \mathcal{T})$ is the Lie superalgebra of all holomorphic vector fields on $(M, \mathcal{O})$, $H^1(M, \mathcal{T})$ is the set of all infinitesimal deformations of $(M, \mathcal{O})$, $H^2(M, \mathcal{T})$ contains the obstructions to the existence of analytic deformations (see [2, 8]). In the case when $(M, \mathcal{O})$ is split, these cohomology groups can sometimes be calculated, using the well elaborated methods of complex analytic geometry, since $\mathcal{T}$ is a locally free analytic sheaf of a finite rank on the complex manifold M. In the nonsplit case, these methods cannot be applied directly. In what follows, we shall construct a spectral sequence that helps to reduce the calculation of the tangent sheaf cohomology to the split case.

To describe the cohomology $H^q(M, \mathcal{A})$, where $\mathcal{A}$ is a sheaf of groups on M, we use Čech cochains with values in $\mathcal{A}$. Let $\mathfrak{U} = (U_i)_{i \in I}$ be an open cover of M, and let $C^q(\mathfrak{U}, \mathcal{A})$ denote the group of q-cochains of $\mathfrak{U}$ with values in $\mathcal{A}$ (without any conditions of skew-symmetricity) and $d : C^{q-1}(\mathfrak{U}, \mathcal{A}) \to C^q(\mathfrak{U}, \mathcal{A})$ the coboundary operator. As usual, we define $Z^q(\mathfrak{U}, \mathcal{A}) = \mathrm{Ker}\,d$ and $H^q(\mathfrak{U}, \mathcal{A}) = Z^q(\mathfrak{U}, \mathcal{A})/dC^{q-1}(\mathfrak{U}, \mathcal{A})$, so that $H^*(\mathfrak{U}, \mathcal{A}) = \bigoplus_{q \geq 0} H^q(\mathfrak{U}, \mathcal{A})$ is the cohomology of the cochain complex $C^*(\mathfrak{U}, \mathcal{A}) = \bigoplus_{q \geq 0} C^q(\mathfrak{U}, \mathcal{A})$. Suppose that the cover $\mathfrak{U}$ is Stein, i.e., that all the open subsets U_i (and hence all their finite intersections) are Stein. If $\mathcal{A}$ is a coherent analytic sheaf on M or on $(M, \mathcal{O})$, then any Stein cover is acyclic relative to $\mathcal{A}$ (see [2, 8]), and, by the well-known theorem of Leray (see [3]), the natural mapping $H^*(\mathfrak{U}, \mathcal{A}) \to H^*(M, \mathcal{A}) = \bigoplus_{q \geq 0} H^q(M, \mathcal{A})$ is an isomorphism of graded groups.

Now we make some remarks concerning the tangent sheaf cohomology $H^*(M, \mathcal{T})$.

First, the $\mathbb{Z}_2$-grading of $\mathcal{T}$ gives rise to the $\mathbb{Z}_2$-gradings in $C^*(\mathfrak{U}, \mathcal{T})$ and $H^*(M, \mathcal{T})$ given by

$$
\begin{aligned}
C_{\bar{0}}(\mathfrak{U}, \mathcal{T}) &= \bigoplus_{q \geq 0} C^{2q}(\mathfrak{U}, \mathcal{T}_{\bar{0}}) \oplus \bigoplus_{q \geq 0} C^{2q+1}(\mathfrak{U}, \mathcal{T}_{\bar{1}}), \\
C_{\bar{1}}(\mathfrak{U}, \mathcal{T}) &= \bigoplus_{q \geq 0} C^{2q}(\mathfrak{U}, \mathcal{T}_{\bar{1}}) \oplus \bigoplus_{q \geq 0} C^{2q+1}(\mathfrak{U}, \mathcal{T}_{\bar{0}}). \\
H_{\bar{0}}(M, \mathcal{T}) &= \bigoplus_{q \geq 0} H^{2q}(M, \mathcal{T}_{\bar{0}}) \oplus \bigoplus_{q \geq 0} H^{2q+1}(M, \mathcal{T}_{\bar{1}}), \\
H_{\bar{1}}(M, \mathcal{T}) &= \bigoplus_{q \geq 0} H^{2q}(M, \mathcal{T}_{\bar{1}}) \oplus \bigoplus_{q \geq 0} H^{2q+1}(M, \mathcal{T}_{\bar{0}}).
\end{aligned}
\tag{8}
$$

If $(M, \mathcal{O})$ is split, then we actually have the bigradings

$$
C^*(\mathfrak{U}, \mathcal{T}) = \bigoplus_{p,q} C^p(\mathfrak{U}, \mathcal{T}_q), \quad H^*(M, \mathcal{T}) = \bigoplus_{p,q} H^p(M, \mathcal{T}_q),
$$

and the $\mathbb{Z}_2$-gradings (8) are compatible with the complete degree of these two bigradings.

Secondly, the bracket (3) on $\mathcal{T}$ gives rise to bilinear operations in the vector spaces $C^*(\mathfrak{U}, \mathcal{T})$ and $H^*(M, \mathcal{T})$ which we denote by $[\,,\,]$ as well (see [3]). The operation in $C^*(\mathfrak{U}, \mathcal{T})$ is given by

$$
[c, c']_{i_0, \ldots, i_{p+q}} = [c_{i_0, \ldots, i_p}, c'_{i_{p+1}, \ldots, i_{p+q}}],
\tag{9}
$$

where $c \in Z^p(\mathfrak{U}, \mathcal{T})$, $c' \in Z^q(\mathfrak{U}, \mathcal{T})$, and the operation in $H^*(M, \mathcal{T})$ is inherited from that in $C^*(\mathfrak{U}, \mathcal{T})$. Clearly, $C^*(\mathfrak{U}, \mathcal{T})$ and $H^*(M, \mathcal{T})$ are $(\mathbb{Z} \times \mathbb{Z}_2)$-graded algebras, and they are bigraded algebras whenever $(M, \mathcal{O})$ is split. They also are superalgebras with respect to the $\mathbb{Z}_2$-gradings (8) ($H^*(M, \mathcal{T})$ is actually a Lie superalgebra).

Now we recall some known facts concerning the classification of the nonsplit supermanifolds having the given associated split manifold.

Let $(M, \mathcal{O})$ be a supermanifold. Consider the sheaf $\mathcal{A}ut\,\mathcal{O}$ of automorphisms of the structure sheaf $\mathcal{O}$ (as usually, any automorphism is even and maps each stalk $\mathcal{O}_x$, $x \in M$, onto itself). This is a sheaf of groups. Its sections are the automorphisms of $(M, \mathcal{O})$ that are identical on M. Clearly, any $a \in \mathcal{A}ut\,\mathcal{O}_x$, $x \in M$, maps $\mathcal{J}_x$ onto itself, and hence preserves the filtration (2) and induces a germ of an automorphism of $\mathrm{gr}\,\mathcal{O}$. By definition, a induces the identity mapping on $\mathcal{F}_x = \mathcal{O}_x / \mathcal{J}_x$. Consider the filtration

$$
\mathcal{A}ut\,\mathcal{O} = \mathcal{A}ut_{(1)}\mathcal{O} \supset \ldots \supset \mathcal{A}ut_{(p)}\mathcal{O} \supset \ldots,
\tag{10}
$$

where

$$
\mathcal{A}ut_{(p)}\mathcal{O}_x = \{a \in \mathcal{A}ut\,\mathcal{O}_x \mid a(f) - f \in \mathcal{J}_x^p \text{ for all } f \in \mathcal{O}_x\}, \quad x \in M.
$$

Clearly, any $\mathcal{A}ut_{(p)}\mathcal{O}$ is a subsheaf of invariant subgroups of $\mathcal{A}ut\,\mathcal{O}$. It is the subsheaf of trivial subgroups whenever $p \geq m + 1$, where, as above, m denotes the odd dimension of $(M, \mathcal{O})$. We also notice that the group $\mathrm{Aut}(M, \mathcal{O})$ of automorphisms of the supermanifold $(M, \mathcal{O})$ acts on $\mathcal{A}ut\,\mathcal{O}$ by automorphisms of this sheaf, leaving any $\mathcal{A}ut_{(p)}\mathcal{O}$ invariant.

We apply the sheaves of automorphisms to our classification problem. Let $\mathbf{E}$ be a holomorphic vector bundle over a complex manifold M, and $\mathcal{E}$ the sheaf of holomorphic sections of $\mathbf{E}$. Consider the split supermanifold $(M, \mathcal{O}_{\mathrm{gr}})$, where $\mathcal{O}_{\mathrm{gr}} = \bigwedge \mathcal{E}$. We want to describe the family of all supermanifolds $(M, \mathcal{O})$ such that $\mathrm{gr}\,\mathcal{O} = \mathcal{O}_{\mathrm{gr}}$. Let $\mathrm{Aut}\,\mathbf{E}$ be the group of all holomorphic automorphisms of the vector bundle $\mathbf{E}$. Since $\mathrm{Aut}\,\mathbf{E} \subset \mathrm{Aut}(M, \mathcal{O}_{\mathrm{gr}})$, this group acts on the sheaf $\mathcal{A}ut_{(2)}\mathcal{O}_{\mathrm{gr}}$, and hence on its 1-cohomology set. The following assertion was proved in [4].

Proposition 2 *To any supermanifold $(M, \mathcal{O})$ satisfying $\mathrm{gr}\,\mathcal{O} = \mathcal{O}_{\mathrm{gr}}$ there corresponds an element of the set $H^1(M, \mathcal{A}ut_{(2)}\mathcal{O}_{\mathrm{gr}})$. This correspondence gives rise to a bijection between the isomorphism classes of supermanifolds, satisfying the above condition, and the orbits of the group $\mathrm{Aut}\,\mathbf{E}$ on $H^1(M, \mathcal{A}ut_{(2)}\mathcal{O}_{\mathrm{gr}})$.*

Let us describe the correspondence mentioned in Proposition 2. Let $(M, \mathcal{O})$ be a supermanifold such that $\mathrm{gr}\,\mathcal{O} = \mathcal{O}_{\mathrm{gr}}$. We can choose an open cover $\mathfrak{U} = (U_i)_{i \in I}$ of M such that there exist sheaf isomorphisms $f_i : \mathcal{O}|U_i \to \mathcal{O}_{\mathrm{gr}}|U_i$, $i \in I$, inducing the identity isomorphisms of $\mathrm{gr}\,\mathcal{O}|U_i$ onto $\mathcal{O}_{\mathrm{gr}}|U_i$. Setting $g_{ij} = f_i \circ f_j^{-1}$, we get a 1-cocycle $g = (g_{ij}) \in Z^1(\mathfrak{U}, \mathcal{A}ut_{(2)}\mathcal{O}_{\mathrm{gr}})$. Its cohomology class $\gamma \in H^1(M, \mathcal{A}ut_{(2)}\mathcal{O}_{\mathrm{gr}})$ does not depend of the choice of f_i.

Clearly, the given split supermanifold $(M, \mathcal{O}_{\mathrm{gr}})$ corresponds to the distinguished point $e \in H^1(M, \mathcal{A}ut_{(2)}\mathcal{O}_{\mathrm{gr}})$ which is fixed under the action of $\mathrm{Aut}\,\mathbf{E}$.

Now we use the even items $\mathcal{A}ut_{(2p)}\mathcal{O}_{\mathrm{gr}}$ of the filtration (10) of $\mathcal{O}_{\mathrm{gr}}$, $p = 1, \ldots, [m/2] + 1$, where m is the odd dimension of $(M, \mathcal{O}_{\mathrm{gr}})$, which is equal to the rank of $\mathbf{E}$. Denoting by $H_{(2p)}$ the image of the natural mapping $H^1(M, \mathcal{A}ut_{(2p)}\mathcal{O}_{\mathrm{gr}}) \to H^1(M, \mathcal{A}ut_{(2)}\mathcal{O}_{\mathrm{gr}})$, we get the $(\mathrm{Aut}\,\mathbf{E})$-invariant filtration

$$H^1(M, \mathcal{A}ut_{(2)}\mathcal{O}_{\mathrm{gr}}) = H_{(2)} \supset \ldots \supset H_{(2p)} \supset \ldots \supset H_{(2([\frac{m}{2}]+1))} = \{e\}. \quad (11)$$

Take $\gamma \in H^1(M, \mathcal{A}ut_{(2)}\mathcal{O}_{\mathrm{gr}})$. Following [7], we define the *order* of γ (denoted $o(\gamma)$) as the maximal of the numbers $2p$ such that $\gamma \in H_{(2p)}$. The order of a supermanifold $(M, \mathcal{O})$ such that $\mathrm{gr}\,\mathcal{O} = \mathcal{O}_{\mathrm{gr}}$ is, by definition, the order of the corresponding cohomology class.

There exists a natural relationship between automorphisms and derivations of $\mathcal{O}_{\mathrm{gr}}$. Let $\mathcal{T}_{\mathrm{gr}} = \mathcal{D}er\,\mathcal{O}_{\mathrm{gr}}$ be the tangent sheaf of $(M, \mathcal{O}_{\mathrm{gr}})$. Then we have the filtration (see (5))

$$(\mathcal{T}_{gr})_{\bar{0}} \supset (\mathcal{T}_{gr})_{(2)\bar{0}} \supset \cdots \supset (\mathcal{T}_{gr})_{(2p)\bar{0}} \supset \cdots,$$

where, by (7),

$$(\mathcal{T}_{gr})_{(2p)\bar{0}} = \bigoplus_{r \geq p} (\mathcal{T}_{gr})_{2r}.$$

Using the exponential series, define the mapping

$$\exp : (\mathcal{T}_{gr})_{(2)\bar{0}} \to \mathcal{A}ut_{(2)}\mathcal{O}_{gr}.$$

Since any derivation from $(\mathcal{T}_{gr})_{(2)\bar{0}}$ is nilpotent, this series is actually a polynomial. One proves that exp is an isomorphism of sheaves of sets (but not of sheaves of groups, in general) and maps $(\mathcal{T}_{gr})_{(2p)\bar{0}}$ onto $\mathcal{A}ut_{(2p)}\mathcal{O}_{gr}$ [7].

   Choose a class $\gamma \in H^1(M, \mathcal{A}ut_{(2)}\mathcal{O}_{gr})$. Clearly, $o(\gamma) = 2p$ if and only if γ can be represented by a cocycle $g \in Z^1(\mathfrak{U}, \mathcal{A}ut_{(2p)}\mathcal{O}_{gr})$. We have $g = \exp z$, where $z = z_{2p} + z_{2p+2} + \cdots$, $z_{2r} \in C^1(\mathfrak{U}, (\mathcal{T}_{gr})_{2r})$. One proves (see [7]) that $z_{2r} \in Z^1(\mathfrak{U}, (\mathcal{T}_{gr})_{2r})$ for all $r < 2p$. In particular, the cocycle z_{2p} gives rise to a cohomology class $\lambda_{2p}(\gamma) \in H^1(M, (\mathcal{T}_{gr})_{2p})$ which is uniquely determined by γ.

3. The spectral sequence

Now we pass to the construction of a spectral sequence for the cohomology of the tangent sheaf $\mathcal{T}$ of an arbitrary supermanifold $(M, \mathcal{O})$ of dimension $n|m$. We fix an open Stein cover $\mathfrak{U} = (U_i)_{i \in I}$ of M and consider the corresponding Čech cochain complex $C^*(\mathfrak{U}, \mathcal{T}) = \bigoplus_{p \geq 0} C^p(\mathfrak{U}, \mathcal{T})$. The filtration (4) gives rise to the filtration

$$C^*(\mathfrak{U}, \mathcal{T}) = C_{(-1)} \supset C_{(0)} \supset \cdots \supset C_{(p)} \supset \cdots \supset C_{(m+1)} = 0 \qquad (12)$$

of this complex by the subcomplexes

$$C_{(p)} = C^*(\mathfrak{U}, \mathcal{T}_{(p)}).$$

Denoting by $H(M, \mathcal{T})_{(p)}$ the image of the natural mapping $H^*(M, \mathcal{T}_{(p)}) \to H^*(M, \mathcal{T})$, we get the filtration

$$H^*(M, \mathcal{T}) = H(M, \mathcal{T})_{(-1)} \supset \cdots \supset H(M, \mathcal{T})_{(p)} \supset \cdots . \qquad (13)$$

Note that (12) is a filtration of the graded differential algebra $C^*(\mathfrak{U}, \mathcal{T})$ (under the bracket (9)) by graded differential subalgebras, and hence (13) is a filtration of the graded algebra $H^*(M, \mathcal{T})$ by graded subalgebras. Denote by $\operatorname{gr} H^*(M, \mathcal{T})$ the bigraded algebra associated with the filtration (13); its bigrading is given by

$$\operatorname{gr} H^*(M, \mathcal{T}) = \bigoplus_{\substack{p \geq -1 \\ q \geq 0}} \operatorname{gr}_p H^q(M, \mathcal{T}).$$

By the general procedure invented by J. Leray, the filtration (12) gives rise to a spectral sequence of bigraded algebras E_r converging to $E_\infty \simeq \operatorname{gr} H^*(M, \mathcal{T})$. It is constructed in the following way.

For any $p \geq -1, r \geq 0$, define the vector spaces

$$C_r^p = \{ c \in C_{(p)} \,|\, dc \in C_{(p+r)} \}.$$

Then, for a fixed p, consider

$$C_{(p)} = C_0^p \supset \ldots \supset C_r^p \supset C_{r+1}^p \supset \ldots .$$

The r-th term of the spectral sequence is defined by

$$E_r = \bigoplus_{r=-1}^{m} E_r^p, \ r \geq 0, \ \text{where} \ E_r^p = C_r^p / (C_{r-1}^{p+1} + dC_{r-1}^{p-r+1}).$$

The bracket (9) gives rise to a structure of the graded algebra in E_r. Since $d(C_r^p) \subset C_r^{p+r}$, d induces a derivation d_r of E_r of degree r such that $d_r^2 = 0$. Then E_{r+1} is naturally isomorphic to the homology algebra $H(E_r, d_r)$. Denoting $Z_r = \operatorname{Ker} d_r$, we have the natural mapping $\kappa_{r+1}^r : Z_r \to E^{r,r+1}$. For any $s > r$, denote $\kappa_s^r = \kappa_s^{s-1} \circ \ldots \circ \kappa_{r+1}^r$ (this composition is not defined on the entire kernel Z_r).

The $\mathbb{Z}_2$-grading (8) in $C^*(\mathfrak{U}, \mathcal{T})$ gives rise to certain $\mathbb{Z}_2$-gradings in C_r^p and E_r^p, turning E_r into a superalgebra. Clearly, the coboundary operator d in $C^*(\mathfrak{U}, \mathcal{T})$ is odd. It follows that the coboundary operator d_r is odd for any $r \geq 0$. The superalgebras E_r are also endowed with a second $\mathbb{Z}$-grading. Namely, for any $q \in \mathbb{Z}$, set

$$C_r^{p,q} = C_r^p \cap C^{p+q}(\mathfrak{U}, \mathcal{T}), \ E_r^{p,q} = C_r^{p,q} / (C_{r-1}^{p+1,q-1} + dC_{r-1}^{p-r+1,q+r-2}).$$

Then

$$E_r = \bigoplus_{p,q} E_r^{p,q}.$$

Clearly,

$$d_r(E_r^{p,q}) \subset E_r^{p+r,q-r+1} \tag{14}$$

for any r, p, q.

One sees easily that $C_r^{p,q} = 0$ for all p, r if $q \leq -(m+1)$. Therefore, for a fixed q, we have $d(C_r^{p,q}) = 0$ for all $p \geq -1$ and all $r \geq q - m + 2$.

This implies that $\kappa^r_{r+1} : E^{p,q}_r \to E^{p,q}_{r+1}$ is an isomorphism for all p and $r \geq r_0(q) = q - m + 2$. Setting $E^{p,q}_\infty = E^{p,q}_{r_0(q)}$, we get the bigraded superalgebra

$$E_\infty = \bigoplus_{p,q} E^{p,q}_\infty.$$

Now we prove certain properties of the spectral sequence (E_r). Some of them are well known and are valid in a more general situation.

Proposition 3 *The first three terms of the spectral sequence (E_r) can be identified with the following bigraded algebras:*

$$E_0 = C^*(\mathfrak{U}, \mathcal{T}_{\mathrm{gr}}), \ E_1 = E_2 = H^*(M, \mathcal{T}_{\mathrm{gr}}).$$

Here

$$E^{p,q}_0 = C^{p+q}(\mathfrak{U}, (\mathcal{T}_{\mathrm{gr}})_p), \ E^{p,q}_1 = E^{p,q}_2 = H^{p+q}(M, (\mathcal{T}_{\mathrm{gr}})_p).$$

We have $d_{2k+1} = 0$ and, hence, $E_{2k+1} = E_{2k+2}$ for all $k \geq 0$.

Proof. By definition, we have

$$E^p_0 = C_{(p)}/C_{(p+1)}, \ p \geq -1,$$

where the coboundary operator d_0 of degree 0 is induced by $d : C_{(p)} \to C_{(p)}$. On the other hand, the exact sequence

$$0 \to \mathcal{T}_{(p+1)} \to \mathcal{T}_{(p)} \xrightarrow{\sigma_p} (\mathcal{T}_{\mathrm{gr}})_p \to 0 \tag{15}$$

(see Proposition 1) and Theorem B for Stein supermanifolds (see [2, 8]) yield the exact sequence

$$0 \to \mathcal{T}_{(p+1)}(U) \to \mathcal{T}_{(p)}(U) \xrightarrow{\sigma_p} (\mathcal{T}_{\mathrm{gr}})_p(U) \to 0$$

for any Stein open subset $U \subset M$. Therefore

$$C^*(\mathfrak{U}, (\mathcal{T}_{\mathrm{gr}})_p) \simeq C_{(p)}/C_{(p+1)} = E^p_0, \ p \geq -1.$$

One sees easily that this is an isomorphism of complexes and that the resulting isomorphism $C^*(\mathfrak{U}, \mathcal{T}_{\mathrm{gr}}) \simeq E_0$ is an isomorphism of bigraded algebras. It follows that

$$E_1 \simeq H(E_0, d_0) \simeq H^*(\mathfrak{U}, \mathcal{T}_{\mathrm{gr}}) \simeq H^*(M, \mathcal{T}_{\mathrm{gr}}).$$

We see that

$$(E_0)_{\bar{0}} = \bigoplus_{p,q} E^{p,2q}_0, \ (E_0)_{\bar{1}} = \bigoplus_{p,q} E^{p,2q+1}_0.$$

Hence, the parity of elements lying in $E^{p,q}_0$ is the same as the parity of q. Passing to homology, we see that this is true for any E_r, $r \geq 0$. This implies, by (14), that the coboundary operator d_{2k+1} is even. Since it is odd, we have $d_{2k+1} = 0$ for any $k \geq 0$. $\square$

Proposition 4 *One has the following identification of bigraded algebras:*

$$E_\infty = \operatorname{gr} H^*(M, \mathcal{T}), \quad \text{where } E_\infty^{p,q} = \operatorname{gr}_p H^{p+q}(M, \mathcal{T}).$$

Proof. Clearly, for $r \geq r_0(q)$ we have $C_r^{p,q} = Z^{p+q}(\mathfrak{U}, \mathcal{T}_{(p)})$. It follows that

$$
\begin{aligned}
E_\infty^{p,q} &= Z^{p+q}(\mathfrak{U}, \mathcal{T}_{(p)})/(Z^{p+q}(\mathfrak{U}, \mathcal{T}_{(p+1)}) + dC^{p+q-1}(\mathfrak{U}, \mathcal{T}) \cap Z^{p+q}(\mathfrak{U}, \mathcal{T}_{(p)})) \\
&= H^{p+q}(M, \mathcal{T})_{(p)}/(Z^{p+q}(\mathfrak{U}, \mathcal{T}_{(p+1)})/(dC^{p+q-1}(\mathfrak{U}, \mathcal{T}) \cap Z^{p+q}(\mathfrak{U}, \mathcal{T}_{(p+1)}))) \\
&= H^{p+q}(M, \mathcal{T})_{(p)}/H^{p+q}(M, \mathcal{T})_{(p+1)} = \operatorname{gr}_p H^{p+q}(M, \mathcal{T}).
\end{aligned}
$$

$\square$

Corollary 1 *If M is compact, then*

$$\dim H^k(M, \mathcal{T}) = \sum_{p+q=k} \dim E_\infty^{p,q}.$$

Proof. In fact, if M is compact, then all cohomology groups with values in a coherent analytic sheaf on $(M, \mathcal{O})$ or M are of finite dimension. $\square$

Consider the cohomology exact sequence

$$H^{p+q}(M, \mathcal{T}_{(p+1)}) \to H^{p+q}(M, \mathcal{T}_{(p)}) \overset{\sigma_p^*}{\to} H^{p+q}(M, (\mathcal{T}_{\mathrm{gr}})_p) = E_2^{p,q}$$

associated with (15). We would like to interpret the vector subspace $\operatorname{Im} \sigma_p^* \subset H^{p+q}(M, (\mathcal{T}_{\mathrm{gr}})_p)$ by means of our spectral sequence. An element $a \in E_2^{p,q}$ shall be called a *permanent cocycle* if $d_2 a = 0$, $d_4(\kappa_4^2 a) = 0$, $d_6(\kappa_6^2 a) = 0$, etc. Let us denote by $Z_\infty^{p,q}$ the subspace of permanent cocycles. Any $a \in Z_\infty^{p,q}$ determines an element $a^* \in E_\infty^{p,q}$, and hence we have a linear mapping $\kappa_\infty^2 : Z_\infty^{p,q} \to H_\infty^{p,q}$.

Note that $\operatorname{Im} \sigma_p^* \subset Z_\infty^{p,q}$. In fact, if $a = \sigma_p^*(b)$, where $b \in H^{p+q}(M, \mathcal{T}_{(p)})$, then b is represented by a cocycle $z \in Z^{p+q}(\mathfrak{U}, \mathcal{T}_{(p)})$, and a is the coset of the same z regarded as an element of $C_2^{p,q}$. Since $dz = 0$, a is a permanent cocycle of the spectral sequence.

Proposition 5 *We have the commutative diagram*

$$
\begin{array}{ccc}
H^{p+q}(M, \mathcal{T}_{(p)}) & \overset{\sigma_p^*}{\longrightarrow} & Z_\infty^{p,q} \subset E_2^{p,q} = H^{p+q}(M, (\mathcal{T}_{\mathrm{gr}})_p) \\
\downarrow & & \downarrow \kappa_\infty^2 \\
H^{p+q}(M, \mathcal{T})_{(p)} & \longrightarrow & E_\infty^{p,q} = \operatorname{gr}_p H^{p+q}(M, \mathcal{T}) \longrightarrow 0,
\end{array}
$$

where the bottom sequence is exact, and the vertical mappings are natural surjections.

In the case when $p + q = 0$, we have $\sigma_p(H^0(M, \mathcal{T}_{(p)})) = Z_\infty^{p,-p}$, and $\kappa_\infty^2 : Z_\infty^{p,-p} \to E_\infty^{p,-p} = \operatorname{gr}_p H^0(M, \mathcal{T})$ is an isomorphism.

Proof. The commutativity of the diagram follows easily from the definitions. Clearly, $H^0(M, \mathcal{T})_{(p)} = H^0(M, \mathcal{T}_{(p)})$, which implies that κ^2_∞ is bijective for $q = -p$. Since the bottom line is exact, σ_p is surjective. $\square$

Now we prove our main result concerning the first nonzero coboundary operators among d_2, d_4, ... As usually, we denote by $\operatorname{ad} x$ the adjoint operator determined by an element x of an algebra L with the product denoted by $[\ ,\ ]$, i.e., $(\operatorname{ad} x)y = [x, y]$, $y \in L$. We may suppose that for each $i \in I$ there exists an isomorphism of sheaves $f_i : \mathcal{O}|U_i \to \mathcal{O}_{gr}|U_i$, inducing the identity isomorphism $(\operatorname{gr} \mathcal{O})|U_i \to \mathcal{O}_{gr}|U_i$. By Proposition 2, $(M, \mathcal{O})$ corresponds to the cohomology class γ of the 1-cocycle $g = (g_{ij}) \in Z^1(\mathfrak{U}, \mathcal{A}ut_{(2)}\mathcal{O}_{gr})$, where $g_{ij} = f_i \circ f_j^{-1}$. If $o(\gamma) = 2k$, then we may choose f_i, $i \in I$, in such a way that $g \in Z^1(\mathfrak{U}, \mathcal{A}ut_{(2k)}\mathcal{O}_{gr})$. We can write $g = \exp z$, where $z \in C^1(\mathfrak{U}, (\mathcal{T}_{gr})_{(2)\bar{0}})$.

We shall identify the differential algebras (E_0, d_0) and $(C^*(\mathfrak{U}, \mathcal{T}_{gr}), d)$ via the isomorphism of Proposition 3. On the other hand, the isomorphism f_i, $i \in I$ gives rise to the injective homomorphism of sheaves $\psi_i : (\mathcal{T}_{gr})_p|U_i \to \mathcal{T}_{(p)}|U_i$ given by $\psi_i(v) = f_i \circ v \circ f_i^{-1}$. Clearly, $\psi_i : (\mathcal{T}_{gr})_{(p)}|U_i = \bigoplus_{r \geq p}(\mathcal{T}_{gr})_r|U_i \to \mathcal{T}_{(p)}|U_i$ is an isomorphism of sheaves for any $i \in I$, $p \geq -1$. These local sheaf isomorphisms permit us to define an isomorphism of graded cochain groups

$$\psi : C^*(\mathfrak{U}, \mathcal{T}_{gr}) \to C^*(\mathfrak{U}, \mathcal{T})$$

such that

$$\psi : C^*(\mathfrak{U}, (\mathcal{T}_{gr})_{(p)}) \to C^*(\mathfrak{U}, \mathcal{T}_{(p)}), \quad p \geq -1.$$

We define it by setting

$$\psi(c)_{i_0 \ldots i_q} = f_{i_0} \circ c_{i_0 \ldots i_q} \circ f_{i_0}^{-1}$$

for any $(i_0, \ldots, i_q)$ such that $U_{i_0} \cap \ldots \cap U_{i_q} \neq \emptyset$. In general, ψ is not an isomorphism of complexes. Nevertheless, we can express explicitly the coboundary d of the complex $C^*(\mathfrak{U}, \mathcal{T})$ by means of d_0 and g.

Proposition 6 *For any $c \in C^q(\mathfrak{U}, \mathcal{T}_{gr}) = E_0^q$, we have*

$$(\psi^{-1}(d\psi(c)))_{i_0 \ldots i_{q+1}} = (d_0 c)_{i_0 \ldots i_{q+1}} + \sum_{l \geq 1} \frac{1}{l!}(\operatorname{ad} z_{i_0 i_1})^l c_{i_1 \ldots i_{q+1}}.$$

Proof. We can write

$$(d\psi(c))_{i_0 \ldots i_{q+1}} = \sum_{\alpha=0}^{q+1}(-1)^\alpha \psi(c)_{i_0 \ldots \hat{i}_\alpha \ldots i_{q+1}}$$

$$= \sum_{\alpha=1}^{q+1}(-1)^{\alpha}\psi(c)_{i_0...\hat{i}_{\alpha}...i_{q+1}} + \psi(c)_{i_1...i_{q+1}}$$

$$= f_{i_0} \circ (\sum_{\alpha=1}^{q+1}(-1)^{\alpha}c_{i_0...\hat{i}_{\alpha}...i_{q+1}}) \circ f_{i_0}^{-1} + f_{i_1} \circ c_{i_1...i_{q+1}} \circ f_{i_1}^{-1}$$

$$= f_{i_0} \circ ((d_0 c)_{i_0...i_{q+1}} - c_{i_1...i_{q+1}}) \circ f_{i_0}^{-1} + f_{i_1} \circ c_{i_1...i_{q+1}} \circ f_{i_1}^{-1}.$$

Therefore

$$(\psi^{-1}(d\psi(c)))_{i_0...i_{q+1}} = f_{i_0}^{-1} \circ (d\psi(c))_{i_0...i_{q+1}} f_{i_0}$$

$$= (d_0 c)_{i_0...i_{q+1}} - c_{i_1...i_{q+1}} + g_{i_0 i_1} \circ c_{i_1...i_{q+1}} \circ g_{i_0 i_1}^{-1}$$

$$= (d_0 c)_{i_0...i_{q+1}} - c_{i_1...i_{q+1}} + (\exp \text{ad } z_{i_0 i_1})(c_{i_1...i_{q+1}})$$

$$= (d_0 c)_{i_0...i_{q+1}} + \sum_{l \geq 1}\frac{1}{l!}(\text{ad } z_{i_0 i_1})^l c_{i_1...i_{q+1}}.$$

This implies our assertion. $\square$

This proposition allows to calculate the spectral sequence (E_r) whenever d_0 and the cochain z are known. Now we find the explicit form of certain coboundary operators d_r, $r \geq 2$.

Theorem 1 *Suppose that the supermanifold $(M, \mathcal{O})$ has order $2k$ and $\gamma \in H_{(2k)}$ denote the cohomology class corresponding to $(M, \mathcal{O})$ by Proposition 2. Then $d_r = 0$ for $r = 1, \ldots, 2k-1$, and $d_{2k} = \text{ad } \lambda_{2k}(\gamma)$.*

Proof. Take a cocycle $c \in E_0^{p,q-p}$, $d_0 c = 0$, and denote by c^* its cohomology class in $E_1^{p,q-p}$. Clearly, c and c^* are represented by the cochain $\psi(c) \in C_0^p$. By Proposition 6,

$$(\psi^{-1}(d\psi(c)))_{i_0...i_{q+1}} = \sum_{l \geq 1}\frac{1}{l!}(\text{ad } z_{i_0 i_1})^l c_{i_1...i_{q+1}}.$$

Writing $z = z^{(2k)} + z^{(2k+2)} + \ldots$, where $z^{(2l)} \in C^1(\mathfrak{U}, (\mathcal{T}_{\text{gr}})_{2l})$, we see that

$$(\psi^{-1}(d\psi(c)))_{i_0...i_{q+1}} = [z_{i_0 i_1}^{(2k)}, c_{i_1...i_{q+1}}] + u_{i_0...i_{q+1}},$$

where $u \in C_{(2k+2)}$. This means that

$$\psi^{-1}(d\psi(c)) = [z^{(2k)}, c] + u,$$

whence $d_1 = d_2 = \ldots = d_{(2k-1)} = 0$. Identifying E_{2k} with E_1, we also see that $d_{2k}c^*$ is represented by the cochain $\psi([z^{(2k)}, c])$. It follows that

$$d_{2k}c^* = [\lambda_{2k}(\gamma), c^*],$$

since $z^{(2k)}$ is a cocycle representing $\lambda_{2k}(\gamma)$. $\square$

Theorem 1 and Proposition 5 imply the following assertion, proved in [1], for $k = 1$ and $p = -1, 0$, in a different way.

Corollary 2 *If* $o(\gamma) = k$ *and a vector field* $v \in H^0(M, \mathcal{T}_p)$ *lies in* $\operatorname{Im}\sigma_p$, *then* $[\lambda_{2k}(\gamma), v] = 0$. *In the case when* $m \le 2k+1$, *the converse is also true.*

4. An example

As an application of our spectral sequence, we are going to calculate here the tangent sheaf cohomology of a series of homogeneous supermanifolds considered in [1]. The reduction of these supermanifolds is the projective line $M = \mathbb{CP}^1$. Namely, let $(M, \mathcal{O}) = Q_{(q)}^{1|m}$ be the superquadric in $\mathbb{CP}^{2|m}$, determined by a nondegenerate skew-symmetric bilinear form ω of rank $q \le m/2$ in $\mathbb{C}^m$. In the homogeneous coordinates $z_0, z_1, z_2, \zeta_1, \ldots, \zeta_m$, it is given by the equation

$$z_0^2 - z_1 z_2 + \omega(\zeta) = 0, \tag{16}$$

where we may suppose that

$$\omega(\zeta) = \sum_{j=1}^{q} \zeta_j \zeta_{r+j}.$$

Introducing the local coordinates

$$x = \frac{z_0}{z_1}, \; \xi_j = \frac{\zeta_j}{z_1} \; \text{ in } \; U = \{z_1 \ne 0\};$$

$$y = \frac{z_0}{z_2}, \; \eta_j = \frac{\zeta_j}{z_2} \; \text{ in } \; V = \{z_2 \ne 0\},$$

we get the following transition functions in $U \cap V$:

$$y = x(x^2 + \omega(\xi))^{-1}, \; \eta_j = (x^2 + \omega(\xi))^{-1}\xi_j, \; j = 1, \ldots, m.$$

It follows that the corresponding split supermanifold is given by the transition functions

$$y = x^{-1}, \; \eta_j = x^{-2}\xi_j, \; j = 1, \ldots, m,$$

and hence corresponds to the vector bundle $\mathbf{E} = m\mathbf{L}_{-2}$ over M, where we denote by $\mathbf{L}_k$ the holomorphic line bundle over $\mathbb{CP}^1$ of degree $k \in \mathbb{Z}$. The cohomology class $\gamma \in H^1(M, \mathcal{A}ut_{(2)}\mathcal{O}_{\mathrm{gr}})$ (see Proposition 2) is given by the cocycle $g \in \mathcal{A}ut_{(2)}\mathcal{O}_{\mathrm{gr}}(U \cap V)$ such that

$$\begin{aligned} g(x^{-1}) &= x(x^2 + \omega(\xi))^{-1}, \\ g(x^{-2})g(\xi_j) &= (x^2 + \omega(\xi))^{-1}\xi_j, \; j = 1, \ldots, m. \end{aligned}$$

It follows that

$$g(x) = x + x^{-1}\omega(\xi),$$
$$g(\xi_j) = (1 + x^{-2}\omega(\xi))\xi_j, \quad j = 1, \ldots, m.$$

Therefore the 2-component $z^{(2)}$ of the cochain $z = \log g$ has the form

$$z^{(2)} = x^{-1}\omega(\xi)\frac{\partial}{\partial x} + x^{-2}\omega(\xi)\varepsilon, \tag{17}$$

where $\varepsilon = \sum_{j=1}^m \xi_j \partial/\partial \xi_j$. This cocycle determines a cohomology class $\zeta \in H^1(M, (\mathcal{T}_{\mathrm{gr}})_2)$. It is nonzero (see [1]), and hence $o(\gamma) = 2$, and $\zeta = \lambda_2(\gamma)$.

Now we must compute the groups $E_2^{p,q-p} = H^q(M, (\mathcal{T}_{\mathrm{gr}})_p)$. To do this, we use the well-known theorem of Bott on the cohomology of homogeneous vector bundles combined with the calculation of the supertangent bundle performed in [1]. We regard $\mathbb{CP}^1$ as the homogeneous space of the Lie group $G = \mathrm{SL}_m(\mathbb{C}) \times \mathrm{SL}_2(\mathbb{C})$ with the stabilizer $H = \mathrm{SL}_m(\mathbb{C}) \times B_-$ of the point $x_0 = \{x = 0\} \in U$, where B_- is the Borel subgroup corresponding to the (unique) negative root $-\lambda$ of $\mathrm{SL}_2(\mathbb{C})$. Let us denote by $\mu_1, \ldots, \mu_m$ the weights of the standard representation ρ of $\mathrm{SL}_m(\mathbb{C})$, by Ad the adjoint representation of this group, and by σ the character of B_- satisfying $d\sigma = \lambda$. Let φ_s, $s \geq 0$, be the irreducible representation of $\mathrm{SL}_2(\mathbb{C})$ (of dimension $s + 1$) with the highest weight $s\lambda$.

It is well known that there is an exact sequence

$$0 \to \mathcal{A}_{p+1} \to (\mathcal{T}_{\mathrm{gr}})_p \to \mathcal{B}_p \to 0, \quad p \geq 1, \tag{18}$$

where

$$\mathcal{A}_p = \mathcal{E}^* \otimes \bigwedge^p \mathcal{E}, \quad \mathcal{B}_p = \Theta \otimes \bigwedge^p \mathcal{E},$$

Θ being the tangent sheaf of M.

Clearly, $\mathbf{E} = m\mathbf{L}_{-2}$ can be regarded as the G-homogeneous vector bundle over $\mathbb{CP}^1$ determined by the representation $\rho \otimes \sigma^{-2}$ of H. Then $\mathcal{A}_p$ and $\mathcal{B}_p$ are determined by the representations $\rho^* \bigwedge^p \rho \otimes \sigma^{2(1-p)}$ and $\bigwedge^p \rho \otimes \sigma^{2(1-p)}$ of H, respectively, since the isotropy representation for $\mathbb{CP}^1$ is $1 \otimes \sigma^2$. Using the theorem of Bott, we easily get the following result.

Proposition 7 *The nonzero groups $H^q(M, \mathcal{A}_p)$, $H^q(M, \mathcal{B}_p)$ with the induced representations of G are indicated in the following table:*

$H^0(M, \mathcal{A}_0)$	$H^0(M, \mathcal{A}_1)$	$H^1(M, \mathcal{A}_p), \ 2 \leq p \leq m$
$\rho^* \otimes \varphi_2$	$1 + \mathrm{Ad} \otimes \varphi_0$	$\rho^* \bigwedge^p \rho \otimes \varphi_{2p-4}$
$H^0(M, \mathcal{B}_0)$	$H^0(M, \mathcal{B}_1)$	$H^1(M, \mathcal{B}_p), \ 2 \leq p \leq m$
$1 \otimes \varphi_2$	$\rho \otimes \varphi_0$	$\bigwedge^p \rho \otimes \varphi_{2p-4}$

Note that for $p \leq m - 1$ we have

$$\rho^* \overset{p}{\bigwedge} \rho \simeq \overset{p-1}{\bigwedge} \rho + \psi_p, \tag{19}$$

where ψ_p is the irreducible representation of $\mathrm{SL}_m(\mathbb{C})$ with the highest weight $\mu_1 + \ldots + \mu_p - \mu_m$.

Using (18) and some results of [1], we can deduce from this the following description of the cohomology of $\mathcal{T}_{\mathrm{gr}}$.

Proposition 8 *In the following table the nonzero groups* $H^q(M, (\mathcal{T}_{\mathrm{gr}})_p)$ *with the induced representations of G are indicated:*

$H^0(M, (\mathcal{T}_{\mathrm{gr}})_{-1})$	$H^0(M, (\mathcal{T}_{\mathrm{gr}})_0)$
$\rho^* \otimes \varphi_2$	$1 + \mathrm{Ad} \otimes \varphi_0 + 1 \otimes \varphi_2$
$H^1(M, (\mathcal{T}_{\mathrm{gr}})_1),\ m \geq 3$	$H^1(M, (\mathcal{T}_{\mathrm{gr}})_p),\ \ 2 \leq p \leq m$
$\psi_2 \otimes \varphi_0$	$\bigwedge^p \rho \otimes \varphi_{2p-4} + \bigwedge^p \rho \otimes \varphi_{2p-2} + \psi_{p+1} \otimes \varphi_{2p-2}$

Proof. For $p = -1, 0$ or $p = 2$, the assertion follows immediately from Proposition 7. In the case $p = 1$, we deduce from (18) and Proposition 7 the following exact sequence

$$0 \to H^0(M, (\mathcal{T}_{\mathrm{gr}})_1) \to H^0(M, \mathcal{B}_1) \overset{\partial^*}{\to} H^1(M, \mathcal{A}_2) \to H^1(M, (\mathcal{T}_{\mathrm{gr}})_1) \to 0.$$

By Proposition 18 of [1],

$$(\mathcal{T}_{\mathrm{gr}})_1 \simeq 2\mathcal{F}(-1) \oplus \frac{1}{2}(m - 2)(m^2 - m + 2)\mathcal{F}(-2),$$

where we denote by $\mathcal{F}(k)$ the sheaf of holomorphic sections of the bundle $\mathbf{L}_k$. Hence $H^0(M, (\mathcal{T}_{\mathrm{gr}})_1) = 0$ for all m and $H^1(M, (\mathcal{T}_{\mathrm{gr}})_1) = 0$ for $m = 2$. Thus, ∂^* is injective, and our assertion follows from Proposition 7 and (19). $\square$

For simplicity, we consider now the case when $m = 2q$ or $2q+1$, i.e., when ω has the maximal possible rank. Denote the corresponding superquadric by $\mathbb{Q}^{1|m}$. Let $\mathfrak{osp}_{3|m}(\mathbb{C})$ be the subalgebra of $\mathfrak{gl}_{3|m}(\mathbb{C})$ consisting of linear transformations that annihilate the left-hand side of (16). If $m = 2q$ is even, then this is a simple Lie superalgebra, belonging to the orthosymplectic series (see [5]). For odd m, it is not simple. Our main result here is as follows.

Theorem 2 *For the superquadric* $(M, \mathcal{O}) = \mathbb{Q}^{1|m}$, $m \geq 2$, *we have*

$$H^0(M, \mathcal{T}) \simeq \mathfrak{osp}_{3|m}(\mathbb{C})$$

(*as Lie superalgebras*);

$$H^1(M, \mathcal{T}_{\bar{0}}) \simeq \bigoplus_{p \geq 1} H^1(M, \mathcal{A}_{2p+1}) \oplus \bigoplus_{p \geq 2} H^1(M, \mathcal{B}_{2p}),$$

$$H^1(M, \mathcal{T}_{\bar{1}}) \simeq H^1(M, (\mathcal{T}_{gr})_1) \oplus \bigoplus_{p \geq 2} H^1(M, \mathcal{A}_{2p}) \oplus \bigoplus_{p \geq 1} H^1(M, \mathcal{B}_{2p+1}).$$

Proof. By Proposition 3, $E_2^{p,q} = H^{p+q}(M, (\mathcal{T}_{gr})_p)$. By Proposition 8, we must calculate $d_2 = \operatorname{ad} \zeta$ (see Theorem 1), where ζ is represented by the cocycle (17), on

$$H^0(M, \mathcal{T}_{gr}) = H^0(M, (\mathcal{T}_{gr})_0) \oplus H^0(M, (\mathcal{T}_{gr})_{-1}).$$

Consider the natural projective action of $\mathfrak{gl}_{3|m}(\mathbb{C})$ on $\mathbb{CP}^{2|m}$. One verifies that $\mathfrak{osp}_{3|m}(\mathbb{C})$ leaves invariant the ideal sheaf of the submanifold $\mathbb{Q}^{1|m}$ of $\mathbb{CP}^{2|m}$, and so this action induces an action of $\mathfrak{osp}_{3|m}(\mathbb{C})$ on $\mathbb{Q}^{1|m}$, i.e., a homomorphism $\alpha : \mathfrak{osp}_{3|m}(\mathbb{C}) \to H^0(M, \mathcal{T})$. One verifies easily that α is injective. Since $\dim(\mathfrak{osp}_{3|m}(\mathbb{C}))_{\bar{1}} = \dim H^0(M, (\mathcal{T}_{gr})_{-1})$, we see that α is an isomorphism on the odd parts and, by Proposition 5, that $E_2^{-1,1} = H^0(M, (\mathcal{T}_{gr})_{-1})$ consists of permanent cocycles.

It is also clear (see [1]), that $E_2^{0,0} = H^0(M, (\mathcal{T}_{gr})_0)$ can be identified with

$$\mathfrak{gl}_m(\mathbb{C}) \oplus \mathfrak{sl}_2(\mathbb{C}) = \mathfrak{sl}_m(\mathbb{C}) \oplus \langle \varepsilon \rangle \oplus \mathfrak{sl}_2(\mathbb{C}).$$

The class $d_2\varepsilon = [\zeta, \varepsilon]$ is represented by the cocycle $[z^{(2)}, \varepsilon] = -2z^{(2)}$. Thus,

$$d_2\varepsilon = -2\zeta.$$

The mapping $d_2 : \mathfrak{sl}_m(\mathbb{C}) \oplus \mathfrak{sl}_2(\mathbb{C}) \to E_2^{2,-1} = H^1(M, (\mathcal{T}_{gr})_2)$ is determined by the representation of G on this latter cohomology space described in Proposition 8. It follows that

$$E_2^{0,0} \cap \operatorname{Ker} d_2 = \mathfrak{osp}_{3|m}(\mathbb{C}).$$

Identifying $H^1(M, (\mathcal{T}_{gr})_2)$ with $H^1(M, \mathcal{A}_3) \oplus H^1(M, \mathcal{B}_2)$, we see that $\operatorname{Im} d_2$ coincides with the tangent space at 0 to the orbit $G(\zeta')$ in $H^1(M, \mathcal{B}_2) \simeq \bigwedge^2 \mathbb{C}^m$, where ζ' is determined by the cocycle $x^{-1}\partial/\partial x \otimes \omega$. Since ω has maximal rank, this orbit is open, and hence $\operatorname{Im} d_2 = H^1(M, \mathcal{B}_2)$.

Therefore only the following groups $E_3^{p,q}$ are nonzero:

$$E_3^{-1,1} = H^0(M, (\mathcal{T}_{gr})_{-1}) \simeq (\mathfrak{osp}_{3|m})_{\bar{1}}, \ E_3^{0,0} \simeq (\mathfrak{osp}_{3|m})_{\bar{0}},$$

$$E_3^{1,0} = H^1(M, (\mathcal{T}_{gr})_1), \ E_3^{2,-1} \simeq H^1(M, \mathcal{A}_3),$$

$$E_3^{p,1-p} \simeq H^1(M, (\mathcal{T}_{gr})_p) \simeq H^1(M, \mathcal{A}_{p+1}) \oplus H^1(M, \mathcal{B}_p), \ 3 \leq p \leq m.$$

This implies that α is an isomorphism, and that the elements of $E_3^{0,0}$ are permanent cocycles. Therefore $d_3 = d_4 = \ldots = 0$, and $E_\infty = E_3$. The assertion concerning $H^1(M, \mathcal{T})$ now follows from Proposition 4. □

References

1. Bunegina, V.A. and Onishchik, A.L.: Homogeneous supermanifolds associated with the complex projective line, *J. Math. Sci.*, **82** (1996) no. 4, 3503–3527.
2. Flenner, H. and Sundararaman, D.: Analytic geometry of complex superspaces, *Trans. Amer. Math. Soc.*, **330** (1992) 1–40.
3. Godement, R.: *Topologie Algébrique et Théorie des Faisceaux*. Hermann, Paris, 1958.
4. Green, P.: On holomorphic graded manifolds, *Proc. Amer. Math. Soc.*, **85** (1982) 587–590.
5. Kac, V.G.: Lie superalgebras, *Adv. Math.* **26** (1977) 8–96.
6. Manin, Y.I.: *Gauge Field Theory and Complex Geometry*, Nauka, Moscow, 1984 (in Russian); English transl.: Springer-Verlag, Berlin–Heidelberg–New York, 1988.
7. Rothstein, M.J.: Deformations of complex supermanifolds, *Proc. Amer. Math. Soc.*, **95** (1985) 255–260.
8. Vajntrob, A.Yu.: Deformations of complex superspaces and coherent sheaves on them, in *Curr. Problems in Math. Recent Achievements*, **32**, VINITI, Moscow, 1988, 27–70 (in Russian); English transl.: *J. Soviet Math.* **51** (1990) 2069–2083.

ON A DUALITY OF VARIETIES OF REPRESENTATIONS OF TERNARY LIE AND SUPER LIE SYSTEMS

YU.P. RAZMYSLOV
Department of Mechanics and Mathematics,
Moscow State University, 119899 Moscow, Russia

Abstract. A duality between the varieties of representations of ternary Lie algebras and superalgebras is established. The isomorphism of the lattices of subvarieties of ternary Lie algebras and superalgebras is induced by this duality. If a variety of (representations of) ternary Lie (super)algebras satisfies an identity of Young symmetry type D, then the dual variety satisfies an identity of Young symmetry type D^*, and D is dual to D^*. Finite basis systems of identities of ternary Lie algebras and superalgebras of infinite dimension as well as of representations of such algebras and superalgebras in their universal enveloping algebras are given in explicit form.

Mathematics Subject Classification (1991): 17B65, 17A30, 17A40, 17A70, 16W55.

Key words: ternary Lie algebra, ternary Lie superalgebra, identity, variety, representation.

1. Ternary algebras, their representations and identities

Let us fix the following notation. K is the ground field. $G \overset{\text{def}}{=} G_0 \oplus G_1$ is a Grassmann algebra with a countable set of generators, G_0, G_1 being its even and odd components respectively, $\mathbf{H}$ is an absolutely free associative algebra with the countable set $X = \{x_1, x_2, \ldots\}$ of free generators.

A keystone of my considerations is the well-known correspondence between bosons and fermions, that is to say, between Lie and super Lie algebras, namely the following one.

B. P. Komrakov et al. (eds.), Lie Groups and Lie Algebras, 217–230.
© 1998 *Kluwer Academic Publishers. Printed in the Netherlands.*

Let $\mathcal{G} = \mathcal{G}_0 \oplus \mathcal{G}_1$ be an arbitrary Lie superalgebra, then its Grassmann envelope $\mathcal{G}^{\sharp} \stackrel{\text{def}}{=} G_0 \otimes_K \mathcal{G}_0 + G_1 \otimes_K \mathcal{G}_1$ is a bi-graded Lie algebra. This sentence remains valid if we replace the words "Lie superalgebra" by the words "bi-graded Lie algebra".

I am going to establish the correspondence $\sharp$ in its proper natural framework by means of the language of identities and varieties connected with Lie algebras, Lie superalgebras and their representations. I treat the term "variety" in the sense of the theory of Universal Algebras. A *variety* is a class of algebras that can be given by a certain set of identities. For example, the class of all Lie algebras is a variety because it can be defined by two identities: the skew-symmetric identity and the Jacobi identity.

I prefer speaking in terms of ternary Lie and super Lie algebras and their representations.

Exact definitions.

Let $\mathcal{T}$ be the class of all ternary algebras with a ternary operation $[\ ,\ ,\ ]$

$$[\ ,\ ,\ ] : T \otimes_K T \otimes_K T \to T ,$$

defined by the following two identities

(i) $[x, y, z] + [y, z, x] + [z, x, y] = 0$ (Jacobi identity)

(ii) $[x, y, [u, v, z]] = [[x, y, u], v, z] + [u, [x, y, v], z] +$

$$[u, v, [x, y, z]] \qquad \text{(derivation identity)}$$

The second one admits an absolutely trivial interpretation:

(a) each linear transformation $[x, y] : T \to T$ that transforms an arbitrary element $z \in T$ into the result of the ternary operation $[x, y, z]$ is a derivation of our algebra T, hence $[x, y] \in \mathrm{Der}_K T$;

(b) the linear subspace $\mathcal{G}_0$ spanned in $\mathrm{End}_K T$ by all these linear transformations is closed under the usual Lie bracket, hence $\mathcal{G}_0$ is a Lie subalgebra in the Lie algebra $(\mathrm{End}_K T)^{(-)}$ of all endomorphisms of the linear space T.

Now we are in a position to introduce a new bi-graded object $\mathcal{G} \stackrel{\text{def}}{=} \mathcal{G}_0 \oplus T$ with a natural operation $[\ ,\ ] : \mathcal{G} \otimes_K \mathcal{G} \to \mathcal{G}$. The only composition that has yet to be defined is $\mathcal{G}_0 \otimes_K \mathcal{G}_1 \to \mathcal{G}_1$,

$$[g_0, g_1] \stackrel{\text{def}}{=} -[g_1, g_0] = g_0 \times g_1, \quad \text{where } g_0 \in \mathcal{G}_0,\ g_1 \in T.$$

All these ternary algebras may be conveniently divided into two classes $\mathcal{T}_+$ and $\mathcal{T}_-$.

$\mathcal{T}_+$ is the subvariety of $\mathcal{T}$ defined by a symmetric identity

(iii$_+$) $[x, y, z] = [y, x, z]$

$\mathcal{T}_-$ is the subvariety of $\mathcal{T}$ defined by a skew-symmetric identity

(iii$_-$) $[x, y, z] = -[y, x, z]$

We will call them the varieties of ternary super Lie and Lie algebras respectively.

The main examples are presented in the Table.

Comments.

C.1. Example 2 is absolutely universal: each ternary algebra, Lie or super Lie, may be realized as the odd component for a suitable bi-graded Lie or super Lie algebra.

C.2. Example 1 enables us to define such a notion as a representation of an arbitrary ternary algebra T in an associative algebra A through the homomorphism either $\rho : T \to A^{[+]}$, if $T \in \mathcal{T}_+$, or $\rho : T \to A^{[-]}$, if $T \in \mathcal{T}_-$.

C.3. Since any bi-graded Lie or super Lie algebra $\mathcal{G} = \mathcal{G}_0 \oplus \mathcal{G}_1$ admits an faithful realization in its universal associative enveloping algebra $U(\mathcal{G})$, then due to previous two comments an arbitrary ternary (whether Lie or super Lie) algebra T has a faithful representation $\rho : T \to U(T)$ in its universal associative enveloping algebra $U(T)$. But we must warn the reader that $U(T)$ may be different from the universal enveloping algebra of the bi-graded object $\mathcal{G} \overset{\text{def}}{=} T \otimes \mathcal{G}_0$ constructed above for any ternary algebra $T \in \mathcal{T}$.

C.4. Example 4 permits us to treat every polynomial $f(x_1, \ldots, x_l)$ in the ternary signatures $\mathcal{T}_+$, $\mathcal{T}_-$ as being an associative polynomial of $\mathcal{M}_+$ or $\mathcal{M}_-$ respectively. Moreover, for any faithful representation $\rho : T \to A$ in an associative algebra A the formula $f = 0$ is an identity of the ternary algebra T if and only if in A the associative polynomial f vanishes on $\rho(T)$. This observation enables us to generalize such a notion as a polynomial identity over representation ρ and to make the theory of identities and subvarieties of $\mathcal{T}_+$, $\mathcal{T}_-$ a particular case of that in the class of all representations of ternary algebras in associative ones.

Definition 1 Let $f(x_1, \ldots, x_l)$ be a polynomial of the absolutely free associative algebra $\mathbf{H}$, and $\rho : T \to A$ an arbitrary representation of ternary algebra T in an associative algebra A. The formula $f = 0$ is called a *polynomial dentity of ρ* if and only if for all $t_1, \ldots, t_l \in T$

$$f(\rho(t_1), \ldots, \rho(t_l)) = 0$$

in the algebra A.

Examples	
$\mathcal{T}_+ : [x, y, z] = [y, x, z]$	$\mathcal{T}_- : [x, y, z] = -[y, x, z]$
1$^+$. Let A be an associative algebra. Then $[,,]_+ : A^{[+]} \otimes_K A^{[+]} \otimes_K A^{[+]} \to A^{[+]}$, where $A^{[+]} \overset{\text{def}}{=} A$ as a linear space and $[x, y, z]_+ \overset{\text{def}}{=} [xy + yx, z]$ for all $x, y, z \in A$, is a ternary Lie superalgebra of the variety $\mathcal{T}_+$.	**1$^-$**. Let A be an associative algebra. Then $[,,]_- : A^{[-]} \otimes_K A^{[-]} \otimes_K A^{[-]} \to A^{[-]}$, where $A^{[-]} \overset{\text{def}}{=} A$ as a linear space and $[x, y, z]_- \overset{\text{def}}{=} [[x, y], z]$ for all $x, y, z \in A$, is a ternary Lie algebra of the variety $\mathcal{T}_-$.
2$^+$. Let $\mathcal{G}_F = \mathcal{G}_0 \oplus \mathcal{G}_1$ be a Lie superalgebra. Then $[,,]_+ : \mathcal{G}_1 \otimes_K \mathcal{G}_1 \otimes_K \mathcal{G}_1 \to \mathcal{G}_1$, where for all $x, y, z \in \mathcal{G}_1$ $[x, y, z]_+ \overset{\text{def}}{=} [[x, y], z]$, is a ternary Lie superalgebra of the variety $\mathcal{T}_+$.	**2$^-$**. Let $\mathcal{G}_B = \mathcal{G}_0 \oplus \mathcal{G}_1$ be a bigraded Lie algebra. Then $[,,]_- : \mathcal{G}_1 \otimes_K \mathcal{G}_1 \otimes_K \mathcal{G}_1 \to \mathcal{G}_1$, where for all $x, y, z \in \mathcal{G}_1$ $[x, y, z]_- \overset{\text{def}}{=} [[x, y], z]$, is a ternary Lie algebra of the variety $\mathcal{T}_-$.
3$^+$. Let $\langle , \rangle : V \otimes_K V \to K$ be a skew-symmetric nondegenerate bilinear form on a linear space V. Then $[,,]_+ : V^{\langle , \rangle} \otimes_K V^{\langle , \rangle} \otimes_K V^{\langle , \rangle} \to V^{\langle , \rangle}$, where $[x, y, z]_+ \overset{\text{def}}{=} \langle y, z \rangle x - \langle z, x \rangle y$ for all $x, y, z \in V$ and $V^{\langle , \rangle} \overset{\text{def}}{=} V$, is a ternary Lie superalgebra of the variety $\mathcal{T}_+$.	**3$^-$**. Let $(,) : V \otimes_K V \to K$ be a symmetric nondegenerate bilinear form on a linear space V. Then $[,,]_- : V^{(,)} \otimes_K V^{(,)} \otimes_K V^{(,)} \to V^{(,)}$, where $[x, y, z]_- \overset{\text{def}}{=} (y, z) x - (z, x) y$ for all $x, y, z \in V$ and $V^{(,)} \overset{\text{def}}{=} V$, is a ternary Lie algebra of the variety $\mathcal{T}_-$.
4$^+$. Let $\mathcal{M}_+$ be the ternary Lie subsuperalgebra of $H^{[+]}$ generated in $H^{[+]}$ by generators $x_1, x_2, \ldots$. Then $\mathcal{M}_+$ is an absolutely free ternary Lie superalgebra of $\mathcal{T}_+$ with free generators $x_1, x_2, \ldots$.	**4$^-$**. Let $\mathcal{M}_-$ be the ternary Lie subalgebra of $H^{[-]}$ generated in $H^{[-]}$ by generators $x_1, x_2, \ldots$. Then $\mathcal{M}_-$ is an absolutely free ternary Lie algebra of $\mathcal{T}_-$ and $x_1, x_2, \ldots$ are its free generators.

Definition 2 Let $\{f_j | j \in J\}$ be an arbitrary subset of polynomials in the free associative algebra **H**. The class of all representations ρ of ternary algebras in associative ones such that all these polynomials vanish on $\rho(T)$ in A is called a *variety* given by the set of identities $\{f_j = 0 | j \in J\}$.

Both definitions are according to the mode in which one looks at the corresponding notions in the theory of ternary $\mathcal{T}_+$, $\mathcal{T}_-$-algebras.

The following results present a delicate little touch on the great canvas of the uniform theory of Lie algebras, Lie superalgebras and their representations.

Theorem 1 *Let $\mathcal{L}_+$ be the lattice of all subvarieties of the variety $\mathcal{T}_+$, and $\mathcal{L}_-$ the lattice of all subvarieties of the variety $\mathcal{T}_-$. Then there exists an isomorphism $\natural : \mathcal{L}_+ \to \mathcal{L}_-$ of both lattices.*

Theorem 2 *Let $\mathcal{R}_+$ be the lattice of all subvarieties of representations $\mathcal{T}_+$-algebras, and $\mathcal{R}_+$ the lattice of all subvarieties of representations $\mathcal{T}_-$-algebras in associative ones. Then there exists an isomorphism $\natural : \mathcal{R}_+ \longleftrightarrow \mathcal{R}_-$ of both lattices.*

Explanations.

The second result admits a natural interpretation on the three different levels: *identities, objects, ideals*. We will suggest that the ground field K is of zero characteristic. In that case any set of polynomial identities is well-known to be equivalent to a set of multilinear ones.

E.1. The duality $\natural$ on the level of identities. Let $\mathcal{R}_{\mathcal{F}}$ be the variety of representations of ternary Lie superalgebras in associative ones given by a set $\mathcal{F} = \{f_j = 0 | j \in J\}$, where each f_j is an associative multilinear polynomial in **H**, and $\mathcal{R}_{\mathcal{F}}^{\natural} \stackrel{\text{def}}{=} \natural(\mathcal{R}_{\mathcal{F}})$ a corresponding subvariety of representations of ternary Lie algebras.

We can consider that every multilinear polynomial f in $\mathcal{F}$ has a form

$$f(x_1, \ldots, x_n) = \sum_{\sigma \in S_n} \beta_\sigma x_{\sigma(1)} \cdots x_{\sigma(n)},$$

where $\beta_\sigma \in K$ and the sum is taken over all permutations of the set $\{1, \ldots, n\}$. Denoting $\mathcal{F}^{\natural} \stackrel{\text{def}}{=} \{f_j^{\natural} = 0 | j \in J\}$ and

$$f^{\natural}(x_1, \ldots, x_n) \stackrel{\text{def}}{=} \sum_{\sigma \in S_n} \text{sgn}(\sigma) \beta_\sigma x_{\sigma(1)} \cdots x_{\sigma(n)},$$

where $\text{sgn}(\sigma)$ is a sign of σ, we claim: the *dual variety* $\mathcal{R}_{\mathcal{F}}^{\natural}$ in $\mathcal{L}_-$ is defined by the set of identities $\mathcal{F}^{\natural}$. As a matter of fact we can now change in our exposition above the words "Lie superalgebras" for the words "Lie algebras", and vise versa, because of the involutory property of the mapping $\natural$.

E.2. The duality $\natural$ on the level of objects. Let T be an arbitrary ternary Lie subsuperalgebra in $A^{[+]}$ (see Example 1^+), and $\widehat{T}$ a linear subspace $G_1 \otimes T$ in the tensor product $\widehat{A} = G \otimes A$ considered as a natural associative algebra. One easily verifies that $\widehat{T}$ is closed under the ternary composition $[\ ,\ ,\ ]_-$ on $\widehat{A}^{[-]}$. Hence $\widehat{T}$ is a ternary Lie subalgebra of $\widehat{A}^{[-]}$ (see Example 1^-), and we are in a position to introduce the following two natural representations

$$\mathbf{id} : T \to A \ , \widehat{\mathbf{id}} : \widehat{T} \to \widehat{A} \ ,$$

where $\mathbf{id}$ is an identity imbedding, of the ternary Lie and super Lie algebras T, $\widehat{T}$ in the associative algebras A, $\widehat{A}$ respectively. Let $\mathcal{F}_{(A,T)}$ be the set of all the identities that hold in $\mathbf{id} : T \to A$, and $\mathcal{B}_{(\widehat{A},\widehat{T})}$ the set of all the identities that hold in $\widehat{\mathbf{id}} : \widehat{T} \to \widehat{A}$. By denoting $\mathrm{var}(A,T)$, $\mathrm{var}(\widehat{A},\widehat{T})$ the subvarieties given in $\mathcal{R}_+$, $\mathcal{R}_-$ by $\mathcal{F}_{(A,T)}$, $\mathcal{B}_{(\widehat{A},\widehat{T})}$ respectively, I claim

Proposition 1 $(\mathrm{var}(A,T))^{\natural} = \mathrm{var}(\widehat{A},\widehat{T})$.

By the result of Explanation 1 it means that *every multilinear polynomial identity of the pair* $\mathrm{var}(\widehat{A},\widehat{T})$ *has the form* $f^{\natural} = 0$, *where* $f = 0$ *is a multilinear identity of the pair* $\mathrm{var}(A,T)$, *and vise versa*. This fact may be easily verified by direct computation of values $f^{\natural}|_{\widehat{T}}$, $f|_T$ using basic definitions.

As a matter of fact our assertion remains valid when T is an arbitrary ternary Lie subalgebra in $A^{[-]}$. In that case we have merely to change in our previous exposition the words "ternary Lie algebra" for the words "ternary Lie subalgebra".

E.3 The duality $\natural$ of $\mathcal{R}_+$, $\mathcal{R}_-$ on the levels of ideals. Let $\rho : T \to A$ be a representation of the ternary Lie superalgebra T in an associative algebra A, and $f(x_1, \ldots, x_l) = 0$ ($f \in \mathbf{H}$) an identity of ρ. Then, one easily sees, for all $g_1, \ldots, g_l \in \mathcal{M}_+$, $a, b \in \mathbf{H}$ the identity $af(g_1, \ldots, g_l)b = 0$ holds also for the representation ρ. It means that polynomials from the left-hand side of all the identities of ρ *form* an ideal of $\mathbf{H}$ closed under all homomorphisms $\varphi : \mathbf{H} \to \mathbf{H}$ such that $\varphi(\mathcal{M}_+) \subseteq \mathcal{M}_+$. Moreover, if an ideal I is stable under all homomorphisms $\varphi : \mathbf{H} \to \mathbf{H}$ each of which transforms $\mathcal{M}_+$ into $\mathcal{M}_+$, then every identity of the pair $(\mathbf{H}/I, (\mathcal{M}_+ + I)/I)$, that is to say, of the identical representation $(\mathcal{M}_+ + I)/I \to \mathbf{H}/I$, has a form $f = 0$, where f ranges over I. We call all these ideals I *Fermi-ideals* , or shortly **F**-ideals, of the free associative algebra $\mathbf{H}$. Replacing in the previous consideration $\mathcal{M}_+$ by $\mathcal{M}_-$, as well as reducing the prefix *super-* in the word *superalgebra*, one forces an entrance to the following definition of *Bose-ideals*: an ideal I of $\mathbf{H}$ is called a **B**-ideal if $\varphi(I) \subseteq I$ for all homomorphisms $\varphi : \mathbf{H} \to \mathbf{H}$ such that $\varphi(\mathcal{M}_-) \subseteq \mathcal{M}_-$ (or, *which is just the same*, for all $f(x_{i_1}, \ldots, x_{i_l}) \in I$, $g_1, \ldots, g_l \in \mathcal{M}_-$ the polynomial $f(g_1, \ldots, g_l)$ lies in I). A natural one-to-one correspondence pointed out above between the lattice of all **F**-ideals and the $\mathcal{R}_+$, as well as that between the lattice of all **B**-ideals and the $\mathcal{R}_-$, provides due to Explanation 1 the following description of the mapping $\sharp : I \to I^\sharp$, where I, $I^\sharp$ range simultaneously over all **F**- and **B**-ideals respectively: *for all multilinear polynomial $f \in$ **H**

$$f \in I \iff f^\sharp \in I^\sharp$$

(see Explanation 1). But in reality $f^\sharp$ admits a natural definition for multihomogeneous polynomials $f \in$ **H**. Let us introduce it and consider our situation from that point of view.

For any subset $\lambda = \{x_{i_1}, \ldots, x_{i_n}\}$ of the set X of free generators of the K-algebra **H** and an arbitrary array $\mathbf{r} = (r_1, r_2, \ldots, r_n)$, where $r_j \in \mathbf{N} \cup \{0\}$, we denote by $_\mathbf{r}\mathbf{H}_\lambda$ the set of all multihomogeneous polynomials $f(x_{i_1}, \ldots, x_{i_n})$ of polydegree $\mathbf{r}$

$$f(\beta_1 \cdot x_{i_1}, \ldots, \beta_n \cdot x_{i_n}) \overset{\text{def}}{=} \beta_1^{r_1} \cdots \beta_n^{r_n} f(x_{i_1}, \ldots, x_{i_n})$$

for any $\beta_1, \ldots, \beta_n \in K$. It is evident that **H** is a direct sum of its multihomogeneous components, that is,

$$\mathbf{H} = \bigoplus_{\mathbf{r}, \lambda} {}_\mathbf{r}\mathbf{H}_\lambda.$$

Thus, in the case of an infinite field K an identity $g = 0$ is equivalent to a certain finite set of multihomogeneous identities. In the particular case of $\mathbf{r} = (1, \ldots, 1)$ we denote the subspace $_\mathbf{r}\mathbf{H}_\lambda$ by $\mathbf{H}_\lambda$ and call it by a space of multilinear polynomials over λ. Applying Vandermonde determinant, in the case of an infinite field K one can easily prove that if a polynomial g belongs to a **F**-ideal I (or a **B**-ideal I), then all multihomogeneous components of g are in I, that is

$$I = \bigoplus_{\mathbf{r}, \lambda} I \cap {}_\mathbf{r}\mathbf{H}_\lambda .$$

We fix a natural order $x_1 < x_2 < \cdots < x_n < \cdots$ on the set X. We establish a linear mapping $\pi : \mathbf{H}_\lambda \to \mathbf{H}_\lambda$ by the formula

$$\pi(f) \overset{\text{def}}{=} \sum_{\sigma \in S_\lambda} \operatorname{sgn} \sigma \cdot \beta_\sigma x_{\sigma(i_1)} \cdots x_{\sigma(i_n)} , \qquad (1)$$

where $\lambda = \{i_1, \ldots, i_n\}$, $i_1 < \cdots < i_n$, and

$$f = \sum_{\sigma \in S_\lambda} \beta_\sigma x_{\sigma(i_1)} \cdots x_{\sigma(i_n)} .$$

Let $\widehat{\mathbf{K}}$ be a Clifford algebra generated by $\widehat{t}_1, \widehat{t}_2, \ldots$ and presented by the following relations

$$\widehat{t}_i^2 = 1, \ \widehat{t}_i \cdot \widehat{t}_j = -\widehat{t}_j \cdot \widehat{t}_i \ (i, j = 1, 2, \ldots, \ i \neq j).$$

Denote $\widehat{\mathbf{H}}$ the subalgebra of the tensor product $\widehat{\mathbf{K}} \otimes_K \mathbf{H}$ generated by elements $\widehat{x}_i \overset{\text{def}}{=} \widehat{t}_i \otimes x_i \ (i = 1, 2, \ldots)$. It is evident that for any word $w = w(x_1, \ldots, x_l)$ the following equality holds in $\widehat{\mathbf{K}} \otimes_K \mathbf{H}$

$$w(\widehat{x}_1, \ldots, \widehat{x}_l) = w(\widehat{t}_1, \ldots, \widehat{t}_l) \otimes w(x_1, \ldots, x_l) .$$

This equality shows that $\widehat{\mathbf{H}}$ is an absolutely free associative K-algebra and a mapping $\varphi : x_i \to \widehat{x}_i$ establishes an isomorphism of $\mathbf{H}$ and $\widehat{\mathbf{H}}$. From the defining relations it follows that in the Clifford algebra $\widehat{\mathbf{K}}$ the word $w(\widehat{t}_1, \ldots, \widehat{t}_l)$ equals up to sign to $\widehat{t}_1^{r_i} \cdots \widehat{t}_l^{r_l}$, where r_i is the degree of the word w with respect to x_i. Consequently, a linear mapping $\pi : \mathbf{H} \to \mathbf{H}$, given on words by the formula

$$w(\widehat{x}_1, \ldots, \widehat{x}_l) \overset{\text{def}}{=} \widehat{t}_1^{r_i} \cdots \widehat{t}_l^{r_l} \otimes \pi(w)$$

coincides on multilinear polynomials $f \in \mathbf{H}_\lambda$ with the mapping (1) and $\pi(w)$ equals either w or $-w$.

Corollary 1 *The mapping $\pi : \mathbf{H} \to \mathbf{H}$ has the following properties*
   a) $\pi^2 = \mathbf{id}$;
   b) $\pi([x_1, x_2, x_3]) = \{x_1, x_2, x_3\}$;
   c) $\pi : \mathbf{F}_\lambda \to \mathbf{B}_\lambda$, *where* $\mathbf{F}_\lambda \overset{\text{def}}{=} \mathbf{F} \cap \mathbf{H}_\lambda$, $\mathbf{B}_\lambda \overset{\text{def}}{=} \mathbf{B} \cap \mathbf{H}_\lambda$, *and this mapping is a one-to-one correspondence.*

Theorem 3 *If a multilinear identity $f = 0$ of Young symmetry type D, where D is a Young tableau, holds in a variety (of representations) $\mathcal{N}$ of ternary Lie (respectively super Lie) systems, then the multilinear identity $\pi(f) = 0$ holds in the dual variety (of representations) $\mathcal{N}^\sharp$ of ternary super Lie (respectively Lie) systems, and Young symmetry type of $\pi(f) = 0$ is D^*, where D^* is the transpose of a Young tableau D.*

2.   Ternary algebras associated with bilinear forms

In this section we consider Example 3 in detail.

2.1.   TERNARY SYSTEMS OF SYMMETRIC AND SKEW-SYMMETRIC BILINEAR FORMS

Let $\mathbf{b} : V \otimes_K V \to K$ be an arbitrary bilinear form. On the linear space V one can establish a new ternary operation $\{,,\}_\mathbf{b} : V \otimes_K V \otimes_K V \to V$ by the formula

$$\{x, y, z\}_{\mathbf{b}} = x \cdot \mathbf{b}(y, z) - y \cdot \mathbf{b}(z, x) \tag{2}$$

for any $x, y, z \in V$.

This ternary algebra is known as a ternary system of a bilinear form $\mathbf{b}$ and is denoted by $V^{\mathbf{b}}$. It can be immediately verified that the algebra $V^{\mathbf{b}}$ satisfies the Jacobi identity

$$\{x, y, z\}_{\mathbf{b}} + \{y, z, x\}_{\mathbf{b}} + \{z, x, y\}_{\mathbf{b}} = 0 \ .$$

In the particular case of a nondegenerate symmetric bilinear form $\mathbf{b}$ the ternary system $V^{\mathbf{b}}$ is well-known. It is called a ternary Lie system, and its identities and identities of the subvarieties of the varieties generated by this ternary system $V^{\mathbf{b}}$ are studied [1, 2, 4]. In this case let us denote $\mathbf{b}(x, y)$ by (x, y) and its ternary Lie system $V^{\mathbf{b}}$ by B_n, if $\dim_K V = n$, or B_∞, if $\dim_K V = \infty$. It is well-known that the bi-graded Lie algebra $\mathcal{G}_{B_n} = \mathcal{G}_0 \oplus \mathcal{G}_1$, associated with the ternary Lie system B_n is isomorphic to $so(n + 1, K)$, its Lie subalgebra $\mathcal{G}_0$ can be identified with the subalgebra $so(n, K)$, and the universal enveloping associative algebra $U(\mathcal{G}_{B_n})$ of the Lie algebra $\mathcal{G}_{B_n}$ is isomorphic to the universal enveloping associative algebra of the ternary Lie system $\mathcal{G}_1 = B_n$.

Theorem of S.Yu. Vasilovsky [3] states that the following identities

$$\sum_{\sigma \in S_3} \mathrm{sgn}(\sigma)[[x, z_{\sigma(1)}, z_{\sigma(2)}], z_{\sigma(3)}, y] = \sum_{\sigma \in S_3} \mathrm{sgn}(\sigma)[[x, y, z_{\sigma(1)}], z_{\sigma(2)}, z_{\sigma(3)}] \ ,$$

$$\tag{3}$$
$$[[[x, y, u], u, z], u, v] = [[[x, y, u], z, u], u, v] \tag{4}$$

form a basis of identities of B_∞ over fields of characteristic not equal to 2,3,5,7.

If a bilinear form $\mathbf{b}$ is nondegenerate and skew-symmetric, then the ternary algebra $V^{\mathbf{b}}$ is a ternary super Lie system, that is, it satisfies identities (i), (ii), (iii$_+$) from Section 1. We denote this ternary super Lie system by F_n, if $\dim_K V = n = 2m$, and F_∞, if $\dim_K V = \infty$.

We denote a skew-symmetric bilinear form $\mathbf{b}(x, y)$ by $\langle x, y \rangle$. The Specht property of the varieties var F_n generated the ternary super Lie system F_n and the varieties of the representations $\mathcal{R}_{F_n}$ generated by the all representations $\rho : F_n \to A$ of F_n in associative algebras was announced by A.G. Loginov [1]. In that paper it was also pointed out that the Lie superalgebra $\mathcal{G}_{F_{2m}} = \mathcal{G}_0 \oplus \mathcal{G}_1$, associated with the ternary super Lie system F_{2m}, is isomorphic to the superalgebra of type $B(1, 2m)$, i.e., the Lie algebra $\mathcal{G}_0$ is isomorphic to the symplectic Lie algebra $sp(2m, K)$, and the $\mathcal{G}_0$-module $\mathcal{G}_1$ is isomorphic to the natural $sp(2m, K)$-module of dimension $2m$. It can be easily shown that the universal enveloping associative algebra $U(\mathcal{G}_{F_{2m}})$

of the Lie superalgebra $\mathcal{G}_{F_{2m}}$ is the universal enveloping associative algebra of the ternary super Lie system F_{2m}. It follows from the fact that the dimension of the vector space $\widetilde{\mathcal{G}}_0$ generated in $U(F_{2m})$ by elements of the form $xy + yx$ $(x, y \in \mathcal{G}_1 = F_{2m})$, is not greater than $\dim_K \mathcal{G}_0$. Since the Lie algebra $\mathcal{G}_0$ is simple, the Lie algebra $\widetilde{\mathcal{G}}_0$ with respect to the operation $[,]$ is isomorphic to $\mathcal{G}_0$.

Here we state the following theorems.

Theorem 4 $(\mathrm{var}\, B_\infty)^\sharp = \mathrm{var}\, F_\infty$ over fields of characteristic zero.

Corollary 2 *All the multilinear identities of the ternary super Lie system F_∞ of a skew-symmetric form $\langle\,,\,\rangle$ are equivalent to the following identities*

$$\sum_{\sigma \in S_3} (\{\{x, z_{\sigma(1)}, z_{\sigma(2)}\}, z_{\sigma(3)}, y\} + \{\{x, y, z_{\sigma(1)}\}, z_{\sigma(2)}, z_{\sigma(3)}\}) = 0 ,$$

$$\sum_{\sigma \in S_3} \mathrm{sgn}(\sigma)(\{\{\{x, y, z_{\sigma(1)}\}, z_{\sigma(2)}, u\}, z_{\sigma(3)}, v\} +$$

$$\{\{\{x, y, z_{\sigma(1)}\}, u, z_{\sigma(2)}\}, z_{\sigma(3)}, v\}) = 0$$

over fields of characteristic not equal to 2,3,5,7.

Theorem 5 *Let $U(B_\infty)$, $U(F_\infty)$ be the universal enveloping associative algebras of infinite-dimensional ternary systems B_∞, F_∞ of symmetric and skew-symmetric forms respectively. Then over fields of characteristic zero the varieties of pairs $\mathrm{var}(U(B_\infty), B_\infty)$, $\mathrm{var}(U(F_\infty), F_\infty)$ (by the representations $B_\infty \hookrightarrow U(B_\infty)$, $F_\infty \hookrightarrow U(F_\infty)$) are dual to each other in the sense of Theorems 2, 3, that is, $(\mathrm{var}(U(B_\infty), B_\infty))^\sharp = \mathrm{var}(U(F_\infty), F_\infty)$.*

2.2. THE VARIETY OF THREE-SUPPORT ALGEBRAS GIVEN BY THE IDENTITY (2)

Lemma 1 *Any multilinear identity of the ternary algebra B_∞ (respectively F_∞) is a consequence of the generalized identity (2), where $\mathbf{b}$ is a nondegenerate symmetric (respectively skew-symmetric) bilinear form.*

Proof. We shall identify the algebra B_∞ with the subalgebra of its universal enveloping associative algebra $U(B_\infty)^{[-]}$ with respect to the operation $[,,]$ (cf. Section 1). Let $f(x_0, x_1, \ldots, x_m) = 0$ be an arbitrary multilinear identity of the ternary Lie system B_∞. Then $m = 2q$ and

$$f(x_0, x_1, \ldots, x_m) = \sum_{\sigma \in S_m} \beta_\sigma [x_0, x_{\sigma(1)}, \ldots, x_{\sigma(m)}] \ (\beta_\sigma \in K) .$$

Using the equality (2) we are able to represent in $U(B_\infty)$ any left-normed commutator in variables $x_0, x_1, \ldots, x_m \in B_\infty$ in the form of a linear combination of generalized monomials of the form $x_t \cdot (x_{i_1}, x_{j_1}) \cdots (x_{i_q}, x_{j_q})$, where the set $\{t, i_1, j_1, \ldots, i_q, j_q\}$ is equal to the set $\{0, 1, \ldots, m\}$.

Since $(x_{i_\nu}, x_{j_\nu}) \in K$, we may assume that for any such monomial $x_{i_1} < \cdots < x_{i_q}$, $x_{i_\nu} > x_{j_\nu}$. Then any multilinear combination

$$\sum \beta_{t,i_1,j_1,\ldots,i_q,j_q} x_t \cdot (x_{i_1}, x_{j_1}) \cdots (x_{i_q}, x_{j_q}), \quad (\beta_{t,i_1,j_1,\ldots,i_q,j_q} \in K) \quad (5)$$

where $i_1 < \cdots < i_q$, $i_1 > j_1, \ldots, i_q > j_q$, vanishes in the algebra $U(B_\infty)$ for any $x_0, x_1, \ldots, x_m \in B_\infty$ if and only if all the coefficients $\beta_{t,i_1,j_1,\ldots,i_q,j_q}$ are equal to zero in the field K. Let E_ν be the vector space spanned by the vectors x_{i_ν}, x_{j_ν} and $(x_{i_\nu}, x_{j_\nu}) = 1$. We may assume that $(E_\mu, E_\nu) = (E_\nu, x_t) = 0$ for any $\mu, \nu \in \{1, \ldots, q\}$, $\mu \neq \nu$. Then the terms of the sum (5) are equal to zero except for $\beta_{t,i_1,j_1,\ldots,i_q,j_q} \cdot x_t \cdot (x_{i_1}, x_{j_1}) \cdots (x_{i_q}, x_{j_q})$. Since $(x_{i_\nu}, x_{j_\nu}) = 1$ and $x_t \neq 0$, we have $\beta_{t,i_1,j_1,\ldots,i_q,j_q} = 0$. Thus, a multilinear identity $f(x_0, x_1, \ldots, x_m) = 0$ holds true in B_∞ if and only if the expression of the form (5) obtained from the multilinear polynomial f by using the generalized identity (2) equals to zero.

This proves Lemma 1 for the case of B_∞. One can obtain a proof of Lemma 1 for the case of the ternary super Lie system F_∞ by changing $(,)$, $[,,]$, B_∞, $U(B_\infty)^{[-]}$ for $\langle,\rangle$, $\{,,\}$, F_∞, $U(F_\infty)^{[+]}$. $\square$

Proof of Theorem 4. Let E, A and **b** be either $B_\infty, U(B_\infty), (,)$ or $F_\infty, U(F_\infty), \langle,\rangle$ with respect to ternary system in question. We construct new objects $\widehat{A} = G \otimes_K A$, $\widehat{E} = G_1 \otimes_K A$ and establish a bilinear G_0-form $\widehat{\mathbf{b}} : \widehat{E} \otimes_{G_0} \widehat{E} \to G_0$ by the formula $\widehat{\mathbf{b}}(\eta_1 \otimes e_1, \eta_2 \otimes e_2) = \eta_1 \eta_2 \cdot \mathbf{b}(e_1, e_2)$. It is evident that for a symmetric (respectively skew-symmetric) form **b** the form $\widehat{\mathbf{b}}$ is skew-symmetric (respectively symmetric). It can be easily verified that the following equalities hold true in $\widehat{A}$

$$\{\eta_1 \otimes e_1, \eta_2 \otimes e_2, \eta_3 \otimes e_3\}_{\widehat{\mathbf{b}}} = \eta_1 \eta_2 \eta_3 \otimes \{e_1, e_2, e_3\}_{\mathbf{b}} =$$

$$\eta_1 \eta_2 \eta_3 \otimes (e_1 \cdot \mathbf{b}(e_2, e_3) - e_2 \cdot \mathbf{b}(e_3, e_1)) =$$

$$\eta_1 \otimes e_1 \cdot \widehat{\mathbf{b}}(\eta_2 \otimes e_2, \eta_3 \otimes e_3) - \eta_2 \otimes e_2 \cdot \widehat{\mathbf{b}}(\eta_3 \otimes e_3, \eta_1 \otimes e_1).$$

This shows that the equality (2) holds true in $\widehat{A}$ for any $x, y, z \in \widehat{E}$ with respect to the bilinear form $\widehat{\mathbf{b}} : \widehat{E} \otimes_{G_0} \widehat{E} \to G_0$. Thus, from Lemma 1 and the proof of Lemma 1 it follows for $E = B_\infty$ that any identity of F_∞ is an identity of the ternary Lie super system $\widehat{B}_\infty$, and for $E = F_\infty$ any identity of B_∞ is an identity of the ternary Lie system $\widehat{F}_\infty$. It follows from Proposition 1 of Section 1 that $(\text{var } E)^\sharp = \text{var } \widehat{E}$, and $(\text{var } B_\infty)^\sharp \subseteq \text{var } F_\infty$, $(\text{var } F_\infty)^\sharp \subseteq \text{var } B_\infty$. Applying the involutory property of the mapping $\sharp$ we have that $(\text{var } B_\infty)^\sharp \supseteq \text{var } F_\infty$, $(\text{var } F_\infty)^\sharp \supseteq \text{var } B_\infty$. Thus, $(\text{var } B_\infty)^\sharp = \text{var } F_\infty$, and Theorem 4 is proved. $\square$

Lemma 2 *Any multilinear identity of the pair* $(U(F_\infty), F_\infty)$ *(respectively* $(U(B_\infty), B_\infty))$ *follows from the generalized identity* (2), *where* **b** *is a non-degenerate skew-symmetric (respectively symmetric) bilinear form.*

Proof. We identify the universal enveloping associative algebra $U(F_\infty)$ of the ternary super Lie system F_∞ with the universal enveloping Lie superalgebra $\mathcal{F}_\infty = \mathcal{G}_0 \oplus \mathcal{G}_1$. We identify $\mathcal{G}_0$ with the linear vector space in $U(F_\infty)$ spanned by elements of the form $\{g_i \circ g_j\} \overset{\text{def}}{=} g_i \cdot g_j + g_j \cdot g_i$ $(i \geq j)$, where $g_0 < g_1 < g_2 < \cdots$ is an arbitrary partially ordered basis of the vector space $\mathcal{G}_1 = F_\infty$. Then the elements of the form

$$g_{r_1} \cdots g_{r_i} \{g_{t_1} \circ g_{s_1}\}^{n_1} \cdot \cdots \cdot \{g_{t_m} \circ g_{s_m}\}^{n_m}$$

$$(r_1 < \cdots < r_l, \ t_\nu \geq s_\nu, \ (t_1, s_1) < \cdots < (t_m, s_m))$$

form a basis of the universal enveloping associative algebra $U(\mathcal{G}_{F_\infty}) = U(F_\infty)$. In particular, all the elements of the form

$$g_{r_1} \cdots g_{r_i} \{g_{t_1} \circ g_{s_1}\} \cdot \cdots \cdot \{g_{t_m} \circ g_{s_m}\}, \qquad (6)$$

$$(r_1 < \cdots < r_l, \ t_\nu \geq s_\nu, \ t_1 < \cdots < t_m),$$

that is linear in each variable, are linearly independent. Let $f(x_1, \ldots, x_n) = 0$ be an arbitrary identity of the pair $(U(F_\infty), F_\infty)$. Since any super Lie commutator of an odd length can be represented in the form (5), where $(,)$ is changed for $\langle,\rangle$, by using generalized identity (2), then a multilinear associative polynomial $f(x_1, \ldots, x_n)$ can be represented as a linear combination of generalized monomials of the form

$$x_{r_1} \cdots x_{r_i} \{x_{t_1} \circ x_{s_1}\} \cdots \{x_{t_m} \circ x_{s_m}\} \cdot \langle x_{i_1}, x_{j_1}\rangle \cdots \langle x_{i_q}, x_{j_q}\rangle, \qquad (7)$$

where

$$r_1 < \cdots < r_l, \ t_\nu < s_\nu, \ t_1 < \cdots < t_m; i_\nu < j_\nu, \ i_1 < \cdots < i_q,$$

and the set $\{1, 2, \ldots, n\}$ is a disjoint union of all the indices of type r, t, s, i, j. A multilinear expression $\sum_w b_w \cdot w$, where the sum is taken over different generalized monomials (7), is equal to zero in $U(F_\infty)$ if and only if all the coefficients $b_w \in K$ are equal to zero. For any substitution of the variables $x_1, x_2, \ldots, x_n$ for linearly independent elements $g_1, g_2, \ldots, g_n \in F_\infty$ monomials w of the forms (7) are mapped up to element of the base field K to elements $\overline{w}$ of the form (6). These monomials are linearly independent in $U(F_\infty)$. One may choose an arbitrary generalized monomial w_0 of the form (7). One may assume that the two-dimensional vector subspace E_ν

$(\nu = 1, 2, \ldots, q)$, spanned by $x_{i_\nu} = g_{i_\nu}$, $x_{j_\nu} = g_{j_\nu}$, are orthogonal to each other, and $\langle g_{i_\nu}, g_{j_\nu} \rangle = 1$. Then

$$\sum_w \beta_w \cdot \overline{w} = \beta_{w_0} \cdot \overline{w}_0 + \sum_{\overline{w} \neq \overline{w}_0} \alpha_{\overline{w}} \cdot \overline{w}$$

and $\sum_w \beta_w \cdot \overline{w} = 0$ only if $\beta_{w_0} = 0$. This shows that a multilinear identity $f = 0$ holds in the pair $(U(F_\infty), F_\infty)$ if and only if the linear combination of different generalized monomials of the form (7) constructed by multiple using of the generalized identity (2), where $\mathbf{b} = \langle, \rangle$, has zero coefficients, that is, Lemma 2 is proved for the case of the ternary super Lie system of a skew-symmetric bilinear form. The case of the ternary Lie system of a symmetric bilinear form is treated in a similar way. In this case we consider linearly independent elements of $U(B_\infty)$ of the form

$$g_{r_1} \cdots g_{r_l} \cdot [g_{t_1}, g_{s_1}] \cdot \ldots \cdot [g_{t_m}, g_{s_m}], \tag{8}$$

$$(r_1 < \cdots < r_l, \ t_\nu \geq s_\nu, \ t_1 < \cdots < t_m),$$

instead of elements of the form (6), and generalized monomials of the form

$$x_{r_1} \cdots x_{r_l} [x_{t_1}, x_{s_1}] \cdots [x_{t_m}, x_{s_m}] \cdot (x_{i_1}, x_{j_1}) \cdots (x_{i_q}, x_{j_q}), \tag{9}$$

where

$$r_1 < \cdots < r_l, \ t_\nu < s_\nu, \ t_1 < \cdots < t_m; i_\nu < j_\nu, \ i_1 < \cdots < i_q,$$

instead of elements of the form (7). $\square$

Proof of Theorem 5. Let E, A, $\mathbf{b}$, $\widehat{E}$, $\widehat{A}$, $\widehat{\mathbf{b}}$ be the same as in the proof of Theorem 4. Since the pair $(\widehat{A}, \widehat{E})$ satisfies the generalized identity (2) with respect to the G_0-linear form $\widehat{\mathbf{b}}$, then it follows from Lemma 2 that any identity $f = 0$ of the pair $(U(F_\infty), F_\infty)$ (respectively $(U(B_\infty), B_\infty)$) holds in the pair $(U(\widehat{B_\infty}), B_\infty)$ (respectively $(U(\widehat{B_\infty}), B_\infty)$). Consequently,

$$\mathrm{var}(U(\widehat{B_\infty}), B_\infty) \subseteq \mathrm{var}(U(F_\infty), F_\infty),$$

$$\mathrm{var}(U(\widehat{F_\infty}), F_\infty) \subseteq \mathrm{var}(U(B_\infty), B_\infty).$$

It follows from Proposition 1 that $(\mathrm{var}(A, E))^\sharp = \mathrm{var}(\widehat{A}, \widehat{E})$ and

$$(\mathrm{var}(U(B_\infty), B_\infty))^\sharp \subseteq \mathrm{var}(U(F_\infty), F_\infty),$$

$$(\mathrm{var}(U(F_\infty), F_\infty))^\sharp \subseteq \mathrm{var}(U(B_\infty), B_\infty).$$

Since the involutory property of the mapping $\sharp$ we have the inverse inclusions, and Theorem 5 is proved. $\square$

Theorem 6 *All the multilinear identities of the pair $(U(B_\infty),\ B_\infty)$ (of the pair $(U(F_\infty), F_\infty))$ are equivalent to the identities of degree ≤ 7 over a base field of characteristic not equal to 2.*

By Theorem 5 the proof of the fact may be established only for the pair $(U(B_\infty), B_\infty)$. In that case now one can easily prove this theorem using the standard B_∞-collecting process for the basis commutator elements.

References

1. Loginov, A.G.: On Spechtness of the varieties of the representations of an odd component of a super Lie algebra of the type $B(1, n)$, *18th All-Union Conference on Algebra. Proceedings. Part 1,* Kishinev, 1985 (in Russian)
2. Trishin, I.M.: *On identities of the representations of the ternary Lie algebra of a bilinear form,* VINITI, N875-B86, Moscow, 1985 (in Russian).
3. Vasilovsky, S.Yu.: *Bases of identities of certain simple non-associative algebras over an infinite field,* Cand. Thesis, Novosibirsk, 1989 (in Russian).
4. Iltyakov, A.V.: The Specht property of ideals of identities of certain simple non-associative algebras, *Algebra i logika* **24** (1985) no. 3, 327–351.
5. Razmyslov, Yu.P.: *Identities of algebras and their representations,* Amer. Math. Soc., Providence, RI, 1994.

VARIOUS ASPECTS AND GENERALIZATIONS
OF THE GODBILLON–VEY INVARIANT

à la mémoire de Claude GODBILLON (1937–1990)
et Jacques VEY (1943–1979)

CLAUDE ROGER

Bat 101, Univ. Claude Bernard,
43 bld du 11 novembre 1918, 69622 Villeurbanne, France.
E-mail: roger@geometrie.univ-lyon1.fr

Abstract. The aim of this article is to stress both the historical importance and the ubiquitous character of the Godbillon–Vey invariant; it is mainly a survey of known results scattered in different places and does not contain really new results except for the computations on Lie superalgebras.

Mathematics Subject Classification (1991): 57R20, 57R30, 57R32, 58H10, 57R57, 17B70, 17B58.

Key words: Godbillon–Vey invariant, Gelfand–Fuks cohomology, Virasoro algebra, foliations, supergeometry, quantum invariant.

We first recall the original presentation of this invariant as a characteristic class for codimension one foliations. Its cohomological aspects are then developed, first of all as three-dimensional cohomology class of the Lie algebra of formal vector fields, and also as cohomology class of the Lie algebra of tangent vector fields on the circle: it appeared as the very first example of explicit computation of Gelfand–Fuks cohomology. One shall then use the link with Virasoro class to obtain further generalizations; there exists several quantum analogues of the Godbillon–Vey class, using the algebra of symbols of pseudodifferential operators and its q-analogue. There exists also analogous classes in supergeometry in the cohomology of superconformal Lie algebras. Finally, computations about Witt algebras in characteristic p due to F. Weinstein, will also give classes analogous to the Godbillon–Vey invariant.

B. P. Komrakov et al. (eds.), Lie Groups and Lie Algebras, 231–247.
© 1998 *Kluwer Academic Publishers. Printed in the Netherlands.*

The aim of this article is to stress both the historical importance and the ubiquitous character of the Godbillon–Vey invariant; it is mainly a survey of known results scattered in different places and does not contain really new results except for the computations on Lie superalgebras in Section 6.

1. Presentation of the original the Godbillon–Vey invariant, [10]

The original construction of Godbillon and Vey (1971) works as follows: consider $\mathbb{F}$, a codimension one foliation on a 3-dimensional manifold V defined by the Pfaff equation $w = 0$. The 1-form $w \in \Omega^1(V)$ is then integrable, one has $dw = \alpha \wedge w$ for some $\alpha \in \Omega^1(V)$. One deduces $d\alpha \wedge w = 0$; so there exists $\beta \in \Omega^1(V)$ such that $d\alpha = w \wedge \beta$. Let $\eta = \alpha \wedge d\alpha$; it satisfies $d\eta = d\alpha \wedge d\alpha = w \wedge \beta \wedge w \wedge \beta = 0$, so it is closed. Let us replace α by $\alpha' = \alpha + fw$; we shall also have $dw = \alpha' \wedge w$. One has $d\alpha' = d\alpha + d(fw)$ and $\alpha' \wedge d\alpha' = \alpha \wedge d\alpha + \alpha \wedge d(fw) + fw \wedge d\alpha$. Finally if one sets $\eta' = \alpha' \wedge d\alpha'$, then $\eta' - \eta = d(\alpha \wedge fw)$. So the closed form η does not depend on the particular choice of α, modulo an exact form. Besides, the 1-form w which defines the foliation $\mathbb{F}$ is certainly not uniquely defined; are can replace it by φw, where φ is a never vanishing function. The calculation gives $d(\varphi w) = \varphi dw + d\varphi \wedge w$, so $d(\varphi w) = \alpha_\varphi \wedge (\varphi w)$, where $\alpha_\varphi = \alpha + d\text{Log} |\varphi|$. So $d\alpha_\varphi - d\alpha = w \wedge \beta(\varphi w) \wedge (\varphi^{-1}\beta)$. The corresponding form η is then replaced by $\eta_\varphi = \alpha_\varphi \wedge d\alpha_\varphi = \alpha \wedge d\alpha + d(\alpha\text{Log}(\varphi))$.

So the cohomology class of η does not depend on the choice of the Pfaff equation defining $\mathbb{F}$, it is an invariant of the foliation itself and belongs to the third de Rham cohomology group of V. This is the Godbillon–Vey invariant of $\mathbb{F}$, and it is denoted by $GV(\mathbb{F})$. Different dynamical and topological properties of this invariant have been obtained; one can see the survey by E. Ghys [8] for a very clear and complete description of these.

There exists another construction of this invariant, which makes use of an atlas of transition functions defining the foliation, and gives the same class in Čech cohomology, rather than de Rham cohomology:

Let us consider $W(1)$, the Lie algebra of formal vector fields in one variable; explicitly each element of $W(1)$ can be written as $p\partial/\partial x$, where $p \in \mathbb{R}[[x]]$ is a formal power series with real coefficients. The Lie bracket is then naturally defined by

$$[p\frac{\partial}{\partial x}, q\frac{\partial}{\partial x}] = (pq' - qp')\frac{\partial}{\partial x}.$$

If one sets $e_n = x^{n+1}\partial/\partial x$, one obtains a (topological) basis $\{e_n | n \geq -1\}$ with the following bracket $[e_n, e_m] = (n - m)e_{n+m}$. We shall make use later of $C^*(W(1))$, the cohomological complex of the Lie algebra $W(1)$; recall that $C^p(W(1))$ is the space of antisymmetric mappings on $W(1)$ with values in $\mathbb{R}$, continuous with respect to m-adic topology and $d : C^p(W(1)) \rightarrow$ $C^{p+1}(W(1))$ is Chevalley–Eilenberg differential for Lie algebra cohomology. That cohomology will be denoted by $H^*(W(1))$.

Suppose now that our codimension one foliation $\mathbb{F}$ is given by a locally finite open covering $U(u_i)_{i \in I}$ with submersions $f_i : u_i \subset V \rightarrow \mathbb{R}$ and transition functions that are local diffeomorphisms:

$$\gamma_{ij} : V_{ij} = f_j(u_i \cap u_j) \rightarrow V_{ji} = f_i(u_i \cap u_j)$$

such that $\gamma_{ij} \circ f_j = f_i$ for every (i, j). The leaves of $\mathbb{F}$ will then be made from the slices $f_i^{-1}(t)$ for $t \in \mathbb{R}$.

To such data, one can associate a characteristic mapping for the foliation. Let us denote by $\check{C}(U)$ the Čech complex on V with coefficients in the locally constant sheaf, with respect to the covering $U = (u_i)$. One can then construct a characteristic mapping

$$\mathbb{F}^U : C^*(W(1)) \rightarrow \check{C}^*(U)$$

The explicit formula for $\mathbb{F}^U$ is rather complicated and makes use of the infinite order jet of transition functions γ_{ij}. The map $\mathbb{F}^U$ is a morphism of complexes, and when one replaces covering U by an equivalent one U', maps $\mathbb{F}^U$ and $\mathbb{F}^{U'}$ induce the same map in cohomology. This construction exists for foliations of every codimension and was obtained by Bott and Haefliger in the early seventies, and independently by Bernstein and Rosenfeld (see Haefliger [11]). There exists a well-defined characteristic mapping in cohomology denoted by $\mathbb{F}$,

$$\mathbb{F} : H^*(W(1)) \rightarrow \check{H}^*(V, \mathbb{R}) = H^*_{DR}(V).$$

So one is lead to the computation of $H^*(W(1))$. This is the first known result (1968) of the Gelfand–Fuks program of understanding and computing cohomology of infinite-dimensional Lie algebras, mainly Lie algebras of vector fields on manifolds and their formal analogues.

One can give a very elementary computation for $W(1)$. Let ε^n be the dual basis of e_n. So $C^k(W(1))$ is then generated by monomials of the type $\varepsilon^{i_1} \wedge \ldots \wedge \varepsilon^{i_k}$, where $-1 \leq i_1 < i_2 < \ldots < i_k$. Vector e_0 corresponding to homothety gives $[e_0, e_m] = m e_m$ and for coadjoint action one deduces $e_0 \bullet \varepsilon^n = -n \varepsilon^n$. The action on cochains $C^k(W(1))$ is as follows:

$$e_0 \bullet (\varepsilon^{i_1} \wedge \ldots \wedge \varepsilon^{i_k}) = -(\sum_{l=1}^{k} i_\ell) \varepsilon^{i_1} \wedge \ldots \wedge \varepsilon^{i_k}.$$

Note now that this action of e_0 commutes with the Chevalley–Eilenberg differential for the cochains, and then $C^*(W(1))$ splits into the direct sum of subcomplexes that are eigenspaces for the action of e_0.

$$C^*(W(1)) = \oplus_{n \in \mathbb{N}} C^*_{(-n)}(W(1)).$$

Suppose now $C \in C^*_{(-n)}(W(1))$ with $n \neq 0$ is a cocycle; then $dc = 0$. From $e_0 \bullet C = (-n)C$ and the well known formula $\text{ad}e_0 = d \circ i(e_0) + i(e_0) \circ d$, one deduces $d[i(e_0)C] = -nC$. So $C = d(-i(e_0)C/n)$ is then a coboundary. So all the subcomplexes $C^*_{(-n)}(W(1))$ for $n \neq 0$ are cohomologically trivial. In other words, $H^*(W(1)) = H^*(C^*_{(0)}(W(1)))$.

It is now very easy to identify explicitly all the terms of $C^*_{(0)}(W(1))$: the only k-uples $(i_1, \ldots, i_k)$ such that $i^1 + \ldots i^k = 0$ are (0), $(-1, 1)$, $(-1, 0, 1)$. One now checks immediately that $d(\varepsilon^{-1} \wedge \varepsilon^1) = 2\varepsilon^0$. The only remaining class is then $\varepsilon^{-1} \wedge \varepsilon^0 \wedge \varepsilon^1$, and it generates all the cohomology in positive degrees. One has:

Theorem 1 (Gelfand and Fuks) $H^k(W(1)) = 0$ for $k \neq 0, 3$ and $H^3(W(1))$ is one-dimensional.

One can give a more explicit formula for this cocycle. From $e_n = x^{n+1} \partial / \partial x$ one has

$$\langle \varepsilon^n, p \frac{\partial}{\partial x} \rangle = \frac{1}{(n+1)!} \frac{d^{n+1} p}{dx^{n+1}}(0).$$

If we denote $2\varepsilon^{-1} \wedge \varepsilon^0 \wedge \varepsilon^1$ by gv, we obtain the following formula:

$$gv(p \frac{\partial}{\partial x}, q \frac{\partial}{\partial x}, r \frac{\partial}{\partial x}) = \begin{vmatrix} p(0) & p'(0) & p''(0) \\ q(0) & q'(0) & q''(0) \\ r(0) & r'(0) & r''(0) \end{vmatrix}$$

This can be considered as the formal Godbillon–Vey invariant, since one has $\mathbb{F}^*(gv) = GV(\mathbb{F})$ (Bott–Haefliger). The proof makes use of the bicomplex of Čech cochains with coefficients in differential forms and goes along the lines of the proof of the isomorphism between de Rham and Čech cohomology. We shall now specialize to the case of foliations transverse to a circle bundle.

2. Foliations transverse to circle bundles and group of diffeomorphisms

Let us consider the following geometric situation: one has a locally trivial fibre bundle $E \xrightarrow{\pi} V$ whose fibres are diffeomorphic to S^1 and a codimension one foliation on the total space E transverse to the fibres; so for every $e \in E$ one has a direct sum decomposition $T_e E = T_e \mathbb{F} \oplus \text{Ker}\,(T_e \pi)$, $\text{Ker}\,(T_e \pi)$ being the vertical bundle tangent to the fibres of π and $T_e \mathbb{F}$ the tangent space to the leaf of $\mathbb{F}$ which contains e. Following well known results due to Ehresmann on flat bundles, one has an alternative description of such a bundle with transverse foliation. It can be constructed through an open covering $U = (u_i)$ of V with a one cocycle $g_{ij} : u_i \cap u_j \mapsto \mathrm{Diff}\,(S^1)$ with values in the group of diffeomorphisms of the circle such that g_{ij} is a locally constant function. This data enables us to construct the foliation on E by sticking together the $u_i \times S^1$ with trivial foliation by $u_i \times \{\sigma\}(\theta \in S^1)$, using the g_{ij} (here the locally constant hypothesis is explicitly needed). Such an object will be called an S^1 foliated bundle. We shall denote by $A(S^1)$ the Lie algebra of tangent vector fields on the circle; one has once more the formula

$$[f\frac{\partial}{\partial\theta}, g\frac{\partial}{\partial\theta}] = (fg' - f'g)\frac{\partial}{\partial\theta}$$

but here f and g are functions on the circle instead of formal power series. The cohomology of this algebra will give characteristic classes for foliations transverse to circle bundles (see Haefliger [11]). One has a characteristic mapping, which is in some sense a globalization of the above defined one:

$$H^*(A(S^1), \mathbb{R}) \xrightarrow{C_{\mathbb{F}}} H^*_{DR}(E).$$

Besides, one has a natural action of rotation group $SO(2)$ on both sides; on the left hand side, $SO(2)$ acts on the circle S^1 by rotation and thus on $A(S^1)$ through the adjoint action; on the right, the action is obtained by rotation along the fibres of the principal S^1 bundle. The map $C_{\mathbb{F}}$ is equivariant for those $SO(2)$ actions, so it can be quotiented through the basic cohomology, and one obtains a commutative diagram.

$$
\begin{array}{ccc}
H^k(A(S^1); \mathbb{R}) & \xrightarrow{\ C_{\mathbb{F}}\ } & H^k_{DR}(E) \\
{\scriptstyle I'}\downarrow & & \downarrow{\scriptstyle I} \\
H^{k-1}(A(S^1)SO(2); \mathbb{R}) & \xrightarrow{\ C'_{\mathbb{F}}\ } & H^{k-1}_{DR}(V)
\end{array}
$$

The mapping I is induced by integration of forms along the fibres. Further, $H^*(A(S^1), SO(2); \mathbb{R})$ denotes $SO(2)$ basic cohomology: this is the cohomology of the subcomplex of $C^*(A(S^1))$ made from $SO(2)$-invariant cochains that vanish on the constant vector field $\partial/\partial\theta$, representing the Lie algebra of $SO(2)$. Those cohomologies are well known since the work of Gelfand and Fuks ([7]) Interesting things happen when $k = 3$. The space $H^3(A(S^1); \mathbb{R})$ is one-dimensional, generated by the Godbillon–Vey invariant on S^1. One can use the following cocycle:

$$gv_{S^1}(f\frac{\partial}{\partial\theta}, g\frac{\partial}{\partial\theta}, h\frac{\partial}{\partial\theta}) = \int_{S^1} \begin{vmatrix} f(\theta) & g(\theta) & h(\theta) \\ f'(\theta) & g'(\theta) & h'(\theta) \\ f''(\theta) & g''(\theta) & h''(\theta) \end{vmatrix} d\theta.$$

One checks now that the following mapping $A(S^1) \xrightarrow{J} W(1)$ defined by

$$J(f\frac{\partial}{\partial\theta}) = J_x^\infty(t)\frac{\partial}{\partial x},$$

where $J_x^\infty(f)$ stands for the infinite jet of f at x, is a Lie algebra morphism and induces the 3-cohomology mapping $J^* : H^3(W(1)) \to H^3(A(S^1))$ such that $J^*(gv) = gv_{S^1}$. So here one can deduce $C_\mathbb{F}(gv_{S^1}) = GV(\mathbb{F})$, the Godbillon–Vey class of the codimension one foliation on the total space E. The space $H^2(A(S^1), SO(2); \mathbb{R})$ is again one-dimensional, generated by the 2-cocycle C, obtained from gv_{S^1} by inner product with the constant field $\partial/\partial\theta$.

Explicitly:

$$C(f\frac{\partial}{\partial\theta}, g\frac{\partial}{\partial\theta}) = \int_{S^1} \begin{vmatrix} f'(\theta) & g'(\theta) \\ f''(\theta) & g''(\theta) \end{vmatrix} d\theta.$$

Its image $C'_\mathbb{F}(C) \in H^2_{DR}(V)$ is then the Godbillon–Vey invariant integrated on the fibers. To summarize, we have just constructed a characteristic class in two-dimensional cohomology of the basis of a S^1 foliated bundle. On can easily checks its functoriality for natural transformations of foliated bundles.

3. Analogous constructions for groups

We can give a similar construction using the cocycle with value in the group $\text{Diff}(S^1)$. Recall first that for any group G, one has the notion of group cocycle with values in a trivial representation, say $\mathbb{R}$; it leads to the famous Eilenberg–MacLane cohomology for groups. The k-dimensional cochains are mapping from G^k into $\mathbb{R}$ and the differential is defined by:

$$d\gamma(g_0,\ldots,g_k) = \gamma(g_1,\ldots,g_k) + \sum_{i=0}^{k-1}(-1)^{i+1}\gamma(\ldots,g_ig_{i+1},\ldots,g_k)$$
$$+(-1)^{k+1}\gamma(g_0,\ldots,g_{k-1}).$$

The space of k-cochains is denoted by $C^k(G,\mathbb{R})$. A 1-cocycle is then a mapping $\gamma : G \to \mathbb{R}$ such that $\gamma(g_1g_2) = \gamma(g_1) + \gamma(g_2)$, i.e., a group homomorphism from G into $\mathbb{R}$. Then if $G = \text{Diff}(S^1)$ we shall use the cocycle $\{g_{ij}\}$ defining the foliated bundle $E \xrightarrow{\pi} V$ to construct a characteristic mapping:

$$C^k(G,\mathbb{R}) \xrightarrow{\bar{C}_F} \check{C}^k(U,\mathbb{R}),$$

where $\bar{C}_F(\gamma)$ is Čech k-cocycle on the covering U defined by

$$\bar{C}_F(\gamma)_{i_0,\ldots,i_k} = \gamma(g_{i_0i_1}(x),\ldots,g_{i_{k-1}i_k}(x)).$$

The fact that the functions $g_{ij}(x)$ are locally constant implies that the cocycle $\bar{C}_F(\gamma)$ is actually a Čech cocycle with constant coefficients. The fact that $\bar{C}_F$ commutes with the Eilenberg–MacLane and Čech differentials is checked by a straightforward computation. One has a characteristic mapping in cohomology, again denoted by $\bar{C}_F$:

$$\bar{C}_F : H^k(G, \mathbb{R}) \longrightarrow \check{H}^k(V, \mathbb{R}).$$

For $k = 2$, a well known cocycle has been constructed by Thurston (see Bott [3]). To every $f \in \text{Diff}(S^1)$ one associates a function $\mu_f : S^1 \longrightarrow \mathbb{R}$ with positive values by setting $f^*(d\theta) = \mu_f d\theta$. The Thurston cocycle is then defined as follows

$$\wp(f, g) = \int_{S^1} \text{Log}\, \mu_f d\text{Log}\, \mu_{f \circ g}.$$

One has $\bar{C}_F(\wp) = C'_F(C)$ for the Godbillon–Vey invariant integrated along the fibers. It gives a group-theoretical construction of this characteristic class in Čech cohomology, and the Thurston cocycle is thus an analogue of the Godbillon–Vey invariant in group cohomology. The significance of this fact lies in a Van Est type isomorphism ([Ve]) between the Lie algebra cohomology of $A(S^1)$ and an appropriate differentiable cohomology of $\text{Diff}(S^1)$. One has:

$$H^k(A(S^1), SO(2); \mathbb{R}) = H^k_{\text{diff}}(\text{Diff}(S^1), \mathbb{R})$$

(see Haefliger [12] for details).

4. Relationship with the Virasoro algebra

One can give an algebraic description of $A(S^1)$ by using the Lie derivative: if one considers $N = C^\infty(S^1)$ the associative algebra of C^∞ functions on the circle, then $A(S^1) = \text{Der}(N)$ the space of derivations of N, by the Lie derivative: the field $f\partial/\partial\theta$ acts on function φ and gives $f\varphi'$. If one takes the complexification $A(S^1) \otimes \mathbb{C}$, then a holomorphic model is useful. Replace N by the space of holomorphic functions on $\mathbb{C}\backslash\{0\}$, then $A(S^1) \otimes \mathbb{C}$ can be identified with the holomorphic vector fields on $\mathbb{C}\backslash\{0\}$. One can use this fact to give an algebraic basis of $A(S^1) \otimes \mathbb{C}$. Set

$$e_n = \frac{\exp(in\theta)}{i}\frac{\partial}{\partial\theta}$$

(the Fourier basis) or

$$e'_n = Z^{n+1}\frac{\partial}{\partial Z}$$

(the Laurent basis) with $n \in \mathbb{Z}$.

Then one has $[e_n, e_m] = (m-n)e_{n+m}$, as well as $[e'_n, e'_m] = (m-n)e'_{n+m}$. If one computes the cocycle C on those bases, one finds:

$$C(e_n, e_m) = -n^3 \delta_{n+m,0}, \quad C(e'_n, e'_m) = (n^3 - n)\delta_{n+m,0}.$$

One can check directly that they are cohomologous. We shall denote by L the Lie algebra of finite linear combinations of e_n(or e'_n). One has $L = \mathrm{Der}\,(C[Z, Z^{-1}])$ and $L \subset A(S^1) \otimes \mathbb{C}$ as an everywhere dense subalgebra. This algebra is sometimes improperly called the Witt algebra. From standard constructions in homological algebra, one can associate to this algebra and to a 2-cocycle with coefficients in a trivial representation, a central extension: this is an exact sequence of Lie algebras:

$$0 \to \mathbb{C} \to \mathrm{Vir} \xrightarrow{\pi} L \to 0$$

such that the terms corresponding to the kernel $\mathbb{C}$ lie in the center of Vir. The Lie bracket in Vir then reads:

$$[(e_n, \lambda), (e_m, \mu)] = ((m - n)e_{n+m}, \tilde{C}n^3 \delta_{n+m,0}),$$

λ and μ being complex numbers and $\tilde{C}$ an arbitrary complex number sometimes called the central charge. This algebra is called the Virasoro algebra, from the works of M. Virasoro [21] in the early seventies, who constructed it within the theory of dual string models in theoretical physics. See the book [9] for an excellent anthology of various articles concerning this algebra and its representations, and their applications in mathematical physics.

From standard considerations of homological algebra (see for example [16]) and the fact that $L = [L, L]$, one can prove that this central extension is in fact the universal one. One has also a real version, an extension of the Lie algebra of vector fields on the circle:

$$0 \to \mathbb{R} \to \hat{A}(S^1) \xrightarrow{\pi} A(S^1) \to 0.$$

One can again embed Vir $\to A(S^1) \otimes \mathbb{C}$ as a dense subalgebra.

There exists a group extension corresponding to this one: remember that $\mathrm{Diff}\,(S^1)$ is a connected group which is in some weak sense the Lie group associated to the Lie algebra $A(S^1)$; to any one-parameter subgroup of $\mathrm{Diff}\,(S^1)$ one associates easily its generator in $A(S^1)$, but in the opposite direction, the "dictionary" does not work so well.

Thurston's cocycle allows one to construct a central extension of groups:

$$0 \to \mathbb{R} \to \widehat{\mathrm{Diff}}\,(S^1) \to \mathrm{Diff}\,(S^1) \to 0.$$

As a set, $\widehat{\mathrm{Diff}}\,(S^1)$ is isomorphic to the product $\mathrm{Diff}\,(S^1) \times \mathbb{R}$, and the group law is as follows:

$$(f,t) \circ (g,s) = (f \circ g, t + s + \wp(f,g)).$$

This group is sometimes called the Virasoro group; see [9] for some of its applications to representation theory of $\mathrm{Diff}\,(S^1)$. Note that a result of A. A. Kirillov implies that there exists no complex infinite-dimensional Lie group corresponding to $A(S^1) \otimes \mathbb{C}$ (see [17]); a fortiori, there does not exist any complexification of the Virasoro group.

As a final remark let us say that the Godbillon–Vey invariant exists in $H^3(\hat{A}(S^1), \mathbb{R})$; in fact the projection π induces an isomorphism: $\pi^* : H^3(A(S^1), \mathbb{R}) \tilde{\rightarrow} H^3(\hat{A}(S^1), \mathbb{R})$.

5. Quantum version of the Godbillon–Vey invariant

Generally speaking, the appearance of the word "quantum" in mathematical literature means introducing some noncommutativity in a commutative world, or some discreteness in a continuous one. But noncommutative theory does not necessarily imply quantized theory: in this respect, the work of Connes about the Godbillon–Vey invariant in noncommutative geometry, extensively developed in [5], remains a classical one. Here we shall introduce quantization by replacing differential operators by finite order difference operators. On the ring of Laurent power series $A = \mathbb{C}[x, x^{-1}]$ one considers the following linear operator $D_q : A \to A$ defined by

$$D_q(f)(x) = \frac{f(qx) - f(x)}{(q - 1)},$$

where q stands for either a complex number or a formal supplementary parameter (one must then replace A by $A[q]$). In any case as $q \to 1$ the operator D_q tends to $x\partial/\partial x$. So it could be tempting to take as a quantized version of $A(S^1)$, the Lie algebra generated by operators of type uD_q for $u \in A$, where $(uD_q)(f) = uD_q(f)$.

This point of view immediately fails, since these operators do not generate a Lie algebra for the bracket of operators. Explicitly one has:

$$[x^n D_q, x^m D_q] = ((m)_q - (n)_q)x^{n+m}D_q + (q^m - q^n)D_q^2,$$

where $(n)_q = (q^n - 1)/(q - 1)$ is the q-analogue of n. So one must enlarge this algebra considerably, by adding powers of operator D_q. One is lead to consider the algebra $\Psi\mathrm{DO}_q$ of q-pseudodifferential operators. An element $A \in \Psi\mathrm{DO}_q$ has the form: $A = \sum_{n=-\infty}^{n=N} a_n(x)D_q^n$. Multiplication of such q-pseudodifferential operators obey the following rule:

$$D_q \circ u = D_q(u) + \delta(u)D_q,$$

where $\delta(u)(x) = u(qx)$. This implies $D_q(fg) = D_q(f)g + \delta(f)D_q(g)$ and in particular $D_q \circ x = x \circ (qD_q(x) + 1)$. The associative algebra ΨDO_q can then be obtained as a twisted loop algebra over $A = \mathbb{C}[D_q, D_q^{-1}]$, using the automorphism $\sigma : A \to A$, where $\sigma(D_q) = qD_q + 1$. Then ΨDO_q is isomorphic to $A_\sigma[x, x^{-1}]$, the loop algebra over A twisted by σ; product of generators is obtained as follows: $(x^n a).(x^m b) = x^{n+m} \sigma^m(a) b$.

This is a kind of Ore algebra; this construction has been developed by V. Kac and A. Radul. When $q \to -1$ the algebra ΨDO_q contracts onto the algebra ΨDO, well known as the algebra of symbols of pseudo-differential operators on the circle; they have the form $A = \sum_{n=-\infty}^{N} a_n(x)\partial^n$, where ∂ acts as $\partial/\partial x$. The multiplication rule is again obtained through $\partial \circ u = \partial(u) + u\partial$, and one has an analogous interpretation in terms of the twisted loop algebra. The associative algebras ΨDO and ΨDO_q are provided with a Lie algebra structure, by commutators, as usual. This algebra admits one further contraction: let $\phi_t : \Psi DO \to \Psi DO$ be defined via

$$\phi_t(a(x)\partial^p) = a(x)t^{p-1}\partial^p, \ t \in]0, 1].$$

It is a family of isomorphisms, so $[A, B]_t = \phi_t^{-1}[\phi_t A, \phi_t B]$ is a Lie bracket equivalent to the original one. But one computes easily $[A, B]_t = \{A, B\} + t\ldots$, where $\{A, B\}$ is the Poisson bracket:

$$\{A, B\}(x, \delta) = \frac{\partial A}{\partial x}\frac{\partial B}{\partial \delta} - \frac{\partial A}{\partial \delta}\frac{\partial B}{\partial x}.$$

So, as $t \to 0, \Psi DO$ contracts on to the Poisson algebra of functions on the cotangent bundle on S^1. Besides, one has the Lie algebra embeddings $i : A(S^1) \to \Psi DO$, where

$$i(a(x)\frac{\partial}{\partial x}) = a(x)\partial,$$

and $i' : A(S^1) \to C^\infty(T^*S^1) \otimes \mathbb{C}$, where

$$i'(a(x)\frac{\partial}{\partial x}) = a(x)\xi,$$

ξ being a linear function on the fiber. So one has a diagram of embeddings of Lie algebras:

$$
\begin{array}{ccc}
 & & \Psi DO \\
 & \nearrow^{i} & \downarrow \\
A(S^1) & \xrightarrow{i'} & C^*(T^*S') \oplus \mathbb{C}
\end{array}
$$

The vertical arrow denotes contraction.

One has the following theorem (cohomology means Lie algebra cohomology).

Theorem 1 (Existence of the quantum Godbillon–Vey invariant) *There exist cohomology classes $C_{GV}^q \in H^3(\Psi DO_q, \mathbb{C})$ and $C_{GV} \in H^3(\Psi DO, \mathbb{C})$ such that:* (1) $i^*(C_{GV}) = gV$ *and* (2) $C_{GV}^q \mapsto C_{GV}$ *as $q \to 1$ (cf. [15], Theorem 4.12).*

One can further quantize the embedding i into i_q defined as follows $i_q : A(S^1) \to \widetilde{\Psi}DO_q$, where $\widetilde{\Psi}DO_q$ denotes ΨDO_q with one formal variable $\mathrm{Log}\,(D_q)$ added. One has $i_q(x^{n+1}\partial/\partial x) = x^n f_q(D_q)$ where

$$
\begin{aligned}
f_q(D_q) &= \frac{\mathrm{Log}\,(1 + D_q(q-1))}{\mathrm{Log}\, q} = \\
&= \frac{1}{\mathrm{Log}\, q}\left[\mathrm{Log}\, D_q + \mathrm{Log}\,(q-1) - \sum_{k=1}^{\infty} \frac{D_q^{-k}}{k(1-q)^k}\right].
\end{aligned}
$$

One easily sees that $f_q(D_q) \to x\partial$ as $q \to 1$, so $i_q \to i$ as $q \to 1$, and one has the commutative diagram:

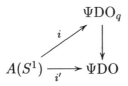

Besides one can show by using explicit formulas for C_{GV}^q that we have $i_q^*(C_{GV}^q) = gv$. So C_{GV}^q fully deserves the name of quantum Godbillon–Vey invariant.

In the proof of the above theorem, Adler's residue plays an essential role; it is defined by $\mathrm{Res}\left(\sum_{i=-\infty}^{N} u_i(x)D_q^i\right) = u_{-1}(x)$ and allows one to construct a symmetric bilinear and nondegenerate invariant form on ΨDO_q and ΨDO. We put $\langle A, B \rangle = \int_{S^1} \mathrm{Res}\,(ABE)$, where

$$
E = \frac{1}{1 + (q-1)D_q} = \sum_{i=1}^{\infty} \frac{(-1)^{i-1}}{(q-1)^i}(D_q)^{-i}.
$$

This form allows one, roughly speaking, to go from cohomology with coefficients in the adjoint representation, to cohomology with trivial coefficients. The two other derivations generating $H^1(\Psi DO_q, \Psi DO_q)$ play an essential role; their generators are respectively $\xi = \mathrm{ad}\,\mathrm{Log}\, x$ and $\eta = \mathrm{ad}\,\mathrm{Log}\, D_q$. The first one can be defined very simply: set $\ell_x(fD_q^a) = afD_q^{a-1}$, and then $\xi(A) = -\ell_x(A)((q-1)D_q+1)$ ([15], Prop. 4.7); the second is more difficult to write down explicitly (see [15], Prop. 4.2).

The corresponding derivations in $H^1(\Psi DO, \Psi DO)$ allowed B. Khesin and O. Kravchenko to construct central extensions of ΨDO generalizing the Virasoro algebra ([14]). Besides, those classes contract into the well known generators of the cohomology of Poisson algebra of T^*S^1. Let us now consider the cup product of ξ and η in $C^2(\Psi\mathrm{DO}_q, \Psi\mathrm{DO}_q)$ and θ the corresponding 3-cochain in $C^3(\Psi\mathrm{DO}_q, \mathbb{C})$:

$$\theta(A, B, C) = \sum_{(\mathrm{cycl})} (\langle \langle \xi(A)\eta(B), C \rangle - \langle \eta(A)\xi(B), C \rangle).$$

Unfortunately, θ is not a cocyle and one has to modify it as follows. Define the cochain a by:

$$a(A, B, C) = \sum_{(\mathrm{cycl})} \mathrm{Tr}(D_q^{-1}ABC).$$

Then $d\theta + da = 0$. So we can define $C_{GV}^q = \theta + a$ and get a 3-cocycle in the Lie algebra cohomology of $\Psi\mathrm{DO}_q$. One can easily deduce an analogous formula for C_{GV} in the 3-cocycles of the Lie algebra cohomology of $\Psi\mathrm{DO}$. For C_{GV}^q, one can compute explicitly:

$$C_{GV}^q(x^l D_q, x^m D_q, x^n D_q) = \frac{1}{2}\delta_0^{l+m+n} \sum_{(\mathrm{alt})} q^{-n}(m)_q(n)_q^2.$$

Here $\sum_{(\mathrm{alt})}$ stands for alternating sum on all permutations of (l, m, n).

6. Superanalogues of the Godbillon–Vey invariant

In that case, the analogy will turn out to be less noticeable. We shall investigate 3-dimensional cohomology of superanalogues of $a(S^1)$ and $W(1)$. Let us first recall the definition of a Lie superalgebra; it is a $\mathbb{Z}/2\mathbb{Z}$ graded vector space $g = g_0 \oplus g_1$ with a graded bracket: $g \times g \xrightarrow{[\cdot,\cdot]} g$, so $[g_i, g_j] \subset g_{i+j}$ with $i, j \in \mathbb{Z}/2\mathbb{Z}$.

This bracket satisfies:
(1) graded anticommutativity $[x, y] = (-1)^{|x||y|+1}[y, x]$;
(2) graded Jacobi identity $\sum_{(\mathrm{alt})}(-1)^{|x||y|}[[x, y], z] = 0$
(here $|x| \in \mathbb{Z}/2\mathbb{Z}$ denotes the degree of $x \in g$).

These axioms imply that g_0 is a Lie algebra; we shall speak of the superisation of a Lie algebra g_0 when one has a Lie superalgebra whose even part contains g_0. For general results and classification theorems concerning Lie superalgebras, the reader is referred to ([6], [13]).

We shall first consider the superalgebras of formal vector fields extending $W(1)$. Define $W(1|n)$ as the superalgebra of vector fields on the superspace of dimension $(1, n)$. Algebraically speaking, take $A(1|n) = \mathbb{R}[[x]] \otimes \wedge^*(\theta_1, \ldots, \theta_n)$, where $\wedge^*(\theta_1, \ldots, \theta_n)$ is the exterior algebra on generators

$(\theta_1, \ldots, \theta_n)$; it is an associative algebra, graded anticommutative for the grading $|x| = 0, |\theta_i| = 1$. Then $W(1|n)$ is the Lie superalgebra of superderivations of $A(1|n)$; a typical element of $W(1|n)$ has the following form:

$$X = F\frac{\partial}{\partial x} + \sum_{i=1}^n \varphi_i \frac{\partial}{\partial \theta_i},$$

where $F, \varphi_i \in A(1|n)$.

So $W(1)$ embeds naturally into the even part of $W(1|n)$. The cohomology of those algebras has been computed by D.B. Fuks, and proofs appear in [2]. For $n \neq 1$, the cohomology of $W(1|n)$ is purely "super" and does not contain anything analogous to the Godbillon–Vey invariant. For $n = 1$, the situation can be described as follows: $H^k(W(1|1)) = 0$ unless $k = 3$, and $H^3(W(1|1))$ is one-dimensional with the following generator S_{gv}. If a field $X \in W(1|1)$ is defined by four formal series (f, g, h, i) in $\mathbb{R}[[x]]$,

$$X = (f + \theta g)\frac{\partial}{\partial x} + (h + \theta_i i)\frac{\partial}{\partial \theta},$$

then:

$$S_{gv}(X_1, X_2, X_3) = \sum_{\sigma \in G_3} \varepsilon(\sigma) g_{\sigma(1)}(0) f_{\sigma(2)}(0) (f'_{\sigma(3)}(0) - i_{\sigma(3)}(0)).$$

But one then sees immediately that this superversion of Godbillon–Vey is completely disconnected from the original one: if i denotes the natural inclusion $W(1) \to W(1|1)$, one immediately checks that $i^*(S_{gv}) = 0$. So one can justify the name super Godbillon–Vey by the fact it is a degree 3 class generating all the cohomology; but it certainly cannot be considered as a superisation of the Godbillon–Vey invariant.

Superanalogues of Lie algebra $A(S^1)$ are known under the name of conformal Lie superalgebras. They have been classified by Feigin, Kac, Leites and Van de Leur; the reader is referred to the excellent survey by Vaintrob ([19]). We shall describe here the best known examples, popular for their physical applications in string theory, the so-called Neveu–Schwarz algebra, denoted by NS. One has $NS = NS_0 \oplus NS_1$, where $NS_0 = A(S^1)$, $NS_1 = F_{-1/2}$, the space of $(-1/2)$ densities on the circle, with the natural action of $NS_0 = A(S^1)$ on it; so the bracket reads:

$$[F\partial, \varphi\sqrt{\partial}] = (F\varphi' - \frac{1}{2}\varphi F)\sqrt{\partial}.$$

The symmetric odd part of the bracket is then

$$[\varphi\sqrt{\partial}, \phi\sqrt{\partial}] = \varphi\phi\partial.$$

One has a natural inclusion $i : A(S^1) \to NS$.

Proposition 1 *One has* $\dim H^2(NS) = 1$ *and* i^* *induces an isomorphism in 2-dimensional cohomology.*

So one has a universal central extension of NS, the Neveu–Schwarz super-algebra $\widehat{NS}$ and the inclusion i extends to $i : \mathrm{Vir} \to \widehat{NS}$. The generating cocycle of $H^2(NS)$ can be written as follows:

$$C(f\delta + \varphi\sqrt{\partial}, g\partial + \phi\sqrt{\partial}) = \int_{S^1} f'''g\,dt + \frac{1}{2}\int_{S^1} \varphi'\phi'\,dt.$$

The fact that i^* is an isomorphism can be seen immediately from this formula. So, concerning the Godbillon–Vey invariant integrated along the fibers, superization works well; it still remains to find reasonably geometric definition of foliations with transverse superstructures and to interpret the above cocycle in that framework.

We shall see that for 3-dimensional cohomology, things are quite different. For superalgebras, one has in fact the following

Proposition 2 *The group* $H^3(NS, \mathbb{R})$ *is one-dimensional.*

Since the above result does not seem to be already known, we give its proof in the appendix.

There exists a strange kind of the Godbillon–Vey invariant in superge-ometry.

7. The Godbillon–Vey invariant in nonzero characteristic

We shall describe here some results deduced from the work of F. Wein-stein [22], where he develops sophisticated techniques for computing difficult cases of Gelfand–Fuks cohomology in characteristic p. The natural generalization of the ring of Laurent polynomials is the ring of truncated polynomials $\mathbb{F}[x]/(x^p)$, where $\mathrm{char}(\mathbb{F}) = p$. The Witt algebra W_p is then the Lie algebra of derivations of this associative algebra. One sees easily that W_p is a p-dimensional vector space on $\mathbb{F}$ with basis $\{e_{-1}, e_0, \ldots, e_{p-2}\}$; each e_i acting on $\mathbb{F}[x]/(x^p)$ as $x^{i+1}\partial/\partial x$. One has almost the same formula for the bracket:

$$[e_i, e_j] = \begin{cases} (j-i)e_{i+j} & \text{for } i+j \le p-2, \\ 0, & \text{if } i+j > p-2. \end{cases}$$

The problem is now to compute $H^*(W_p; \mathbb{F})$; one can use a method analogous to the one described in Section 1 to reduce it to a much smaller computation.

One can use the action of e_0 on cochains to obtains as above a decomposition into a direct sum of eigenspaces:

$$C^*(W_p) = \oplus_n C^*_{(-n)}(W_p).$$

But the above argument does not work completely in the same way. The equation $d(i(e_0)c) = -nc$ allows us to deduce triviality of c only when n is not a multiple of p. So one has: $H^*(W_p, \mathbb{F}) = \oplus_{k \in \mathbb{N}} H^*_{(-kp)}(W_p, \mathbb{F})$.

For $k = 0$, the result is quite identical to the zero characteristic case: $H^3_{(o)}(W_p, \mathbb{F}) = \mathbb{F}_p$ with one generator $\varepsilon^{-1} \wedge \varepsilon^0 \wedge \varepsilon^1$ (see [22], §4.4). This class really deserves the name of the Godbillon–Vey invariant in characteristic p. Cohomology groups for $k > 0$ are more mysterious and much harder to compute; already known results do not show any other 3-dimensional class (see [22]). In particular, when $p \geq 13$, $H^3_{(p)}(W_p, \mathbb{F})$ is trivial. But for $p = 5$, one can do elementary computations: the basis is $\{e_{-1}, e_0, e_1, e_2, e_3\}$; besides $\varepsilon^{-1} \wedge \varepsilon^0 \wedge \varepsilon^1$, one has the element $\varepsilon^0 \wedge \varepsilon^2 \wedge \varepsilon^3$, which is a nontrivial 3-cocycle of weight (-5). One can conjecture that, when p increases, one can find several 3-cocycles of higher weight.

8. Appendix: Proof of the fact that $\dim H^3(NS, \mathbb{R}) = 1$

For brevity we shall denote $A(S^1)$ by L_0 and $F_{-1/2}$ by L_1. So the centerless Neveu–Schwarz algebra is denoted by $NS = L_0 \oplus L_1$, the direct sum of its odd an even parts. In the following, even elements will be denoted by latin letters, and odd ones by greek letters. One has a natural splitting for graded 3-cochains, corresponding to different kinds of parity:

$$\wedge^3_{gr}(NS) = \wedge^3(L_0^*) \oplus (\wedge^2(L_0^*) \otimes L_1^*) \oplus (\wedge^1(L_0^*) \otimes S^2(L_1^*)) \oplus S^3(L_1^*).$$

So any cochain $C \in \wedge^3_{gr}(NS)$ splits according to parity: $C = C_0 + C_1 + C_2 + C_3$, so $C_{even} = C_0 + C_2$, $C_{odd} = C_1 + C_3$. The cochain C_i will be nonvanishing only when evaluated on i odd terms. To study cocycle equation $dC = 0$, one can then split it into five components corresponding again to different parities, and one gets five equations for $C_i = 0$ ($i = 0, 1, 2, 3$).

(1) $dC_0 = 0$, where d is the differential of the cohomology complex for L_0 with trivial coefficients.

(2) $dC_1 = 0$, where d is the differential of the cohomology complex for L_0 with coefficients in L_1^*.

(3) $dC_2 + \tilde{C}_0 = 0$, where d is the differential of the cohomology complex for L_0 with coefficients in $S^2 L_1^*$ and $\tilde{C}_0$ is the 4-cochain of parity $(0, 0, 1, 1)$ defined by: $\tilde{C}_0(X, Y, \alpha, \beta) = C_0(X, Y, \{\alpha, \beta\})$.

(4) $\delta C_1 + I(C_3) = 0$. This equation is among cochains of parity $(0, 1, 1, 1)$ and one has:

$$\delta C_1(X, \alpha, \beta, \gamma) = C_1(X, \{\alpha, \beta\}, \gamma) + C_1(X, \{\alpha, \alpha\}, \beta) + C_1(X, \{\beta, \gamma\}, \alpha)$$

and

$$I(C_3)(X, \alpha, \beta, \gamma) = C_3([X, \alpha], \beta, \gamma) + C_3(\alpha, [X, \beta], \gamma) + C_3(\alpha, \beta, [X, \gamma])$$
$$= [ad_X C_3](\alpha, \beta, \gamma).$$

(5) $\delta C_2 = 0$. This equation is among cochains of parity $(1,1,1,1)$ and reads $\delta C_2(\alpha_1, \alpha_2, \alpha_3, \alpha_4) = \sum_{1 \leq i,j \leq 4} C_2(\{\alpha_i, \alpha_j\}, \alpha_k, \alpha_l)$.

One sees that those equations split into two groups: equations (2) and (4) involve only C_1 and C_3, and, besides, equations (1), (3) and (5), which involve only C_0 and C_2. From equation (2) one sees that $dC_1 = 0$ so C_1 is a 2-cocycle for cohomology of L_0 with coefficients in L_1^*. Since $L_1 = F_{-1/2}$, one has $L_1^* = F_{3/2}$ as an L_0-module. Computations in [18] then show that $H^2(L_0, L_1^*) = 0$, so C_1 is a coboundary, hence one can take $C_1 = 0$ up to cohomology.

Equation $I(C_3) = 0$ then means that there exists an L_0-invariant trilinear symmetric mapping from $F_{-1/2}$ into F_1 (one-forms) denoted $\tilde{C}_3$ such that $C_3(\alpha, \beta, \gamma) = \int_{S^1} \tilde{C}_3(\alpha, \beta, \gamma)$. Considerations developed in the book of Fuks ([6], p. 275) show that there cannot exist such an invariant mapping defined by multidifferential operators. So the odd part of the cohomology must vanish.

Now let us study the even part: from (1) it follows that modulo coboundary $C_0 = gv_{S_1}$. One must now find $C_2 \in \wedge^1(L_0^*) \otimes S^2(L_1^*)$ satisfying the cohomological equation $dC_2 = -gv_{S_1}$. One verifies easily that $C_2(X, \alpha, \beta) = \int_{S^1} X(\alpha\beta)'' dt$ satisfies (3). So we have found a cochain C_2 satisfying (3), and it is uniquely determined up to some cocycle in $Z^1(L_0, S^2(L_1^*))$.

A. Astashkevich pointed out to me that $H^1(L_0, S^2 L_1^*)$ is one-dimensional. So $H^3(NS, \mathbb{R})$ is also one-dimensional, with a generator of type (even, odd, odd). The author thanks A. Astashkevich for correcting a mistake in a previous version of this article.

Remark One could describe the above computation in terms of the Leites spectral sequence for cohomology of superalgebras as described in the book of Fuks ([6] pp. 58–59). The obstruction in equation (3) then comes from the differential $d_2 : E_2^{0,3} \to E_2^{2,2}$.

References

1. Lie, S.: Allgemeine Untersuchung über Differentialgleichnungen, die eine kontinuierliche, endliche Gruppe gestatten, *Math. Ann.* **25** (1985) no. 1, 71–151.
2. Astashkevich, A. and Fuks, D.B.: On the cohomology of the Lie superalgebra $W(m|n)$, *Advances in Soviet Math.* **17** (1993) 1–13.
3. Bott, R.: On the characteristic classes of groups of diffeomorphisms, *L'enseignement mathematique* **23** (1977) no. 3–4, 209–220.
4. Bott, R. and Haefliger, A.: On characteristic classes of Γ-foliations, *Bull. AMS* **78** (1972) no. 6, 1039–1044.
5. Connes, A.: *Non commutative Geometry*, Academic Press, 1994.
6. Fuks, D.B.: *Cohomology of infinite-dimensional Lie algebras*, Consultants Bureau, 1986.
7. Gelfand, I.M. and Fuks, D.B.: The cohomologies of the Lie algebra of vector fields on a circle, *Funkt. Anal and Appl.* **2** (1986) no. 4, 342–343.

8. Ghys, E.: L'invariant de Godbillon–Vey, *(Séminaire Bourbaki. Février 1989)*, Asterisque **177–178** (1989) 155–182.

9. Goddard, P. and Olive, D.: Kac–Moody and Virasoro algebra, *Advanced series in Mathematical Physics* **3** World Scientific, 1988.

10. Godbillon, C. and Vey, J.: Un invariant des feuilletages de codimension 1 *C.R.A.S.* **273** (1971) no. 2, 92–95.

11. Haefliger, A.: *Sur les classes caractéristiques des feuillages*, Séminaire Bourbaki, 1972.

12. A.Haefliger, *Differentiable Cohomology*, Cours au CIME, 1976.

13. Kac, V.: Lie Superalgebras, *Advances in Mathematics* **26** (1977) 8–96.

14. Kravchenko, O. and Khesin, B.: A non trivial central extension of Lie algebra of pseudo-differential symbols on the circle, *Funct. Anal. and Appl.* **25** (1991) no. 2, 83–85.

15. Khesin, B., Lyubachenko, V. and Roger, C.: Extensions and contractions of the Lie algebra of q pseudo differential symbols, Preprint IHES, 1994.

16. Kassel, C. and Loday, L.: Extensions centrales d'algèbres de Lie, *Annales de l'Institut Fourier* **33** (1982) 119–142.

17. Kirillov, A.A.: *The Orbit Method*, Lectures at the University of Maryland, 1990.

18. Ovsienko, V. and Roger, C.: *Generalizations of Virasoro group and Virasoro algebra through extensions by modules of tensor densities on S^1*, Preprint CPT/94, 3024 (in press).

19. Vaintrob, A.: Geometric models and the moduli spaces for string theories, *Lectures Notes in Phys.* **375** (1991) 263–268.

20. Van Est, W.: Group cohomology and Lie algebra cohomology and Lie group, *Indag. Math.* **15** (1953) 484–504.

21. Virasoro, M.: Subsidiary conditions and ghosts in dual-resonance models, *Phys. Rev. D* **1** (1970) 2933–2936.

22. Weinstein, F.: Filtering bases: a tool to compute cohomologies of abstract subalgebras of the Witt algebra, *Advances in Soviet Math.* **17** (1993) 155–216.

VOLUME OF BOUNDED SYMMETRIC DOMAINS AND COMPACTIFICATION OF JORDAN TRIPLE SYSTEMS

GUY ROOS
URA 1322 du CNRS, Département de Mathématiques,
Université de Poitiers, 40, avenue du Recteur Pineau, 86022
POITIERS CEDEX, FRANCE

Abstract. For an irreducible bounded complex circled homogeneous domain, there is a natural normalization of the Euclidean volume, such that this volume is an *integer* of some projective realization of its compact dual.

We give an explanation of this phenomenon in the language of Jordan triple systems. First, we give a simplified version of the projective imbedding of a compactification introduced by O. Loos. We then compute the pullback of the invariant projective volume element by this imbedding; this pullback turns out to be "dual" to the Bergman kernel of the domain. Finally, we prove the equality mentioned above (and more general identities) using some special (real analytic) isomorphism, defined *via* the Jordan structure, between the bounded domain and its ambient vector space.

Mathematics Subject Classification (1991): 17C50, 32M15, 53C35.

Key words: Jordan triple systems, bounded symmetric domains, compactification, Bergman kernel, invariant projective volume.

1. Introduction

1.1. Let D be a bounded symmetric circled domain in a finite-dimensional complex vector space V; assume that D is irreducible (i.e., not the direct product of two such domains).

The classification of these domains is known (E. Cartan, 1935); they fall into four infinite series:

type $\mathrm{I}_{p,q}$ $(1 \le p \le q)$:

B. P. Komrakov et al. (eds.), Lie Groups and Lie Algebras, 249–259.
© *1998 Kluwer Academic Publishers. Printed in the Netherlands.*

(generalized) unit ball of $p \times q$ complex matrices $= \{A\overline{A}' << I_p\}$;
type II_n ($2 \leq n$): unit ball of alternating $n \times n$ complex matrices;
type III_n ($1 \leq n$): unit ball of symmetric $n \times n$ complex matrices;
type IV_n ($n \neq 2$): "Lie ball" in $\mathbb{C}^n$;
and two exceptional types:
type V: dimension 16;   type VI: dimension 27.

1.2. Let h_0 be the Bergman (Hermitian) metric of D at the origin; then V, endowed with h_0, is an Hermitian vector space. The *volume* of D in (V, h_0) has been computed by Hua Lu-keng (1958), for the "classical" domains (the four infinite series). A general formula, using the numerical invariants r (rank of D), a and b (see below) has been established by Koranyi (see [7]).

1.3. Each bounded symmetric domain has a *"compact dual"*, which was initially defined by E. Cartan using the theory of semisimple Lie groups. For example, the compact dual of the domain of type $\mathrm{I}_{p,q}$ is the Grassmann manifold $G_p(\mathbb{C}^{p+q})$, which can be realized as a projective manifold *via* the Plücker imbedding; the compact dual of the Lie ball (type IV_n) is a complex quadric hypersurface in $\mathbb{P}_{n+1}(\mathbb{C})$.

The *degree* of Grassmann manifolds (in the Plücker imbedding) and of other "minuscule flag manifolds" is also well know (Hirzebruch, 1957; see also [3], [11]).

It turns out that, if the volume element in (V, h_0) is suitably normalized, that is if one gives volume 1 to the greatest Hermitian ball of center 0 included in D, then the volume of D is always an *integer*, and this integer is *equal to the degree* of some "natural" projective realization of the compact dual of D. For example, the volume of the $\mathrm{I}_{p,q}$-type domain is equal to the degree of the Plücker imbedding of the corresponding Grassmann manifold; the volume of Lie balls is equal to 2 — the degree of the complex quadrics. A case by case verification shows that the same holds for all other types of Hermitian symmetric domains.

This simple fact, which apparently had not been noticed before, calls for an intrinsic and simple explanation, which will be the subject of this communication.

1.4. An alternative approach to Lie theoretic methods for the study of Hermitian symmetric domains is provided by *Jordan algebras* (for domains of "tube type") and *Jordan triple systems* (for general domains). More precisely, there is an equivalence of categories between the category of bounded symmetric domains and the category of "positive Hermitian Jordan triple systems" (Koecher [6], Loos [9]). This approach leads to a very transparent translation of geometric facts for bounded symmetric domains, such as: boundary components, geodesics of the Bergman metric, invariant polynomials and invariant differential operators, compactification.

We use the compactification of positive Hermitian Jordan triple systems given by Loos [9] — and a simplified version of its canonical projective imbedding — to state and prove the equality between the volume of a bounded symmetric domain and the degree of the canonical projective imbedding of its compact dual.

In Section 2, we will recall basic facts about positive Hermitian Jordan triple systems. New results are gathered in Section 3.

2. Jordan triple systems and bounded symmetric domains

We recall here the basic facts about Jordan triple systems (see Loos [9], Satake [13], Nomura [10]).

2.1. An *Hermitian Jordan triple system* (Hermitian JTS) is a (finite-dimensional) complex vector space equipped with a $\mathbb{C}$-antilinear involution $z \mapsto \bar{z} \colon V \mapsto V$ (called *conjugation*) and a $\mathbb{C}$-trilinear mapping (called *triple product*) $\{\,,\,,\,\} \colon V \times V \times V \longrightarrow V$ such that

$$\{u, v, w\} = \{w, v, u\}, \tag{2.1}$$

$$\{x, y, \{u, v, w\}\} - \{u, v, \{x, y, w\}\} = \{\{x, y, u\}, v, w\} - \{u, \{v, x, y\}, w\}. \tag{2.2}$$

The operators $D(x, y)$, $Q(x, z)$, $Q(x)$ are then defined by

$$D(x, y)z = D(x, z)y = \{x, y, z\}, \qquad 2Q(x) = Q(x, x). \tag{2.3}$$

An Hermitian JTS is called *positive* if

$$(u \mid v) = \operatorname{tr} D(u, \bar{v}) \tag{2.4}$$

is positive definite on V.

2.2. Let V be a positive Hermitian JTS. An element $c \in V$ is called *tripotent*, if

$$D(c, \bar{c})c = \{c, \bar{c}, c\} = 2c. \tag{2.5}$$

For a tripotent element c, the operator $D(c, \bar{c})$ (which is self-adjoint with respect to $(\,\mid\,)$) has its eigenvalues in $\{0, 1, 2\}$, giving rise to the *Peirce decomposition* of V with respect to c:

$$V = V_0(c) \oplus V_1(c) \oplus V_2(c), \tag{2.6}$$

where $V_j(c) = \{x \in V; D(c, \bar{c})x = jx\}$.

Two tripotents c_1, c_2 are called (strongly) *orthogonal* if $D(c_1, \bar{c}_2) = 0$; then $c_1 + c_2$ is also a tripotent. If $(c_1, \cdots, c_p)$ is a family of mutually orthogonal tripotents, $(D(c_j, \bar{c}_j))$ is a commutative family of self-adjoint operators, giving rise to the *simultaneous Peirce decomposition*

$$V = \sum_{0 \le i \le j \le p}^{\oplus} V_{ij}, \tag{2.7}$$

where

$$V_{ij} = \{x \in V; D(c_k, \bar{c}_k)x = (\delta_i^k + \delta_j^k)x, \ \forall k, \ l \le k \le p\}. \tag{2.8}$$

2.3. A positive Hermitian JTS is always *semisimple* and from now on *we will assume that V is simple* (i.e., not the direct sum of two non trivial subsystems with component-wise triple product).

A nonzero tripotent $c \in V$ is called *primitive* if it is not the sum of two nonzero orthogonal tripotents; a maximal family of mutually orthogonal tripotents is called a *frame*. All frames $(c_1, \cdots, c_r)$ of V have the same number r of elements which is the *rank* of V.

If $(c_1, \cdots, c_r)$ is a frame of V and if $V = \sum_{0 \le i \le j \le r}^{\oplus} V_{ij}$ is the corresponding Peirce decomposition, then $V_{00} = 0$, $V_{ii} = \mathbb{C}c_i$ $(1 \le i \le r)$; all V_{ij} have the same dimension, denoted by a:

$$a = \dim V_{ij} \qquad (1 \le i < j \le r); \tag{2.9}$$

also all V_{0i} have the same dimension, denoted by b:

$$b = \dim V_{0i} \qquad (1 \le i \le r). \tag{2.10}$$

The number

$$g = 2 + a(r - 1) + b \tag{2.11}$$

is called the *genus* of V; if $n = \dim V$, we have also

$$n = r + ar(r - 1)/2 + br.$$

2.4. Each element x of V has a (unique) *spectral decomposition*

$$x = \lambda_1 c_1 + \cdots + \lambda_p c_p, \tag{2.12}$$

where $(c_1, \cdots, c_p)$ is a family of mutually orthogonal tripotents and $\lambda_1 > \cdots > \lambda_p > 0$. The map $x \mapsto \lambda_1$ is a norm on V, called the *spectral norm*. The unit ball of this norm is an *irreducible bounded symmetric domain* D; conversely, the triple product $\{ \ , \ , \ \}$ on V can be recovered from the Bergman kernel and the Bergman metric of D, providing an equivalence of categories between (simple) positive Hermitian JTS and (irreducible) bounded complex symmetric domains.

2.5. The integer p in (2.12) is called the *rank* of x. The elements x such that $p = r$ (the rank of V) are called *regular* and form an open dense subset V_{reg} of V; the map

$$((c_1, \cdots, c_p), (\lambda_1, \cdots, \lambda_p)) \mapsto \lambda_1 c_1 + \cdots + \lambda_p c_p \qquad (2.13)$$

is a diffeomorphism $\Phi : \mathcal{F} \times \{\lambda_1 > \cdots > \lambda_p > 0\} \to V_{\text{reg}}$; here $\mathcal{F}$ is the "manifold of frames" of V.

The *automorphism group* $\text{Aut} V$ is the group of complex linear isomorphisms $f : V \to V$ verifying

$$f\{u, \overline{v}, w\} = \{fu, \overline{fv}, fw\}; \qquad (2.14)$$

$\text{Aut} V$ is a compact Lie group and we denote by K its identity component. Then K acts transitively on the manifold of frames $\mathcal{F}$ (recall that we assume that V is simple).

2.6. The *Bergman operator* of a positive Hermitian JTS V is defined by

$$B(x, y) = \text{id}_V - D(x, y) + Q(x)Q(y). \qquad (2.15)$$

It is closely related to the Bergman kernel and Bergman metric of D; if $k(x, \overline{y})$ is the Bergman kernel of D with respect to the flat volume form α^n (see 3.3) and h_z the Bergman metric at $z \in V : h_z(u, \overline{v}) = \partial_u \overline{\partial}_{\overline{v}} \log k(z, \overline{z})$, then

$$h_z(u, \overline{v}) = h_0(B(z, \overline{z})^{-1} u, \overline{v}), \qquad h_0(u, \overline{v}) = \text{tr } D(u, \overline{v}), \qquad (2.16)$$

$$k(x, \overline{y}) = (\text{vol } D)^{-1} \det B(x, \overline{y})^{-1} \qquad (x, y \in D). \qquad (2.17)$$

If $x \in V_{\text{reg}}$ is a regular element with spectral decomposition

$$x = \lambda_1 c_1 + \cdots + \lambda_r c_r$$

and if $V = \sum_{0 \leq i \leq j \leq r}^{\oplus} V_{ij}$ is the Peirce decomposition relative to the frame $(c_1, \cdots, c_r)$, then

$$B(x, \overline{x})y = (1 - \lambda_i^2)(1 - \lambda_j^2)y, \qquad \text{if } y \in V_{ij} \qquad (1 \leq i \leq j \leq n), \quad (2.18)$$

where $\lambda_0 = 0$. It follows immediately from (2.18) that

$$\det B(x, \overline{x}) = \left(\prod_{i=1}^{r} (1 - \lambda_i^2) \right)^g, \qquad (2.19)$$

where g is the genus defined by (2.11). The expression $\prod_{i=1}^{r}(1-\lambda_i^2)$ defines a K-invariant polynomial (on the underlying real space $V_{\mathbb{R}}$), called the *generic norm* and denoted by $m(x,\overline{x})$, which extends to a complex polynomial $m(x,y)$ on $V \times V$, also called generic norm. The generic norm $m(x,y)$ has the expansion

$$m(x,y) = 1 - m_1(x,y) + \cdots + (-1)^j m_j(x,y) + \cdots + (-1)^r m_r(x,y), \quad (2.20)$$

where the m_j's are polynomials, homogeneous of bidegree (i,j) and $m_j(x,\overline{x})$ > 0 for $x \neq 0$.

2.7. FUNCTIONAL CALCULUS ON A POSITIVE HERMITIAN JTS (LOOS, [9], 3.18).

For $x \in V$, the *odd powers* $x^{(2k+1)}$ $(k \geq 0)$ are defined by

$$x^{(1)} = x, \qquad x^{(2k+1)} = Q(x)\overline{x}^{(2k-1)}; \qquad (2.21)$$

if $x = \lambda_1 c_1 + \cdots + \lambda_p c_p$ is the spectral decomposition of x, then

$$x^{(2k+1)} = \lambda_1^{2k+1} c_1 + \cdots + \lambda_p^{2k+1} c_p.$$

Let $f(t)$ be an odd, complex valued, analytic function of the real variable t, defined for $|t| < \rho$:

$$f(t) = \sum_{k=0}^{\infty} a_k t^{2k+1};$$

then the series

$$f(x) = \sum_{k=0}^{\infty} a_k x^{(2k+1)} \qquad (2.22)$$

converges for $\|x\| < \rho$ ($\|x\|$ = spectral norm of x) and defines a real analytic function, also named f, from $\{\|x\| < \rho\}$ to V. If x has the spectral decomposition $x = \lambda_1 c_1 + \cdots + \lambda_p c_p$, $\rho > \lambda_1 > \cdots > \lambda_p$, then

$$f(x) = f(\lambda_1)c_1 + \cdots + f(\lambda_p)c_p. \qquad (2.23)$$

Of particular interest are the following special cases:

1) th : $V \to D$, associated to th : $\mathbb{R} \to]-1,1[$, which is the *exponential map* at 0 of the Riemannian space D (endowed with the Riemannian metric $\mathrm{Re}h_z$);

2) the *quasi-inverse* map $x \mapsto x^{\overline{x}}$, associated to $t \mapsto t/(1-t^2)$, defined by

$$x^{\overline{x}} = \sum_{k=0}^{\infty} x^{(2k+1)}, \qquad x \in D \qquad (2.24)$$

3) the maps $\vartheta : V \to D$, $\psi : D \to V$, inverse of each other, defined by

$$\begin{aligned}
\vartheta(x) &= B(x, -\bar{x})^{-1/4} x & \text{for } x \in V, \\
\psi(x) &= B(x, -\bar{x})^{-1/4} x & \text{for } x \in D,
\end{aligned} \qquad (2.25)$$

respectively associated to $t \mapsto t(1 + t^2)^{-1/2}$, $t \mapsto t(1 - t^2)^{-1/2}$, which will be used below to prove the main result.

3. Compactification of a positive Hermitian Jordan triple system

3.1. STRUCTURE OF THE GENERIC NORM

Proposition 3.1 *Let V be a simple positive Hermitian JTS, $m(x,y)$ its generic norm. There exist maps $\sigma_j : V \to V^{(j)}$ $(0 \le j \le r = \text{rank of } V)$ with the following properties:*

1) $V^0 = \mathbb{C}$, $V^{(1)} = V$, $V^{(2)}, \cdots, V^{(r)}$ *are (finite-dimensional) complex vector spaces, with conjugation $z \mapsto \bar{z}$ and complex inner product $(:)$;*
2) $\sigma_0 = 1$, $\sigma_1 = \text{id}_V$, $\sigma_2, \cdots, \sigma_r$ *are homogeneous polynomial maps of degree j, such that $\sigma_j(\bar{x}) = \sigma_j(x)^-$ and $\sigma_j(V)$ spans $V^{(j)}$;*
3) *the identity*

$$m(x, y) = 1 - (\sigma_1 x : \sigma_1 y) + \cdots + (-1)^r (\sigma_r x : \sigma_r y) \qquad (3.1)$$

holds in $V \times V$.

Examples.
   1) **Type I_{pq}** $(1 \le p \le q)$. $V = \mathcal{M}_q^p(\mathbb{C}) \cong \mathcal{H}am(\mathbb{C}^q, \mathbb{C}^p)$;
      $\{xyz\} = xy'z + zy'x$ $(y' : \text{transposed of } y)$;
      $\bar{x} = \text{conjugate matrix of } x$;
      $m(x, y) = \det(I_p - xy')$;
      $m_j(x, y) = \text{tr}\, (\Lambda^i x \Lambda^i y')$ $(0 < i \le p)$;
      $V^{(i)} = \mathcal{H}am(\Lambda^i \mathbb{C}^q, \Lambda^i \mathbb{C}^p)$ $(0 < i \le p)$;
      $\sigma_i x = \Lambda^i x$.
   2) **Type IV_n** $(n \ne 2)$. $V = \mathbb{C}^n$; $m(x, y) = 1 - q(x, y) + q(x)q(y)$, with
      $q(x, y) = 2 \sum x_i y_i$, $q(x) = 2 \sum x_i^2$;
      $V^{(1)} = V$, $V^{(2)} = \mathbb{C}$; $\sigma_1 x = x$, $\sigma_2 x = q(x)$.
   3) **Type V** (see [12] for the definitions of $\mathcal{O}_\mathbb{C}$, $\mathcal{J}'$ and x^*).
      $V = \mathcal{J}' \cong \mathcal{O}_\mathbb{C} \otimes \mathcal{O}_\mathbb{C} \cong \mathbb{C}^{16}$, $m(x, y) = 1 - (x : y) + (x^* : y^*)$;
      $V^{(1)} = V$, $V^{(2)} = \mathbb{C} \otimes \mathbb{C} \otimes \mathcal{O}_\mathbb{C} \cong \mathbb{C}^{10}$ (type IV_{10});
      $\sigma_1 x = x$, $\sigma_2 x = x^*$.

Remark. At this time, we are only able to prove Proposition 3.1 through a case by case verification. In addition to the facts stated in Proposition 3.1, it turns out that the $V^{(j)}$'s are actually simple positive Hermitian JTS.

We hope to be soon able to give an intrinsic characterization of the maps $\sigma_j : V \to V^j$.

3.2. COMPACTIFICATION

Here we give a simplified version of a compactification described by Loos ([9], §7).

The definition of the *quasi-inverse* x^y extends the definition (2.26) of $x^{\bar{x}}$: if $B(x,y)$ is invertible (as a linear operator), the pair (x,y) is said to be invertible and the quasi-inverse x^y is then defined by

$$x^y = B(x,y)^{-1} (x - Q(x)y). \tag{3.2}$$

An equivalence relation $\sim$ is defined in $V \times V$ as follows:

$$(x,y) \sim (x',y') \Leftrightarrow ((x, y - y') \text{ is invertible and } x' = x^{y-y'}).$$

The equivalence class of (x,y) is denoted $[x : y]$ and the quotient space is named X. For each $v \in V$, let $X_v = \{[x : v]; x \in V\}$. Then there exists ([9]) on X a unique structure of smooth algebraic variety such that X_v is an open affine subvariety and $[x : v] \mapsto x$ is an isomorphism of X_v on V.

The injection $x \mapsto [x : 0]$ of V in X will be called the *canonical compactification* of V. It has the following realization:

Proposition 3.2 *Let V be a simple positive Hermitian JTS of rank* r *and let $\sigma_1, \ldots, \sigma_r$ as in Proposition 3.1. Let*

$$W = \mathbb{C} \oplus V^{(1)} \oplus \cdots \oplus V^{(r)}$$

and let $\sigma : V \to \mathbb{P}(W)$ be defined by

$$\sigma(x) = [1, \sigma_1 x, \ldots, \sigma_r x]. \tag{3.3}$$

Then the closure of $\sigma(V)$ is an algebraic submanifold $\widetilde{X}$ of $\mathbb{P}(W)$, which is isomorphic to X, and $\sigma : V \to \mathbb{P}(W)$ is isomorphic to the canonical compactification above.

From now on, we will identify $\widetilde{X}$ with X.

Examples.

1) **Type I_{pq}** $(1 \leq p \leq q)$. $V = \mathcal{M}_q^p(\mathbb{C}) \cong \mathcal{H}am(\mathbb{C}^q, \mathbb{C}^p)$;
   $W = \sum_{0 \leq i \leq p}^{\oplus} \mathcal{H}am(\Lambda^i \mathbb{C}^q, \Lambda^i \mathbb{C}^p) \cong \Lambda^p(\mathbb{C}^p \oplus \mathbb{C}^q)$;
   $\sigma_i x = \Lambda^i x$, $\sigma : V \to \mathbb{P}(W)$ is the Plücker imbedding;
   $D = \{x; I_p - xx' \gg 0\}$, $X = \text{Grass}_p(\mathbb{C}^{p+q})$.

2) **Type IV_n** $(n \neq 2)$. The compactification imbedding
   $\sigma : V \to \mathbb{P}(W) = \mathbb{P}(\mathbb{C} \oplus V \oplus \mathbb{C})$

is defined by
$$\sigma x = [1, x, q(x)];$$
the compact dual $X = \overline{\sigma(V)}$ is the complex quadric hypersurface of homogeneous equation
$$t_0 t_{n+1} = q(t_1, \cdots, t_n).$$

3) Type V.
$$V^{(1)} = V, \quad V^{(2)} = \mathbb{C} \oplus \mathbb{C} \oplus \mathcal{O}_{\mathbb{C}} \cong \mathbb{C}^{10} \text{ (type IV}_{10});$$
$$W \cong \mathcal{J} \cong \mathbb{C}^{27};$$
$$\sigma : V \to \mathbb{P}(W) = \mathbb{P}(\mathcal{J}) \text{ is defined by } \sigma x = [1, x, x^*];$$
$$X = \{[u]; u \in \mathcal{J}, u^* = 0\}.$$

3.3. VOLUME ELEMENTS

In V, we choose the Hermitian scalar product
$$(x : \overline{y}) = m_1(x, \overline{y}) = \frac{1}{g} h_0(x, \overline{y}); \tag{3.4}$$

let α be the associated Kähler form
$$\alpha = (i/2\pi)\partial\overline{\partial} m_1(x, \overline{x}). \tag{3.5}$$

If V is endowed with the volume form α^n ($n = \dim V$), the unit ball of the norm associated to m_1 has then a volume equal to 1.

On $W = \mathbb{C} \oplus V^{(1)} \oplus \cdots \oplus V^{(r)}$, we put the Hermitian scalar product $(: ^-)$ which is the direct sum of the Hermitian products $(: ^-)$ on the $V^{(j)}$, obtained from Proposition 3.1. Let $F(z) = (z : \overline{z})$ for $z \in W$; then
$$\beta = (i/2\pi)\partial\overline{\partial} \log F \tag{3.6}$$

defines a form on $\mathbb{P}(W)$. If Z is a smooth submanifold of pure dimension d in $\mathbb{P}(W)$, the *degree* of Z in $\mathbb{P}(W)$ is given by
$$\deg Z = \int_Z \beta^d.$$

In particular, the degree of the compactification X of V, described in 3.2, is
$$\deg X = \int_X \beta^n = \int_V \sigma^*\beta^n, \tag{3.7}$$

as $\sigma(V)$ is an open dense subset of X. From the definition of σ and by Proposition 3.1, it is clear that
$$\sigma^*\beta = (i/2\pi)\partial\overline{\partial} \log(1 + m_1(x, \overline{x} + \cdots + m_r(x, \overline{x})),$$

which can also be written

$$\sigma^* \beta = (i/2\pi) \partial \bar{\partial} \log m(x, -\bar{x}). \tag{3.8}$$

The following proposition allows us to compute the pull-back $\sigma^* \beta^n$ in (3.7):

Proposition 3.3 *The following identity holds in the Hermitian positive Jordan triple system V:*

$$(i/2\pi)^n (\partial \bar{\partial} \log m(x, -\bar{x}))^n = \det B(x, -\bar{x})^{-1} \alpha^n. \tag{3.9}$$

3.4. COMPUTATION OF α^N IN THE SPECTRAL DECOMPOSITION

Proposition 3.4 *Let*

$$\Phi : \mathcal{F} \times \{\lambda_1 > \lambda_2 > \cdots > \lambda_r > 0\} \longrightarrow V_{\text{reg}}$$

be the diffeomorphism defined by

$$\Phi((c_1, \cdots, c_r), (\lambda_1, \cdots, \lambda_r)) = \sum \lambda_j c_j,$$

where $\mathcal{F}$ is the manifold of frames of V. Then the pull-back of α^n by Φ is given by

$$\Phi^* \alpha^n = \Theta \wedge \prod_{j=1}^{r} \lambda_j^{2b+1} \prod_{j>k} (\lambda_j^2 - \lambda_k^2)^a d\lambda_1 \wedge \cdots \wedge d\lambda_r, \tag{3.10}$$

where a, b are the numerical invariants of V defined in 2.3 and Θ is a K-invariant volume form on $\mathcal{F}$.

3.5. THE EFFECT OF ϑ AND ψ ON VOLUME ELEMENTS

Recall the definition of ϑ and ψ (inverse of each other) given in 2.7: if $x = \lambda_1 e_1 + \cdots + \lambda_p e_p$ is the spectral decomposition of x, then

$$\begin{aligned}
\vartheta(x) &= B(x, -\bar{x})^{-1/4} x & \text{for } x \in V, \\
\psi(x) &= B(x, -\bar{x})^{-1/4} x & \text{for } x \in D,
\end{aligned} \tag{2.25}$$

Proposition 3.5 *The following relations hold in an Hermitian positive Jordan triple system V:*

$$\psi^* (m(x, -\bar{x})^\gamma \alpha^n) = m(x, \bar{x})^{-g-\gamma} \alpha^n, \tag{3.11}$$

or equivalently

$$\psi^* (\det B(x, \bar{x})^s \alpha^n) = \det B(x, \bar{x})^{-1-s} \alpha^n; \tag{3.11'}$$

$$\vartheta^*(m(x,\overline{x})^\gamma \alpha^n) = m(x,\overline{x})^{-g-\gamma}\alpha^n; \qquad (3.12)$$

or equivalently

$$\vartheta^*(\det B(x,\overline{x})^s \alpha^n) = \det B(x,\overline{x})^{-1-s}\alpha^n. \qquad (3.12')$$

As a consequence, taking $s = -1$ in $(3.11')$ and integrating over V, we get by (3.7) and (3.9) the desired result:

Theorem

$$\deg X = \int_D \alpha^n. \qquad (3.13)$$

References

1. Cartan, E.: Sur les domaines homogénes bornés de l'espace de n variables complexes, *Abh. Math. Sem. Univ. Hamburg* **11** (1935) 116–162.
2. Helgason, S.: *Differential Geometry, Lie Groups and Symmetric Spaces*. Academic Press, XV+634 pp., 1978.
3. Hiller, H.: The Geometry of Coxeter groups, *Research Notes in Math.* **54** (1982) Pitman Advanced Publishing Program, Boston, 213 pp.
4. Hirzebruch, F.: *Characteristic numbers of homogeneous domains, Seminar on Analytic Functions*, Institute for Advanced Study, Princeton, N.J., 1957.
5. Hua Lu-keng: *Harmonic Analysis of Functions of Several Complex Variables in Classical Domains*. Science Press, Beijing, 1958 (in Chinese).
6. Koecher, M.: *An elementary approach to bounded symmetric domains*. Lect. Notes, Rice Univ., Houston, 1969.
7. Korányi, A.: Analytic invariants of bounded symmetric domains, *Proc. Amer. Math. Soc.* **19** (1968) 279–284.
8. Loos, O.: Jordan pairs, *Lect. Notes in Math.* **460** (1975) Springer–Verlag.
9. Loos, O.: *Bounded Symmetric Domains and Jordan Pairs*, Math. Lect., Univ. of California, Irvine, 1977.
10. Nomura, T.: Algebraically independent generators of invariant differential operators on a bounded symmetric domain, *J. Math. Kyoto Univ.* **31** (1991) 265–279.
11. Proctor, R. A.: The geometry of Coxeter groups, by H. Hiller (Book review), *Bull. Amer. Math. Soc. (New Series)*, **10** (1984) 142–150.
12. Roos, G.: Algèbres de composition, Systémes triples de Jordan exceptionnels. *In G. Roos, J.P. Vigue, Systémes triples de Jordan et domaines symétriques*, Travaux en cours, 43, Hermann, Paris, 1992, 1–84.
13. Satake, I.: *Algebraic Structures of Symmetric Domains*, Iwanami Shoten & Princeton University Press, 1980, 320 pp.

ASYMPTOTIC BEHAVIOR OF THE POISSON TRANSFORM ON A HYPERBOLOID OF ONE SHEET

A.A. ARTEMOV
Tambov State University,
Internatsionalnaya 33, Tambov 392622, Russia.
E-mail: artemov@math-univ.tambov.su

Abstract. In this work we study the asymptotic expansion of the Poisson transform at infinity for the hyperboloid of one sheet $SO_0(1,2)/SO_0(1,1)$.

Mathematics Subject Classification (1991): 22E30, 22E46, 43A32, 43A85.

Key words: Poisson transform, asymptotic expansion, symmetric spaces.

Introduction

In this work we study the asymptotic expansion of the Poisson transform at infinity for the hyperboloid of one sheet $SO_0(1,2)/SO_0(1,1)$. It is the simplest but also the key example of a semisimple symmetric space G/H with noncompact H. We consider arbitrary, not only K-finite, functions and study the expansion not only in exponents but also in powers of some other functions.

In order to keep the size of the paper within reasonable limits, we restrict ourselves to a generic case: the parameter σ (see Section 2) is not half integer ($\sigma \notin 1/2 + \mathbb{Z}$). For σ half integer, formulas for asymptotic series become rather cumbersome.

Here are some notations and formulas, used in the work.

$\mathbb{N} = \{0, 1, 2, ...\}$, $\mathbb{Z}$, $\mathbb{R}$, $\mathbb{C}$ are integer, real, complex numbers, respectively.

If M is a manifold, then $\mathcal{D}(M)$ denotes the Schwartz space of complex valued infinitely differentiable functions on M with the compact support

B. P. Komrakov et al. (eds.), Lie Groups and Lie Algebras, 261–284.
© *1998 Kluwer Academic Publishers. Printed in the Netherlands.*

endowed with usual topology, and $\mathcal{D}'(M)$ denotes the space of distributions on M, i.e., the space of the antilinear continuous functionals on $\mathcal{D}(M)$. We shall denote the value of the functional $F \in \mathcal{D}'(M)$ on $f \in \mathcal{D}(M)$ by (F, f).

The symbol $\equiv$ denotes the reduction modulo 2.

We use the distributions on the real line $x_{\pm}^{\lambda}$, $(x \pm i0)^{\lambda}$, $|x|^{\lambda}$, $|x|^{\lambda}\mathrm{sgn}x$ from [3]. For brevity we shall use the notation

$$x^{\lambda,\varepsilon} = |x|^{\lambda} \, \mathrm{sgn}^{\varepsilon} x = x_{+}^{\lambda} + (-1)^{\varepsilon} \, x_{-}^{\lambda} \quad (\lambda \in \mathbb{C}, \varepsilon = 0, 1). \qquad (0.1)$$

The Dirac delta-function and its k-th derivative are denoted by $\delta(x)$ and $\delta^{(k)}(x)$.

$\Gamma(x)$ is the Euler gamma-function, $F(a, b; c; x)$ is the Gauss hypergeometric function, see [1], Ch. 1, 2.

The generalized powers are

$$a^{(m)} = a(a-1)...(a-m+1), \qquad a^{[m]} = a(a+1)...(a+m-1).$$

Some formulas of the differential calculus, see [4] 0.433(1), 0.431(1):

$$\left(\frac{d}{xdx}\right)^{r} = \sum_{j=0}^{r-1} c_{rj} \, x^{-r-j} \left(\frac{d}{dx}\right)^{r-j}, \qquad (0.2)$$

where

$$c_{rj} = (-1)^{j} \, \frac{(r-1+j)!}{2^{j}j!(r-1-j)!}. \qquad (0.3)$$

If $xy = 1$, then (we use the standard notation for binomial coefficients)

$$\left(\frac{d}{dx}\right)^{r} = (-1)^{r} \sum_{k=0}^{r-1} \binom{r}{k} (r-1)^{(k)} \left(\frac{d}{dy}\right)^{r-k}. \qquad (0.4)$$

1. Group $SO_0(1,2)$ and its representations

In this Section we recall some facts about representations of the group $G = SO_0(1,2) = PSL(2, \mathbb{R})$, see [7], [2]. Let us consider the bilinear form

$$[x, y] = -x_1 y_1 + x_2 y_2 + x_3 y_3 \qquad (1.1)$$

in $\mathbb{R}^3$. The group G is the connected group of all linear transformations of $\mathbb{R}^3$ preserving the form (1.1). We shall suppose that G acts on $\mathbb{R}^3$ from the right: $x \mapsto xg$, hence we shall write the vector x in row form: $x = (x_1, x_2, x_3)$.

Let us denote by H, A, K the subgroups of G consisting of the matrices, respectively $(t \in \mathbb{R})$:

$$h_t = \begin{pmatrix} \text{ch}t & \text{sh}t & 0 \\ \text{sh}t & \text{ch}t & 0 \\ 0 & 0 & 1 \end{pmatrix}, a_t = \begin{pmatrix} \text{ch}t & 0 & \text{sh}t \\ 0 & 1 & 0 \\ \text{sh}t & 0 & \text{ch}t \end{pmatrix}, k_t = \begin{pmatrix} 1 & 0 & 0 \\ 0 & \text{cos}t & -\text{sin}t \\ 0 & \text{sin}t & \text{cos}t \end{pmatrix}.$$
(1.2)

The basis in the Lie algebra $\mathfrak{g}$ of the group G consists of the elements L_H, L_A, L_K of $\mathfrak{g}$, corresponding to the subgroups (1.2), respectively.

The element

$$\Delta_{\mathfrak{g}} = -L_H^2 - L_A^2 + L_K^2 \qquad (1.3)$$

from the universal enveloping algebra is a generator in the center of this algebra.

Let X_0 denote the cone $[x, x] = 0, x \neq 0$, in $\mathbb{R}^3$. The group G acts on it, transitively on every sheet $x_1 > 0$ and $x_1 < 0$. Let S be the section of the cone by the plane $x_1 = 1$, so that S consists of the points

$$s = (1, \sin\alpha, \cos\alpha), \quad \alpha \in \mathbb{R}. \qquad (1.4)$$

The operator $\Delta_S = d^2/d\alpha^2$ is the Laplace–Beltrami operator on S corresponding to the metric $d\alpha^2$. Sometimes we shall write functions $\varphi(s)$ on S as $\varphi(\alpha)$, where s and α are connected by (1.4). Let us take the usual inner product of functions on S:

$$(\psi, \varphi) = \int_S \psi(s)\overline{\varphi(s)}ds = \int_0^{2\pi} \psi(\alpha)\overline{\varphi(\alpha)}d\alpha. \qquad (1.5)$$

The basis in $\mathcal{D}(S)$ consisting of the functions

$$\psi_m(\alpha) = e^{im\alpha}, \quad m \in \mathbb{Z}, \qquad (1.6)$$

is an eigenbasis for Δ_S,

$$\Delta_S \psi_m = \lambda_m \psi_m, \quad \lambda_m = -m^2. \qquad (1.7)$$

Let σ be a complex number. The representation T_σ of the group G acts on $\mathcal{D}(S)$ in the following way:

$$T_\sigma(g)\varphi(s) = \varphi\left(\frac{sg}{(sg)_1}\right)(sg)_1^\sigma. \qquad (1.8)$$

It is continuous and infinitely differentiable. The restriction of the representation T_σ to the subgroup K does not depend on σ, it is the representation T by rotations in $\mathcal{D}(S)$:

$$T(k)\varphi(s) = \varphi(sk), \quad k \in K. \qquad (1.9)$$

We denote

$$\sigma^* = -\sigma - 1. \tag{1.10}$$

The form (1.5) is invariant with respect to the pair $(T_\sigma, T_{\bar{\sigma}^*})$, so that

$$(T_\sigma(g)\psi, \varphi) = (\psi, T_{\bar{\sigma}^*}(g^{-1})\varphi). \tag{1.11}$$

If σ is not integer, then T_σ is irreducible. If σ is integer, then the sub-spaces $V_\sigma^+ = \{\psi_m, m \geq -\sigma\}$ and $V_\sigma^- = \{\psi_m, m \leq \sigma\}$ are invariant; for $\sigma < 0$ both are irreducible, for $\sigma \geq 0$ their intersection E_σ is irreducible and has the dimension $2\sigma + 1$.

The element (1.3) in the representation T_σ turns into a scalar operator:

$$T_\sigma(\Delta_\mathfrak{g}) = \sigma^* \sigma E, \tag{1.12}$$

where E is the identity operator in $\mathcal{D}(S)$.

Define the operator A_σ on $\mathcal{D}(S)$ as follows:

$$(A_\sigma \varphi)(s) = \int_S \left(-[s, \tilde{s}] \right)^{\sigma^*} \varphi(\tilde{s}) d\tilde{s} = \int_0^{2\pi} [1 - \cos(\alpha - \tilde{\alpha})]^{\sigma^*} \varphi(\tilde{\alpha}) d\tilde{\alpha}. \tag{1.13}$$

Here the integral converges absolutely for $\operatorname{Re}\sigma < -1/2$ and can be extended to other σ as an analytic function. The operator A_σ is a continuous operator in $\mathcal{D}(S)$. It is a meromorphic function of σ, so the operator

$$\tilde{A}_\sigma = \frac{A_\sigma}{\Gamma(-\sigma - \frac{1}{2})} \tag{1.14}$$

is an entire nonvanishing function of σ.

The basis ψ_m is an eigenbasis for the operator A_σ:

$$A_\sigma \psi_m = a(\sigma, m)\psi_m, \tag{1.15}$$

where

$$a(\sigma, m) = (-1)^m 2^{-\sigma} \sqrt{\pi} \frac{\Gamma(-\sigma)\Gamma(-\sigma - \frac{1}{2})}{\Gamma(-\sigma + m)\Gamma(-\sigma - m)}. \tag{1.16}$$

The operator A_σ intertwines the representations T_σ and T_{σ^*}:

$$T_{\sigma^*} A_\sigma = A_\sigma T_\sigma.$$

By (1.16) we have

$$A_{\sigma^*} A_\sigma = \gamma(\sigma) \, E, \tag{1.17}$$

where

$$\gamma(\sigma) = \frac{4\pi \operatorname{tg}\sigma\pi}{2\sigma + 1}. \tag{1.18}$$

Thus if σ is not integer, then the representations T_σ and T_{σ^*} are equivalent. For integer σ, there is a partial equivalence.

The operator A_σ interacts with the form (1.5) as follows:

$$(A_\sigma \psi, \varphi) = (\psi, A_{\bar{\sigma}} \varphi). \qquad (1.19)$$

For $\sigma \in \mathbb{C}$ we define by (1.11) the representation of G on $\mathcal{D}'(S)$ which we denote by the same symbol T_σ. Now in (1.11) ψ is a distribution in $\mathcal{D}'(S)$, φ is a function in $\mathcal{D}(S)$, and (ψ, φ) denotes the value of the distribution ψ at function φ. This representation is an extension of T_σ, provided we attach to a function $\psi \in \mathcal{D}(S)$ the functional $\varphi \mapsto (\psi, \varphi)$ in $\mathcal{D}'(S)$ by means of (1.5). Similarly we can extend the operator A_σ to $\mathcal{D}'(S)$ by (1.19).

2. The eigenfunctions of the Laplace–Beltrami operator on the hyperboloid of one sheet in $\mathbb{R}^3$

Let us consider the hyperboloid X of one sheet in $\mathbb{R}^3$ defined by the equation $[x, x] = 1$. It is a homogeneous space G/H. Indeed, H is the stabilizer of the point $x^0 = (0, 0, 1)$ in X. The metric $[dx, dx]$ on X is G-invariant. It gives rise to the G-invariant measure dx and the Laplace–Beltrami operator Δ on X.

Let us introduce on X the "polar" coordinates t, α ($t \in \mathbb{R}, \alpha \in [0, 2\pi]$):

$$x = x^0 a_t k_\alpha = (\mathrm{sh}\, t, \ \mathrm{ch}\, t \ \sin\alpha, \ \mathrm{ch}\, t \ \cos\alpha). \qquad (2.1)$$

In these coordinates we have

$$dx = \mathrm{ch}\, t \ dt \ d\alpha, \qquad \Delta = -\frac{\partial^2}{\partial t^2} - \mathrm{th}\, t \frac{\partial}{\partial t} + \frac{1}{\mathrm{ch}^2 t} \frac{\partial^2}{\partial \alpha^2}. \qquad (2.2)$$

Notice that $x/\mathrm{ch}\, t \to (\pm 1, \sin\alpha, \cos\alpha)$ as $t \to \pm\infty$, so the set $S \cup (-S)$ is the boundary of the hyperboloid X (in the sense of Karpelevich).

Let U denote the representation of G on $C^\infty(X)$ by translations:

$$(U(g)f)(x) = f(xg).$$

It is continuous. The basis elements L_H, L_A, L_K go to some differential operators. Let us write one of them:

$$U(L_A) = \cos\alpha \frac{\partial}{\partial t} - \mathrm{th}\, t \ \sin\alpha \frac{\partial}{\partial \alpha}. \qquad (2.3)$$

The image of the element $\Delta_\mathfrak{g}$ (see (1.3)) is the operator Δ:

$$\Delta = U(\Delta_\mathfrak{g}). \qquad (2.4)$$

For $\sigma \in \mathbb{C}, \varepsilon = 0, 1$ we denote by $\mathcal{H}_{\sigma,\varepsilon}$ the subspace of $C^\infty(X)$ consisting of functions $f(x)$ satisfying

$$\Delta f = \sigma^* \sigma \; f, \qquad\qquad f(-x) = (-1)^\varepsilon \; f(x).$$

It is closed. Clearly $\mathcal{H}_{\sigma,\varepsilon} = \mathcal{H}_{\sigma^*,\varepsilon}$. Let $U_{\sigma,\varepsilon}$ denote the restriction of U on $\mathcal{H}_{\sigma,\varepsilon}$.

Let us take the function $f(x) = f(t, \alpha)$ from $\mathcal{H}_{\sigma,\varepsilon}$ and decompose it into the Fourier series of α:

$$f(t, \alpha) = \sum_{m \in \mathbb{Z}} c_m(t) \; e^{im\alpha}, \tag{2.5}$$

where

$$c_m(t) = \frac{1}{2\pi} \int_0^{2\pi} f(t, \alpha) \; e^{-im\alpha} d\alpha. \tag{2.6}$$

Theorem 2.1. *The series* (2.5) *converges in the topology of the space* $C^\infty(X)$. *In particular, we can differentiate equation* (2.5) *with respect to* t *and* α *as many times as desired.*

Proof. For $\tau > 0$, $k, r \in \mathbb{N}$, we denote $M_\tau(k, r) = \max \left| \partial_t^k \partial_\alpha^r f \right|$, where we take the maximum over the compact set $X_\tau : |t| \leq \tau, \alpha \in [0, 2\pi]$. If we integrate by parts the expression (2.6) r times, we obtain the following estimate:

$$\left| \partial_t^k c_m(t) \right| \leq |m|^{-r} M_\tau(k, r).$$

This estimate allows to prove that after the application of the operator $\partial_t^k \partial_\alpha^p$ the series (2.5) converges uniformly on X_τ. Indeed, the convergent majorant for such series is the series with terms $M_\tau(k, p + 2)|m|^{-2}$ for $m \neq 0$. $\square$

The Fourier coefficients $c_m(t)$ from (2.5) satisfy the differential equation

$$\frac{d^2 y}{dt^2} + \mathrm{th}t \, \frac{dy}{dt} + \left(\frac{m^2}{\mathrm{ch}^2 t} + \sigma^* \sigma \right) y = 0 \tag{2.7}$$

and have the parity $\varepsilon + m$: $c_m(-t) = (-1)^{\varepsilon + m} c_m(t)$.

Let us write explicit expressions of the basis solutions of (2.7). They are power series in powers of some functions of t. We use the following functions: $\mathrm{th}^2 t, 1 - \mathrm{th}t, (\mathrm{ch}t)^{-2}, e^{-2t}$. Different functions are suitable for different aims. For example, the first of them is used for study of the structure of $\mathcal{H}_{\sigma,\varepsilon}$ (reducibility and irreducibility, etc.), and others are used to study the behavior at infinity ($t \to \pm\infty$).

At first we take $z = \mathrm{th}^2 t$ and change the desired function:

$$y(t) = (\mathrm{ch}t)^\sigma \; (\mathrm{th}t)^l \; v(z),$$

whÑre $l = 0, 1$ respectively for even ($l = 0$) or odd ($l = 1$) function $y(t)$. Then equation (2.7) gives for $v(z)$ the hypergeometric equation (see [1] 2.1) with the parameters $a = (l - \sigma + m)/2$, $b = (l - \sigma - m)/2$, $c = l + 1/2$.

Thus, basis of solutions of (2.7) consists of the following two ($l = 0, 1$) functions of parity l:

$$R(\sigma, l, m; t) = (\mathrm{ch}\, t)^\sigma \, (\mathrm{th}\, t)^l \, F\left(\frac{l - \sigma + m}{2}, \frac{l - \sigma - m}{2}; l + \frac{1}{2}; \mathrm{th}^2 t\right). \quad (2.8)$$

These functions are defined for all $\sigma \in \mathbb{C}$ and all $m \in \mathbb{Z}$. Note that by [1] 2.1(23) $R(\sigma, l, m; t) = R(\sigma^*, l, m; t)$.

Thus

$$c_m(t) = \alpha_m \, R(\sigma, l, m; t), \quad l \equiv \varepsilon + m,$$

where α_m are some numbers.

The functions (recall (2.1))

$$h(\sigma, \varepsilon, m; x) = R(\sigma, l, m; t) \, e^{im\alpha}, \quad l \equiv \varepsilon + m, \quad (2.9)$$

form a basis in the space $\mathcal{H}_{\sigma, \varepsilon}$. Sometimes we shall denote them $h_m(x)$.

For an integer $\sigma \in \mathbb{Z}$ we denote by $\mathcal{H}^+_{\sigma, \varepsilon}$ the subspace of $\mathcal{H}_{\sigma, \varepsilon}$ spanned by h_m for which $m \geq -\sigma$, if $\sigma \equiv \varepsilon$, and $m \geq -\sigma^*$, if $\sigma \equiv \varepsilon + 1$. We denote by $\mathcal{H}^-_{\sigma, \varepsilon}$ the subspace of $\mathcal{H}_{\sigma, \varepsilon}$ spanned by h_m with $m \leq \sigma$ for $\sigma \equiv \varepsilon$ and $m \leq \sigma^*$ for $\sigma \equiv \varepsilon + 1$. These subspaces are well-defined: they are invariant with respect to the replacement of σ by σ^*, i.e., $\mathcal{H}^\pm_{\sigma, \varepsilon} = \mathcal{H}^\pm_{\sigma^*, \varepsilon}$.

Theorem 2.2. *If σ is not integer, then the representation $U_{\sigma, \varepsilon}$ is irreducible. If σ is integer, then the two subspaces $\mathcal{H}^\pm_{\sigma, \varepsilon}$ of $\mathcal{H}_{\sigma, \varepsilon}$ are invariant. For $\sigma \geq 0, \varepsilon \equiv \sigma + 1$ and for $\sigma < 0, \varepsilon \equiv \sigma$ these two subspaces are irreducible. For $\sigma \geq 0, \varepsilon \equiv \sigma$ and for $\sigma < 0, \varepsilon \equiv \sigma + 1$ these subspaces have a finite-dimensional intersection $\mathcal{E}_{\sigma, \varepsilon}$ with dimensions $2\sigma + 1$ and $2\sigma^* + 1$, respectively.*

Proof. The operator $U(L_A)$ (see (2.3)) acts on the basis functions h_m in the following way:

$$U(L_A)h_m = -\frac{1}{2}(\sigma + m)(\sigma^* + m)h_{m-1} - \frac{1}{2}(\sigma - m)(\sigma^* - m)h_{m+1}, \quad (2.10)$$

if $m \equiv \varepsilon$, and

$$U(L_A)h_m = \frac{1}{2}h_{m-1} + \frac{1}{2}h_{m+1}, \quad (2.11)$$

if $m \equiv \varepsilon + 1$ (the proof will be given in Section 4). It follows that on the axis m the barrier of the motion from the left to right is the point σ for $\sigma \geq 0, \varepsilon \equiv \sigma$ and the point σ^* for $\sigma < 0, \varepsilon \equiv \sigma + 1$ and for the motion from the right to left it is the point $-\sigma$ for $\sigma \geq 0, \varepsilon \equiv \sigma$ and the point $-\sigma^*$ for $\sigma < 0, \varepsilon \equiv \sigma + 1$. All this implies the statements of the theorem. $\square$

Now we write the basis solutions of the equation (2.7) as series in powers of $1 - \text{th}t$, $(\text{ch}t)^{-2}$, e^{-2t}:

$$R(\sigma, l, m; t) = g(\sigma, l, m)(\text{ch}t)^\sigma V(\sigma, m; \xi) + g(\sigma^*, l, m)(\text{ch}t)^{\sigma^*} V(\sigma^*, m; \xi), \tag{2.12}$$

$$R(\sigma, l, m; t) = g(\sigma, l, m)(\text{ch}t)^\sigma W(\sigma, m; \eta) + g(\sigma^*, l, m)(\text{ch}t)^{\sigma^*} W(\sigma^*, m; \eta), \tag{2.13}$$

$$R(\sigma, l, m; t) = g(\sigma, l, m)2^{-\sigma} e^{\sigma t} U(\sigma, m; \zeta) + g(\sigma^*, l, m)2^{-\sigma^*} e^{\sigma^* t} U(\sigma^*, m; \zeta), \tag{2.14}$$

where

$$V(\sigma, m; \xi) = F\left(-\sigma + m, -\sigma - m; -\sigma + \frac{1}{2}; \frac{1 - \xi}{2}\right), \tag{2.15}$$

$$W(\sigma, m; \eta) = F\left(\frac{-\sigma + m}{2}, \frac{-\sigma - m}{2}; -\sigma + \frac{1}{2}; \eta\right), \tag{2.16}$$

$$U(\sigma, m; \zeta) = (1 - \zeta)^m F\left(-\sigma + m, \frac{1}{2} + m; \frac{1}{2} - \sigma; \zeta\right), \tag{2.17}$$

$$g(\sigma, l, m) = \frac{\Gamma(\sigma + \frac{1}{2})\Gamma(l + \frac{1}{2})}{\Gamma\left(\frac{\sigma + l - m + 1}{2}\right)\Gamma\left(\frac{\sigma + l + m + 1}{2}\right)}, \tag{2.18}$$

$$\xi = \text{th}t, \quad \eta = \frac{1}{\text{ch}^2 t}, \quad \zeta = -e^{-2t},$$

so

$$\eta = 1 - \xi^2 \quad \frac{1 - \xi}{2} = \frac{\zeta}{\zeta - 1}. \tag{2.19}$$

Formula (2.13) is obtained from (2.8) with the help of the transform [1] 2.10(1) and [1] 2.1(23) for $l = 1$. Then the formula (2.12) is obtained from (2.13) with the help of the quadratic transform [1] 2.1(27). At last, formula (2.14) comes from (2.12) with the help of the transform [1] 2.10(6). The functions V and W coincide, if $t \geq 0$, i.e.,

$$V(\sigma, m; \xi) = W(\sigma, m; \eta), \quad \eta = 1 - \xi^2, \quad 0 \leq \xi \leq 1, \tag{2.20}$$

see [1] 2.1.5. To find the connection between V and W, if $t < 0$, we must use the following symmetry relation for V, arising from [1] 2.10(1), 2.9(2):

$$(\text{ch}t)^\sigma V(\sigma, m; \xi) =$$
$$= A(\sigma, m)(\text{ch}t)^\sigma V(\sigma, m; -\xi) + B(\sigma, m)(\text{ch}t)^{\sigma^*} V(\sigma^*, m; -\xi),$$

where

$$A(\sigma, m) = \frac{(-1)^m}{\cos \sigma \pi}, \quad B(\sigma, m) = 2^{\sigma^* - \sigma} \frac{\Gamma(-\sigma + \frac{1}{2})\Gamma(-\sigma - \frac{1}{2})}{\Gamma(-\sigma + m)\Gamma(-\sigma - m)}.$$

The functions V, W, U are defined for $\sigma \notin 1/2 + \mathbb{N}$. So the expansions (2.12)—(2.14) are true for $\sigma \notin 1/2 + \mathbb{Z}$.

Let us denote by $v_r(\sigma, m)$, $w_r(\sigma, m)$ and $u_r(\sigma, m)$ the coefficients of the power series (2.15), (2.16), (2.17) of the degrees of $1 - \xi, \eta, \zeta$, respectively, i.e.,

$$V(\sigma, m; \xi) = \sum_{r=0}^{\infty} v_r(\sigma, m)(1 - \xi)^r, \qquad (2.21)$$

$$W(\sigma, m; \eta) = \sum_{r=0}^{\infty} w_r(\sigma, m)\eta^r, \qquad (2.22)$$

$$U(\sigma, m; \zeta) = \sum_{r=0}^{\infty} u_r(\sigma, m)\zeta^r. \qquad (2.23)$$

Lemma 2.3. The coefficients w_r, u_r are expressed by means of the coefficients v_r in the following way:

$$w_r(\sigma, m) = \frac{1}{2^r r!} \sum_{j=0}^{r-1} (-1)^j c_{rj} (r - j)! v_{r-j}(\sigma, m), \qquad (2.24)$$

$$u_r(\sigma, m) = (-1)^r \sum_{k=0}^{r} 2^k \binom{\sigma - k}{r - k} v_k(\sigma, m), \qquad (2.25)$$

where c_{rj} are the numbers defined by (0.3); it is assumed in (2.24) that $r \geq 1$, and there is $w_0 = u_0 = 1$ for $r = 0$.

Proof. Relations (2.19), (0.2), (0.4) imply the following expansions of derivatives:

$$\left(\frac{d}{d\eta}\right)^r = \left(-\frac{1}{2}\right)^r \sum_{j=0}^{r-1} c_{rj} \, \xi^{-r-j} \left(\frac{d}{d\xi}\right)^{r-j}, \qquad (2.26)$$

$$\left(\frac{d}{d\zeta}\right)^r = \sum_{j=0}^{r-1} (-1)^j \, 2^{r-j} \binom{r}{j} (r - 1)^{(j)} (\zeta - 1)^{-2r+j} \left(\frac{d}{d\xi}\right)^{r-j}. \qquad (2.27)$$

Comparing (2.12) with (2.14), we see that

$$U(\sigma, m; \zeta) = (1 - \zeta)^\sigma \, V(\sigma, m; \xi). \qquad (2.28)$$

We apply the operator (2.26) to (2.20), the operator (2.27) to (2.28), use (2.21)—(2.23) and put $\xi = 1$, then $\eta = 0$, $\zeta = 0$. In the first case we obtain (2.24). In the second case we obtain

$$u_r = (-1)^r \frac{\sigma^{(r)}}{r!} v_0 + \frac{1}{r!} \sum_{s=0}^{r-1} \binom{r}{s} \sigma^{(s)}(-1)^s \cdot$$

$$\cdot(-1)^{r-s}\sum_{j=0}^{r-s-1}2^{r-s-j}\binom{r-s}{j}(r-s-1)^{(j)}(r-s-j)!(-1)^j v_{r-s-j}.$$

Let us introduce here the variable $k = r - s - j$ instead of j and change the summation order over s and k. We obtain:

$$u_r = (-1)^r \frac{\sigma^{(r)}}{r!} v_o +$$

$$+ \sum_{k=1}^{r}(-1)^k 2^k v_k \sum_{s=0}^{r-k}(-1)^s \frac{1}{s!(r-k-s)!}\sigma^{(s)}(r-s-1)^{(r-k-s)}.$$

The last factor is equal to $(-1)^{r-k-s}(-k)^{(r-k-s)}$, so according to the binomial theorem for generalized powers, the last sum is equal to

$$(-1)^{r-k}\frac{(\sigma - k)^{(r-k)}}{(r-k)!},$$

which implies (2.25). $\square$

Let us write the explicit form of the coefficients v_r, w_r, u_r. For the two first ones it follows from (2.15), (2.16), see [1] 2.1(2), for u_r we exploit (2.25). Further we shall need the expression of the coefficient v_r in terms of the eigenvalues of the intertwining operator, see (1.16). We also give an interesting expression of the coefficient u_r in terms of the hypergeometric function $_3F_2$. Thus, we have

$$v_r(\sigma, m) = 2^{-r}\frac{(-\sigma + m)^{[r]}(-\sigma - m)^{[r]}}{(-\sigma + \frac{1}{2})^{[r]}r!} = (-1)^r\binom{\sigma}{r}\frac{a(\sigma^* + r, m)}{a(\sigma^*, m)}, \quad (2.29)$$

$$w_r(\sigma, m) = \frac{\left(\frac{-\sigma + m}{2}\right)^{[r]}\left(\frac{-\sigma - m}{2}\right)^{[r]}}{(-\sigma + \frac{1}{2})^{[r]}r!},$$

$$u_r(\sigma, m) = (-1)^r\sum_{k=0}^{r}\binom{\sigma - k}{r - k}\frac{(-\sigma + m)^{[k]}(-\sigma - k)^{[k]}}{(-\sigma + \frac{1}{2})^{[k]}k!} =$$

$$= (-1)^r\binom{\sigma}{r}{}_3F_2(-r, -\sigma + m, -\sigma - m; -\sigma + \frac{1}{2}, -\sigma; 1).$$

It follows from these formulas that the coefficients v_r, w_r, u_r are the even polynomials in m of degree $2r$. So, they are polynomials of degree r in $\lambda_m = -m^2$, see (1.7). Let us normalize these polynomials so that the leading coefficient (i.e., the coefficient of λ_m^r) is equal to 1. We denote these normalized polynomials in λ by $\overset{o}{v}_r(\sigma, \lambda)$, $\overset{o}{w}_r(\sigma, \lambda)$, $\overset{o}{u}_r(\sigma, \lambda)$. Therefore,

$$v_r(\sigma, m) = (-1)^r \frac{1}{2^r r! (\sigma - \frac{1}{2})^{(r)}} \overset{o}{v}_r(\sigma, \lambda), \tag{2.30}$$

$$w_r(\sigma, m) = (-1)^r \frac{1}{2^{2r} r! (\sigma - \frac{1}{2})^{(r)}} \overset{o}{w}_r(\sigma, \lambda), \tag{2.31}$$

$$u_r(\sigma, m) = \frac{1}{r! (\sigma - \frac{1}{2})^{(r)}} \overset{o}{u}_r(\sigma, \lambda), \tag{2.32}$$

where $\lambda = -m^2$. Here is the explicit form of the polynomials $\overset{o}{v}_r$, $\overset{o}{w}_r$, $\overset{o}{u}_r$:

$$\overset{o}{v}_r(\sigma, \lambda) = \prod_{j=0}^{r-1} \left[(\sigma - j)^2 + \lambda \right],$$

$$\overset{o}{w}_r(\sigma, \lambda) = \prod_{j=0}^{r-1} \left[(\sigma - 2j)^2 + \lambda \right],$$

$$\overset{o}{u}_r(\sigma, \lambda) = \sum_{k=0}^{r} (-1)^{r-k} \binom{r}{k} (\sigma - k)^{(r-k)} \left(\sigma - k - \frac{1}{2} \right)^{(r-k)} \overset{o}{v}_k(\sigma, \lambda).$$

3. H-invariants

Let us specify the elements invariant with respect to the subgroup H in the representations of G described in Section 1. As a rule, these invariants are distributions (functionals). They belong to $\mathcal{D}'(S)$ or to its subfactors. The material of this paragraph is taken from [5], [6].

Theorem 3.1. *The space of H-invariant elements in the representation T_σ has dimension 2 for $\sigma \neq -1, -2, \dots$ and dimension 3 for $\sigma = -1, -2, \dots$. The basis of this space consists of two distributions for $\sigma \neq -1, -2, \dots$:*

$$\theta_{\sigma, \varepsilon} = s_3^{\sigma, \varepsilon} = (\cos\alpha)^{\sigma, \varepsilon}, \quad \varepsilon = 0, 1,$$

and of three distributions

$$(\cos\alpha)^{-n-1}, \quad \delta^{(n)}(\cos\alpha) \, Y(\sin\alpha), \quad \delta^{(n)}(\cos\alpha) \, Y(-\sin\alpha)$$

for $\sigma = -n - 1$, $n \in \mathbb{N}$. Here Y is the Heaviside function: $Y(t) = 1$ for $t > 0$, $Y(t) = 0$ for $t < 0$.

The operator A_σ takes $\theta_{\sigma, \varepsilon}$ to $\theta_{\sigma^*, \varepsilon}$ with a factor:

$$A_\sigma \, \theta_{\sigma, \varepsilon} = j(\sigma, \varepsilon) \, \theta_{\sigma^*, \varepsilon}, \tag{3.1}$$

where

$$j(\sigma, \varepsilon) = 2^{-\sigma} \, \pi^{-\frac{1}{2}} \, \Gamma(\sigma + 1) \, \Gamma\left(-\sigma - \frac{1}{2}\right) [1 - (-1)^\varepsilon \cos \sigma \pi]. \qquad (3.2)$$

It follows from (1.17) and (3.1) that (see γ in (1.18))

$$j(\sigma, \varepsilon) \, j(\sigma^*, \varepsilon) = \gamma(\sigma). \qquad (3.3)$$

The distribution

$$\tilde{\theta}_{\sigma, \varepsilon} = \frac{\theta_{\sigma, \varepsilon}}{\Gamma\left(\frac{\sigma + 1 + \varepsilon}{2}\right)}$$

is defined for all $\sigma \in \mathbb{C}$ and is invariant with respect to $T_\sigma(H)$. For $\sigma \neq -1, -2, \dots$ it differs from $\theta_{\sigma, \varepsilon}$ by a factor only.

4. Poisson transform

Let us define the Poisson transform $P_{\sigma, \varepsilon}$ associated with the H-invariant $\theta_{\sigma, \varepsilon}$ as follows

$$\left(P_{\sigma, \varepsilon} \, \varphi\right)(x) = \int_S \left(T_\sigma(g^{-1}) \theta_{\sigma, \varepsilon}\right)(s) \, \varphi(s) \, ds = \qquad (4.1)$$

$$= \int_S [x, s]^{\sigma, \varepsilon} \, \varphi(s) \, ds, \qquad (4.2)$$

where $x = x^0 g$. It is a linear continuous operator from $\mathcal{D}(S)$ to $C^\infty(X)$. It intertwines T_{σ^*} with U:

$$U(g) \, P_{\sigma, \varepsilon} = P_{\sigma, \varepsilon} \, T_{\sigma^*}(g), \quad g \in G. \qquad (4.3)$$

Hence (see (1.12), (2.4)):

$$\Delta \circ P_{\sigma, \varepsilon} = \sigma^* \sigma \, P_{\sigma, \varepsilon}. \qquad (4.4)$$

Further, it interacts with the operator A_σ as follows

$$P_{\sigma, \varepsilon} \, A_\sigma = j(\sigma, \varepsilon) \, P_{\sigma^*, \varepsilon}. \qquad (4.5)$$

Moreover, $(P_{\sigma, \varepsilon} \varphi)(x)$ has the parity ε:

$$\left(P_{\sigma, \varepsilon} \varphi\right)(-x) = (-1)^\varepsilon \left(P_{\sigma, \varepsilon} \varphi\right)(x). \qquad (4.6)$$

The image of $P_{\sigma, \varepsilon}$ is contained in $\mathcal{H}_{\sigma, \varepsilon}$ (this follows from (4.4), (4.6)). The transform $P_{\sigma, \varepsilon}$ is a meromorphic function of σ with poles at the points $\sigma = -1 - \varepsilon - 2k, k \in \mathbb{N}$.

In order to include, all the values of σ, we can consider the transform $\widetilde{P}_{\sigma,\varepsilon}$ corresponding to the H-invariant $\widetilde{\theta}_{\sigma,\varepsilon}$, see Section 3, that is

$$\widetilde{P}_{\sigma,\varepsilon} = \frac{P_{\sigma,\varepsilon}}{\Gamma\left(\frac{\sigma+1+\varepsilon}{2}\right)}. \tag{4.7}$$

However, we prefer to consider first $\sigma \neq -1 - \varepsilon - 2k$ (the generic case) and the transform $P_{\sigma,\varepsilon}$, and then the excluded points.

Let us write the Poisson transform in polar coordinates (2.1), (2.2):

$$\left(P_{\sigma,\varepsilon}\varphi\right)(t,\alpha) = \int_0^{2\pi} \left[- \text{sh}\, t + \text{ch}\, t \, \cos(\alpha - \beta)\right]^{\sigma,\varepsilon} \, \varphi(\beta) \, d\beta. \tag{4.8}$$

In particular, for the basis functions ψ_m, see (1.6), we have from (4.8):

$$(P_{\sigma,\varepsilon}\,\psi_m)(t,\alpha) = \Psi(\sigma,\varepsilon,m;t) \, e^{im\alpha}, \tag{4.9}$$

where

$$\Psi(\sigma,\varepsilon,m;t) = \int_0^{2\pi} (-\text{sh}\,t + \text{ch}\,t\,\cos\gamma)^{\sigma,\varepsilon} \, e^{-im\gamma} \, d\gamma. \tag{4.10}$$

Notice that the last function is even in m and has the parity $\varepsilon + m$ in t.

We see that the Fourier series expansion (2.5) of $P_{\sigma,\varepsilon}\,\psi_m$ consists of one term, it is the right-hand side of (4.9), and the corresponding Fourier coefficient is the function $\Psi(\sigma,\varepsilon,m;t)$. It differs from the function R by a factor only, see (2.8):

$$\Psi(\sigma,\varepsilon,m;t) = \chi(\sigma,\varepsilon,m) \, R(\sigma,l,m;t), \tag{4.11}$$

where $l \equiv \varepsilon + m$. Let us call this factor the *Poisson coefficient*. Let us find it. Substituting (2.12) into (4.11) we obtain:

$$\Psi(\sigma,\varepsilon,m;t) = \chi(\sigma,\varepsilon,m)g(\sigma,l,m)(\text{ch}\,t)^\sigma V(\sigma,m;\xi)+$$

$$+\chi(\sigma,\varepsilon,m)g(\sigma^*,l,m)(\text{ch}\,t)^{\sigma^*} V(\sigma^*,m;\xi). \tag{4.12}$$

Assume that $\text{Re}\,\sigma > -1/2$. Then the main term of the asymptotics of the right-hand side of (4.12) as $t \to +\infty$ is produced from the first term, it is equal to $(\text{ch}\,t)^\sigma\chi(\sigma,\varepsilon,m) \, g(\sigma,l,m)$. On the other hand, from (4.10) we have as $t \to +\infty$:

$$\Psi(\sigma,\varepsilon,m;t) \sim (\text{ch}\,t)^\sigma (-1)^\varepsilon \int_0^{2\pi} (1 - \cos\gamma)^{\sigma,\varepsilon} \, e^{-im\gamma} \, d\gamma =$$

$$= (\text{ch}\,t)^\sigma (-1)^\varepsilon \, a(\sigma^*,m),$$

see (1.13), (1.15), (1.16). Equating these asymptotics, we find

$$\chi(\sigma, \varepsilon, m) g(\sigma, l, m) = (-1)^\varepsilon a(\sigma^*, m) =$$

$$= (-1)^{\varepsilon + m} 2^{\sigma + 1} \sqrt{\pi} \, \frac{\Gamma(\sigma + 1)\Gamma(\sigma + \frac{1}{2})}{\Gamma(\sigma + 1 + m)\Gamma(\sigma + 1 - m)}. \tag{4.13}$$

Using (2.18), we obtain χ

$$\chi(\sigma, \varepsilon, m) = (-1)^l 2^{1-\sigma} \pi^{\frac{3}{2}} \frac{\Gamma(\sigma + 1)}{\Gamma\left(l + \frac{1}{2}\right)\Gamma\left(\frac{\sigma - l - m + 2}{2}\right)\Gamma\left(\frac{\sigma - l + m + 2}{2}\right)}.$$

The coefficient of the second term in (4.12) is equal to $(-1)^\varepsilon j(\sigma, \varepsilon)$, see (3.2). So we have obtained

$$\Psi(\sigma, \varepsilon, m; t) =$$

$$= (-1)^\varepsilon \left\{ a(\sigma^*, m)(\mathrm{cht})^\sigma V(\sigma, m; \xi) + j(\sigma, \varepsilon)(\mathrm{cht})^{\sigma^*} V(\sigma^*, m; \xi) \right\}. \tag{4.14}$$

Incidentally, we can now prove formulas (2.10), (2.11). According to (4.3), the operator $U(L_A)$, see (2.3), acts on the functions $\Psi(\sigma, \varepsilon, m; t) \, e^{im\alpha}$ in the same way as the operator $T_{\sigma^*}(L_A)$ acts on the functions ψ_m, i.e.,

$$U(L_A)\left[\Psi(\sigma, \varepsilon, m; t) \, e^{im\alpha}\right] =$$

$$\frac{-\sigma - 1 + m}{2} \Psi(\sigma, \varepsilon, m - 1; t) e^{i(m-1)\alpha} + \frac{-\sigma - 1 - m}{2} \Psi(\sigma, \varepsilon, m + 1; t) e^{i(m+1)\alpha}.$$

Substituting (4.11) and using (2.9), we get the action of the operator $U(L_A)$ on h_m, defined by (2.10), (2.11).

Now let us determine the kernel and the closure of the image of the Poisson transform. In order to include the integer negative σ, we consider the transform $\widetilde{P}_{\sigma,\varepsilon}$ instead of $P_{\sigma,\varepsilon}$, see (4.7), which is defined for all $\sigma \in \mathbb{C}, \varepsilon = 0, 1$.

Theorem 4.1. For every σ, ε, except for $\sigma \in \mathbb{Z}$, $\sigma \equiv \varepsilon$, the kernel of $\widetilde{P}_{\sigma,\varepsilon}$ is $\{0\}$ and the closure of the image coincides with $\mathcal{H}_{\sigma,\varepsilon}$. If $\sigma \geq 0, \sigma \equiv \varepsilon$, then the kernel is $V_{\sigma^*}^+ + V_{\sigma^*}^-$ and the closure of the image is $\mathcal{E}_{\sigma,\varepsilon}$. If $\sigma < 0, \sigma \equiv \varepsilon$, then the kernel is E_{σ^*} and the closure of the image is $\mathcal{H}_{\sigma,\varepsilon}^+ + \mathcal{H}_{\sigma,\varepsilon}^-$.

Proof. It suffices to find out for which m the function of τ:

$$\widetilde{\chi}(\tau, \varepsilon, m) = \frac{1}{\Gamma(\frac{\tau + 1 + \varepsilon}{2})} \chi(\tau, \varepsilon, m) =$$

$$= (-1)^l 2\pi^{\frac{3}{2}} \frac{\Gamma(\frac{\tau + 1 + \varepsilon}{2})}{\Gamma\left(l + \frac{1}{2}\right)\Gamma\left(\frac{\tau - l - m + 2}{2}\right)\Gamma\left(\frac{\tau - l + m + 2}{2}\right)}$$

vanishes at $\tau = \sigma$. $\square$

5. Asymptotic behavior of the Poisson transform

First we write the expansion of the Poisson transform in powers of $1 - \text{th}t$, $(\text{ch}t)^{-2}$ and e^{-2t} for the K-finite functions φ from $\mathcal{D}(S)$, i.e., for the functions φ with the finite Fourier series.

For $\sigma \in \mathbb{C}$, $r \in \mathbb{N}$ we define the differential operators $L_{\sigma,r}$, $M_{\sigma,r}$, $K_{\sigma,r}$ on S as follows. For $r > 0$ we set

$$L_{\sigma,r} = \overset{o}{v}_r(\sigma^*, \Delta_S), \quad M_{\sigma,r} = \overset{o}{w}_r(\sigma^*, \Delta_S), \quad K_{\sigma,r} = \overset{o}{u}_r(\sigma^*, \Delta_S).$$

(See Δ_S in Section 1, the polynomials $\overset{o}{v}_r$, $\overset{o}{w}_r$, $\overset{o}{u}_r$ are defined in Section 2). For $r = 0$ we put these operators equal to 1.

Theorem 5.1. *Let σ be generic: $\sigma \notin 1/2 + \mathbb{Z}$, $\sigma \notin -1 - \varepsilon - 2\mathbb{N}$. For a K-finite function $\varphi \in \mathcal{D}(S)$, its Poisson transform $\left(P_{\sigma,\varepsilon}\varphi\right)(t, \alpha)$ has the following expansion in powers of $1 - \text{th}t$:*

$$\left(P_{\sigma,\varepsilon}\varphi\right)(t, \alpha) = (\text{ch}t)^\sigma \sum_{r=0}^\infty x_r(\sigma, \varepsilon)\left(A_{\sigma^*+r}\varphi\right)(\alpha)\,(1 - \text{th}t)^r +$$

$$+(\text{ch}t)^{\sigma^*} \sum_{r=0}^\infty y_r(\sigma, \varepsilon)\left(L_{\sigma,r}\varphi\right)(\alpha)\,(1 - \text{th}t)^r, \tag{5.1}$$

where

$$x_r(\sigma, \varepsilon) = (-1)^{\varepsilon+r}\binom{\sigma}{r}, \tag{5.2}$$

$$y_r(\sigma, \varepsilon) = (-1)^{\varepsilon+r}j(\sigma, \varepsilon)\frac{1}{2^r r!(\sigma + \frac{3}{2})^{[r]}} = \tag{5.3}$$

$$= (-1)^r 2^{-\sigma-r}\pi^{-\frac{1}{2}}\frac{1}{r!}\Gamma(\sigma + 1)\Gamma\left(-\frac{1}{2} - \sigma - r\right)\left[(-1)^\varepsilon - \cos\sigma\pi\right].$$

Both series in (5.1) converge absolutely.

Proof. It is sufficient to consider the case when $\varphi = \psi_m$, see (1.6). Then (4.9) holds. We expand both functions V in (4.14) into hypergeometric series (2.15), (2.21). We use formula (2.29) in the first case and (2.30) in the second one. We obtain

$$\Psi(\sigma, \varepsilon, m; t) = (\text{ch}t)^\sigma \sum_{r=0}^\infty x_r(\sigma, \varepsilon)\,a(\sigma^* + r, m)\,(1 - \text{th}t)^r +$$

$$+(\text{ch}t)^{\sigma^*} \sum_{r=0}^\infty y_r(\sigma, \varepsilon)\,\overset{o}{v}_r(\sigma^*, \lambda_m)\,(1 - \text{th}t)^r, \tag{5.4}$$

where x_r, y_r are defined by formulas (5.2), (5.3), respectively. Formula (5.4) is (5.1) for $\varphi = \psi_m$. The absolute convergence follows from the absolute convergence of the hypergeometric series. $\square$

Equation (4.5) gives two relations for the operators occurring in (5.1):

$$L_{\sigma,r}\, A_\sigma = j(\sigma,\varepsilon)\, \frac{x_r(\sigma^*,\varepsilon)}{y_r(\sigma,\varepsilon)}\, A_{\sigma+r}, \tag{5.5}$$

$$A_{\sigma^*+r}\, A_\sigma = j(\sigma,\varepsilon)\, \frac{y_r(\sigma^*,\varepsilon)}{x_r(\sigma,\varepsilon)}\, L_{\sigma^*,r}. \tag{5.6}$$

We denote the coefficient in (5.5) by $w_r(\sigma)$ (it does not depend on ε):

$$w_r(\sigma) = j(\sigma,\varepsilon)\, \frac{x_r(\sigma^*,\varepsilon)}{y_r(\sigma,\varepsilon)} = 2^r\, (\sigma+1)^{[r]} \left(\sigma + \frac{3}{2}\right)^{[r]}. \tag{5.7}$$

We can go from formula (5.5) to formula (5.6), and back with the help of (1.17) and (3.3).

Theorem 5.2. *Under the hypotheses of Theorem 5.1 the Poisson transform* $\left(P_{\sigma,\varepsilon}\varphi\right)(t,\alpha)$ *of the K-finite function φ has for $t > 0$ the following expansion in powers of* $(\mathrm{ch}\,t)^{-2}$:

$$\left(P_{\sigma,\varepsilon}\varphi\right)(t,\alpha) = (\mathrm{ch}\,t)^\sigma \sum_{r=0}^\infty 2^{-r}\, x_r(\sigma,\varepsilon)\left(B_{\sigma,r}\varphi\right)(s)\, (\mathrm{ch}\,t)^{-2r} +$$

$$+ (\mathrm{ch}\,t)^{\sigma^*} \sum_{r=0}^\infty 2^{-r}\, y_r(\sigma,\varepsilon)\left(M_{\sigma,r}\varphi\right)(s)\, (\mathrm{ch}\,t)^{-2r}, \tag{5.8}$$

where

$$B_{\sigma,r} = \frac{1}{w_r(\sigma^*)}\, M_{\sigma^*,r}\, A_{\sigma^*} = \tag{5.9}$$

$$= \sum_{j=0}^{r-1} \frac{c_{rj}}{(\sigma - r + 1)^{[j]}}\, A_{\sigma^*+r-j}, \tag{5.10}$$

x_r, y_r, w_r, c_{rj} *are defined by the formulas* (5.2), (5.3), (5.7), (0.3). *Both series in* (5.8) *converge absolutely and uniformly with respect to $t > 0$, $s \in S$.*

Proof. It is sufficient to consider the case when $\varphi = \psi_m$. Then (4.9) holds. According to (2.20), for $t \geq 0$ we may replace V by W. Then we decompose both functions W into the hypergeometric series (2.16), (2.22) and use for the first of them formulas (2.24), (2.29), and for the second one formula (2.31). We obtain

$$\Psi(\sigma,\varepsilon,m;t) = (\mathrm{ch}\,t)^\sigma \sum_{r=0}^\infty 2^{-r}\, x_r(\sigma,\varepsilon) b_r(\sigma,m)\, \eta^r +$$

$$+(\mathrm{ch} t)^{\sigma^*} \sum_{r=0}^{\infty} 2^{-r} \, y_r(\sigma, \varepsilon) \overset{o}{w}_r(\sigma^*, \lambda_m) \, \eta^r, \tag{5.11}$$

where

$$b_r(\sigma, m) = \sum_{j=0}^{r-1} c_{rj} \frac{\sigma^{(r-j)}}{\sigma^{(r)}} \, a(\sigma^* + r - j, m) =$$

$$= \sum_{j=0}^{r-1} \frac{c_{rj}}{(\sigma - r + 1)^{[j]}} \, a(\sigma^* + r - j, m). \tag{5.12}$$

Denote by $B_{\sigma,r}$ the operator in $\mathcal{D}(S)$, which has $b_r(\sigma, m)$ as its eigenvalues. Then (5.11) and (5.12) give formula (5.8) for $\varphi = \psi_m$, and formula (5.10).

Formula (4.5) implies that the following relations for the operators B and M from (5.8) hold (similarly to (5.5), (5.6)):

$$M_{\sigma,r} \, A_\sigma = w_r(\sigma) \, B_{\sigma^*,r}, \tag{5.13}$$

$$B_{\sigma,r} \, A_\sigma = j(\sigma, \varepsilon) \, \frac{y_r(\sigma^*, \varepsilon)}{x_r(\sigma, \varepsilon)} \, M_{\sigma^*,r}.$$

Now (5.9) is obtained from (5.3) by replacing σ by σ^*. □

Theorem 5.3. *Under the hypotheses of Theorem 5.1, the Poisson transform $P_{\sigma,\varepsilon}\varphi(t, s)$ of the K-finite function $\varphi \in \mathcal{D}(S)$ has the following expansion in powers of $\zeta = -e^{-2t}$:*

$$\left(P_{\sigma,\varepsilon}\varphi\right)(t, s) = \left(\frac{e^t}{2}\right)^\sigma \sum_{r=0}^{\infty} 2^r(-1)^r \, x_r(\sigma, \varepsilon)\left(C_{\sigma,r}\varphi\right)(s) \, \zeta^r +$$

$$+\left(\frac{e^t}{2}\right)^{\sigma^*} \sum_{r=0}^{\infty} 2^r(-1)^r \, y_r(\sigma, \varepsilon)\left(K_{\sigma,r}\varphi\right)(s) \, \zeta^r, \tag{5.14}$$

where

$$C_{\sigma,r} = \frac{1}{w_r(\sigma^*)} \, K_{\sigma^*,r} \, A_{\sigma^*} = \tag{5.15}$$

$$= (-1)^r \, 2^{-r} \sum_{k=0}^{r} (-1)^k \, 2^k \binom{r}{k} A_{\sigma^*+k}, \tag{5.16}$$

x_r, y_r, w_r *are defined by formulas* (5.2), (5.3), (5.7). *Both series in* (5.14) *converge absolutely and uniformly with respect to* $t \in \mathbb{R}$, $s \in S$.

Proof. It suffices to consider the case in which $\varphi = \psi_m$. Then we have (4.9). Comparing (2.12) and (2.14), we obtain the following formula instead of (4.14):

$$\Psi(\sigma, \varepsilon, m; t) =$$

$$= (-1)^\varepsilon \left\{ \left(\frac{e^t}{2}\right)^\sigma a(\sigma^*, m) U(\sigma, m; \zeta) + \left(\frac{e^t}{2}\right)^{\sigma^*} j(\sigma, \varepsilon) U(\sigma^*, m; \zeta) \right\}.$$

Let us expand both functions U in this equation into the series (2.23) and use formulas (2.25), (2.29) for the first of them, and formula (2.32) for the second one. We obtain

$$\Psi(\sigma, \varepsilon, m; t) = \left(\frac{e^t}{2}\right)^{\sigma} \sum_{r=0}^{\infty} (-1)^r\, 2^r\, x_r(\sigma, \varepsilon)\, c_r(\sigma, m)\, \zeta^r +$$

$$+\left(\frac{e^t}{2}\right)^{\sigma^*} \sum_{r=0}^{\infty} (-1)^r\, 2^r\, y_r(\sigma, \varepsilon)\, \overset{o}{u}_r(\sigma^*, \lambda_m)\, \zeta^r, \qquad (5.17)$$

where

$$c_r(\sigma, m) = (-1)^r\, 2^{-r} \sum_{k=0}^{r} (-1)^k\, 2^k \binom{r}{k}\, a(\sigma^* + k, m).$$

Denote by $C_{\sigma, r}$ the operator $\mathcal{D}(S)$, which has the eigenvalues $c_r(\sigma, m)$ in the basis ψ_m. Then (5.17) is (5.14) for $\varphi = \psi_m$. Moreover, we have proved (5.16). Formula (5.15) is proved similarly to (5.9). $\square$

Using (4.5), we obtain the following relations for the operators C and K from (5.14) (similarly to (5.5), (5.6)):

$$K_{\sigma, r}\, A_{\sigma} = \omega_r(\sigma)\, C_{\sigma^*, r}, \quad C_{\sigma, r}\, A_{\sigma} = j(\sigma, \varepsilon)\, \frac{y_r(\sigma^*, \varepsilon)}{x_r(\sigma, \varepsilon)}\, K_{\sigma^*, r}.$$

We may remove the restriction $\sigma \notin -1 - \varepsilon - 2\mathbb{N}$ in Theorems 5.1, 5.2, 5.3 by passing from $P_{\sigma, \varepsilon}$ to $\tilde{P}_{\sigma, \varepsilon}$. Then the coefficients x_r and y_r must be changed, respectively, to

$$\tilde{x}_r(\sigma, \varepsilon) = \frac{x_r(\sigma, \varepsilon)}{\Gamma\left(\frac{\sigma+1+\varepsilon}{2}\right)}, \quad \tilde{y}_r(\sigma, \varepsilon) = \frac{y_r(\sigma, \varepsilon)}{\Gamma\left(\frac{\sigma+1+\varepsilon}{2}\right)}.$$

Let us take an *arbitrary* function $\varphi \in \mathcal{D}(S)$ (not necessarily K-finite) and decompose it into its Fourier series:

$$\varphi(\alpha) = \sum_{m \in \mathbb{Z}} d_m\, e^{im\alpha}. \qquad (5.18)$$

It is well known that its Fourier-coefficients d_m decrease faster than any power function, i.e., for every $h = 1, 2, \ldots$ there exists a constant D such that

$$|d_m| \leq D\, |m|^{-h}. \qquad (5.19)$$

By the continuity of the Poisson transform, we have

$$(P_{\sigma, \varepsilon}\varphi)(t, \alpha) = \sum_{m} d_m\, \Psi(\sigma, \varepsilon, m; t)\, e^{im\alpha}.$$

Therefore by formula (4.14) the Poisson transform $(P_{\sigma,\varepsilon}\varphi)(t,\alpha)$ is

$$(P_{\sigma,\varepsilon}\varphi)(t,\alpha) = (\text{cht})^{\sigma}\left(P^{+}_{\sigma,\varepsilon}\varphi\right)(\xi,\alpha) + (\text{cht})^{\sigma^{*}}\left(P^{-}_{\sigma,\varepsilon}\varphi\right)(\xi,\alpha), \qquad (5.20)$$

where $\xi = \text{tht}$, and the operators $P^{\pm}_{\sigma,\varepsilon}$ are defined in the following way:

$$\left(P^{+}_{\sigma,\varepsilon}\varphi\right)(\xi,\alpha) = \sum_{m} d_{m}\,(-1)^{\varepsilon}\,a(\sigma^{*},m)\,V(\sigma,m;\xi)\,e^{im\alpha}, \qquad (5.21)$$

$$\left(P^{-}_{\sigma,\varepsilon}\varphi\right)(\xi,\alpha) = \sum_{m} d_{m}\,(-1)^{\varepsilon}\,j(\sigma,\varepsilon)\,V(\sigma^{*},m;\xi)\,e^{im\alpha}. \qquad (5.22)$$

Theorem 5.4. Let $\sigma \notin 1/2 + \mathbb{Z}$ and $\sigma \neq -1 - \varepsilon - 2k,\; k \in \mathbb{N}$. For $\varphi \in \mathcal{D}(S)$, the functions $P^{\pm}_{\sigma,\varepsilon}\varphi$ have the following asymptotic expansions in powers of $1 - \xi$ as $t \to +\infty$:

$$\left(P^{+}_{\sigma,\varepsilon}\varphi\right)(\xi,\alpha) \sim \sum_{r=0}^{\infty} x_{r}(\sigma,\varepsilon)\left(A_{\sigma^{*}+r}\varphi\right)(\alpha)\,(1-\xi)^{r}, \qquad (5.23)$$

$$\left(P^{-}_{\sigma,\varepsilon}\varphi\right)(\xi,\alpha) \sim \sum_{r=0}^{\infty} y_{r}(\sigma,\varepsilon)\left(L_{\sigma,r}\varphi\right)(\alpha)\,(1-\xi)^{r}. \qquad (5.24)$$

So the Poisson transform of φ has the following asymptotic expansion as $t \to +\infty$:

$$\left(P_{\sigma,\varepsilon}\varphi\right)(t,\alpha) \sim (\text{cht})^{\sigma} \sum_{r=0}^{\infty} x_{r}(\sigma,\varepsilon)\left(A_{\sigma^{*}+r}\varphi\right)(\alpha)\,(1-\xi)^{r} +$$

$$+ (\text{cht})^{\sigma^{*}} \sum_{r=0}^{\infty} y_{r}(\sigma,\varepsilon)\left(L_{\sigma,r}\varphi\right)(\alpha)\,(1-\xi)^{r}.$$

The expansions (5.23) and (5.24) mean the following: for example, formula (5.23) means that for every $N \in \mathbb{N}$ there exists a constant C such that

$$\left|\left(P^{+}_{\sigma,\varepsilon}\varphi\right)(\xi,\alpha) - \sum_{r=0}^{N} u_{r}(\sigma,\varepsilon)\,A_{\sigma^{*}+r}\varphi(\alpha)\,(1-\xi)^{r}\right| \le C\,(1-\xi)^{N+1}, \qquad (5.25)$$

for $\xi \in [0,1], \alpha \in [0,2\pi]$.

To prove the theorem we need some lemmas.

Denote

$$\Phi(\sigma,\varepsilon,m;\xi) = \int_{0}^{2\pi} (-\xi + \cos\alpha)^{\sigma,\varepsilon}\,e^{-im\alpha}\,d\alpha, \qquad (5.26)$$

so, see (4.10),

$$\Psi(\sigma, \varepsilon, m; t) = (\mathrm{ch}t)^{\sigma} \; \Phi(\sigma, \varepsilon, m; \mathrm{th}t). \tag{5.27}$$

For every $\xi \in [-1, 1]$ the integral (5.26) converges absolutely for $\mathrm{Re}\sigma > -1/2$ and extends to the whole σ-plane as a meromorphic function (with poles at the points $\sigma = -(1+k)/2$, $k \in \mathbb{N}$). The function $\Phi(\sigma, \varepsilon, m; \xi)$ is even with respect to m and has parity $\varepsilon + m$ with respect to ξ.

Lemma 5.5. *For $\mathrm{Re}\sigma > -1/2$ the estimate*

$$\left| \Phi(\sigma, \varepsilon, m; \xi) \right| \le C \tag{5.28}$$

holds for all $m \in \mathbb{Z}$ and all $\xi \in [0, 1]$.

(Here and below $C, C_1, C_2, \ldots$ are constants which depend on σ only. The constants C may be different in different formulas).

Proof. Denote $\rho = \mathrm{Re}\sigma$. For $\rho > -1/2$ we have from (5.26):

$$\left| \Phi(\sigma, \varepsilon, m; \xi) \right| \le \int_0^{2\pi} |-\xi + \cos\alpha|^{\rho} \, d\alpha. \tag{5.29}$$

From this inequality for $\rho \ge 0$ we obtain (5.28) for all $\xi \in [-1, 1]$.

Now let $-1/2 < \rho < 0$. The right-hand side of (5.29) is $\Phi(\rho, 0, 0; \xi)$. By (5.27), (4.12), (4.13) this function equals to

$$2^{\rho+1} \; \sqrt{\pi} \; \frac{\Gamma(\rho + \frac{1}{2})}{\Gamma(\rho + 1)} \; F\left(-\rho, -\rho; -\rho + \frac{1}{2}; \frac{1-\xi}{2} \right) + \tag{5.30}$$

$$+ \left(1 - \frac{1}{\cos\rho\pi} \right) 2^{-\rho} \; \sqrt{\pi} \; \frac{\Gamma(\rho + 1)}{\Gamma(\rho + \frac{3}{2})} \; (\mathrm{ch}t)^{-2\rho-1} \; F\left(\rho + 1, \rho + 1; \rho + \frac{3}{2}; \frac{1-\xi}{2} \right).$$

When ξ ranges over $[0, 1]$, the argument of both hypergeometric functions ranges over $[0, 1/2]$. They are both bounded on this interval, so the function (5.30) for $0 \le \xi \le 1$ is bounded by a constant depending only on ρ. $\square$

Let us denote

$$X(\sigma, m; \xi) = a(\sigma^*, m) \; V(\sigma, m; \xi),$$

see (2.15) for V.

Lemma 5.6. *The function X can be expressed in terms of Φ, see (5.26), (5.27), in the following way:*

$$X(\sigma, m; \xi) = \sum_{\varepsilon=0,1} \frac{1}{2}\left((-1)^{\varepsilon} + \frac{1}{\cos \sigma\pi} \right) \Phi(\sigma, \varepsilon, m; \xi). \tag{5.31}$$

Proof. From the two ($\varepsilon = 0, 1$) equalities (4.14), we express $V(\sigma, m; \xi)$ in terms of Ψ and then substitute (5.27). $\square$

Lemma 5.7. *For the function X the following recursive relation holds*

$$X(\sigma, m; \xi) = X(\sigma, m - 2; \xi) + \frac{2(m-1)}{\sigma + 1} X(\sigma + 1, m - 1; \xi). \qquad (5.32)$$

Proof. We make the integration by parts in (5.26):

$$\Phi(\sigma, \varepsilon, m; \xi) = \frac{\sigma}{im} \int_0^{2\pi} (-\xi + \cos\alpha)^{\sigma - 1, \varepsilon - 1} \sin\alpha \, e^{im\alpha} \, d\alpha =$$

$$= -\frac{\sigma}{2m} \Big[\Phi(\sigma - 1, \varepsilon - 1, m + 1; \xi) - \Phi(\sigma - 1, \varepsilon - 1, m - 1; \xi) \Big].$$

If we replace σ, ε, m by $\sigma + 1, \varepsilon + 1, m - 1$, respectively, we get

$$\Phi(\sigma, \varepsilon, m; \xi) = \Phi(\sigma, \varepsilon, m - 2; \xi) - \frac{2(m-1)}{\sigma + 1} \Phi(\sigma + 1, \varepsilon + 1, m - 1; \xi).$$

Substituting this expression into (5.31), we obtain (5.32). $\square$

Lemma 5.8. *For every $\sigma \notin 1/2 + \mathbb{Z}, \sigma \notin -1 - \mathbb{N}$, there exist constants C and R depending on σ only such that*

$$\Big| X(\sigma, m; \xi) \Big| \leq C \, (|m| + 1)^R \qquad (5.33)$$

for all $m \in \mathbb{Z}$ and all $\xi \in [0, 1]$.

Proof. For $\rho > -1/2$ from lemmas 5.5 and 5.6 we obtain

$$\Big| X(\sigma, m; \xi) \Big| \leq C \qquad (5.34)$$

for all $m \in \mathbb{Z}$, $\xi \in [0, 1]$. It is (5.33) with $R = 0$.

Now we consider the strip $-3/2 < \rho \leq -1/2$. We can assume that $m \geq 0$ because X is even by m. We apply formula (5.32) to $X(\sigma, m; \xi)$ so many times as to get $X(\sigma, 0; \xi)$ or $X(\sigma, 1; \xi)$, depending on the parity of m:

$$X(\sigma, m; \xi) = X(\sigma, d; \xi) - \frac{2}{\sigma + 1} \sum_{k=0}^{[\frac{m-2}{2}]} (m - 1 - 2k) X(\sigma + 1, m - 1 - 2k; \xi),$$
$$(5.35)$$

where $d = 0$ or 1 for m even or odd, respectively. We may apply (5.34) to the items with parameter $\sigma + 1$. Then from (5.35) we obtain

$$\Big| X(\sigma, m; \xi) \Big| \leq \Big| X(\sigma, d; \xi) \Big| + C m^2. \qquad (5.36)$$

Both of the two $(d = 0, 1)$ functions $X(\sigma, d; \xi)$ are bounded for $\xi \in [0, 1]$, therefore from (5.36) we have

$$\left| X(\sigma, m; \xi) \right| \leq C_1 + Cm^2. \tag{5.37}$$

So we obtain estimate (5.33) with $R = 2$.

The case $-5/2 < \rho \leq -3/2$ is similar. Namely, from (5.35) and (5.37), proved already, replacing σ by $\sigma + 1$, we obtain the estimate

$$\left| X(\sigma, m; \xi) \right| \leq C_1 + C_2 m^4.$$

Hence we obtain (5.33) with $R = 4$, etc. $\square$

Lemma 5.9. Let $\sigma \notin 1/2 + \mathbb{Z}, \sigma \notin -1 - \mathbb{N}$. For every $k \in \mathbb{N}$ there exist constants C and R (depending on σ only) such that

$$\left| \left(\frac{\partial}{\partial \xi} \right)^k X(\sigma, m; \xi) \right| \leq C \left(|m| + 1 \right)^R,$$

for all $m \in \mathbb{Z}$ and $\xi \in [0, 1]$.

Proof. According to [6] 2.1(7), we have

$$\left(\frac{\partial}{\partial \xi} \right)^k X(\sigma, m; \xi) = \sigma^{(k)} X(\sigma - k, m; \xi).$$

Now the lemma follows from Lemma 5.8. $\square$

Lemma 5.10. *Both series (5.21) and (5.22) and all their derivatives with respect to ξ converge uniformly on $\xi \in [0, 1]$, $\alpha \in [0, 2\pi]$. Hence, it is possible to differentiate them with respect to ξ term by term and to pass to the limit as $\xi \to 1$.*

Proof. Let us consider, for example, the series (5.21). From Lemma 5.9, formula (4.13) and estimate (5.19), it follows that there exist constants C, R, h such that the series

$$\sum_m C \left(|m| + 1 \right)^{R-h} \tag{5.38}$$

is a majorant of series (5.21) differentiated k times with respect to ξ. We take the constant h such that $h > R + 1$. Then this majorant converges. $\square$

This lemma allows us to extend the functions $(P_{\sigma, \varepsilon}^{\pm} \varphi)(\xi, \alpha)$ of ξ from $(-1, 1)$, where they are defined by equation (5.20) originally, to $(-1, 1]$, in particular, to $[0, 1]$, interesting for us. These functions and all their derivatives with respect to ξ are continuous on $[0, 1]$, in particular, at the point $\xi = 1$.

Proof of Theorem 5.4. Let us prove (5.23). From Lemma 5.10 (and from the subsequent remark) it follows that the Taylor polynomial

$$\sum_{k=0}^{N} \frac{1}{k!} \left(\frac{\partial}{\partial \xi}\right)^k P_{\sigma,\varepsilon}^+ \varphi(\xi, \alpha)\Big|_{\xi=1} (\xi - 1)^k \qquad (5.39)$$

of the function $\left(P_{\sigma,\varepsilon}^+ \varphi\right)(\xi, \alpha)$ in powers of $\xi - 1$ is obtained by summation over m of the Taylor polynomials of $\left(P_{\sigma,\varepsilon}^+ \psi_m\right)(\xi, \alpha)$ multiplied by d_m. But the last polynomials are the partial sums of the first of two series (5.1). So, the polynomial (5.39) is equal to

$$\sum_{m} d_m \sum_{r=0}^{N} x_r(\sigma, \varepsilon) \left(A_{\sigma^*+r} \psi_m\right)(\alpha) (1 - \xi)^r.$$

Here it is possible to transpose the summation over m with the summation over r and to insert it under the symbol of the operator A_{σ^*+r} (the latter is continuous). We obtain that the Taylor polynomial (5.39) is equal to the partial sum $\sum_{r=0}^{N}$ of the series (5.23). By the Taylor formula with remainder term in the Lagrange form, the difference between the function $\left(P_{\sigma,\varepsilon}^+ \varphi\right)(\xi, \alpha)$ and its Taylor polynomial (5.39) is

$$\frac{1}{(N+1)!} \left(\frac{\partial}{\partial \xi}\right)^{N+1} \left(P_{\sigma,\varepsilon}^+ \varphi\right)(\xi, \alpha)\Big|_{\xi=\eta} (\xi - 1)^{N+1} =$$

$$= \frac{1}{(N+1)!} \left\{(-1)^\varepsilon \sum_{m} d_m \left(\frac{\partial}{\partial \xi}\right)^{N+1} X(\sigma, m; \xi)\Big|_{\xi=\eta} e^{im\alpha}\right\} (\xi - 1)^{N+1},$$

where $\eta \in (\xi, 1)$. The terms of the series between braces were estimated above. Its majorant for $\xi \in [0, 1]$, $\alpha \in [0, 2\pi]$ is (5.38) with $k = N + 1$. So the difference mentioned above does not exceed $C (1 - \xi)^{N+1}$ in absolute value. It proves (5.25), and hence (5.23).

Formula (5.24) is proved similarly. □

We may remove the restriction $\sigma \notin -1 - \mathbb{N}$ by passing from $P_{\sigma,\varepsilon}$ to $\widetilde{P}_{\sigma,\varepsilon}$.

Similar theorems are true for the expansions $\eta = (\text{ch} t)^{-2}$ and $\zeta = -e^{-2t}$. We shall not formulate them here. The case of *half integer* σ differs from the general case by the appearance of the factor $\ln \text{ch} t$ in the expansions.

References

1. Erdelyi, A., Magnus, W., Oberhettinger, F. and Tricomi, F.: *Higher Transcendental Functions, I*, McGraw–Hill, New York, 1953.
2. Gelfand, I.M., Graev, M.I. and Vilenkin, N.Ya.: *Integral Geometry and Representation Theory*, Fizmatgiz, Moscow, 1962. (Engl. transl.: Academic Press, New York, 1966.)

3. Gelfand, I.M. and Shilov, G.E.: *Generalized Functions and Operations on Them*, Fizmatgiz, Moscow, 1958. (Engl. transl.: Academic Press, New York, 1964.)
4. Gradshtein, I.S. and Ryzhik, I.M.: *The Tables of Integrals, Sums, Series and Derivatives*, Fizmatgiz, Moscow, 1963.
5. Molchanov, V.F.: Decomposition of the tensor square of a representation of the supplementary series of the unimodular group of real second order matrices, *Sib. Mat. Zh.* **18** (1977) no. 1, 174–188. (Engl. transl.: *Sib. Math. J.* **18** (1977) 128–138.)
6. Molchanov, V.F.: The Plancherel formula for pseudo-Riemannian symmetric spaces of the universal covering of the group $SL(2, \mathbb{R})$, *Sib. Mat. Zh.* **25** (1984) no. 6, 89–105. (Engl. transl.: *Sib. Math. J.* **25** (1984) 903–917.)
7. Vilenkin, N.Ya.: *Special Functions and the Theory of Group Representations*, Nauka, Moscow 1965. (Engl. transl.: *Transl. Math. Monographs* **22**, Amer. Math. Soc., Providence RI 1968.)

MAXIMAL DEGENERATE REPRESENTATIONS, BEREZIN KERNELS AND CANONICAL REPRESENTATIONS

G. VAN DIJK AND S.C. HILLE
Department of Mathematics, Leiden University,
P.O Box 9512, 2300 RA Leiden, The Netherlands.
E-mail: dijk@wi.leidenuniv.nl, shille@wi.leidenuniv.nl

Abstract. In this work we present a new context of the canonical representations which have been introduced by Berezin, Gel'fand, Graev and Vershik for simple Lie groups G of Hermitian type. We discuss maximal-degenerate representations of the complexification of G and the decomposition of the canonical representations into irreducible parts.

Mathematics Subject Classification (1991): 22E30, 22E46, 43A80, 43A85, 53C35

Key words: canonical representations, Berezin kernels, group complexification, degenerate representations, Hermitian symmetric spaces.

Introduction

Canonical representations, being a special type of reducible unitary representations, were introduced by Vershik, Gel'fand and Graev for the group $SU(1,1)$ in [11]. These are the same representations as those introduced by Berezin from a completely other point of view, namely related to quantization [1]. The groups which are considered are the simple Lie groups G with maximal compact subgroup K such that G/K is a Hermitian symmetric space. The main problem discussed up to now is the decomposition of the canonical representations into irreducible constituents. This is not an easy task. It has been done by Berezin for the classical groups, and recently, for all G (from the above class) by Upmeier and Unterberger [10], who applied techniques from the theory of Jordan algebras. There are however, in both

285

B. P. Komrakov et al. (eds.), Lie Groups and Lie Algebras, 285–298.
© *1998 Kluwer Academic Publishers. Printed in the Netherlands.*

treatments, conditions on the set of parameters of the representations: only large parameters are allowed. For small parameters (see [2]) an interesting new phenomenon occurs: complementary series representations take part in the decomposition.

In this note we present a new context for the canonical representations. We hope that the reader will be convinced once more in this way of the relevance of these representations. We start therefore with the introduction of so-called maximal degenerate representations of the complexification G_c of G. Maximal degenerate representations have been discussed recently for various groups and in various contexts ([3], [6], [7], [9]). We then formulate a few interesting problems, which we shall discuss. The canonical representations occur by restriction of the maximal degenerate representations of G_c to G. We shall treat the case $G = \mathrm{SU}(1, n)$ in detail. An analogous theory of canonical representations has recently been developed by Molchanov and the author for the class of para-Hermitian symmetric spaces, see [8].

1. Irreducible Hermitian symmetric spaces

Let $\mathfrak{g}$ be a noncompact simple real Lie algebra with complexification $\mathfrak{g}_c$. Let $\mathfrak{g} = \mathfrak{k} + \mathfrak{p}$ be a Cartan decomposition and let θ denote the corresponding Cartan involution. We assume that $\mathfrak{k}_s = [\mathfrak{k}, \mathfrak{k}] \neq \mathfrak{k}$. The center $\mathfrak{z}$ of $\mathfrak{k}$ is one-dimensional and there is an element $Z_0 \in \mathfrak{z}$ such that $(\mathrm{ad}\, Z_0)^2 = -1$ on $\mathfrak{p}$. Fixing i a square root of -1, $\mathfrak{p}_c = \mathfrak{p} + i\mathfrak{p} = \mathfrak{p}_+ + \mathfrak{p}_-$ where $\mathrm{ad}\, Z_0|_{\mathfrak{p}_+} = i$, $\mathrm{ad}\, Z_0|_{\mathfrak{p}_-} = -i$. Then $\mathfrak{g}_c = \mathfrak{k}_c + \mathfrak{p}_+ + \mathfrak{p}_-$, $[\mathfrak{p}_\pm, \mathfrak{p}_\pm] = 0$, $[\mathfrak{p}_+, \mathfrak{p}_-] = \mathfrak{k}_c$ and $[\mathfrak{k}_c, \mathfrak{p}_\pm] = \mathfrak{p}_\pm$. Let G_c be a connected, simply connected Lie group with Lie algebra $\mathfrak{g}_c$ and K_c, P_+, P_-, G, K, K_s, $Z(K)$ the analytic subgroups corresponding to $\mathfrak{k}_c$, $\mathfrak{p}_+$, $\mathfrak{p}_-$, $\mathfrak{g}$, $\mathfrak{k}$, $\mathfrak{k}_s$, $\mathfrak{z}$ respectively. Then $K_c P_-$ and $K_c P_+$ are maximal parabolic subgroups of G_c with split component $A = \exp i\mathbb{R}Z_0$. If K_s^c is the analytic subgroup corresponding to $\mathfrak{k}_s^c$, then $K_c = K_s^c Z(K) A = M_c A$ with $M_c = K_s^c Z(K)$. Observe that $M_c A P_\pm$ is the Langlands decomposition of $K_c P_\pm$. G is closed in G_c and U, the analytic subgroup corresponding to $\mathfrak{u} = \mathfrak{k} + i\mathfrak{p}$, is a maximal compact subgroup of G_c. The space $X = G/K$ is an irreducible Hermitian symmetric space of the noncompact type. Here is a classification (up to local isomorphy).

Irreducible Hermitian symmetric spaces

Type	Noncompact	Compact	Rank	Dimension
AIII	$SU(p,q)/S(U_p \times U_q)$	$SU(p+q)/S(U_p \times U_q)$	$\min(p,q)$	$2pq$
DIII	$SO^*(2n)/U(n)$	$SO(2n)/U(n)$	$[1/2\,n]$	$n(n-1)$
BDI	$SO_0(p,2)/$	$SO(p+2)/$	$\min(p,2)$	$2p$
$(q=2)$	$SO(p) \times SO(2)$	$SO(p) \times SO(2)$		
CI	$Sp(n,\mathbb{R})/U(n)$	$Sp(n)/U(n)$	n	$n(n+1)$
EIII	$e_{6(-14)}/so(10)+\mathbb{R}$	$e_{6(-78)}/so(10)+\mathbb{R}$	2	32
EVII	$e_{7(-25)}/e_6+\mathbb{R}$	$e_{7(-133)}/e_6+\mathbb{R}$	3	54

Tube domains: $AIII : p = q$, $DIII : n$ even, BDI $(q = 2)$, CI, $EVII$.

2. Maximal degenerate series of G_c

Let $P^\pm$ be the two (standard) maximal parabolic subgroups of G_c given by $K_c P_\pm$. Let $n = \dim_{\mathbb{C}} \mathfrak{p}_\pm$. For $\mu \in \mathbb{C}$, define the character ω_μ of $P^\pm$ by the formula:

$$\omega_\mu(p) = |\det \mathrm{Ad}\,(k)|_{\mathfrak{p}_+}|^\mu, \qquad (2.1)$$

if $p = k n$ ($k \in K_c$, $n \in P_\pm$). Here "det" is the determinant over $\mathbb{C}$.

Consider the representations $\pi_\mu^\pm$ of G_c induced from $P^\pm$:

$$\pi_\mu^\pm = \mathrm{Ind}\,\omega_{\mp\mu}. \qquad (2.2)$$

Let us describe these representations in the "compact picture".

One has the following decompositions

$$G_c = U P^+, \qquad (2.3)$$

$$= U P^-, \qquad (2.4)$$

which we call the Iwasawa type and the anti-Iwasawa type decomposition, respectively.

For the corresponding decompositions $g = u p$ of an element $g \in G_c$ the factors p and u are defined up to an element of the subgroup $U \cap P^+ = U \cap P^- = U \cap K_c = K$. The coset spaces $G_c/P^\pm$ can be identified with the coset space U/K. Now $K = K_s Z(K)$. Put $S = U/K_s$. $Z(K)$ acts on S from the right.

Let us denote by V_0 the vector space of C^∞-functions φ on S satisfying

$$\varphi(s\lambda) = \varphi(s) \qquad (2.5)$$

for all $\lambda \in Z(K)$.

V_0 can be seen as the representation space of both π_μ^+ and π_μ^-. In fact, $\pi_\mu^+ = \pi_\mu^- \circ \tau$, where τ is the Cartan involution of G_c with respect to U.

The group G_c acts on S; denote it by $g \cdot s$ ($g \in G_c, s \in S$). If e_0 is the "origin" of S and $s = u\,e_0$, write $gu = u'man$ with $u' \in U, m \in K_s^c, a \in$

$A, n \in P_+$. Then u' is unique modulo K_s and $g \cdot s = u' e_0$. We have for $\varphi \in V_0$:

$$\pi_\mu^-(g)\,\varphi(s) = \varphi(g^{-1} \cdot s)\,\|g^{-1}(s)\|^\mu. \qquad (2.6)$$

Here $\|g^{-1}(s)\|$ means the following. Choose a representative $u \in U$ for s (modulo K_s). Write as above

$$g^{-1}u = u'man$$

with $u' \in U$, $m \in K_s^c$, $a \in A$, $n \in P_+$. Then we define

$$\|g^{-1}(s)\| = \det \mathrm{Ad}\,(a)|_{\mathfrak{p}_+}. \qquad (2.7)$$

Observe that $\|s\| = 1$ for $s \in S$.

In a similar way we have

$$\pi_\mu^+(g)\,\varphi(s) = \varphi(\tau(g^{-1}) \cdot s)\|\tau(g^{-1})(s)\|^\mu. \qquad (2.8)$$

Let $(\ |\ )$ be the standard inner product on $L^2(S)$,

$$(\varphi|\psi) = \int_S \varphi(s)\overline{\psi(s)}\,ds. \qquad (2.9)$$

Here ds is a normalized U-invariant measure on S. This measure ds is transformed by the action of $g \in G_c$ as follows:

$$d\tilde{s} = \|g(s)\|^{-2}\,ds, \quad \tilde{s} = g \cdot s. \qquad (2.10)$$

It implies that the Hermitian form (2.9) is invariant with respect to the pairs

$$(\pi_\mu^-, \pi_{-\bar\mu-2}^-) \quad \text{and} \quad (\pi_\mu^+, \pi_{-\bar\mu-2}^+).$$

Therefore, if $\mathrm{Re}\,\mu = -1$, then the representations $\pi_\mu^\pm$ are unitarizable, the inner product being (2.9).

3. Intertwining operators and irreducibility

U acts on $i\mathfrak{k} + \mathfrak{p}$ irreducibly, so the smallest real subalgebra of $\mathfrak{g}_c$ containing $\mathfrak{u}$ and $i\,Z_0$ is precisely $\mathfrak{g}_c$. For the study of the irreducibility of $\pi_\mu^\pm$ it should be useful to consider their U-types. So we have to decompose V_0 into U-types, which can be done multiplicity free. Let $L_0 = i\,Z_0$. Then a crucial role is played by the operators $\pi_\mu^\pm(L_0)$. One will notice that these operators send spherical functions to spherical functions, since L_0 commutes with $\mathfrak{k}$. We will not pursue the irreducibility problem here. It will turn out that $\pi_\mu^\pm$ is at least irreducible if $\mu \notin \mathbb{R}$. Let us turn to intertwining operators.

$P_- K_c P_+$ is an open dense subset of G_c, and $U' = U \cap P_- K_c P_+$ is open and dense in U. For $u \in U'$ write $u = n_- k_u n_+$ with $n_- \in P_-, k \in K_c$ and $n_+ \in P_+$. If $s = u e_0, t = v e_0$ with $u, v \in U$ such that $v^{-1} u \in U'$, then we set

$$(s, t) = \det \mathrm{Ad}\, (k_{v^{-1} u})|_{\mathfrak{p}_+}. \tag{3.1}$$

This is a well-defined expression. One has: $(s, t) = \overline{(t, s)}$ and (s, t) is invariant with respect to the diagonal action of U. If $\lambda \in Z(K)$, then

$$(s\lambda, t) = (s, t\lambda^{-1}) = (\det \mathrm{Ad}\, (\lambda)|_{\mathfrak{p}_+}) \cdot (s, t). \tag{3.2}$$

In order to get an explicit expression for (s, t) and to be able to study its behavior on $S \times S$, it is enough to study the expression $(u\, e_0, e_0)$ for u in a maximal "split" torus of U, i.e., a torus contained in $\exp i\mathfrak{p}$. Therefore we need some additional structure theory.

Let $\mathfrak{h}$ be a Cartan subalgebra of $\mathfrak{g}$ contained in $\mathfrak{k}$. Choose an ordering on the roots of $(\mathfrak{g}_c, \mathfrak{h}_c)$ such that $\mathfrak{p}_+ = \sum_{\alpha \in \Delta_n^+} \mathfrak{g}_\alpha$, $\mathfrak{p}_- = \sum_{\alpha \in \Delta_n^+} \mathfrak{g}_{-\alpha}$, where Δ_n^+ denotes the set of positive noncompact roots. Let r be the split rank of $\mathfrak{g}$. Then one can select r linearly independent positive noncompact roots $\beta_1, \dots, \beta_r$ such that $\beta_i \pm \beta_j$ is not a root ($i \neq j$). Let $X \to \overline{X}$ be the complex conjugation of $\mathfrak{g}_c$ with respect to $\mathfrak{g}$. Set

$$\begin{aligned}
H_k &= i[X_{\beta_k}, \overline{X}_{\beta_k}], \\
X_k &= X_{\beta_k} + \overline{X}_{\beta_k}, \\
Y_k &= -i(X_{\beta_k} - \overline{X}_{\beta_k},)
\end{aligned}$$

where $X_{\beta_k} \in \mathfrak{g}_{\beta_k}$ is chosen such that $[iH_k, X_{\beta_k}] = -2X_{\beta_k}$. Then H_k, X_k, Y_k are all real and satisfy

$$[H_k, X_k] = -2Y_k, \quad [H_k, Y_k] = 2X_k, \quad [X_k, Y_k] = 2H_k.$$

The correspondence

$$\begin{pmatrix} i & 0 \\ 0 & -i \end{pmatrix} \leftrightarrow H_k, \quad \begin{pmatrix} 0 & 1 \\ 1 & 0 \end{pmatrix} \leftrightarrow X_k, \quad \begin{pmatrix} 0 & i \\ -i & 0 \end{pmatrix} \leftrightarrow Y_k$$

gives a copy of $\mathfrak{su}(1,1)$ in $\mathfrak{g}$. The r correspondences commute. Let $\mathfrak{b} = \{i \sum s_j Y_j \,|\, s_j \in \mathbb{R}\}$ and $B = \exp \mathfrak{b}$. B is a maximal "split" torus of U: $\mathfrak{b} \in i\mathfrak{p}$.

Set $i\mathfrak{h}^- = \mathrm{span}_{\mathbb{R}}(iH_1, \dots, iH_r)$. By Cayley transformation this space is mapped onto $\mathfrak{a} = \mathrm{span}_{\mathbb{R}}(X_1, \dots, X_r)$, which is a maximal abelian subspace of $\mathfrak{p}$. The restricted roots of $i\mathfrak{h}^-$ are $\pm\frac{1}{2}(\beta_j \pm \beta_k)$ ($j < k$), $\pm\beta_j$ and $\pm\frac{1}{2}\beta_j$ with root multiplicities $a, 1$ and $2b$ respectively. If G/K is of positive type, then $b = 0$. The nonzero restrictions of the positive noncompact roots are $\frac{1}{2}(\beta_j + \beta_k)$ $(j < k)$, β_j and $\frac{1}{2}\beta_j$ with root multiplicities $a, 1$ and b respectively. Set $p = (r-1)a + b + 2$, the *genus* of G/K.

Now an easy computation in $\mathfrak{su}(1,1)$ shows that

$$((\exp i \sum s_j Y_j)e_0, e_0) = (\cos s_1)^p \ldots (\cos s_r)^p \tag{3.3}$$

provided $\cos s_i \neq 0$ for $i = 1, \ldots r$. So (s,t) can be extended to a C^∞-function on $S \times S$.

Define the operator A_μ on V_0 by the formula

$$A_\mu \varphi(s) = \int_S |(s,t)|^{-\mu-2} \varphi(t)\, dt. \tag{3.4}$$

This integral converges absolutely for $\operatorname{Re}(\mu) < -2$ (at least) and can be analytically extended to the whole μ-plane as a meromorphic function. It is easily checked that A_μ is an intertwining operator:

$$A_\mu \pi_\mu^\pm(g) = \pi_{\mu'}^\mp(g)\, A_\mu, \tag{3.5}$$

where $\mu' = -\mu - 2$. For the proof of the intertwining property we apply the relation:

$$(g \cdot s, \tau(g) \cdot t) = (s,t)\, \|\tau(g)t\|^{-1}\, \|g(s)\|^{-1} \quad (g \in G_c; s, t \in S). \tag{3.6}$$

It is an interesting problem to determine those μ for which $\pi_\mu^\pm$ is unitarizable.

4. Restriction to G

One has, according to Harish-Chandra, $G \subset P_-K_cP_+$, $G \cap K_cP_+ = K$. Even GK_cP_+ is open in $P_-K_cP_+$ and G/K is thus identified with a bounded open domain $\mathcal{D}$ in $\mathfrak{p}_-$ (via the exponential map), G acting on it by means of holomorphic transformations. On the other hand G/K_s can be identified with an open subset of S by means of the map

$$g \to g \cdot e_0. \tag{4.1}$$

If we identify s and $s\lambda$ for $\lambda \in Z(K)$, then this set can be identified with G/K. Let $\mathcal{O} = G \cdot e_0$. For $s \in \mathcal{O}$, $s = g \cdot e_0$ $(g \in G)$, set

$$J s = \tau(g) \cdot e_0. \tag{4.2}$$

This is a well-defined (analytic) map.

For any $\varphi \in V_0$ with $\operatorname{Supp} \varphi \in \mathcal{O}$, put

$$E\varphi(s) = \varphi(Js) \qquad (s \in S).$$

Then

$$\pi_\mu^-(g)\, E\varphi(s) = E(\pi_\mu^+(g)\varphi)(s) \tag{4.3}$$

for all $s \in \mathcal{O}$ and $g \in G$.

We shall write

$$[s,t] = (s, J\,t) \tag{4.4}$$

for $s,t \in \mathcal{O}$. Clearly $[s,s] > 0$ for $s \in \mathcal{O}$.

Notice that

$$[g \cdot s,\, g \cdot t] = [s,t]\, \|g(s)\|^{-1} \|g(t)\|^{-1} \tag{4.5}$$

for all $g \in G$, $s,t \in \mathcal{O}$. Let φ be a C^∞-function with compact support on $\mathcal{O}$, satisfying (2.5). Set

$$\psi(s) = \varphi(s)\,[s,s]^{-\mu/2}. \tag{4.6}$$

Then ψ satisfies the same condition (2.5). Moreover

$$\begin{aligned}
\psi(g^{-1} \cdot s) &= \varphi(g^{-1} \cdot s)\,[s,s]^{-\mu/2}\, \|g^{-1}(s)\|^\mu \\
&= \pi_\mu^-(g)\,\varphi(s) \cdot [s,s]^{-\mu/2}.
\end{aligned}$$

So the linear map $\varphi \to \psi$ ($\varphi \in V_0$, $\operatorname{Supp}\varphi \subset \mathcal{O}$) intertwines the restriction of π_μ^- to G with the left regular representation of G on $\mathcal{D}(G/K)$.

A G-invariant measure on $\mathcal{O}$ is given by

$$d\nu(s) = \frac{ds}{[s,s]}, \tag{4.7}$$

(cf. (2.10)).

So, if we provide $V_0 \cap \mathcal{D}(\mathcal{O})$ with the inner product given by

$$(\varphi_1, \varphi_2) = \int_S \varphi_1(s)\, \overline{\varphi_2(s)}\, [s,s]^{-\operatorname{Re}\mu - 1}\,ds, \tag{4.8}$$

then π_μ^- becomes unitary, if we restrict it to G. Call this (double) restriction of π_μ^- to G the representation R_μ. The intertwining operator becomes

$$A_\mu\,\varphi(s) = \int_S |[s,t]|^{-\mu-2} \varphi(t)\,dt \quad (s \in \mathcal{O}). \tag{4.9}$$

Observe that $|[s,t]| > 0$ for $s,t \in \mathcal{O}$. Indeed, it suffices to notice that $|(g \cdot e_0, e_0)| > 0$ for $g \in G$ and this is clear from $G \subset P_- K_c P_+$. The intertwining operator (4.9), with $\varphi \in V_0 \cap \mathcal{D}(\mathcal{O})$, is defined for all μ, provided $s \in \mathcal{O}$. Then $A_\mu\varphi$ is a C^∞-function on $\mathcal{O}$, nonnecessarily with compact support. Let μ be *real* from now on.

For φ_1, $\varphi_2 \in V_0 \cap \mathcal{D}(\mathcal{O})$ consider the Hermitian form

$$(\varphi_1 \mid A_\mu \varphi_2) = \int_S \int_S |[s, t]|^{-\mu-2} \varphi_1(s) \overline{\varphi_2(t)} \, ds dt. \tag{4.10}$$

This form is clearly invariant with respect to R_μ. Applying the linear transformation (4.6) on $\mathcal{O}$, we get the following:

$$\langle \psi_1, \psi_2 \rangle = \int_S \int_S \psi_1(s) \overline{\psi_2(t)} \left\{ \frac{[s, s][t, t]}{[s, t][t, s]} \right\}^{\frac{\mu}{2}+1} d\nu(s) d\nu(t). \tag{4.11}$$

Let $s, t \in \mathcal{O}$. Then

$$B_\lambda(s, t) = \left\{ \frac{[s, s][t, t]}{[s, t][t, s]} \right\}^\lambda \tag{4.12}$$

is called the Berezin kernel on $\mathcal{D}(G/K)$.

B_λ is clearly G-invariant. The function

$$\varphi_\lambda(g) = B_\lambda(g \cdot e_0, e_0)$$

is equal to

$$\varphi_\lambda(g) = Q(g)^{-\lambda}, \tag{4.13}$$

where $Q(g) = \|[g \cdot e_0, e_0]\|^2 \cdot \|g(e_0)\|^2$. φ_λ is a bi-K-invariant function on G.

If $\mathcal{K}(z, w)$ is the Bergman kernel of G/K ($z, w \in \mathcal{D}$), where $\mathcal{D}$ is the bounded domain corresponding to G/K in $\mathfrak{p}_-$, then $\mathcal{K}(z, z) = Q(g)^{-1}$ if z corresponds to gK.

5. Remarks

(a) In general there are $\frac{1}{2}(r + 1)(r + 2)$ G-orbits on $\widetilde{S} = U/K$, and $r + 1$ open orbits. There is one closed orbit: the Silov boundary of G/K. The orbit $\mathcal{O} \simeq G/K$ is a Riemannian manifold, all other open orbits are pseudo-Riemannian. In case $G = \mathrm{SU}(1, n)$, where r, the split rank of G, is equal to 1, the second open orbit is isomorphic to $G/\mathrm{S}(\mathrm{U}(1, n-1) \times \mathrm{U}(1))$. A extension of the theory of the previous sections to these orbits is immediate. The Berezin kernel is however by no way positive-definite but remains Hermitian and G-invariant. For the G-orbit structure of $\widetilde{S}$ we refer to Wolf [12].

(b) Taking instead of ω_μ

$$\omega_{\mu,l}(p) = |\det \mathrm{Ad}(k)|_{\mathfrak{p}_+}|^\mu \cdot \left(\frac{\det \mathrm{Ad}(k)|_{\mathfrak{p}_+}}{|\det \mathrm{Ad}(k)|_{\mathfrak{p}_+}|} \right)^l$$

if $p = kn$ ($k \in K_c$, $n \in P_\pm$), so $\omega_\mu = \omega_{\mu,0}$, one can develop a completely similar, but more general theory. It leads to the kernels:

$$B_{\lambda,l}(s, t) = B_\lambda(s, t) \left(\frac{|[s, t]|}{[s, t]} \right)^l \qquad (s, t \in \mathcal{O})$$

defined on $\mathcal{D}(G/K, l)$, the space of C_c^∞-functions φ on G satisfying

$$\varphi(gk) = \omega_{0,l}(k)\varphi(g)$$

($g \in G$, $k \in K$). Details will appear in future publications.
(c) Let as before, p denote the genus and r the rank of G/K. Recall that a is the root multiplicity of $\pm(\beta_j \pm \beta_k)$, ($j < k$). It is well-known that B_λ is a positive-definite kernel for λ in the "Berezin–Wallach set", i.e, for $\lambda > \frac{a}{2}(r-1)$ and $\lambda = \frac{a}{2}j$ with $0 \leq j \leq r-1$ (j an integer). This is a sufficient condition, cf [5]. The function

$$\psi_\lambda(g) = \varphi_{\frac{\lambda}{p}}(g)$$

is therefore a positive-definite bi-K-invariant function for λ as above. The unitary representation associated with ψ_λ is called a *canonical representation*. In order to decompose it into irreducible components, one computes the spherical Fourier transform of ψ_λ for λ so large that ψ_λ is an integrable function on G. For the remaining λ analytic continuation and residue calculus seems to be a good idea. This latter step has only been performed for $G = \mathrm{SU}(1, n)$ so far, see the next section. For large λ there is a general result, obtained by Berezin and Upmeier–Unterberger, see [10]. We shall describe it briefly. Let $X_1, \ldots, X_r$ be the basis of the maximal abelian subspace of $\mathfrak{p}$, constructed in Section 3. Let $\nu \in \mathfrak{a}^*$, $\nu_j = \nu(X_j)$ for $j = 1, \ldots, r$. Let ρ be half the sum of the positive roots, $\rho_j = \rho(X_j)$ for $j = 1, \ldots, r$. Then the spherical Fourier transform of ψ_λ is the function on $\mathfrak{a}^* \simeq \mathbb{R}^r$, given by

$$\pi^n \prod_{j=1}^r \frac{\Gamma(\lambda + i\frac{\nu_j}{2} - \frac{p-1}{2})\Gamma(\lambda - i\frac{\nu_j}{2} - \frac{p-1}{2})}{\Gamma(\lambda + \frac{\rho_j}{2} - \frac{p-1}{2})\Gamma(\lambda - \frac{\rho_j}{2} + \frac{n}{r} - \frac{p-1}{2})},$$

with a suitable normalization of Haar measures and a suitable parametrization of the spherical functions, see [10].

6. An example: $\mathrm{SU}(1, n)$

(a) Irreducibility, equivalence and unitarizability

Let $G = \mathrm{SU}(1, n)$, $K = \mathrm{S}(\mathrm{U}(1) \times \mathrm{U}(n))$. Then $G_c = \mathrm{SL}(n+1, \mathbb{C})$,

$$Z_0 = \mathrm{diag}\,(\frac{n}{n+1}, \frac{-1}{n+1}, \ldots, \frac{-1}{n+1})$$

and $P^\pm$ are the groups of upper and lower block triangular matrices respectively:

$$P^+ : \begin{pmatrix} a & b \\ 0 & c \end{pmatrix}, \qquad P^- : \begin{pmatrix} a & 0 \\ b & c \end{pmatrix},$$

where $a \in \mathbb{C}^{\times}$, $c \in \mathrm{GL}\,(n,\mathbb{C})$, $a \cdot \det c = 1$ and b is a row (column) vector in $\mathbb{C}^n$.

For $\mu \in \mathbb{C}$, ω_μ is the character of $P^{\pm}$ given by

$$\omega_\mu(p) = |a|^{(n+1)\mu}.$$

Consider thus the representations $\pi_\mu^{\pm}$ of G_c induced by ω_μ from $P^{\pm}$, as in (2.2). A complete analysis of these representations has recently been given by one of the authors in [6]. Actually a more general situation is considered there, taking

$$\omega_{\mu,l}(p) = \omega_\mu(p) \left(\frac{a}{|a|} \right)^l \quad (l \in \mathbb{Z})$$

instead. We shall confine ourselves here to a brief description of the results obtained for the case $l = 0$.

The representations $\pi_\mu^{\pm}$ can be seen as acting on the subspace of the Hilbert space $L^2(S)$, where $S = \{z \in \mathbb{C}^n : ||z|| = 1\}$, consisting of functions φ satisfying

$$\varphi(s\lambda) = \varphi(s) \quad (s \in S)$$

for $\lambda \in \mathbb{C}, |\lambda| = 1$.

This space actually splits under $U = \mathrm{SU}(n+1)$ as a direct sum

$$\bigoplus_{a \in \mathbb{N}_0} \mathcal{H}_a,$$

where $\mathcal{H}_a$ is the space of harmonic polynomials on $\mathbb{C}^{n+1}$, homogeneous of degree a in z and $\bar{z}$.

By the method sketched in Section 3, we obtain the following.

• **Irreducibility**

Set $\nu = (n+1)\mu$. Then

(i) $\pi_\mu^{\pm}$ is irreducible for $\nu \notin (-2(n+1) - 2\mathbb{N}_0) \cup 2\mathbb{N}_0$.

(ii) If $\nu = 2k$, k a nonnegative integer, then there is one irreducible finite-dimensional subspace for $\pi_\mu^{\pm}$, namely

$$\bigoplus_{a=0}^{k} \mathcal{H}_a.$$

The quotient representation is irreducible.

(iii) If $\nu = -2k - 2 - 2n$, k a nonnegative integer, then there is one infinite-dimensional irreducible subspace for $\pi_\mu^{\pm}$, namely

$$\bigoplus_{a=k+1}^{\infty} \mathcal{H}_a.$$

The quotient representation is irreducible and finite-dimensional.

• **Equivalence**

Let $\varepsilon_1, \varepsilon_2 \in \{+, -\}$. A continuous intertwining operator between $\pi_{\mu_1}^{\varepsilon_1}$ and $\pi_{\mu_2}^{\varepsilon_2}$ exists only if

(i) $\mu_1 = \mu_2$ and $\varepsilon_1 = \varepsilon_2$, or

(ii) $\mu_1 = -\mu_2 - 2$ and $\varepsilon_1 \neq \varepsilon_2$, or

(iii) $n = 1$; $\mu_1 = \mu_2$, ε_1 and ε_2 arbitrary.

• **Unitarizability**

$\pi_\mu^\pm$ is unitarizable if and only if

(i) $n \geq 1$, $\operatorname{Re} \mu = -1$; the inner product is the usual inner product of $L^2(S)$;

(ii) $n = 1$, $\mu \in (-2, 0)$: one obtains the usual complementary series of $\mathrm{SL}(2, \mathbb{C})$.

(b) The spectral decomposition of the canonical representations

The main reference here is [2].

Consider the Hermitian form on $\mathbb{C}^{n+1}$ given by

$$[x, y] = y_0 \, \overline{x}_0 - y_1 \, \overline{x}_1 - \cdots - y_n \, \overline{x}_n.$$

$G = \mathrm{SU}(1, n)$ is the subgroup of the orthogonal group of this Hermitian form consisting of elements with determinant equal to one. $X = G/K$ can be identified with an open subset of the projective space $P_n(\mathbb{C})$. If π denotes the natural projection map $\mathbb{C}^{n+1} \backslash \{0\} \to P_n(\mathbb{C})$, then X is the image under π of the open set

$$\{x \in \mathbb{C}^{n+1} : [x, x] > 0\}.$$

In this model the function $Q(g)$ from Section 4, see (4.13), which is bi-K-invariant, is given by

$$Q(x) = \frac{|x_0|^{2(n+1)}}{[x, x]^{n+1}}.$$

X has also a model as a bounded domain in $\mathbb{C}^n$ (the Harish-Chandra's realization).

On $\mathbb{C}^n$ we have the usual inner product

$$(x, y) = \overline{y}_1 \, x_1 + \cdots + \overline{y}_n \, x_n$$

with norm $\|x\| = (x, x)^{1/2}$. Let

$$B(\mathbb{C}^n) = \{x \in \mathbb{C}^n : \|x\| < 1\}$$

be the unit ball in $\mathbb{C}^n$. The map from $\{x \in \mathbb{C}^{n+1} : [x,x] > 0\}$ to $\mathbb{C}^n$, given by

$$x \to y \quad \text{with} \quad y_p = x_p x_0^{-1}$$

defines, by going to the quotient space, a real analytic bijection of X onto $B(\mathbb{C}^n)$. G acts on $B(\mathbb{C}^n)$ transitively by fractional linear transformations. If $g \in G$ is of the form

$$g = \begin{pmatrix} a & b \\ c & d \end{pmatrix}$$

with matrices a (1×1), b $(1 \times n)$, c $(n \times 1)$ and d $(n \times n)$, then

$$g \cdot y = (dy + c)(\langle b, y \rangle + a)^{-1},$$

where y and c are regarded as column vectors and

$$\langle b, y \rangle = b_1 y_1 + \cdots + b_n y_n.$$

Clearly $K = \text{Stab}(o)$. On $B(\mathbb{C}^n)$ the function Q takes the form

$$Q(y) = (1 - ||y||^2)^{-(n+1)}.$$

The invariant measure on $B(\mathbb{C}^n)$ is $(1 - ||y||^2)^{-(n+1)} \, dy$. The function ψ_λ, defined in Section 5, is thus equal to

$$\psi_\lambda(g) = (1 - ||y||^2)^\lambda,$$

if $y = g \cdot o$, since the genus of X is equal to $n + 1$.

Clearly ψ_λ is integrable for $\lambda > \rho$ (even for $\text{Re } \lambda > 0$), where $\rho = n$. The split rank of G is one. Let A be the subgroup of the matrices

$$a_t = \begin{pmatrix} \cosh t & 0 & \sinh t \\ 0 & I_{n-1} & 0 \\ \sinh t & 0 & \cosh t \end{pmatrix}$$

$(t \in \mathbb{R})$. Then $G = KA_+K$, where $A_+ = \{a_t : t \geq 0\}$. We normalize the Haar measure dg on G such that

$$\int_G f(g) \, dg = \int_K \int_0^\infty \int_K f(ka_t k') \, \delta(t) \, dk \, dt \, dk'$$

with $\dot{\delta}(t) = 2\frac{\pi^n}{\Gamma(n)}(\sinh t)^{2(n-1)}\left(\frac{\sinh 2t}{2}\right)$.

The spherical functions have an explicit expression in terms of hypergeometric functions:

$$\varphi_s(a_t) = {_2F_1}\left(\frac{s+\rho}{2}, \frac{-s+\rho}{2}; n; -\sinh^2 t\right).$$

Notice that $\psi_\lambda(a_t) = (\cosh t)^{-2\lambda}$.

Harish-Chandra's c-function takes the form:

$$c(s) = \Gamma(n)\, 2^{\rho - s} \frac{\Gamma(s)}{\Gamma(\frac{s+\rho}{2})^2}.$$

For $\operatorname{Re}\lambda > \rho$, we then have

$$\psi_\lambda(g) = \frac{1}{2\pi K} \int_0^\infty a_\lambda(\mu)\, \varphi_{i\mu}(g) \frac{d\mu}{|c(i\mu)|^2},$$

where a_λ is the spherical Fourier transform of ψ_λ:

$$a_\lambda(\mu) = \int_G \psi_\lambda(g)\, \varphi_{-i\mu}(g)\, dg$$

and

$$K = 2^{1-2\rho} \frac{\pi^n}{\Gamma(n)}.$$

The density $a_\lambda(\mu)$ can easily be computed in this case, using the explicit expressions for $\psi_\lambda(a_t)$ and $\varphi_{-i\mu}(a_t)$, and applying Erdélyi [4] 20.2 (9). We get

$$a_\lambda(\mu) = \pi^n \frac{\Gamma(\lambda + \frac{i\mu - \rho}{2})\, \Gamma(\lambda + \frac{-i\mu - \rho}{2})}{\Gamma(\lambda)^2}.$$

Clearly ψ_λ is a positive-definite function for $\lambda > \rho$. But it is easily seen, using the explicit expressions of ψ_λ on $B(\mathbb{C}^n)$:

$$\psi_\lambda(g_1^{-1} g_2) = \left\{ \frac{(1 - \|y\|^2)\,(1 - \|z\|^2)}{[1 - (y, z)]\,[1 - (z, y)]} \right\}^\lambda,$$

if $z = g_1 \cdot o$, $y = g_2 \cdot o$, that ψ_λ is positive-definite for $\lambda > 0$: one simply has to expand $[1 - (y, z)]^{-\lambda}$ and $[1 - (z, y)]^{-\lambda}$ into a power series. So the canonical representation π_λ, defined by ψ_λ, should be decomposed for $0 < \lambda \le \rho$ too. We apply for this purpose a method of analytic continuation: we first shift the integration line $\operatorname{Re} s = 0$ in the inversion formula for the spherical Fourier transform, then we compute the inverse spherical Fourier transform of ψ_λ and finally we shift back. For technical details we refer to [2].

We finally obtain:

$$(\psi_\lambda, f) = \sum_{l, s_l > 0} r_l(\lambda)(f, \varphi_{s_l}) + \frac{1}{2\pi K} \int_0^\infty a_\lambda(i\mu)\, (f, \varphi_{i\mu}) \frac{d\mu}{|c(i\mu)|^2}$$

for $f \in \mathcal{D}(G/K)$, $\lambda > 0$.

Here

$$s_l = \rho - 2\lambda - 2l, \; l = 0, 1, 2, \ldots$$

and

$$r_l(\lambda) = \frac{2^{2\lambda + 2l}}{c(s_l)} \cdot \frac{(1 - \lambda - l)_l^2}{(\rho + 1 - 2\lambda - 2l)_l \, l!},$$

where $(\alpha)_l = \alpha(\alpha + 1) \ldots (\alpha + l - 1)$. So, in particular, we pick up complementary series representations in $s = s_l$. In [2] an explicit description is given of the embedding of the (finitely many) complementary series representations, corresponding to φ_{s_l}, into π_λ.

References

1. Berezin, F.A.: Quantization in complex symmetric spaces, *Math. USSR Izv.* **9** (1975) 341–379.
2. Van Dijk, G. and Hille, S.C.: *Canonical representations related to hyperbolic spaces*, Report no. 3, Institut Mittag Leffler, Djursholm, 1995/96.
3. Van Dijk, G. and Molchanov, V.F.: Tensor products of maximal degenerate representations of SL(n, $\mathbb{R}$), in preparation.
4. Erdélyi, A., et al.: *Tables of Integral Transforms*, **II**, McGraw-Hill, New York, 1954.
5. Faraut, J. and Koranyi, A.: Function spaces and reproducing kernels on bounded symmetric domains, *J. Funct. Anal.* **88** (1990) 64–89.
6. Hille, S.C.: Maximal degenerate representations of SL($n + 1$, $\mathbb{C}$), Preprint Leiden University, 1996.
7. Johnson, K.D.: Degenerate principal series representations and tube domains, *Contemporary Math.* **138** (1992) 175–187.
8. Molchanov, V.F.: On quantization on para-Hermitian symmetric spaces, In: *Adv. in Math. Sciences*, Amer. Math. Soc. (1996).
9. Ørsted, B. and Zhang, G.: Generalized principal series representations and tube domains, *Duke Math. J.* **78** (1995) 335–358.
10. Unterberger, A. and Upmeier, H.: The Berezin transform and invariant differential operators, Preprint, 1994.
11. Vershik, A.M., Gel'fand, I.M. and Graev, M.I.: Representations of the group SL(2, **R**) where **R** is a ring of functions, In: *Representation Theory*. Cambridge University Press, Cambridge, 1982.
12. Wolf, J.A.: The action of a real semisimple group on a complex flag manifold. I: orbit structure and holomorphic arc components, *Bull. Amer. Math. Soc.* **75** (1969) 1121–1236.

ASYMPTOTIC REPRESENTATION
OF DISCRETE GROUPS

ALEXANDER S. MISHCHENKO
Department of Mechanics and Mathematics,
Moscow State University, 119899 Moscow, Russia

AND

NOOR MOHAMMAD
Quaid-i-Azam University, Islamabad, Pakistan

Abstract. If one has a unitary representation $\rho : \pi \to U(H)$ of the fundamental group $\pi_1(M)$ of the manifold M, then one can do many useful things:

1) construct a natural vector bundle over M;
2) construct the cohomology groups with respect to the local system of coefficients;
3) construct the signature of manifold M with respect to the local system of coefficients;

and others. In particular, one can write the Hirzebruch formula, which compares the signature with the characteristic classes of the manifold M, further, based on this, find the homotopy invariant characteristic classes (i.e., the Novikov conjecture). Taking into account that the family of known representations is not sufficiently large, it would be interesting to extend this family to some larger one. Using the ideas of P. de la Harpe and M. Karoubi ([1]) and A. Connes, M. Gromov and H. Moscovici ([2]), a proper notion of asymptotic representation is defined.

Mathematics Subject Classification (1991): 22A25, 20C99, 46L89, 43A65, 55Q05, 55R40, 57R57.

Key words: asymptotic representations, discrete groups, Hirzebruch formula, characteristic classes, fundamental groups, C^*-algebras.

B. P. Komrakov et al. (eds.), Lie Groups and Lie Algebras, 299–312.

1. Introduction

Our interest concentrates on the problem of developing the representation theory for groups and C^*-algebras, especially for the group C^*-algebras of discrete groups. This problem is stimulated by the following very well known phenomenon. Let M be a closed oriented non simply connected manifold with fundamental group π. Let $B\pi$ be the classifying space for the group π and let f_M be a map inducing the isomorphism of fundamental groups. Then if one has a finite-dimensional unitary representation ρ of the finitely represented group π,

$$\rho : \pi \to U(H), \tag{1}$$

then one can construct in a natural way a vector bundle ξ_ρ over the classifying space $B\pi$ with the fiber H and, as a consequence, over an arbitrary manifold M with fundamental group π. The advantage of this construction lies for instance in the fact that one can write a generalization of the Hirzebruch formula for the signature of the manifold in terms of characteristic classes. In fact, if

$$\rho : \pi_1(M) \to U(n) \tag{2}$$

is a unitary representation, the cohomology with respect to the local system of coefficients generated by the representation ρ admits a nondegenerate quadratic form, and the signature, $\mathrm{sign}_\rho(M)$, is defined. One can verify that

$$\mathrm{sign}_\rho M^{4k} = 2^{2k} \langle \mathrm{ch}\rho L(M^{4k}), [M^{4k}] \rangle = \sigma_{\mathrm{ch}\rho}(M^{4k}), \tag{3}$$

where $\mathrm{ch}\rho$ is the Chern character of representation ρ. Since $\mathrm{sign}_\rho M^{4k}$ is a homotopy invariant, the right-hand side of equation (3), $\sigma_{\mathrm{ch}\rho}(M^{4k})$, is also homotopy invariant. Unfortunately the Chern character $\mathrm{ch}\rho$ is trivial for any finite-dimensional unitary representation ρ. Therefore, the way to construct nontrivial examples of $\mathrm{ch}\rho$ lies in considering indefinite metrics on H or infinite-dimensional (Fredholm) representations. In ([2]) A. Connes, M. Gromov and H. Moscovici ([2]) defined a new concept valid in the category of finite-dimensional representations. They introduced the notion of an almost flat bundle on the manifold M. Namely, let $\alpha \in K^0(M)$ be an element of the K-theory on a Riemannian manifold M, and let $\alpha = (E^+, \nabla^+) - (E^-, \nabla^-)$, where E^+, E^- are Hermitian complex vector bundles with connections ∇^+, ∇^-. One can say that the element α is *almost flat*, if for any $\varepsilon > 0$ there exists a representation $\alpha = (E^+, \nabla^+) - (E^-, \nabla^-)$ such that

$$\|(E^+, \nabla^+)\| \le \varepsilon, \ \|(E^-, \nabla^-)\| \le \varepsilon, \tag{4}$$

where

$$\|(E, \nabla)\| = \mathrm{Sup}_{x \in M}\{\|\theta_x(X \wedge Y)\| : \|X \wedge Y\| \le 1\}, \tag{5}$$

and $\theta = \nabla^2$ is the curvature form.

Then for almost flat element $\alpha \in K^0(M)$ and any $\varepsilon > 0$ and a finite subset $\pi^0 \subset \pi$, there exists a representation $\alpha = (E^+, \nabla^+) - (E^-, \nabla^-)$ with

$$\|(E^+, \nabla^+)\| \leq \varepsilon, \quad \|(E^-, \nabla^-)\| \leq \varepsilon, \tag{6}$$

such that the corresponding quasi-representations constructed via the holonomy on the family of paths which form elements of the fundamental group

$$\sigma^+ : \pi \to U(N^+), \quad \sigma^- : \pi \to U(N^-) \tag{7}$$

have the following property:

$$\|\sigma^+\|_{\pi^0} \leq \varepsilon, \quad \|\sigma^-\|_{\pi^0} \leq \varepsilon, \tag{8}$$

where

$$\|\sigma\|_{\pi^0} = \operatorname{Sup}\{\|\sigma(ab) - \sigma(a)\sigma(b)\| : a, b \in \pi^0\}. \tag{9}$$

This notion has a shortcoming, because the choice of quasi-representation depends on the property of almost flatness which in turn depends on the smooth structure of the manifold M. It would be interesting to construct a natural inverse correspondence from the family of quasi-representations of discrete group to the family of almost flat bundles.

Another problem related to quasi-representations was discussed in [3] by P. Halmos and later by Voiculescu ([4]), Loring ([5]), Exel and Loring ([6],[7]). Namely, whether a pair of unitary matrices A and B which almost commute (in the sense that the operator norm $\|AB - BA\|$ is small) can always be slightly perturbed in order to yield a pair of commuting unitary matrices. The answer is known to be false and the proof depends on the second cohomology of the two-torus. We shall show that, in fact, the problem of almost commuting unitary matrices is a special case of quasi-representation of group π when $\pi = Z \oplus Z$ and $B\pi$ is two-torus.

So the general problem can be formulated as follows: whether an almost representation can be perturbed in order to yield a classical representation. Then the obstruction to the existence of such a perturbation can be expressed as a characteristic class of a vector bundle, which can be constructed in a natural way (see Loring [5]). For certain reasons, described below, we prefer to consider the so-called asymptotic representation due to A. Connes.

2. Definitions of asymptotic representations

2.1. FINITE-DIMENSIONAL CASE: DISCRETE VERSION

There are at least two ways to define asymptotical representations. The first is the following:

Definition 1 Let $\varepsilon > 0$ and let $\pi^0 \subset \pi$ be a finite subset and

$$\sigma : \pi \to U(N) \tag{10}$$

a map such that

$$\sigma(g^{-1}) = \sigma(g)^{-1}, \tag{11}$$

$$\|\sigma\|_{\pi^0} = \sup\{\|\sigma(gh) - \sigma(g)\sigma(h)\| : g, h, gh \in \pi^0\} \le \varepsilon. \tag{12}$$

Then σ is called an ε-*almost representation* (with respect to the finite subset π^0).

Further we consider the so-called stable class of ε-almost representations σ. This means that the map σ can be substituted for the composition

$$\sigma \oplus 1 : \pi \to U(N) \to U(N+1). \tag{13}$$

It is evident that

$$\|\sigma\|_{\pi^0} = \|\sigma \oplus 1\|_{\pi^0}. \tag{14}$$

Therefore we shall regard as ε-almost representations maps

$$\sigma : \pi \to U(\infty), \tag{15}$$

which possess the properties (11), (12)

Definition 2 Let

$$\sigma = \{\sigma_n : \pi \to U(\infty)\} \tag{16}$$

be a sequence of maps such that $\sigma_n(g^{-1}) = \sigma_n(g)^{-1}$, and for any $\varepsilon > 0$ and any finite subset $\pi^0 \subset \pi$, there exists a number N_0, such that if $n > N_0$, then

$$\|\sigma_n(a) - \sigma_{n+1}(a)\| \le \varepsilon, \ a \in \pi^0, \tag{17}$$

$$\|\sigma_n\|_{\pi^0} \le \varepsilon. \tag{18}$$

Then the map σ is said to be an *asymptotic representation* of the group π.

There is another approach: Let π be a finitely represented group and let $(A; S)$ be a presentation of π by generators A and relations S. Let $F\langle A \rangle$ denote the free group generated by A.

Definition 3 Let $\sigma : F\langle A \rangle \to SU(\infty)$ be a representation such that $\|\sigma(s) - E\| < \varepsilon$, where $s \in S$. Then σ is called an ε-*almost representation* of the group π.

Definition 4 Let

$$\sigma = \{\sigma_n : F\langle A \rangle \to SU(\infty)\} \tag{19}$$

be a sequence of ε_n-almost representations of the group π such that

$$\lim_{n \to \infty} \varepsilon_n = 0,$$

and for any $a \in A$ one has

$$\lim_{n \to \infty} \|\sigma_n(a) - \sigma_{n+1}(a)\| = 0. \tag{20}$$

Then σ is called an *asymptotic representation* of the group π.

One can easily check that Definitions 2 and 4 are equivalent.

2.2. FINITE-DIMENSIONAL CASE: CONTINUOUS VERSION

Definition 5 Let

$$\sigma = \{\sigma_t : \pi \to U(\infty)\} \tag{21}$$

be a continuous family of maps, $0 \leq t < \infty$, such that $\sigma_t(g^{-1}) = \sigma_t(g)^{-1}$ and for any $\varepsilon > 0$ and any finite subset $\pi^0 \subset \pi$, there exists a number N_0, such that if $t > N_0$, then

$$\|\sigma_t\|_{\pi^0} \leq \varepsilon. \tag{22}$$

Then σ is said to be an *asymptotic representation* of the group π.

There is another definition: Let π be a finitely represented group and let $(A; S)$ be a presentation of π by generators A and relations S. Let $F\langle A \rangle$ denote the free group generated by A.

Definition 6 Let

$$\sigma = \{\sigma_t : F\langle A \rangle \to SU(\infty)\} \tag{23}$$

be a continuous family of ε_t-almost representations of the group π such that $\lim \varepsilon_t = 0$, $t \to \infty$. Then σ is said to be an *asymptotic representation* of the group π.

One can easily check that Definitions 2 and 6 are equivalent. Namely, one has the following

Theorem 1 *Let σ_t be an asymptotic representation of the group π in the sense of Definition 6 Then there exists a sequence t_n such that σ_{t_n} is asymptotic representation in the sense of the Definition 2.*

Conversely, let σ_n be an asymptotic representation of the group π in the sense of Definition 2. Then there exist an asymptotic representation σ_t in the sense of Definition 6, and a sequence t_n such that $\sigma_n = \sigma_{t_n}$.

2.3. THE RING OF ASYMPTOTIC REPRESENTATIONS

The family of asymptotic representations can be transformed into the Witt group $\mathcal{R}_a(\pi)$. Namely, the family $\{\sigma\}$ admits natural operations, i.e., direct sum and tensor product. Let us introduce two equivalence relations between two asymptotic representations. The first equivalence relation is just a homotopy between the two asymptotic representations. Let $\{\sigma_t\}$,

$0 \leq t \leq 1$, be a family of asymptotic representations such that for any $g \in \pi$, the functions $\{\sigma_{t,n}(g)\}$ are continuous and the following property of uniformity holds:

$$\lim_{n \to \infty} \max_{\{t\}} \|\sigma_{t,n}\|_{\pi^0} = 0. \tag{24}$$

Then σ_0 and σ_1 are said to be *homotopic*.

The second relation admits change of basis in the space. Namely, the representations $\{\sigma_n\}$ and $\{U_n \circ \sigma_n \circ U_n^{-1}\}$ are called *equivalent*, if U_n is any sequence of unitary operators from $U(\infty)$ satisfying the property

$$\lim_{n \to \infty} \|U_n - U_{n+1}\| = 0.$$

One can prove that the second relation follows from the first one.

Then the Witt group generated by $\{\sigma\}$ modulo the equivalence relations, as defined above, will be denoted by $\mathcal{R}_a(\pi)$.

3. Construction of vector bundle

Now we want to construct a natural homomorphism

$$\phi : \mathcal{R}_a(\pi) \to K(B\pi). \tag{25}$$

Let M be a finite CW-complex with fundamental group π, i.e.,

$$\pi_1(M) = \pi, \tag{26}$$

and let

$$f_M : M \to B\pi \tag{27}$$

be the map f_M defined previously. Let $\tilde{M}$ be the universal covering of M, and $p : \tilde{M} \to M$ the projection. Let $\bar{M} \subset \tilde{M}$ be a fundamental domain, that is, a finite closed subcomplex such that $p(\bar{M}) = M$. Put

$$\pi^0 = \{g \in \pi : g(\bar{M}) \cap \bar{M} \neq \emptyset\}. \tag{28}$$

One can construct elements α of the group $K(B\pi)$ by using their restrictions to the spaces like M. More precisely, consider the category $\mathcal{B}\pi$ whose objects are CW-complexes M with fundamental group π, and morphisms are the homotopy classes of maps $g : M_1 \to M_2$ such that the diagram

$$\begin{array}{ccc} M_1 & \xrightarrow{f_{M_1}} & B\pi \\ \downarrow{g} & & \downarrow{=} \\ M_2 & \xrightarrow{f_{M_2}} & B\pi \end{array} \tag{29}$$

commutes. Consider the natural correspondence α which associates to any space $M \in B\pi$ a vector bundle $\alpha(M) \in K(M)$ such that for any morphism $g : M_1 \to M_2$ one has

$$\alpha(M_1) = g^*(\alpha(M_2)). \tag{30}$$

Then it is clear that there is a one-to-one correspondence between the set $\{\alpha\}$ and the set of vector bundles over $B\pi$. For the sake of simplicity, we shall restrict our considerations only to finite CW-complexes M from $B\pi$.

To construct a vector bundle over M, we proceed as follows. Consider the trivial vector bundle over $\tilde{M}$, $\xi = \tilde{M} \times C^n$, and an action of the group π which is compatible with the action on the base $\tilde{M}$. This action can be described by a matrix function

$$T_g(x) : \xi_x \to \xi_{gx}, \; x \in \tilde{M}, \; g \in \pi, \tag{31}$$

satisfying the condition

$$T_g(hx) = T_{gh}(x) \circ T_h^{-1}(x), \; x \in \tilde{M}, \; g, h \in \pi. \tag{32}$$

The function (31) with the property (32) is called a *transition function.* From (32) one has

$$T_e(x) = E, \; x \in \tilde{M}, \tag{33}$$

where $e \in \pi$ is the neutral element, and E is the identity matrix. It is clear that the condition (32) is related to separate orbits $\pi(x_0)$ of the action of π on the space $\tilde{M}$. Moreover, the function (31) determines its value only at one fixed point $x_0 \in \pi(x_0)$ by using condition (33). In fact, if the function (31) is defined at the point $x_0 \in \pi(x_0)$ for all $g \in \pi$ and satisfies condition (33), then one can extend the function $T_g(x)$ to an arbitrary point $x \in \pi(x_0)$ using the formula (32):

$$T_g(x) = T_{gu}(x_0) \circ T_u^{-1}(x_0), \; x \in \tilde{M}, \; g, u \in \pi, \; x = ux_0. \tag{34}$$

If $x = x_0$, then in (34) one has $u = e$, and hence $T_g(x) = T_{ge}(x_0) \circ E = T_g(x_0)$. We must verify the condition (32) for $x = ux_0$ and $y = hx = hux_0$. We have

$$T_g(hx) = T_g(hux_0) = T_{ghu}(x_0) \circ T_{hu}^{-1}(x_0), \tag{35}$$
$$T_{gh}(x) = T_{ghu}(x_0) \circ T_u^{-1}(x_0), \tag{36}$$
$$T_h(x) = T_{hu}(x_0) \circ T_u^{-1}(x_0). \tag{37}$$

Therefore

$$
\begin{aligned}
T_g(hx) &= T_{ghu}(x_0) \circ T_{hu}^{-1}(x_0) = \\
&= T_{ghu}(x_0) \circ T_u^{-1}(x_0) \circ (T_u^{-1}(x_0))^{-1} \circ (T_{hu}(x_0))^{-1} = \\
&= T_{ghu}(x_0) \circ T_u^{-1}(x_0) \circ (T_{hu}(x_0) \circ T_u^{-1}(x_0))^{-1} = \\
&= T_{gh}(x) \circ T_h(x).
\end{aligned} \tag{38}
$$

Let $\bar{M} \subset \tilde{M}$ be the fundamental domain and let π^0 be as given in (28). Then one can consider the restriction

$$\bar{T}_g(x) = T_g(x), \ g \in \pi^0, \ x \in \bar{M}, gx \in \bar{M}. \tag{39}$$

The function (39) satisfies the condition (32) for all admissible x, g, h, where

$$x, hx, ghx \in \bar{M}, \ g, h \in \pi^0. \tag{40}$$

The function (39) with property (40) defines a vector bundle over M as well, and will also be called transition function. Therefore one can define the transition function only for a fixed point $x_0 \in (\bar{M} \cap \pi(x_0))$ and for all $g \in \pi$ such that $gx_0 \in \bar{M}$. If $x, gx \in \bar{M} \cap \pi(x_0)$, then $x = hx_0$, and hence $h, g, gh \in \pi^0$. One can use formula (32) to define the values of the transition function for all admissible x, g. Therefore for the construction of the vector bundle $\phi(\sigma)$ as a family of the transition functions, one should start from the zero-dimensional skeleton. Put

$$\bar{T}_g(x_0) = \sigma(g) \tag{41}$$

for a representative x_0 of each zero-dimensional orbit. To extend the transition function from the zero-dimensional skeleton to simplices of higher dimension, one can use the property that the action of the group π is free and the property (12) for $\varepsilon = 1$. Indeed, in the general case one should do the following.

Let us choose representatives $\{a_\alpha\}$ in each orbit of the set $[\tilde{M}]_0$ of vertices. Then the set $[\bar{M}]_0 = [\tilde{M}]_0 \cap \bar{M}$ has the property that $g(\bar{M}) \cap \bar{M} \neq \emptyset$ is equivalent to $g([\bar{M}]_0) \cap [\bar{M}]_0 \neq \emptyset$. Therefore there are elements $g_a \in \pi^0$ such that for any element $a \in [\bar{M}]_0$, there is an α_a such that

$$a = g_a(a_{\alpha_a}). \tag{42}$$

Consider a simplex
$$\sigma = (b_0, b_1, \ldots, b_n) \subset \bar{M}.$$

This means that $\{b_0, b_1, \ldots, b_n\} \subset [\bar{M}]_0$. Then, according to (28), one has

$$b_i = g_i(a_{\alpha_i}), \ i = 0, 1, \ldots, n; \ g_i \in \pi^0, \tag{43}$$

and as before,

$$T_g(b_i) = T_{gg_i}(a_{\alpha_i}) \circ T_{g_i}^{-1}(a_{\alpha_i}) = \sigma(gg_i) \circ \sigma^{-1}(g_i), gg_i \in \pi^0. \tag{44}$$

Therefore for any i

$$\|T_g(b_i) - \sigma(g)\| = \|\sigma(gg_i) \circ \sigma(g_i^{-1}) - \sigma(g)\| < \varepsilon. \tag{45}$$

Let $x \in \sigma$ be a point with the barycentric coordinates $\lambda_0, \lambda_1, \ldots, \lambda_n$,

$$x = \sum_i \lambda_i b_i. \quad \lambda_i \geq 0, \quad \sum_i \lambda_i = 1. \tag{46}$$

We put

$$T_g(x) = \sum_{i=0}^{n} \lambda_i T_g(b_i). \tag{47}$$

According to (48) we have

$$\|F_x \circ G - 1\| < 1, \quad (\varepsilon = 1(!)). \tag{48}$$

The next problem is to establish when two ε-almost representations give the same vector bundles over $B\pi$.

4. The first Chern class of a vector bundle

Definition 4 is more suitable for calculating the first Chern class

$$c_1(\phi(\sigma)) \in H^2(B\pi; R).$$

Namely, the first Chern class can be described as a two-dimensional cocycle, that is, as a function defined on the family of two-dimensional cells of the space $B\pi$. Let X be a finite CW-complex with fundamental group π. Let the presentation (A,S) be induced from 2-skeleton of some simplicial structure of X.

Theorem 2 *The first Chern class $c_1(\phi(\sigma))$ can be described as the cocycle*

$$c_1(s) = \frac{\mathrm{tr}(\log(\sigma(s)))}{2\pi i} \in Z, \tag{49}$$

where s runs through all 2-dimensional cells.

As a matter of fact, the construction of the vector bundle $f_M^*(\phi(\sigma))$ over a compact manifold with fundamental group π depends only on some sufficiently small ε_n, and coincides with the classical one for $\varepsilon = 0$. The magnitude of admissible ε can be estimated by $\exp(-\dim(M))$.

5. Asymptotic representations of $Z \oplus Z$.

Consider the example of fundamental group that seems to be the simplest nontrivial case for which there exists an asymptotic representation giving a nontrivial vector bundle over the classifying space. Let $\pi = Z \oplus Z$, and let a, b $\in \pi$ be generators of the group such that $ab = ba$. To construct an asymptotic representation, it suffices to define two continuous matrix-valued functions, $A(t)$, $B(t) \in U(\infty)$, $0 \le t \le \infty$ such that

$$\lim_{t\to\infty} \|A(t)B(t) - B(t)A(t)\| = 0. \tag{50}$$

First of all we shall construct a discrete series of unitary matrices

$$A_n, \ B_n \in U(n). \tag{51}$$

Let A_n be a matrix of cyclic permutation of the orthonormal basis in C^n:

$$A_n = \begin{pmatrix} 0 & 0 & 0 & \cdots & 0 & 1 \\ 1 & 0 & 0 & \cdots & 0 & 0 \\ 0 & 1 & 0 & \cdots & 0 & 0 \\ \vdots & \vdots & \vdots & & \vdots & \vdots \\ 0 & 0 & 0 & \cdots & 0 & 0 \\ 0 & 0 & 0 & \cdots & 1 & 0 \end{pmatrix}, \tag{52}$$

and let B_n be the diagonal matrix

$$B_n = \begin{pmatrix} \lambda_1 & 0 & 0 & \cdots & 0 & 0 \\ 0 & \lambda_2 & 0 & \cdots & 0 & 0 \\ 0 & 0 & \lambda_3 & \cdots & 0 & 0 \\ \vdots & \vdots & \vdots & & \vdots & \vdots \\ 0 & 0 & 0 & \cdots & \lambda_{n-1} & 0 \\ 0 & 0 & 0 & \cdots & 0 & \lambda_n \end{pmatrix}. \tag{53}$$

Then the commutator has the following form

$$A_n B_n - B_n A_n =$$

$$= \begin{pmatrix} 0 & 0 & 0 & \cdots & 0 & \lambda_n \\ \lambda_1 & 0 & 0 & \cdots & 0 & 0 \\ 0 & \lambda_2 & 0 & \cdots & 0 & 0 \\ \vdots & \vdots & \vdots & & \vdots & \vdots \\ 0 & 0 & 0 & \cdots & 0 & 0 \\ 0 & 0 & 0 & \cdots & \lambda_{n-1} & 0 \end{pmatrix} - \begin{pmatrix} 0 & 0 & 0 & \cdots & 0 & \lambda_1 \\ \lambda_2 & 0 & 0 & \cdots & 0 & 0 \\ 0 & \lambda_3 & 0 & \cdots & 0 & 0 \\ \vdots & \vdots & \vdots & & \vdots & \vdots \\ 0 & 0 & 0 & \cdots & 0 & 0 \\ 0 & 0 & 0 & \cdots & \lambda_n & 0 \end{pmatrix} =$$

$$= \begin{pmatrix} 0 & 0 & 0 & \cdots & 0 & \lambda_n - \lambda_1 \\ \lambda_1 - \lambda_2 & 0 & 0 & \cdots & 0 & 0 \\ 0 & \lambda_2 - \lambda_3 & 0 & \cdots & 0 & 0 \\ \vdots & \vdots & \vdots & & \vdots & \vdots \\ 0 & 0 & 0 & \cdots & 0 & 0 \\ 0 & 0 & 0 & \cdots & \lambda_{n-1} - \lambda_n & 0 \end{pmatrix}. \tag{54}$$

The next step is to frame the matrices A_n and B_n by an additional row and a column:

$$\tilde{A}_n = \begin{pmatrix} A_n & 0 \\ 0 & 1 \end{pmatrix} = \begin{pmatrix} 0 & 0 & 0 & \cdots & 0 & 1 & 0 \\ 1 & 0 & 0 & \cdots & 0 & 0 & 0 \\ 0 & 1 & 0 & \cdots & 0 & 0 & 0 \\ \vdots & \vdots & \vdots & & \vdots & \vdots & \\ 0 & 0 & 0 & \cdots & 0 & 0 & 0 \\ 0 & 0 & 0 & \cdots & 1 & 0 & 0 \\ 0 & 0 & 0 & \cdots & 0 & 0 & 1 \end{pmatrix}, \tag{55}$$

$$\tilde{B}_n = \begin{pmatrix} B_n & 0 \\ 0 & 1 \end{pmatrix} = \begin{pmatrix} \lambda_1 & 0 & 0 & \cdots & 0 & 0 & 0 \\ 0 & \lambda_2 & 0 & \cdots & 0 & 0 & 0 \\ 0 & 0 & \lambda_3 & \cdots & 0 & 0 & 0 \\ \vdots & \vdots & \vdots & & \vdots & \vdots & \vdots \\ 0 & 0 & 0 & \cdots & \lambda_{n-1} & 0 & 0 \\ 0 & 0 & 0 & \cdots & 0 & \lambda_n & 0 \\ 0 & 0 & 0 & \cdots & 0 & 0 & 1 \end{pmatrix}. \tag{56}$$

The commutator for the framed matrices has the form

$$\tilde{A}_n\tilde{B}_n - \tilde{B}_n\tilde{A}_n = \begin{pmatrix} A_nB_n - B_nA_n & 0 \\ 0 & 1 \end{pmatrix} =$$

$$= \begin{pmatrix} 0 & 0 & 0 & \cdots & 0 & \lambda_n - \lambda_1 & 0 \\ \lambda_1 - \lambda_2 & 0 & 0 & \cdots & 0 & 0 & 0 \\ 0 & \lambda_2 - \lambda_3 & 0 & \cdots & 0 & 0 & 0 \\ \vdots & \vdots & \vdots & & \vdots & \vdots & \vdots \\ 0 & 0 & 0 & \cdots & 0 & 0 & 0 \\ 0 & 0 & 0 & \cdots & \lambda_{n-1} - \lambda_n & 0 & 0 \\ 0 & 0 & 0 & \cdots & 0 & 0 & 1 \end{pmatrix}. \tag{57}$$

So the norms of two commutators coincide, i.e.,

$$\|A_nB_n - B_nA_n\| = \|\tilde{A}_n\tilde{B}_n - \tilde{B}_n\tilde{A}_n\| = \max\{|\lambda_1 - \lambda_2|, \ldots, |\lambda_n - \lambda_1|\}. \tag{58}$$

Put

$$\lambda_k = \exp(\frac{2\pi ik}{n}), \; k = 1, \ldots, n. \tag{59}$$

Then $\lambda_n = 1$ and

$$|\lambda_k - \lambda_{k+1}| \le \frac{2\pi}{n}. \tag{60}$$

The next step is to connect the pair $(\tilde{A}_n, \tilde{B}_n)$ and the pair (A_{n+1}, B_{n+1}) by continuous path $(\bar{A}_t, \bar{B}_t)$ satisfying the inequality

$$\|\bar{A}_t\bar{B}_t - \bar{B}_t\bar{A}_t\| \le \frac{2\pi}{n}. \tag{61}$$

Put

$$
\bar{A}_t =
\begin{pmatrix}
0 & 0 & 0 & \cdots & 0 & \cos(2\pi t) & \sin(2\pi t) \\
1 & 0 & 0 & \cdots & 0 & 0 & 0 \\
0 & 1 & 0 & \cdots & 0 & 0 & 0 \\
\vdots & \vdots & \vdots & & \vdots & \vdots & \\
0 & 0 & 0 & \cdots & 0 & 0 & 0 \\
0 & 0 & 0 & \cdots & 1 & 0 & 0 \\
0 & 0 & 0 & \cdots & 0 & -\sin(2\pi t) & \cos(2\pi t)
\end{pmatrix},
\tag{62}
$$

$$
\bar{B}_t = \tilde{B}_n =
\begin{pmatrix}
\lambda_1 & 0 & 0 & \cdots & 0 & 0 & 0 \\
0 & \lambda_2 & 0 & \cdots & 0 & 0 & 0 \\
0 & 0 & \lambda_3 & \cdots & 0 & 0 & 0 \\
\vdots & \vdots & \vdots & & \vdots & \vdots & \vdots \\
0 & 0 & 0 & \cdots & \lambda_{n-1} & 0 & 0 \\
0 & 0 & 0 & \cdots & 0 & 1 & 0 \\
0 & 0 & 0 & \cdots & 0 & 0 & 1
\end{pmatrix}.
\tag{63}
$$

Then one has

$$
\bar{A}_t \bar{B}_t - \bar{B}_t \bar{A}_t =
$$

$$
=
\begin{pmatrix}
0 & 0 & 0 & \cdots & 0 & \cos(2\pi t) & \sin(2\pi t) \\
\lambda_1 & 0 & 0 & \cdots & 0 & 0 & 0 \\
0 & \lambda_2 & 0 & \cdots & 0 & 0 & 0 \\
\vdots & \vdots & \vdots & & \vdots & \vdots & \vdots \\
0 & 0 & 0 & \cdots & 0 & 0 & 0 \\
0 & 0 & 0 & \cdots & \lambda_{n-1} & 0 & 0 \\
0 & 0 & 0 & \cdots & 0 & -\sin(2\pi t) & \cos(2\pi t)
\end{pmatrix}
-
$$

$$
-
\begin{pmatrix}
0 & 0 & 0 & \cdots & 0 & \lambda_1 \cos(2\pi t) & \lambda_1 \sin(2\pi t) \\
\lambda_2 & 0 & 0 & \cdots & 0 & 0 & 0 \\
0 & \lambda_3 & 0 & \cdots & 0 & 0 & 0 \\
\vdots & \vdots & \vdots & & \vdots & \vdots & \vdots \\
0 & 0 & 0 & \cdots & 0 & 0 & 0 \\
0 & 0 & 0 & \cdots & \lambda_n & 0 & 0 \\
0 & 0 & 0 & \cdots & 0 & -\sin(2\pi t) & \cos(2\pi t)
\end{pmatrix}
=
$$

$$
=
\begin{pmatrix}
0 & \cdots & 0 & \nu_n \cos(2\pi t) & \nu_n \sin(2\pi t) \\
\nu_1 & \cdots & 0 & 0 & 0 \\
0 & \cdots & 0 & 0 & 0 \\
\vdots & & \vdots & \vdots & \vdots \\
0 & \cdots & 0 & 0 & 0 \\
0 & \cdots & \nu_{n-1} & 0 & 0 \\
0 & \cdots & 0 & 0 & 0
\end{pmatrix},
\tag{64}
$$

where

$$\nu_k = \lambda_k - \lambda_{k+1}, \; k = 1, \ldots, n; \; \lambda_{n+1} = \lambda_1.$$

When λ_k are given by (59), the inequality (61) holds.

The second part of the path connects the pair $(\bar{A}_1 = A_{n+1}, \; \bar{B}_1 = \tilde{B}_n)$ with the pair $A_{n+1}, \; B_{n+1}$. Let $\bar{A}_t = A_{n+1}$ be constant and let $\bar{B}_t$ be the diagonal matrix, i.e.,

$$\bar{B}_t = \begin{pmatrix} \lambda_1(t) & 0 & 0 & \cdots & 0 & 0 & 0 \\ 0 & \lambda_2(t) & 0 & \cdots & 0 & 0 & 0 \\ 0 & 0 & \lambda_3(t) & \cdots & 0 & 0 & 0 \\ \vdots & \vdots & \vdots & & \vdots & \vdots & \vdots \\ 0 & 0 & 0 & \cdots & \lambda_{n-1}(t) & 0 & 0 \\ 0 & 0 & 0 & \cdots & 0 & \lambda_n(t) & 0 \\ 0 & 0 & 0 & \cdots & 0 & 0 & \lambda_{n+1}(t) \end{pmatrix}, \quad (65)$$

where

$$\lambda_k(t) = \exp 2\pi i k((1 - t)\frac{1}{n} + t\frac{1}{n+1}), \; 1 \le k \le n$$

$$\lambda_{n+1}(t) = 1. \tag{66}$$

Therefore when $t = 1$, one has

$$\bar{B}_1 = B_{n+1}. \tag{67}$$

Thus we have constructed a continuous family of (stable) matrices (A_t, B_t) such that

$$\|A_t B_t - B_t A_t\| \le \frac{2\pi}{[t]}, \tag{68}$$

where A_n, B_n coincide with the discrete sequence. Now let us apply formula (49) to calculate the Chern class of the asymptotic representation constructed above. Since

$$s = aba^{-1}b^{-1}, \tag{69}$$

and hence

$$\sigma_n(s) = A_n B_n A_n^{-1} B_n^{-1}. \tag{70}$$

It is easy to verify that

$$A_n B_n A_n^{-1} B_n^{-1} = \begin{pmatrix} \frac{\lambda_1}{\lambda_2} & 0 & 0 & \cdots & 0 & 0 \\ 0 & \frac{\lambda_2}{\lambda_3} & 0 & \cdots & 0 & 0 \\ 0 & 0 & \frac{\lambda_3}{\lambda_4} & \cdots & 0 & 0 \\ \vdots & \vdots & \vdots & & \vdots & \vdots \\ 0 & 0 & 0 & \cdots & \frac{\lambda_{n-1}}{\lambda_n} & 0 \\ 0 & 0 & 0 & \cdots & 0 & \frac{\lambda_n}{\lambda_1} \end{pmatrix} =$$

$$= \begin{pmatrix} \exp(\frac{-2\pi i}{n}) & 0 & \cdots & 0 & 0 \\ 0 & \exp(\frac{-2\pi i}{n}) & \cdots & 0 & 0 \\ 0 & 0 & \cdots & 0 & 0 \\ \vdots & \vdots & & \vdots & \vdots \\ 0 & 0 & \cdots & \exp(\frac{-2\pi i}{n}) & 0 \\ 0 & 0 & \cdots & 0 & \exp(\frac{-2\pi i}{n}) \end{pmatrix}. \tag{71}$$

Thus

$$c_1(s) = \frac{\operatorname{tr}(\log(\sigma(s)))}{2\pi} = 1. \tag{72}$$

Consequently, one has

Theorem 3 *The homomorphism*

$$\varphi : \mathcal{R}_a(Z \oplus Z) \to K(T^2)$$

is surjective.

References

1. Karoubi, M. and de la Harpe, P.: Representations approchees d'un groupe dans une algebre de Banach, *Manuscripta Math.* **22** (1978) 293–310.
2. Connes, A., Gromov, M. and Moscovoci, H.: Conjecture de Novikov et fibrés presque plats, *C. R. Acad. Sci. Paris, Série I* **310** (1990) 273–277.
3. Halmos, P.R.: Some unsolved problems of unknown depth about operators on Hilbert space, *Proc. Roy. Soc. Edinburgh Sect.* **A 76** (1976) 67–76.
4. Voiculescu, D.: Remarks on the singular extension in the C^*-algebra of the Heisenberg group, *J. Operator Theory* **5** (1981) 147–170.
5. T.A.Loring, K-theory and asymptotically commuting matrices, *Canad. J. Math.* **40** (1988) 197–216.
6. Exel, R. and Loring, T.: Almost commuting unitary matrices, *Proc. Amer. Math. Soc.* **106** (1989) no. 4, 913–915.
7. Exel, R. and Loring, T.: Invariants of almost commuting unitaries, *J. Funct. Analysis* **95** (1991) no. 2, 364–376.
8. Stern, A.: Quasisymmetry. I, *Russian Journal of Math. Phys.* **2** (1993) no. 3, 353–382.

MAXIMAL DEGENERATE SERIES REPRESENTATIONS OF THE UNIVERSAL COVERING OF THE GROUP $SU(n,n)$

V.F. MOLCHANOV
Tambov State University,
Internatsionalnaya 33, 392622 Tambov, Russia.
E-mail: molchanov@math-univ.tambov.su

Abstract. We study representations $T_{\mu,\tau}$ $(\mu, \tau \in \mathbb{C})$ of the universal covering group $\widetilde{G}$ of $G = SU(n,n)$ induced by characters of a maximal parabolic subgroup $\widetilde{P}$ ($\widetilde{G}/\widetilde{P}$ is the Shilov boundary of G/K): composition series, intertwining operators, invariant sesqui-linear forms, realizations on holomorphic functions, etc.

Mathematics Subject Classification (1991): 22E46, 43A85.

Key words: Hermitian symmetric spaces, maximal degenerate series, composition series, intertwining operators, unitarizability, Fock space.

In this paper we study a maximal degenerate series of representations $T_{\mu,\tau}$ $(\mu, \tau \in \mathbb{C})$ of the universal covering group $\widetilde{G}$ of the group $G = SU(n,n)$. They are induced by characters of a maximal parabolic subgroup $\widetilde{P}$ (here $\widetilde{G}/\widetilde{P}$ is the Shilov boundary of G/K). We determine when these representations $T_{\mu,\tau}$ are irreducible and completely describe the structure of the invariant subspaces (composition series) in the reducible case. It is remarkable that this structure can be shown by means of a plane picture. Here a natural instrument is a notion of barrier, which was used by the author earlier for the description of representations of the group $SO_0(p,q)$, see, for example, [7]. Next we find all the intertwining operators for the representations $T_{\mu,\tau}$ and their subfactors and write an integral expression for these operators as well as explicit expressions for the eigenvalues. Further we determine when an invariant sesqui-linear form exists for the pair of representations $T_{\mu,\tau}$, T_{μ_1,τ_1} (or their subfactors), when an invariant Hermitian form exists for the representation $T_{\mu,\tau}$ (or its subfactors), and when

B. P. Komrakov et al. (eds.), Lie Groups and Lie Algebras, 313–336.

the latter form is positive definite (i.e., the corresponding representation is unitarizable). Finally, we indicate which of the representations considered can be realized on spaces (or subfactors) of functions holomorphic on $\Omega = G/K$, the Cartan domain of type I. For this realization we find invariant Hermitian forms (if any) and determine unitary representations.

The problem of describing maximal degenerate series of representations of the group G (or $\widetilde{G}$) for arbitrary Hermitian symmetric spaces G/K was considered by many authors — by different methods and in different generality, see, for example, [1, 11, 10, 12, 8, 2, 4, 6, 9] and also references in [4]. A part of these publications is devoted to studying representations of G (or $\widetilde{G}$) that can be realized on spaces of functions holomorphic on Ω (Fock spaces), the analytic continuation of holomorphic discrete series: [1, 11, 12, 8, 4]. All such unitary representations for Ω of classical type were first found in [1]. They are labelled by a real parameter which ranges over a half line on $\mathbb{R}$ and also over a finite set (n points, $n = \operatorname{rank} G/K$). Incidentally, this finite set, called Wallach set in some papers ([11]), should be called the Berezin–Wallach set, to be fair.

The papers [6, 9] consider spaces G/K of tube type. In [6] representations of G (not $\widetilde{G}$) are studied, and in [9] representations of $\widetilde{G}$ with real parameters (μ, τ in our notation) are considered for the case in which these representations have an invariant Hermitian form.

In [2] the complementary series for the group $SU(n,n)$ itself (also for $SO_0(n,n)$ and $Sp(n,n)$) was treated.

1. The group $SU(n,n)$

The group $G = SU(n,n)$ consists of matrices $g \in SL(2n, \mathbb{C})$ preserving a Hermitian form with the signature (n,n). As such a form, we take the form with the matrix

$$I = \begin{pmatrix} -E & 0 \\ 0 & E \end{pmatrix},$$

where E denotes the identity matrix of order n. Thus, we have $\bar{g}'Ig = I$, where the prime denotes matrix transposition. Let us write the matrix g in $n \times n$ block form:

$$g = \begin{pmatrix} \alpha & \beta \\ \gamma & \delta \end{pmatrix}. \tag{1.1}$$

The Lie algebra $\mathfrak{g}$ of the group G consists of complex zero trace matrices X of order $2n$ satisfying the condition $\bar{X}'I + IX = 0$. Let us denote by $\operatorname{Herm}(n)$ the space of Hermitian matrices of order n. In block form we can write

$$X = \begin{pmatrix} i\xi & \eta \\ \bar{\eta}' & i\zeta \end{pmatrix}, \tag{1.2}$$

where $\xi, \eta \in \mathrm{Herm}(n), \mathrm{tr}(\xi + \zeta) = 0, \eta \in \mathrm{Mat}(n, \mathbb{C})$.

The automorphism θ of G defined by $\theta(g) = IgI$ is a Cartan involution. The subgroup K of fixed points of θ consists of block diagonal matrices:

$$k = \begin{pmatrix} \varphi & 0 \\ 0 & \psi \end{pmatrix}, \tag{1.3}$$

where $\varphi, \psi \in U(n), \det\varphi \cdot \det\psi = 1$; it is a maximal compact subgroup of G. Note that the automorphism θ is inner: indeed, $\theta(g) = (iI)g(iI)^{-1}$.

The Lie algebra $\mathfrak{k}$ of K consists of block diagonal matrices (1.2):

$$X = \begin{pmatrix} i\xi & 0 \\ 0 & i\zeta \end{pmatrix}, \tag{1.4}$$

where $\xi, \zeta \in \mathrm{Herm}(n), \mathrm{tr}(\xi + \zeta) = 0$. It has a one-dimensional center with the basis

$$X_0 = (i/2)\, I. \tag{1.5}$$

The semisimple part $\mathfrak{k}^{(s)} = [\mathfrak{k}, \mathfrak{k}]$ consists of matrices (1.4) for which $\mathrm{tr}\,\xi = \mathrm{tr}\,\zeta = 0$. The corresponding subgroup $K^{(s)}$ of G consists of matrices (1.3) for which $\det\varphi = \det\psi = 1$; it is isomorphic to $SU(n) \times SU(n)$. The center $Z(G)$ of G is a cyclic group of order $2n$ generated by εE, where $\varepsilon = \exp(\pi i/n)$. The intersection $Z^{(s)}(G) = Z(G) \cap K^{(s)}$ is a cyclic group of order n generated by $\varepsilon^2 E$.

Now let us consider the second realization G_1 of the group $SU(n, n)$ (the realization above will be referred as the first one). The group G_1 consists of matrices $g_1 \in SL(2n, \mathbb{C})$ preserving the Hermitian form with the matrix

$$J = \begin{pmatrix} 0 & -iE \\ iE & 0 \end{pmatrix}.$$

The matrices I and J are related by $J = C^{-1}IC$, where

$$C = \frac{1}{\sqrt{2}} \begin{pmatrix} E & iE \\ iE & E \end{pmatrix}, \quad C^{-1} = \bar{C}'.$$

Therefore

$$g_1 = C^{-1}gC \tag{1.6}$$

and, in detail,

$$\begin{pmatrix} \alpha_1 & \beta_1 \\ \gamma_1 & \delta_1 \end{pmatrix} = \frac{1}{2} \begin{pmatrix} \alpha + i\beta - i\gamma + \delta & i\alpha + \beta + \gamma - i\delta \\ -i\alpha + \beta + \gamma + i\delta & \alpha - i\beta + i\gamma + \delta \end{pmatrix}. \tag{1.7}$$

We shall adhere to the following notation system: if A is a subgroup of G, then A_1 is the subgroup of G_1 corresponding to A under (1.6), and vice versa. The same concerns Lie algebras, automorphisms, etc. We denote the identity component of a group by the same symbol with the subscript e.

Let σ_1 be the automorphism of G_1 defined by $\sigma_1(g_1) = I g_1 I$. The fixed point subgroup H_1 is block diagonal:

$$h_1 = \begin{pmatrix} c_1 & 0 \\ 0 & d_1 \end{pmatrix}, \tag{1.8}$$

where $c_1 = \bar{d}_1^{\,-1}$; moreover $\det h_1 = 1$ gives $\det d_1 \in \mathbb{R}$. So H_1 is isomorphic to $SL(n, \mathbb{C}) \cdot \mathbb{R}^*$. The Lie algebra $\mathfrak{h}_1$ of H_1 is block diagonal: $X = \text{diag}\{a, -\bar{a}'\}$, $\text{tr}\, a \in \mathbb{R}$. It has the one-dimensional center with the matrix $(1/2)\, I$ as a basis; in the first realization this basis is

$$Y_0 = (-1/2)\, J. \tag{1.9}$$

Let $L = K \cap H$. This subgroup consists of block diagonal matrices of the following form:

$$l = \begin{pmatrix} w & 0 \\ 0 & w \end{pmatrix}, \tag{1.10}$$

where $w \in U(n)$, $\det w^2 = 1$, so that $\det w = \pm 1$. The map (1.6) preserves matrices l: $l_1 = l$, so that $L_1 = L$. It consists of two connected parts, L_e consists of matrices (1.10) with $\det w = 1$, so that $L_e \subset K^{(s)}$. The group L contains $Z(G)$ and L_e contains $Z^{(s)}(G)$. The space K/L can be identified with the unitary group $U(n)$ by means of the following action of elements (1.3):

$$u \mapsto \tilde{u} = u \cdot k = \psi^{-1} u \varphi \tag{1.11}$$

since L is the stabilizer of the points λE. As the base point, let us take $-iE$, then the projection $K \to U(n)$ is

$$k \mapsto u = -i \psi^{-1} \varphi. \tag{1.12}$$

The Lie algebra $\mathfrak{g}_1$ of G_1 decomposes into the direct sum of ± 1-eigenspaces of σ_1: $\mathfrak{g}_1 = \mathfrak{h}_1 + \mathfrak{q}_1$. Under the adjoint action of H_1, the subspace $\mathfrak{q}_1$ splits into the direct sum of two irreducible subspaces $\mathfrak{q}_1^+$ and $\mathfrak{q}_1^-$ consisting of upper and lower block nilpotent matrices respectively, which are Abelian subalgebras of $\mathfrak{g}_1$.

Let us denote $Q_1^\pm = \exp \mathfrak{q}_1^\pm$. The subgroups $P_1^\pm = H Q_1^\pm = Q_1^\pm H$ are maximal parabolic subgroups of G_1, they consist of upper or lower block triangular matrices. Note that τ_1 takes P_1^+ to P_1^- and inversely. Since τ_1 is inner, the subgroups $P_1^\pm$ are conjugate.

Let us extend the action (1.11) to the whole group G by means of an Iwasawa type decomposition

$$g = qhk \quad (g \in G, q \in Q^+, h \in H, k \in K), \tag{1.13}$$

where h and k are determined up to $l \in L$: $qhk = qhl^{-1} \cdot lk$, namely the action $u \mapsto \tilde{u}$ of G on $U(n)$ is defined as follows: u and $\tilde{u}$ are obtained by the projection respectively from k and $\tilde{k}$, which are connected by the relation $kg = \tilde{q}\tilde{h}\tilde{k}$. This action is given by linear fractional transformations:

$$u \mapsto \tilde{u} = u \cdot g = (u\beta + \delta)^{-1}(u\alpha + \gamma). \tag{1.14}$$

The stabilizer of $-iE$ is P^+ (cf. (1.7)), so that $G/P^+ = U(n)$. To calculate (1.14), we use (1.12) and the following fact: if g, h_1, k for (1.13) are given by (1.1), (1.8), (1.3), then $d_1\varphi = \alpha + i\gamma$, $d_1\psi = \delta - i\beta$.

Let L_X be the action of $\mathfrak{g}$ corresponding to (1.14):

$$(L_X f)(u) = \frac{d}{dt}\bigg|_{t=0} f(u \cdot \exp tX), \quad X \in \mathfrak{g}. \tag{1.15}$$

In particular,

$$L_X(\det u) = -2i(\operatorname{Im}(\operatorname{tr}(u\eta)) + \operatorname{tr}\zeta) \cdot \det u. \tag{1.16}$$

On $U(n)$ let us take the following $U(n)$-invariant measure (see [5] 3.3.2, [13]):

$$du = \prod_{j<k} |e^{i\theta_j} - e^{i\theta_k}|^2 d\theta_1 \ldots d\theta_n \cdot dv, \tag{1.17}$$

where $u = v^{-1}\gamma v$, $\gamma = \operatorname{diag}\{e^{i\theta_1}, \ldots, e^{i\theta_n}\}$, dv being a $U(n)$-invariant measure on $U(n)/\Gamma$, and Γ being the subgroup of diagonal matrices. We normalize dv as in [5]. Then the volume of the whole set $U(n)$ is

$$\operatorname{vol} U(n) = (2\pi)^{n(n+1)/2}/1!2! \ldots (n-1)! \tag{1.18}$$

Under the map (1.14) the measure du is transformed as follows:

$$d\tilde{u} = |\det(u\beta + \delta)|^{-2n} du.$$

Let $\tilde{G}$ denote the universal covering group of G. Let us denote by $\tilde{H}, \tilde{P}^+, \tilde{L}, \overline{K}$ the preimages of H, P^+, L, K respectively under the projection $\pi : \tilde{G} \to G$. The groups $K^{(s)}, H_e, Q^{\pm}$ are simply connected, therefore we can identify them with the analytic subgroups of $\tilde{G}$ with corresponding Lie algebras. Moreover, we also keep the previous notation of subgroups and elements of these subgroups, for example, $E, \varepsilon^2 E, \ldots$.

Let D denote the cyclic subgroup of $\tilde{G}$ generated by $w_0 = \exp \pi W_0$, where

$$W_0 = \operatorname{diag}\{-2i + \frac{i}{n}, \frac{i}{n}, \ldots, \frac{i}{n}\} \in \mathfrak{k}.$$

Note that $\pi(w_0) = \varepsilon E$. The center $Z(\widetilde{G})$ of $\widetilde{G}$ is $Z^{(s)}(G)D$, it is isomorphic to $\mathbb{Z}_n \times \mathbb{Z}$. Notice that the kernel of the projection π is a cyclic group generated by $\varepsilon^{-2}E \cdot w_0^2$. For the groups $\widetilde{H}$, $\widetilde{L}$, and $\widetilde{K}$, we have the following direct decompositions

$$\widetilde{H} = DH_e, \quad \widetilde{L} = DL_e, \quad \widetilde{K} = \exp \mathbb{R}W_0 \cdot K^{(s)}.$$

Therefore, $\widetilde{K}/\widetilde{L} = K/L = U(n)$. The Iwasawa decomposition (1.13) is true for $\widetilde{G}$ too, with $g \in \widetilde{G}, q \in Q^+, h \in \widetilde{H}, k \in \widetilde{K}$.

2. Maximal degenerate series

For $\mu, \tau \in \mathbb{C}$, let us define the following characters ω_μ, ω_τ, $\omega_{\mu,\tau}$ of H_e, $\widetilde{K}$, $\widetilde{P}^+$ respectively:

$$\omega_\mu(h) = |\det d_1|^{2\mu}, \quad \omega_\tau(\exp tW_0 \cdot k) = e^{2\tau it}, \quad \omega_{\mu,\tau}(qw_0^m h) = \omega_\tau(w_0^m)\omega_\mu(h),$$

where $h \in H_e$, $k \in K^{(s)}$, $q \in Q^+$, and d_1 is given by (1.8). Let $T_{\mu,\tau}$ be the representation of $\widetilde{G}$ induced by the character $\omega_{\mu,\tau}$ of $\widetilde{P}^+$. It acts by right translations on the space of functions ψ in $C^\infty(\widetilde{G})$ satisfying the condition $\psi(pg) = \omega_{\mu,\tau}(p)\psi(g)$, $p \in \widetilde{P}^+, g \in \widetilde{G}$.

The representation $T_{\mu,\tau}$ can be realized on the space $V = C^\infty(U(n))$: we put

$$f(u) = \psi(k)\omega_\tau(k)^{-1}, \quad (u \in U(n), k \in \widetilde{K}),$$

where $k \mapsto u$ (notice that the right-hand side is invariant under left translations by $l \in \widetilde{L}$); then

$$(T_{\mu,\tau}(g)f)(u) = f(\widetilde{u})\omega_\tau(\widetilde{h})e^{2\tau i(\widetilde{t}-t)};$$

let us explain this formula: if the elements $k = \exp tW_0 \cdot k_0$ and $\widetilde{k} = \exp \widetilde{t}W_0 \cdot \widetilde{k}_0$ in $\widetilde{K}$ $(k_0, \widetilde{k}_0 \in K^{(s)})$ are projected to u and $\widetilde{u}$ in $U(n)$ respectively, then they are connected by the Iwasawa decomposition: $kg = \widetilde{q}\widetilde{h}\widetilde{k}$, where $\widetilde{q} \in Q^+, \widetilde{h} \in H_e$; in particular $\widetilde{u}$ is given by (1.14).

It is more convenient to deal with the corresponding representation $T_{\mu,\tau}$ of the Lie algebra $\mathfrak{g}$:

$$(T_{\mu,\tau}(X)f)(u) = (L_X f)(u) + 2[\mu \mathrm{Re}\,(\mathrm{tr}(u\eta)) + i\tau \mathrm{Im}\,(\mathrm{tr}(u\eta)) + i\tau \mathrm{tr}\zeta]f(u), \tag{2.1}$$

where $X \in \mathfrak{g}$ is given by (1.2) and L_X is the Lie operator (1.15).

The Hermitian form

$$(f_1, f_2) = \int_{U(n)} f_1(u)\overline{f_2(u)}\,du \tag{2.2}$$

is invariant with respect to the pair $T_{\mu,\tau}, T_{\bar{\mu}^*,\bar{\tau}}$, so that

$$(T_{\mu,\tau}(X)f_1, f_2) = -(f_1, T_{\bar{\mu}^*,\bar{\tau}}(X)f_2); \qquad (2.3)$$

here we denote

$$\mu^* = -\mu - n.$$

Let us indicate the connection of our parameters μ, τ with the parameters from some other papers: h of [1], λ of [8], ν_0 of [6], ε, t of [9]:

$$h = -n/\mu; \ \lambda = -2\mu; \ \nu_0 = -2\mu; \ \varepsilon = -\tau, t = 2\mu + n.$$

3. The structure of maximal degenerate series representations

To study the structure of the representations $T_{\mu,\tau}$ (irreducibility, composition series, etc.) we use the restriction to $\mathfrak{k}$.

Let $X \in \mathfrak{k}$ be given by (1.4). By (2.1), for it we have:

$$(T_{\mu,\tau}(X)f)(u) = (L_X f)(u) + 2i\tau \mathrm{tr}\zeta \cdot f(u). \qquad (3.1)$$

In particular, if $X \in \mathfrak{k}^{(s)}$, i.e., $\mathrm{tr}\xi = \mathrm{tr}\zeta = 0$, we have

$$(T_{\mu,\tau}(X)f)(u) = (L_X f)(u).$$

We observe that the restriction of $T_{\mu,\tau}$ to $\mathfrak{k}^{(s)}$ is the differential of the following representation T of the group $K^{(s)}$ on V:

$$(T(k)f)(u) = f(\psi^{-1}u\varphi), \qquad (3.2)$$

where $k = \mathrm{diag}\{\varphi, \psi\}, \varphi, \psi \in SU(n)$, see (1.3).

Let us recall some facts about representations of the group $U(n)$, see, for example, [13]. An irreducible unitary representation π_ν of $U(n)$ is labelled by the highest weight $\nu = (\nu_1, \ldots, \nu_n)$, a vector in $\mathbb{R}^n$ with $\nu_k \in \mathbb{Z}$ and $\nu_1 \geq \nu_2 \geq \ldots \geq \nu_n$. Let Λ denote the lattice of highest weights. The character χ_ν of π_ν has the following expression (the Weyl formula). Let δ be a diagonal matrix $\mathrm{diag}\{\delta_1, \ldots, \delta_n\}$ and $\rho = (n-1, n-2, \ldots.1, 0)$. Put

$$D_\lambda(\delta) = \det(\delta_j^{\lambda_k}), \ \lambda = (\lambda_1, \ldots, \lambda_n).$$

Then

$$\chi_\nu(\delta) = \frac{D_{\nu+\rho}(\delta)}{D_\rho(\delta)}. \qquad (3.3)$$

The highest weight $\mathbf{1} = (1, 1, \ldots, 1)$ gives the one-dimensional representation $u \mapsto \det u$. We have

$$\pi_\nu(u)(\det u)^m = \pi_{\nu+m\mathbf{1}}(u). \qquad (3.4)$$

Let $\langle\,,\,\rangle$ and $e_1, \ldots, e_n$ denote the standard scalar product and the standard basis in $\mathbb{R}^n$.

Denote by Π_ν the linear span of matrix elements of π_ν. It is invariant with respect to the action (3.2) of $K^{(s)}$.

The restriction of the representation π_ν of $U(n)$ to $SU(n)$ is irreducible. By (3.4), the weights $\nu + m\mathbf{1}$ for all $m \in \mathbb{Z}$ give the same result. The representation (3.2) of the group $K^{(s)} = SU(n) \times SU(n)$ on Π_ν is the restriction of the tensor product $\bar{\pi}_\nu \otimes \pi_\nu$; it is irreducible.

For X_0, see (1.5), the operator (3.1) on Π_ν is a scalar operator (multiplication by a number):

$$(T_{\mu,\tau}(X_0)f)(u) = i(n\tau - \langle \nu, \mathbf{1} \rangle)f(u), \ f \in \Pi_\nu. \tag{3.5}$$

Therefore, Π_ν is irreducible under the restriction of $T_{\mu,\tau}$ to $\mathfrak{k}$ (indeed, in the regular representation of $U(n)$ the center of $U(n)$ acts on Π_ν as a scalar operator).

Thus, the restriction of $T_{\mu,\tau}$ to $\mathfrak{k}$ decomposes into the direct multiplicity-free sum of irreducible representations of $\mathfrak{k}$ on Π_ν. Therefore, any irreducible subfactor of $T_{\mu,\tau}$ acts on a subfactor W of V, which is the direct sum of some Π_ν. Let us denote by $\Lambda(W)$ the set of highest weights $\nu \in \Lambda$ occurring in W. Let us call the dimension of the convex hull of $\Lambda(W)$ the *weight dimension* of W. If this dimension is less than n, we say that the subfactor *exceptional*.

The character χ_ν is the unique up to a factor function in Π_ν invariant under L_e: indeed, for the matrix (1.10), the right hand side of (3.2) is $f(w^{-1}uw)$.

Since Y_0, see (1.9), commutes with L_e, it preserves the linear span of characters. Let us write down the action of Y_0 on characters. Define $2n$ functions β_k, β'_k, $k = 1, \ldots, n$, of ν depending on μ, τ:

$$\beta_k(\mu, \tau; \nu) = \mu + \tau - \nu_k + k - 1, \tag{3.6}$$

$$\beta'_k(\mu, \tau; \nu) = \mu - \tau + \nu_k - k + n. \tag{3.7}$$

There are relations between them:

$$\beta_k(\mu, \tau; \nu) + \beta'_k(\mu^*, \tau; \nu) = -1. \tag{3.8}$$

Theorem 3.1

$$T_{\mu,\tau}(Y_0)\chi_\nu = \frac{i}{2} \sum_{k=1}^{n} \beta_k(\mu, \tau; \nu)\chi_{\nu+e_k} - \frac{i}{2} \sum_{k=1}^{n} \beta'_k(\mu, \tau; \nu)\chi_{\nu-e_k}. \tag{3.9}$$

Proof. By (2.1) we have

$$(T_{\mu,\tau}(Y_0)\chi_\nu)(u) = (L_{Y_0}\chi_\nu)(u) + \frac{i}{2}\{(\mu + \tau)\mathrm{tr}u - (\mu - \tau)\mathrm{tr}\bar{u}'\}\chi_\nu(u). \tag{3.10}$$

Note that

$$\text{tr}u = \chi_{e_1}(u), \quad \text{tr}\bar{u}' = \text{tr}u^{-1} = \chi_{-e_n}(u). \tag{3.11}$$

For $g = \exp tY_0$, formula (1.14) has the form

$$\tilde{u} = (i\sinh(t/2) \cdot u + \cosh(t/2) \cdot E)^{-1}(\cosh(t/2) \cdot u - i\sinh(t/2) \cdot E).$$

If u is a diagonal matrix δ as above, then $\tilde{u} = \tilde{\delta} = \text{diag}\{\tilde{\delta}_1, \ldots, \tilde{\delta}_n\}$, where

$$\tilde{\delta}_j = \frac{\delta_j \cosh(t/2) - i\sinh(t/2)}{\delta_j i\sinh(t/2) + \cosh(t/2)},$$

so that

$$\frac{d}{dt}\bigg|_{t=0} \tilde{\delta}_j^{\lambda_k} = -\frac{i}{2}\lambda_k(\delta_j^{\lambda_k+1} + \delta_j^{\lambda_k-1}).$$

Therefore

$$\frac{d}{dt}\bigg|_{t=0} D_\lambda(\tilde{\delta}) = -\frac{i}{2}\sum_{k=1}^{n}\lambda_k\{D_{\lambda+e_k}(\delta) + D_{\lambda-e_k}(\delta)\}. \tag{3.12}$$

In particular,

$$\frac{d}{dt}\bigg|_{t=0} D_\rho(\tilde{\delta}) = -\frac{i}{2}(n-1)D_{\rho+e_1}(\delta). \tag{3.13}$$

Now we are able to find $(L_{Y_0}\chi_\nu)(\delta)$. By (3.3) and (3.12), (3.13), we have

$$(L_{Y_0}\chi_\nu)(\delta) = \frac{1}{D_\rho(\delta)}\left\{\frac{d}{dt}\bigg|_{t=0} D_{\nu+\rho}(\tilde{\delta}) - \chi_\nu(\delta)\frac{d}{dt}\bigg|_{t=0} D_\rho(\tilde{\delta})\right\} =$$

$$= -\frac{i}{2}\sum_{k=1}^{n}(\nu_k + n - k)(\chi_{\nu+e_k}(\delta) + \chi_{\nu-e_k}(\delta)) + \frac{i}{2}(n-1)\chi_\nu(\delta)\chi_{e_1}(\delta). \tag{3.14}$$

Substituting (3.14) and (3.11) in (3.10) and taking into account the decompositions (see [13])

$$\chi_\nu\chi_{e_1} = \sum_{k=1}^{n}\chi_{\nu+e_k}, \quad \chi_\nu\chi_{-e_n} = \sum_{k=1}^{n}\chi_{\nu-e_k}, \tag{3.15}$$

we obtain (3.9). $\square$

Let us call the hyperplane $\beta_k(\mu, \tau; \nu) = 0$ in the space $\mathbb{R}^n$ of vectors ν a *barrier*, if it intersects the lattice Λ. For that it is necessary and sufficient to have $\mu + \tau \in \mathbb{Z}$. Then the lattice Λ splits into two subsets: $\beta_k \geq 0$ (the interior of the barrier) and $\beta_k < 0$ (the exterior of the barrier). The subspace

$$V_k = V_k(\mu, \tau) = \sum_{\beta_k(\mu,\tau;\nu)\geq 0} \Pi_\nu$$

is invariant under $T_{\mu,\tau}$ (this follows from Theorem 3.1).

Similarly for β'_k one has $\mu - \tau \in \mathbb{Z}$ and the corresponding subspace

$$V'_k = V'_k(\mu, \tau) = \sum_{\beta'_k(\mu,\tau;\nu)\geq 0} \Pi_\nu$$

is invariant.

We have the embeddings:

$$\{0\} \subset V_1 \subset V_2 \subset \ldots \subset V_n \subset V, \tag{3.16}$$

$$\{0\} \subset V'_n \subset V'_{n-1} \subset \ldots \subset V'_1 \subset V. \tag{3.17}$$

To unify the notation, we put $V_0 = V'_{n+1} = \{0\}$, $V_{n+1} = V'_0 = V$.
We have the following

Theorem 3.2 *If both numbers $\mu + \tau$ and $\mu - \tau$ are not integers, then the representation $T_{\mu,\tau}$ is irreducible. If $\mu + \nu \in \mathbb{Z}, \mu - \tau \notin \mathbb{Z}$ (respectively $\mu + \nu \notin \mathbb{Z}, \mu - \tau \in \mathbb{Z}$), then V has an increasing chain (3.16) (respectively (3.17)) of n proper invariant subspaces with irreducible subfactors. If both numbers $\mu + \tau$ and $\mu - \tau$ are integer (then both numbers μ and ν are simultaneously integers or half integers), then the invariant subspaces are V_j and V'_k and everything that can be obtained from them by taking intersections and arithmetic sums.*

The structure of invariant subspaces can be shown by a plane picture as explained below (for typographical reasons we can give no picture but a description only). Let us draw on the plane a rectangle P with horizontal and vertical sides (rather long horizontally). Let us denote its upper and lower sides by a and b respectively. The rectangle P represents the space V. Let us draw two bundles of lines in P. Each bundle has n pairwise disjoint lines.

The lines of the first bundle represent the barriers $\beta_j = 0, j = 1, \ldots, n$. Correspondingly, we enumerate these lines from left to right using the numbers $j = 1, \ldots, n$; for each j let us equip both points where the line j meets the sides a and b with the number j. Let us equip these lines by spikes oriented to the left (inside the barriers). Then the part of P lying to the left of the line j represents V_j.

The lines of the second bundle represent the barriers $\beta'_k = 0, k = 1, \ldots, n$. Correspondingly, we enumerate these lines from left to right by using symbols $k' = 1', \ldots, n'$ and for each k' equip both points where the line k' meets the sides a and b with the symbol k'. Let us equip these lines by spikes oriented to the right (inside the barriers). Then the part of P lying to the right of the line k' represents V'_k.

These two bundles interact as follows.

Let first $\mu < (-n+1)/2$ (recall $\mu \in (1/2)\mathbb{Z}$). On the side a, the points $1, \ldots, n$ and $1', \ldots, n'$ separate each other, namely, $1 < 1' < 2 < 2' < \ldots < n < n'$ ("$\alpha < \beta$" means "α lies to the left of β"). On the side b: if $\mu \leq 1/2 - n$, then all points $1, \ldots, n$ lie to the left of all points $1', \ldots, n'$ (i.e., $1 < \ldots < n < 1' < \ldots n'$), and if $\mu = (1+s)/2 - n$, $s = 1, 2, \ldots, n-1$, then the first s points of the second bundle coincide with the last s points of the first bundle (i.e., $k' = n - s + k$, $k = 1, \ldots, s$).

For $\mu > (-n+1)/2$, the picture is dual to the case $\mu < (-n+1)/2$. On the side a the points $1, \ldots, n$ and $1', \ldots, n'$ separate each other too but in another way: $1' < 1 < 2' < 2 < \ldots n' < n$. On the side b: if $\mu \geq 1/2$, then all points $1', \ldots, n'$ lie to the left of points $1, \ldots, n$ (i.e., $1' < \ldots < n' < 1 < \ldots < n$), and if $\mu = (1-s)/2$, $s = 1, \ldots, n-1$, then the first s of the first bundle coincide with the last s points of the second bundle (i.e., $k = (n - s + k)'$, $k = 1, \ldots, s$).

At last, for $\mu = (-n+1)/2$ both bundles coincide (i.e., the line j is at the same time the line j').

Note that for $\mu < (-n+1)/2$ the line j lies to the left of the line j' and for $\mu > (-n+1)/2$, to the right.

The lines of these two bundles decompose P into several parts (some of them can be degenerate, i.e., be a point or a line, as it happens, for example, for $\mu = (-n+1)/2$ or $\mu = 0$). Any such a part presents some irreducible invariant subfactor of V under $T_{\mu,\tau}$.

From our description of the barrier structure we deduce the following

Theorem 3.3 *In the representation $T_{\mu,\tau}$ any irreducible invariant subfactor is:*

(a) V_{k+1}/V_k, $k = 0, 1, \ldots, n$, if $\mu + \tau \in \mathbb{Z}$, $\mu - \tau \notin \mathbb{Z}$;

(b) V'_k/V'_{k+1}, $k = 0, 1, \ldots, n$, if $\mu + \tau \notin \mathbb{Z}$, $\mu - \tau \in \mathbb{Z}$;

(c) $(V'_k \cap V_{k+p+1})/(V'_{k+1} + V_{k+p})$, $p \in \mathbb{Z}$, if $\mu \pm \tau \in \mathbb{Z}$. $\qquad$ (3.18)

Remarks. The sign "+" denotes the arithmetic sum. For brevity we write U/Z instead of $U/U \cap Z$. Some subfactors (3.18) can be trivial, for example, $V/(V'_j + V_{j+n-r})$ for $\mu = -n + r/2$, $r = 1, \ldots, n$, $j = 1, \ldots, r$. The maximal number of irreducible subfactors is $(n+1)(n+2)/2$. In (3.18) it is sufficient to take $p \geq 0$ for $\mu \leq -n/2$ and $p \leq 0$ for $\mu > -n/2$. In case (c), for $\mu < (-n+1)/2$ there are $n + 1$ irreducible invariant subspaces

$$V_1, \ V'_1 \cap V_2, \ \ldots, \ V'_{n-1} \cap V_n, \ V'_n; \qquad (3.19)$$

for $\mu = -n/2$ the space V splits into the direct sum of the subspaces (3.19); for $\mu = (-n+1)/2$ exceptional irreducible subspaces $V'_j \cap V_j$, $j = 1, \ldots, n$, with weight dimension $n - 1$ appear.

Let U be an invariant subspace of V under $T_{\mu,\tau}$. Denote by $U^\perp$ its orthogonal complement with respect to form (2.2). Then $\Lambda(U^\perp) = \Lambda \backslash \Lambda(U)$

and $U^\perp$ is invariant under $T_{\bar\mu^*,\bar\tau}$, see (2.3). For a subfactor $W = U/Z$ invariant under $T_{\mu,\tau}$, the subfactor $W^* = Z^\perp/U^\perp$ is invariant under $T_{\bar\mu^*,\bar\tau}$. We have $\Lambda(W) = \Lambda(W^*)$. Let us call W^* the *dual* subfactor. In particular, $U^* = V/U^\perp$. For the irreducible subfactors from Theorem 3.3, let us present the dual ones (we use (3.8)):

$$(V_{k+1}/V_k)^* = V'_k/V'_{k+1},$$

$$[(V'_k \cap V_{k+p+1})/(V'_{k+1} + V_{k+p})]^* = (V_{k+1} \cap V'_{k+p})/(V_k + V'_{k+p+1}) \quad (3.20)$$

4. Intertwining operators

First we present some intertwining operators: $A_{\mu,\tau}$ and B, see (4.2) and (4.15) below. Consider the following function on $U(n)$ depending on $\mu, \tau \in \mathbb{C}$:

$$L_{\mu,\tau}(u) = |\det(E - u)|^{2\mu^*} \prod_{k=1}^{n} e^{i\tau(\theta_k - \pi)}, \quad (4.1)$$

where $e^{i\theta_k}$ are eigenvalues of $u \in U(n)$, here $0 \le \theta_k < 2\pi$.

The function (4.1) defines a distribution on $U(n)$, which will be denoted by the same symbol. Let us define the operator $A_{\mu,\tau}$ on V as the convolution with this distribution:

$$(A_{\mu,\tau}f)(u) = \int L_{\mu,\tau}(uv^{-1})f(v)\,dv, \quad (4.2)$$

where dv is defined by (1.17); here and below integrals are taken over $U(n)$ (if some other is not indicated). The integral (4.2) converges absolutely for $\mathrm{Re}\,\mu < 1/2 - n$ and can be extended on other μ as a meromorphic function. The following theorem is proved by direct calculations.

Theorem 4.1 *The operator $A_{\mu,\tau}$ intertwines the representations $T_{\mu,\tau}$ and $T_{\mu^*,\tau}$:*

$$T_{\mu^*,\tau}(X)A_{\mu,\tau} = A_{\mu,\tau}T_{\mu,\tau}(X), \ X \in \mathfrak{g}. \quad (4.3)$$

Theorem 4.2 *On each subspace Π_ν the operator $A_{\mu,\tau}$ is the multiplication by a number:*

$$(A_{\mu,\tau}\varphi)(u) = a(\mu,\tau;\nu)\varphi(u), \ \varphi \in \Pi_\nu. \quad (4.4)$$

This factor has the following explicit expression:

$$a(\mu,\tau;\nu) = (-1)^{\nu_1+\dots+\nu_n}(2\pi)^{\frac{n(n+1)}{2}} \prod_{k=1}^{n} \frac{\Gamma(2\mu^* + k)}{\Gamma(-\beta_k(\mu,\tau;\nu))\Gamma(-\beta'_k(\mu,\tau;\nu))}, \quad (4.5)$$

where β_k, β'_k are the functions given by (3.6), (3.7).

Proof. The restriction of $T_{\mu,\tau}$ to $\mathfrak{k}$ has a multiplicity-free spectrum of irreducible representations acting on Π_ν (see Section 3). Since $A_{\mu,\tau}$ is an intertwining operator, it is the multiplication by a number $a(\mu,\tau;\nu)$ on each Π_ν, i.e., (4.4) holds. The rest of the proof is devoted to the calculation of this factor. It is sufficient to do that for the generic case: $\mu \pm \tau \notin \mathbb{Z}$. For simplicity we write $a(\nu)$ instead of $a(\mu,\tau;\nu)$.

Let us apply (4.3) with $X = Y_0$, see (1.9), to the character χ_ν. By (3.9) we obtain n recurrent relations for $a(\nu)$:

$$a(\nu) = -a(\nu - e_k)\frac{\beta'_k(\mu,\tau;\nu)}{1 + \beta_k(\mu,\tau;\nu)}, \quad k = 1,\dots,n. \qquad (4.6)$$

It is easy to verify that the solution of (4.6) is

$$a(\nu) = a(0) \prod_{k=1}^{n}(-1)^{\nu_k}\frac{\Gamma(-\beta_k(\mu,\tau;0))\Gamma(-\beta'_k(\mu,\tau;0))}{\Gamma(-\beta_k(\mu,\tau;\nu))\Gamma(-\beta'_k(\mu,\tau;\nu))}. \qquad (4.7)$$

So it remains only to compute $a(0)$. In (4.4) let us take $\nu = 0$, $\varphi = \chi_0 = 1$, $u = E$. We obtain

$$a(0) = (A_{\mu,\tau}\chi_0)(E).$$

Assume $\operatorname{Re}\mu < 1/2 - n$, then $a(0)$ is given by an absolutely convergent integral (we have used $d(v^{-1}) = dv$):

$$a(0) = \int L_{\mu,\tau}(v)\, dv\,. \qquad (4.8)$$

Taking the expression (1.17) for the measure dv, we obtain

$$a(0) = C_n \cdot Q_n(\mu^*,\tau), \qquad (4.9)$$

where

$$Q_n(\lambda,\tau) = \int \prod_{k=1}^{n} |1 - e^{i\theta_k}|^{2\lambda} e^{i\tau(\theta_k - \pi)}\omega(\theta)\, d\theta_1\dots d\theta_n, \qquad (4.10)$$

$$\omega(\theta) = \prod_{j<r} |e^{i\theta_j} - e^{i\theta_r}|^2, \quad \theta = (\theta_1,\dots,\theta_n),$$

and the integral is taken over the set $0 \le \theta_k < 2\pi, k = 1,\dots,n$.

For $n = 1$ the integral (4.10), i.e., $Q_1(\lambda.\tau)$, can be calculated by [3] 1.5(29):

$$Q_1(\lambda.\tau) = \int_0^{2\pi} |1 - e^{i\theta}|^{2\lambda} e^{i\tau(\theta - \pi)}d\theta =$$

$$= \frac{2\pi\Gamma(2\lambda + 1)}{\Gamma(\lambda + \tau + 1)\Gamma(\lambda - \tau + 1)}. \qquad (4.11)$$

Let α range over the positive roots $e_j - e_k, j < k$, of the Lie algebra $\mathfrak{gl}(n, \mathbb{C})$ and let δ be their half sum:

$$\delta = ((n-1)/2, (n-3)/2, \ldots, (-n-1)/2).$$

We can write $\omega(\theta)$ as follows

$$\omega(\theta) = (-1)^{\frac{n(n-1)}{2}} \left\{ \prod_\alpha \left(e^{\frac{i}{2}\langle \alpha, \theta \rangle} - e^{-\frac{i}{2}\langle \alpha, \theta \rangle} \right) \right\}^2.$$

It is well-known that the expression in the braces is equal to

$$\sum_{s \in S_n} \mathrm{sgn} s \cdot e^{i\langle s\delta, \theta \rangle},$$

where S_n is the symmetric group of n symbols. Therefore

$$\omega(\theta) = (-1)^{\frac{n(n-1)}{2}} \sum_{s,w \in S_n} \mathrm{sgn}\,(sw) e^{i\langle s\delta + w\delta, \theta \rangle} =$$

$$= (-1)^{\frac{n(n-1)}{2}} \sum_{s \in S_n} \sum_{\sigma \in S_n} \mathrm{sgn}\sigma e^{i\langle \delta + \sigma\delta, s^{-1}\theta \rangle}. \tag{4.12}$$

Substitute (4.12) into (4.10). After integrating, every exponent gives the product

$$\prod_{k=1}^{n} Q_1(\lambda, \tau + \delta_k + (\sigma\delta)_k)$$

independently of $s \in S_n$, hence

$$Q_n(\lambda, \tau) = (-1)^{\frac{n(n-1)}{2}} n! \sum_{\sigma \in S_n} \mathrm{sgn}\sigma \prod_{k=1}^{n} Q_1(\lambda, \tau + \delta_k + (\sigma\delta)_k).$$

The latter sum is none other than a determinant:

$$Q_n(\lambda, \tau) = (-1)^{\frac{n(n-1)}{2}} n! \det(Q_1(\lambda, \tau + \delta_k + \delta_j)) =$$

$$= (-1)^{\frac{n(n-1)}{2}} n! \det(Q_1(\lambda, \tau + n + 1 - j - k)).$$

Denoting $n + 1 - j = l$ and rearranging the columns (or rows) numbered by j correspondingly, we obtain

$$Q_n(\lambda, \tau) = n! \det(Q_1(\lambda, \tau + l - k)).$$

Substituting (4.11) into this expression, we obtain

$$Q_n(\lambda, \tau) = \frac{n!(2\pi)^n \Gamma^n(2\lambda + 1)}{\prod_{k=1}^{n} \Gamma(\lambda + \tau + k)\Gamma(\lambda - \tau + k)} \times$$

$$\times \det\left((\lambda + \tau + n - i)^{(n-j)}(\lambda - \tau + i - 1)^{(j-1)}\right),$$

where

$$a^{(m)} = a(a-1)\ldots(a - m + 1) = \Gamma(a+1)/\Gamma(a - m + 1).$$

The latter determinant can be calculated in an elementary way; it is equal to

$$1!2!\ldots(n-1)!(2\lambda + n - 1)^{(n-1)}(2\lambda + n - 2)^{(n-2)}\ldots(2\lambda + 1),$$

so that

$$Q_n(\lambda, \tau) = 1!2!\ldots n!(2\pi)^n \prod_{k=1}^{n} \frac{\Gamma(2\lambda + k)}{\Gamma(\lambda + \tau + k)\Gamma(\lambda - \tau + k)}.$$

Substituting this into (4.9), we obtain

$$a(0) = D_n \cdot \prod_{k=1}^{n} \frac{\Gamma(2\mu^* + k)}{\Gamma(\mu^* + \tau + k)\Gamma(\mu^* - \tau + k)}, \tag{4.13}$$

where D_n is a number. To find it, let us set $\mu^* = 0, \tau = 0$ here and in (4.8). Then

$$\operatorname{vol} U(n) = D_n/1!2!\ldots(n-1)!$$

Recalling (1.18), we obtain

$$D_n = (2\pi)^{n(n+1)/2}. \tag{4.14}$$

We can rewrite (4.13) replacing $\Gamma(\mu^* - \tau + k)$ by $\Gamma(\mu^* - \tau + n - k + 1)$, so when we substitute (4.13) and (4.14) in (4.7) we obtain (4.5). $\square$

Now consider the following operator B on V:

$$(Bf)(u) = f(u)\det u. \tag{4.15}$$

The following theorem follows from (1.16).

Theorem 4.3 *The operator B intertwines $T_{\mu,\tau-1}$ and $T_{\mu,\tau}$:*

$$T_{\mu,\tau}(X)B = BT_{\mu,\tau-1}(X), \ X \in \mathfrak{g}. \tag{4.16}$$

Thus, the parameter τ can be reduced modulo 1, and we can assume that

$$-1/2 < \operatorname{Re}\tau \le 1/2. \tag{4.17}$$

Let us find nonzero operators that intertwine the representations $T_{\mu,\tau}$ and T_{μ_1,τ_1} (or some their subfactors), i.e.,

$$T_{\mu_1,\tau_1}(X)A = AT_{\mu,\tau}(X), \ X \in \mathfrak{g}. \tag{4.18}$$

By virtue of Theorem 4.3, we can assume that τ and τ_1 lie in the strip (4.17).

Theorem 4.4 *A nonzero operator A on V satisfying (4.18), where τ and τ_1 lie in the strip (4.17), exists only in the following cases:*
  (a) $\mu_1 = \mu, \tau_1 = \tau$,
  (b) $\mu_1 = \mu^*, \tau_1 = \tau$.
In case (a) *the operator A is unique up to a numerical factor except when $\mu = -n/2, \tau \in n/2 + \mathbb{Z}$ (this case is treated in* (b) *below), it is equal to λE (E is the identity operator).*

In case (b), *if $\mu \pm \tau \notin \mathbb{Z}$, then the operator A is unique up to a numerical factor, it is equal to $\lambda A_{\mu,\tau}$. If $\mu + \tau \in \mathbb{Z}$ or $\mu - \tau \in \mathbb{Z}$, then for any irreducible subfactor W for $T_{\mu,\tau}$ there exists a unique (up to a numerical factor) operator A which intertwines the subfactor of $T_{\mu,\tau}$ on W with the subfactor of $T_{\mu^*,\tau}$ on the dual subfactor W^*.*

Proof. Let $A \neq 0$ be an operator on V satisfying (4.18).

First let X range over $\mathfrak{k}^{(s)}$. Then (4.18) implies that A intertwines the representation T of the group $K^{(s)}$, see (3.2), with itself. Restrictions of T to Π_ν are irreducible, but the spectrum is not simple: representations on Π_ν and $\Pi_{\nu'}$ are equivalent if $\nu' = \nu + m\mathbf{1}$ (see Section 3). Therefore,

$$A\Pi_\nu \subset \sum_{m \in \mathbb{Z}} \Pi_{\nu + m\mathbf{1}}.$$

In particular, since A commutes with L_e, we have

$$A\chi_\nu = \sum_{m \in \mathbb{Z}} \alpha(\nu, m)\chi_{\nu + m\mathbf{1}}, \tag{4.19}$$

where $\alpha(\nu, m)$ are some numbers depending on μ, τ, μ_1, τ_1.

Apply (4.18) with $X = X_0$, see (1.5), to χ_ν. Using (3.5) and (4.19), we obtain

$$\alpha(\nu, m)\{n\tau_1 - \langle \nu + m\mathbf{1}, \mathbf{1} \rangle\} = \alpha(\nu, m)\{n\tau - \langle \nu, \mathbf{1} \rangle\}. \tag{4.20}$$

Since $A \neq 0$, we have $\alpha(\nu, m) \neq 0$ at least for one pair (ν, m). Then (4.20) gives $\tau_1 = \tau + m$. By (4.17) we have $\tau_1 = \tau$. Substituting this into (4.20) we obtain $\alpha(\nu, m) = 0$ for $m \neq 0$. This means that A preserves each Π_ν, being the multiplication by a number $a(\nu)$ (depending on μ, μ_1, τ):

$$A\varphi = a(\nu)\varphi, \quad \varphi \in \Pi_\nu.$$

Apply (4.18) with $X = Y_0$, see (1.9), to χ_ν. Using (3.9), we obtain two systems of relations:

$$a(\nu + e_k)\beta_k(\mu, \tau; \nu) = a(\nu)\beta_k(\mu_1, \tau; \nu), \tag{4.21}$$

$$a(\nu - e_k)\beta'_k(\mu, \tau; \nu) = a(\nu)\beta'_k(\mu_1, \tau; \nu), \tag{4.22}$$

where $k = 1, \ldots, n$. Multiplying (4.21) by $\beta'_k(\mu_1, \tau; \nu + e_k)$ and using (4.22) with ν replaced by $\nu + e_k$, we obtain

$$a(\nu)\beta_k(\mu, \tau; \nu)\beta'_k(\mu, \tau; \nu + e_k) = a(\nu)\beta_k(\mu_1, \tau; \nu)\beta'_k(\mu_1, \tau; \nu + e_k). \quad (4.23)$$

Since $A \neq 0$, there exists a $\bar{\nu}$ such that $a(\bar{\nu}) \neq 0$. Putting $\nu = \bar{\nu}$ in (4.23) and dividing by $a(\bar{\nu})$, we obtain

$$\beta_k(\mu, \tau; \bar{\nu})\beta'_k(\mu, \tau; \bar{\nu} + e_k) = \beta_k(\mu_1, \tau; \bar{\nu})\beta'_k(\mu_1, \tau; \bar{\nu} + e_k). \quad (4.24)$$

By the definition of β_k, β'_k (see (3.6), (3.7)), equality (4.24) means that a quadratic trinomial with the sum of roots $-n$ has the same values at points μ and μ_1. This is possible only when $\mu_1 = \mu$ or $\mu_1 + \mu = -n$, i.e., $\mu_1 = \mu^*$. Let $\mu_1 = \mu$. Then by (4.21) and (4.22)

$$a(\nu + e_k) = a(\nu), \quad a(\nu - e_k) = a(\nu) \quad (4.25)$$

for all ν such that $\beta_k(\mu, \tau; \nu) \neq 0, \beta'_k(\mu, \tau; \nu) \neq 0$ respectively. It follows from (4.25) that $a(\nu)$ is constant on the intersection of the line $\nu + \mathbb{R}e_k$ with the lattice Λ except for one case: $\mu = -n/2, \tau \in n/2 + \mathbb{Z}$. For the latter case relations (4.25) do not link the two sets $\beta_k \geq 0$ and $\beta'_k \geq 0$ in Λ. The space V in this case decomposes into the direct sum of $n + 1$ irreducible subspaces (3.19). Thus, if $\mu_1 = \mu$, then A is a scalar operator except for $\mu = -n/2, \tau \in n/2 + \mathbb{Z}$, when A is a scalar operator on each of invariant subspaces (3.19).

Let now $\mu_1 = \mu^*$. Then (4.21) and (4.22) give the same result:

$$a(\nu + e_k)\beta_k(\mu, \tau; \nu) = a(\nu)\beta_k(\mu^*, \tau; \nu), \quad (4.26)$$

If there is no barrier (i.e., $T_{\mu,\tau}$ and $T_{\mu^*,\tau}$ are irreducible), then relations (4.26) define $a(\nu)$ uniquely up to a numerical factor. These relations are precisely relations (4.6) for the eigenvalues of $A_{\mu,\tau}$. Therefore, $A = \lambda A_{\mu,\tau}$.

Let barriers exist ($T_{\mu,\tau}$ and $T_{\mu^*,\tau}$ are reducible). Take some irreducible subfactor W of V with respect to $T_{\mu,\tau}$. As shown in Section 3, W is a coset space, $W = U/U \cap Z$, where U and Z are invariant subspaces such that $\Lambda(U)$ is either Λ or the intersection of the interiors of some barriers (not more than 2) and $\Lambda(Z)$ is either $\emptyset$ or the union of the interiors of some barriers (no more than 2). Relations (4.26) define the function $a(\nu)$ on $\Lambda(U)$ uniquely up to a numerical factor with $a(\nu) = 0$ on $\Lambda(U \cap Z)$. This function $a(\nu)$ defines uniquely up to a factor an operator A on U commuting with $\mathfrak{g}$ (i.e., (4.18) with $\mu_1 = \mu^*, \tau_1 = \tau$ is true) such that $A = 0$ on $U \cap Z$. This operator gives rise to the equivalence of the subfactor $W = U/U \cap Z$ with respect to $T_{\mu,\tau}$ and the dual subfactor $W^* = Z^{\perp}/Z^{\perp} \cap U^{\perp}$ irreducible with respect to $T_{\mu^*,\tau}$. $\square$

5. Unitarity

A sesqui-linear form $H(f, f_1)$ on V is called *invariant* with respect to the pair $T_{\mu,\tau}, T_{\mu_1,\tau_1}$, if for any $X \in \mathfrak{g}$ and any $f, f_1 \in V$ we have

$$H(T_{\mu,\tau}(X)f, f_1) + H(f, T_{\mu_1,\tau_1}(X)f_1) = 0.$$

Similarly we define the invariance of a form on a pair of invariant subfactors U/Z and U_1/Z_1; then $f \in U, f_1 \in U_1$ and $H(f, f_1) = 0$, if $f \in Z$ or $f_1 \in Z_1$.

Let $H(f, f_1)$ be such a form. Then by virtue of (4.16), the forms

$$H_1(f, f_1) = H(f, B^m f_1) \text{ and } H_2(f, f_1) = H(B^k f, f_1)$$

are invariant with respect to the pairs $T_{\mu,\tau}$, T_{μ_1,τ_1-m} and $T_{\mu,\tau-k}, T_{\mu_1,\tau_1}$ respectively. So in studying invariant sesqui-linear forms we can restrict ourselves to the case in which τ and τ_1 lie in the strip (4.17).

The following theorem is proved similarly to Theorem 4.4.

Theorem 5.1 *A nonzero sesqui-linear form $H(f, f_1)$ invariant with respect to the pair $T_{\mu,\tau}, T_{\mu_1,\tau_1}$ on V or on a pair of subfactors (provided that τ and τ_1 lie in the strip (4.17)) exists only in the following cases:*
   (a) $\mu_1 = \bar{\mu}^*, \tau_1 = \bar{\tau}$,
   (b) $\mu_1 = \bar{\mu}, \tau_1 = \bar{\tau}$.
In case (a) the form H coincides up to a factor with the form (2.2) except when $\mu = -n/2, \tau \in n/2 + \mathbb{Z}$. In the latter case the form H coincides up to a factor with the form (2.2) on each of $n + 1$ invariant subspaces (3.19).

In case (b) the form H can be expressed in terms of the form (2.2) and the operator A intertwining $T_{\mu,\tau}$ and $T_{\mu^*,\tau}$ or their subfactors, see Theorem 4.4:

$$H(f, f_1) = (Af, f_1).$$

Now let us determine when the representations $T_{\mu,\tau}$ or their subfactors are unitarizable. For that we have to put $\mu_1 = \mu, \tau_1 = \tau$ in Theorem 5.1 and the form H has to be Hermitian ($H(f, f)$ real) and positive definite. For τ we obtain $\tau \in \mathbb{R}$, and the number τ is reduced modulo 1, so that we can take

$$-1/2 < \tau \leq 1/2. \tag{5.1}$$

For μ there are two possibilities: $\mu = \bar{\mu}^*$ and $\mu = \bar{\mu}$. In the first case we have $\operatorname{Re} \mu = -n/2$. We obtain a series of unitarizable representations $T_{\mu,\tau}$: $\operatorname{Re} \mu = -n/2, -1/2 < \tau \leq 1/2$, the invariant inner product is (2.2), so that the unitary completion acts on $L^2(U(n), du)$. Let us call this series the *continuous* series. Representations of this series are irreducible except one case $\mu = -n/2, \tau \in n/2 + \mathbb{Z}$ when the representation decomposes into the direct sum of $n + 1$ irreducible representations on the subspaces (3.19).

Now let $\mu = \bar{\mu}$, i.e., $\mu \in \mathbb{R}$. First consider the irreducible case.

Theorem 5.2 *If a representation $T_{\mu,\tau}$ with $\mu, \tau \in \mathbb{R}$ is irreducible, then it is unitarizable for pairs (μ, τ) lying in the set*

$$|\mu + n/2| + |\tau| < 1/2 \quad \text{for } n \text{ odd,} \tag{5.2}$$

$$|\mu + n/2| < |\tau| \quad \text{for } n \text{ even,} \tag{5.3}$$

(we keep condition (5.1)). The invariant inner product is $c(\mu, \tau) \cdot (A_{\mu,\tau} f, f_1)$, where $A_{\mu,\tau}$ is the operator from Section 4, $(\cdot, \cdot)$ is the form (2.2), $c(\mu, \tau)$ is some normalizing factor, for example, one can take $c(\mu, \tau) = a(\mu, \tau; 0)^{-1}$.

Let us call the family of representations $T_{\mu,\tau}$ indicated in Theorem 5.2 the *complementary* series.

Proof. We must learn when equations (4.26) have a positive solution $a(\nu)$. For that it is necessary and sufficient that for each $i = 1, \ldots, n$ both functions $\beta_k(\mu, \tau; \nu)$ and $\beta_k(\mu^*, \tau; \nu)$ regarded as functions of ν_k (see (3.6)) are of the same sign on $\mathbb{Z}$. In turn, for that it is necessary and sufficient that the interval $(\mu + \tau, \mu^* + \tau)$ contain no point from $\mathbb{Z}$. Therefore its length has to be less than 1, from where we have $(-n-1)/2 < \mu < (-n+1)/2$. Then (recall (5.1)), the interval mentioned above can contain the following integers only: $(-n \pm 1)/2$ for n odd and $-n/2$ for n even. Removing these ones we obtain (5.2) and (5.3). $\square$

Let us turn now to the reducible case: $\mu + \tau \in \mathbb{Z}$ or $\mu - \tau \in \mathbb{Z}$. From (4.26), we see that if $a(\nu)$ is defined outside some barrier, then $a(\nu) = 0$ inside it. Therefore, we must look for invariant positive definite Hermitian forms on irreducible subfactors.

The proofs of the following two theorems are rather long, but not very complicated, so we omit them.

Theorem 5.3 *Let $\mu + \tau \in \mathbb{Z}, \mu - \tau \notin \mathbb{Z}$. Then a positive definite Hermitian form, invariant under $T_{\mu,\tau}$, exists on the following irreducible subfactors:*

V/V_n for $\mu > (-n-1)/2$,
V_1 for $\mu \le (-n+1)/2$,
$V_1, V_2/V_1, \ldots, V/V_n$ for $(-n-1)/2 < \mu \le (-n+1)/2$

Let $\mu + \tau \notin \mathbb{Z}, \mu - \tau \in \mathbb{Z}$. Then a Hermitian form of the above type exists on the following irreducible subfactors:

V/V'_1 for $\mu > (-n-1)/2$,
V'_n for $\mu \le (-n+1)/2$,
$V'_n, V'_{n-1}/V'_n, \ldots, V/V'_1$ for $(-n-1)/2 < \mu \le (-n+1)/2$

Such a form is unique up to a factor.

Theorem 5.4 *Let $\mu \pm \tau \in \mathbb{Z}$ (then $\mu \in (1/2)\mathbb{Z}$).*

Let $\mu = -n/2$. Then V decomposes into the direct sum of $n+1$ irreducible subspaces (3.19). On each of them there exists a positive definite Hermitian form, invariant under $T_{\mu,\tau}$.

Let $\mu < -n/2$. Then a Hermitian form of the above type exists on the following irreducible subfactors:

(a) subspaces $V'_k \cap V_{k+1}$, $k = 0, 1, \ldots, n$, for all $\mu \leq -n/2$; the weight dimension is n;

(b) coset spaces $V/(V'_{k+1}+V_{k+p})$, $k = 0, 1, \ldots, n-p$, for $\mu = (-n-p)/2$, $p = 1, 2, \ldots, n$; the weight dimension is equal to $n - p$ (exceptional subfactors): coordinates $\nu_{k+1}, \ldots, \nu_{k+p}$ are fixed and are equal to $\mu + \tau + k + p$.

Let $\mu > -n/2$. Then a form of the above type exists on the following irreducible subfactors:

(c) coset spaces $V/(V'_{k+1}+V_k)$, $k = 0, 1, \ldots, n$, for $\mu > -n/2$; the weight dimension is n;

(d) subspaces $V'_{k+q} \cap V_{k+1}$, $k = 0, 1, \ldots, n - q$, for $\mu = (-n+q)/2$, $q = 1, \ldots, n$; the weight dimension is equal to $n-q$ (exceptional subfactors): the coordinates $\nu_{k+1}, \ldots, \nu_{k+q}$ are fixed and are equal to $\mu + \tau + k$.

Such a form is unique up to a numerical factor.

6. Analytic series

Some subfactors of the representations $T_{\mu,\tau}$ can be realized on some spaces of holomorphic functions on the Cartan domain $\Omega = \Omega_{nn}$ of type I. They are subfactors whose sets of highest weights are bounded below, so that the corresponding representations, if unitary, are highest weight representations. Therefore, one of bounding barriers must be $\beta'_n(\mu, \tau; \nu) = 0$, so that the highest weights of our subfactors satisfy $\nu_n \geq \tau - \mu$. From Theorems 3.2 and 3.3, we can extract the following list of subfactors with the barrier $\beta'_n = 0$.

Theorem 6.1 *The following irreducible subfactors have $\beta'_n(\mu, \tau; \nu) = 0$ as one of their bounding barriers:*

(a) V'_n for $\mu - \tau \in \mathbb{Z}$, $\mu + \tau \notin \mathbb{Z}$; unitarizable for $\mu < (-n+1)/2$;

(b) V'_n for $\mu \pm \tau \in \mathbb{Z}$, $\mu < (-n+1)/2$, unitarizable;

(c) $(V'_n \cap V_{k+1})/(V'_n \cap V_k)$, $k = n - q, \ldots, n$, for $\mu = (-n+q)/2$, $q = 1, \ldots, n$, $\mu \pm \tau \in \mathbb{Z}$ (all together $q + 1$ subfactors); exceptional subfactors; the unitarizability is available on the bottom level: the subspace $V'_n \cap V_{n-q+1}$, and on the top level: the coset space $V'_n/(V'_n \cap V_n) \cong V/V_n$.

(d) $(V'_n \cap V_{k+1})/(V'_n \cap V_k)$, $k = 0, 1, \ldots, n$, for $\mu > 0$, $\mu \pm \tau \in \mathbb{Z}$ (all together $n + 1$ subfactors); the unitarizability is available on the top level: the coset space $V'_n/(V'_n \cap V_n) \cong V/V_n$.

We can join unitarizable cases from (a) and (b): V'_n for $\tau - \mu \in \mathbb{Z}$, $\mu < (-n+1)/2$.

Recall some facts concerning Ω, see [5, 1]. It consists of complex matrices z of order n such that $E - z\bar{z}'$ is positive definite. The group $U(n)$ is the

Shilov boundary of Ω. The group $G = SU(n, n)$ acts on Ω as in (1.14): $z \mapsto \tilde{z} = z \cdot g = (z\beta + \delta)^{-1}(z\alpha + \gamma)$. An invariant under G measure on Ω is $[\det(E - z\bar{z}')]^{-2n} dz$, where $z = x + iy$, $x, y \in \mathrm{Mat}(n, \mathbb{R})$, $dz = \prod dx_{pq} dy_{pq}$.

Let $\mathcal{H}(\Omega)$ be the space of functions holomorphic on Ω that are the analytic continuation of functions on V to $U(n)$. Functions in Π_ν can be continued analytically from $U(n)$ to Ω for $\nu_n \geq 0$. Let $\Pi_\nu(\Omega)$ denote the set of these continuations ($\nu_n \geq 0$). For the continuation of the characters χ_ν we preserve the same symbol. Any function holomorphic on Ω decomposes into a series of functions in $\Pi_\nu(\Omega)$. Define the following sesqui-linear form for holomorphic functions on Ω:

$$B_\mu(\psi_1, \psi_2) = \int_\Omega \psi_1(z)\overline{\psi_2(z)}[\det(E - z\bar{z}')]^{2\mu^*} dz. \tag{6.1}$$

The integral (6.1) for functions in $\mathcal{H}(\Omega)$ converges absolutely for $\mathrm{Re}\,\mu < 1/2 - n$ and can be extended to other μ by analyticity as a meromorphic function. For $\psi_1 = \psi_2 = 1 = \chi_0$ the integral (6.1) was calculated in [5]§2.2:

$$B_\mu(1, 1) = \pi^{n^2} \prod_{k=1}^n \frac{\Gamma(2\mu^* + k)}{\Gamma(2\mu^* + n + k)}.$$

Let us return to the subfactors of $T_{\mu, \tau}$. First consider the subspace $V_n'(\mu, \tau)$ for $\mu - \tau \in \mathbb{Z}$. This includes cases (a), (b) and partially (c), (d) from Theorem 6.1. Highest weights of this subspace are specified by the condition $\nu_n \geq \tau - \mu$, the lowest weight is $(\tau - \mu)\mathbf{1}$.

Let us define a transformation $\varphi \mapsto \psi$ of the space V by

$$\psi(u) = \varphi(u)(\det u)^{\mu - \tau}. \tag{6.2}$$

It maps $V_n'(\mu, \tau)$ to the subspace $V^+ \subset V$ defined by $\nu_n \geq 0$. Under (6.2) the representation $T_{\mu, \tau}$ when restricted to $V_n'(\mu, \tau)$ becomes the representation T_μ^+ on V^+ defined by

$$(T_\mu^+(X)\psi)(u) = (L_X\psi)(u) + 2\mu\mathrm{tr}(u\eta + i\zeta) \cdot \psi(u), \tag{6.3}$$

where X is given by (1.2) (to compute (6.3) we have used (1.16)).

By analyticity, T_μ^+ gives rise to the representation T_μ^+ (we preserve the symbol) on $\mathcal{H}(\Omega)$, given by (6.33) with u replaced by z. The form B_μ is invariant with respect to T_μ^+. Normalize B_μ by 1 at $\chi_0 = 1$, i.e., consider the form $F_\mu = B_\mu(1, 1)^{-1}B_\mu$.

This form (as well as B_μ) is invariant with respect to T_μ^+. Therefore, it is just the invariant sesqui-linear form coming from $V'_n(\mu, \tau)$. In particular, subspaces $\Pi_\nu(\Omega)$ are pairwise orthogonal with respect to F_μ. Let us write out $N_\mu(\nu) = F_\mu(\chi_\nu, \chi_\nu)$:

$$N_\mu(\nu) = \prod_{k=1}^{n} \frac{(n-k+1)^{[\nu_k]}}{(-2\mu - k + 1)^{[\nu_k]}}, \qquad (6.4)$$

where $a^{[m]} = a(a+1)\ldots(a+m-1) = \Gamma(a+m)/\Gamma(a)$. This formula is proved by (6.3) with $X = Y_0$, see (1.9), and (3.14), (3.11), (3.15).

Theorem 6.2 *In order that the form F_μ be positive definite on $\mathcal{H}(\Omega)$ it is necessary and sufficient to have $\mu < (-n+1)/2$.*

The theorem follows immediately from (6.4).

If $\mu > (-n+1)/2$ and also $\mu + \tau \in \mathbb{Z}$, then the subspace V'_n is reducible, it contains the invariant subspaces $V'_n \cap V_r$. The map (6.2) takes $V'_n \cap V_r$ to the subspace V_r^+ of V^+ the highest weights that satisfy the inequality $\nu_r \leq 2\mu + r - 1$. The corresponding subspace $\mathcal{H}_{\mu,r}(\Omega)$ of $\mathcal{H}(\Omega)$ is defined by the same condition.

Consider case (c) from Theorem 6.1. The map (6.2) and its analytic continuation take the subfactor $(V'_n \cap V_{k+1})/(V'_n \cap V_k)$ to the subfactor $\mathcal{H}_{\mu,k+1}(\Omega)/\mathcal{H}_{\mu,k}(\Omega)$. The corresponding highest weights satisfy the inequalities

$$\nu_1 \geq \ldots \geq \nu_k \geq k + 2\mu \geq \nu_{k+1} \geq \ldots \geq \nu_n \geq 0. \qquad (6.5)$$

The lowest weight is $\nu(\mu, k) = (k + 2\mu, \ldots, k + 2\mu, 0, \ldots, 0)$, $(n - k$ zeros$)$. The invariant Hermitian form $F_{\mu,k}$ on this subfactor is obtained by analytic continuation with respect to λ of the form $N_\lambda(\nu(\mu, k))^{-1} B_\lambda$ from the domain $\text{Re}\,\lambda < 1/2 - n$ at the point $\lambda = \mu$. Here are its values at the characters:

$$F_{\mu,k}(\chi_\nu, \chi_\nu) = \prod_{i \leq k} \frac{(k + 2\mu + n - i + 1)^{[\nu_i - k - 2\mu]}}{(k - i + 1)^{[\nu_i - k - 2\mu]}} \prod_{j > k} \frac{(n - j + 1)^{[\nu_j]}}{(-2\mu - j + 1)^{[\nu_j]}}. \qquad (6.6)$$

The extreme subfactors ($k = n - q = -2\mu$ and $k = n$) are unitarizable. For $k = n - q = -2\mu$, the subfactor is the subspace $\mathcal{H}_{\mu,n-q+1}(\Omega)$; in this case inequalities (6.5) become $\nu_{k+1} = \ldots = \nu_n = 0$, the lowest weight is $\nu = 0$, so that $F_{\mu,k} = F_\mu$, its values at characters are

$$F_{\mu,k}(\chi_\nu, \chi_\nu) = \prod_{i=1}^{n-q} \frac{(n - i + 1)^{[\nu_i]}}{(-2\mu - i + 1)^{[\nu_i]}}. \qquad (6.7)$$

For $k = n$, the subfactor is $\mathcal{H}(\Omega)/\mathcal{H}_{\mu,n}(\Omega)$; in this case (6.5) becomes $\nu_n \geq n + 2\mu$; the lowest weight is $(n + 2\mu) \cdot \mathbf{1}$. The values of $F_{\mu,n}$ at characters are:

$$F_{\mu,n}(\chi_\nu,\chi_\nu) = \prod_{i=1}^{n} \frac{(2\mu + 2n - i + 1)^{[\nu_i-n-2\mu]}}{(n - i + 1)^{[\nu_i-n-2\mu]}}. \tag{6.8}$$

By (3.20) the latter subfactor for μ is equivalent to the space $\mathcal{H}(\Omega)$ for μ^*.

Case (d) from Theorem 6.1 is treated quite similarly to case (c), the same formulas (6.5), (6.6), (6.8) are true. But unitarizability now holds for $k = n$ only.

For $\mu \in \mathbb{R}$, denote by $\mathcal{F}_\mu(\Omega)$ the set of functions ψ holomorphic on Ω for which $F_\mu(\psi,\psi)$ exists (possibly, as the analytic continuation) and is positive for $\psi \neq 0$ (the Fock space). Then $\mathcal{F}_\mu(\Omega)$ is a Hilbert space [1]. From Theorem 6.3 and formula (6.7) we see that $\mathcal{F}_\mu(\Omega)$ exists for $\mu < (-n+1)/2$ and for $\mu = (-n + q)/2, q = 1, 2, \ldots, n$. In the former case this space is the completion of the whole $\mathcal{H}(\Omega)$ and in the latter case it is the completion of the subspace $\mathcal{H}_{\mu,n-q+1}(\Omega)$. The representation T_μ^+ on $\mathcal{F}_\mu(\Omega)$ is unitary. This family of values of the parameter μ was discovered by Berezin [1], see also [8, 12, 11].

For $\mu \in (1/2)\mathbb{Z}, \mu > -n/2$, the form $F_{\mu,n}$ is positive semi-definite on $\mathcal{H}(\Omega)$ and vanishes on the subspace $\mathcal{H}_{\mu,n}(\Omega)$. As above, we obtain a unitary representation of $\mathfrak{g}$ on the corresponding Hilbert space $\mathcal{F}_{\mu,n}(\Omega)$. It is equivalent to the representation on the Fock space $\mathcal{F}_{\mu^*}(\Omega)$.

Acknowledgments

The author was partially supported by grant 94-01-01603-a from the Russian Foundation for Basic Research and by grant JC 7100 from the International Science Foundation and the Russian Government.

References

1. Berezin, F.A.: Quantization on complex symmetric spaces, *Izv. Akad. nauk SSSR, Ser. Mat.* **39** (1975) 363–402; English transl.: *Math. USSR-Izv.* **9** (1975) 341–379.
2. Caillez, J. and Oberdoerffer, J.: Série complémentaire pour les groupes $SU(n, n, \mathbf{F})$, *C. R. Acad. Sci. Paris, Sér. 1* **297** (1983) no. 5, 279–281.
3. Erdelyi, A. and Magnus, W., Oberhettinger, F. and Tricomi, F.: *Higher transcendental functions, I*, McGraw-Hill, New York, 1953.
4. Faraut, J. and Koranyi, A.: Function spaces and reproducing kernels on bounded symmetric domains, *J. Funct. Anal.* **88** (1990) no. 1, 64–89.
5. Hua, L.K.: *Harmonic Analysis of Functions of Several Complex Variables in the Classical Domains*, Amer. Math. Soc., Providence, RI, 1963.
6. Johnson, K.D.: Degenerate principal series on tube domains, *Contemp. Math.* **138** (1992) 175–187.
7. Molchanov, V.F.: Representations of a pseudo-orthogonal group associated with a cone, *Mat. Sb.* **81** (1970) no. 3, 358–375: English transl.: *Math. USSR–Sb.* **10** (1970) no. 3, 333–347.

8.   Ørsted, B.: Composition series for analytic continuation of holomorphic discrete series of $SU(n, n)$, *Trans. Amer. Math. Soc.* **260** (1980) no. 2, 563–573.

9.   Sahi, S.: Unitary representation on the Shilov boundary of a symmetric tube domain, *Contemp. Math.* **145** (1993) 275–286.

10. Speh, B.: Degenerate series representations of the universal covering group of $SU(2, 2)$, *J. Funct. Anal.* **33** (1979) 95–118.

11. Vergne, M. and Rossi, H.: Analytic continuation of the holomorphic discrete series of a semi-simple Lie group, *Acta Math.* **136** (1976) 1–59.

12. Wallach, N.: The analytic continuation of the discrete series 1,2, *Trans. Amer. Math. Soc.* **251** (1979) 1–17, 19–37.

13. Zhelobenko, D.P.: *Compact Lie Groups and Their Representations*, Nauka, Moscow, 1970; English transl.: Amer. Math. Soc., Providence, RI, 1973.

ALMOST REPRESENTATIONS AND QUASI-SYMMETRY

A.I. SHTERN
Department of Mechanics and Mathematics,
Moscow State University, 119899 Moscow, Russia

Abstract. A quasi-homomorphism T of a semigroup S into a semitopological semigroup Q with uniform structure is a mapping $T : S \to Q$ with uniformly close $T(gh)$ and $T(g)T(h)$, $g, h \in S$. Roughly speaking, a quasi-representation is a quasi-homomorphism with $Q = L(E)$ for a topological vector space E. We discuss conditions on T that guarantee the existence of a representation R of S (in the same space E) for which $T - R$ is "small" in the corresponding uniformity. Such an R exists for bounded representations of amenable locally compact groups (and can depend on the choice of an invariant mean); but in general we can have no close ordinary representations. For this reason we consider a certain class of quasi-representations (the so-called pseudorepresentations) that have additional algebraic properties and approximate any quasi-representation with a reasonable error (depending on the measure of original nearness of $T(gh)$ and $T(g)T(h)$, $g, h \in G$).

Mathematics Subject Classification (1991): 22A25, 22D12, 43A65, 43A40.

Key words: quasi-representations, pseudorepresentations, amenability, approximate diagonal, quasi-characters, pseudocharacters.

1. Introduction

A quasi-homomorphism T of a semigroup S into a uniform semigroup Q is a mapping $T : S \to Q$ with uniformly close $T(gh)$ and $T(g)T(h)$, $g, h \in S$. Roughly speaking, a quasi-representation is a quasi-homomorphism for which Q is $L(E)$ with a topological vector space E. Therefore, quasi-representations form a particular class of the so-called almost represen-

B. P. Komrakov et al. (eds.), Lie Groups and Lie Algebras, 337–358.
© 1998 *Kluwer Academic Publishers. Printed in the Netherlands.*

tations, since the closeness conditions are posed not on a compactum or on a set of generators, but on the whole semigroup. The study of quasi-representations is of physical interest, as explained in Subsection 2.1.

We recall that the existence conditions for a homomorphism close to a given quasi-homomorphism have been studied in various terms for almost 50 years now (Gelfand, 1941; Hyers, 1941; de la Harpe and Karoubi, 1977; Johnson, 1986; Johnson, 1987; Johnson, 1988; Kazhdan, 1982; Lawrence, 1985; Pap, 1988; Shtern, 1980; Shtern, 1983; Shtern, 1982; Shtern, 1991b; Shtern, 1994a); examples of quasi-representations that admit no near-posed representations were given in (Faĭziev, 1987; Faĭziev, 1988; Kazhdan, 1982; Johnson, 1986; Johnson, 1987; Johnson, 1988; Shtern, 1983; Shtern, 1990; Shtern, 1991a; Shtern, 1991b). These examples motivated the definitions of a pseudocharacter (Shtern, 1983) and a pseudorepresentation (Shtern, 1991b). Note that the simplest part of the theory of almost representations, namely, the theory of pseudocharacters (Shtern, 1983; Shtern, 1990; Shtern, 1991b; Shtern, 1994c; Faĭziev, 1987; Faĭziev, 1988) proved its fruitfulness; in particular, this tool was recently applied in the study of bounded cohomology theory (Grigorchuk, 1995) and the theory of weight functions on groups (Grigorchuk, 1996). We also note that general almost representations (mappings that are nearly multiplicative on compacta) recently arose in the C^*-algebra setting, see (Cerry, 1994; Elliott e.a., 1995; Exel and Loring, 1989; Exel and Loring, 1991; Exel, 1993; Exel, 1994; Farsi, 1994).

The paper is organized as follows.

Section 2 contains an exposition of preliminaries on quasi-symmetry. After some discussion in Subsection 2.1 (of introductory nature) we list some general properties of quasi-representations and indicate their connection with almost multiplicative mappings of semigroup and group algebras, in the sense of (Johnson, 1986; Johnson, 1987; Johnson, 1988), see Subsection 2.2, and discuss properties of quasi-representations that contain zero in the closure of their orbits in Subsection 2.3. Here we consider conditions that guarantee the coincidence of a given quasi-representation with an ordinary representation on the whole space of the quasi-representation or on some its invariant subspace. This situation is related to the study of extensions of ordinary representations. In Subsection 2.4 we apply previous results, together with those presented below in Subsection 4.1, to the study of quasi-representations with unbounded orbits. In Subsection 2.5, the structure of finite-dimensional quasi-representations is determined. These results clarify the role of bounded quasi-representations and their unbounded quasi-extensions in the study of algebraic properties of general quasi-representations.

Section 3 is devoted to a special case related to quasi-symmetry, namely, to quasi-characters and pseudocharacters. In Subsection 3.1 we consider the simplest one-dimensional mappings connected with quasi-representations, namely, real quasi-characters (i.e., real functions f on the semigroup S with the property that the family $\{f(s_1 s_2) - f(s_1) - f(s_2), s_1, s_2 \in S\}$ is bounded) and real pseudocharacters (i.e., real quasi-characters whose restrictions to each commutative subsemigroup are ordinary homomorphisms of this subsemigroup); their general properties are studied and certain examples are treated. Among them there is an example connected with the Rademacher symbol (Shtern, 1990). In Subsection 3.2 we describe some results concerning continuous and Borel pseudocharacters on locally compact groups.

In Section 4, conditions for existence of representations close to bounded quasi-representations are specified. In Subsection 4.1, with the help of results of (Johnson, 1972a; Johnson, 1986; Johnson, 1988), the Johnson's amelioration is described and a class of quasi-representations with certain additional algebraic properties is introduced (the class of the so-called pseudorepresentations) in which, for every quasi-representation of a group, a "sufficiently close" mapping can be found; Subsection 4.2 is devoted to the discussion of the notion from the point of view of (de la Harpe and Karoubi, 1977). In Subsection 4.3, examples of quasi-representations of Lie groups and some free products of discrete groups are considered; these examples (whose construction uses the operation of quasi-induction) show that the conditions involved in the definition of pseudorepresentations are substantial.

This research was partially supported by the RFFI Grant 96-01-00276.

2. Preliminaries on quasi-symmetry

2.1. INTRODUCTION

A quasi-homomorphism T of a semigroup S into a uniform semigroup Σ (in more exact form: a U-quasi-homomorphism for a given neighboring U from the uniformity in Σ) is a mapping $T : S \to \Sigma$ with uniformly close elements $T(s_1 s_2)$ and $T(s_1) T(s_2)$ (i.e., $(T(s_1 s_2), T(s_1) T(s_2)) \in U$ for all $s_1, s_2 \in S$). The study of such mappings is related to the unification of the ideas of symmetry and nearness (and the corresponding problem has been posed already in the book (Ulam, 1960) in connection with the results of (Hyers, 1941) and (Hyers and Ulam, 1945)), but it also has a certain physical meaning that might be described as follows. Suppose a group G, the symmetry group for the description of some physical system, possesses quasi-representations (i.e., quasi-homomorphisms into the group of invertible continuous linear operators in a certain topological vector space) with the value of the neighboring U defined by the exactness of measurements but with no "sufficiently near-posed" ordinary representation of the group

G in the same topological vector space; then the symmetry properties of the system are more complicated than for the case in which a near-posed representation does exist, and it can require distinguishing true symmetries (connected with "the laws of nature") from quasi-symmetries.

Our exposition is mainly based on (Johnson, 1972a; Johnson, 1972b; Johnson, 1986; Johnson, 1987; Johnson, 1988; Shtern, 1983; Shtern, 1990; Shtern, 1991a; Shtern, 1991b; Shtern, 1993b; Shtern, 1994a; Shtern, 1994b; Shtern, 1994c; Shtern, 1997). We use the standard notation and well-known facts of representation theory without special comments; the necessary background is covered, for instance, by (Naimark and Shtern, 1982; Zhelobenko and Shtern, 1983; Lyubich, 1985).

2.2. QUASI-REPRESENTATIONS. DEFINITIONS AND MAIN PROPERTIES

2.2.1. *Definition of a quasi-representation. Operations*
Definition 2.2.1. Let S be a semigroup and let E be a topological vector space. A mapping T of the semigroup S to the algebra $L(E)$ of continuous linear operators in E is said to be a *quasi-representation* (to be more precise, a *U-quasi-representation*, where U is a given equicontinuous set in $L(E)$) of the semigroup S in E if the set $U(T)$ defined by

$$U(T) = \{T(s_1 s_2) - T(s_1)T(s_2), \ s_1, s_2 \in S\},$$

is equicontinuous (respectively, if we have $U(T) \subset U$).

If E is a locally convex space with a defining family of seminorms $P = \{p\}$ and $\varepsilon \geq 0$, then the mapping T is said to be an *ε-quasi-representation with respect to the family P*, if

$$p(T(s_1 s_2)x - T(s_1)T(s_2)x) \leq \varepsilon p(x), \qquad s_1, s_2 \in S, \quad x \in E, \quad p \in P.$$

This definition is specialized for locally convex spaces and metrizable spaces. In particular, let E be a linear metric space, let d is be a metric in E, and let $\varepsilon \geq 0$, then the mapping T is said to be *ε-quasi-representation with respect to d* if

$$d(T(s_1 s_2)x, T(s_1)T(s_2)x) \leq \varepsilon d(x, 0), \qquad s_1, s_2 \in S, \quad x \in E.$$

In normed spaces, ε-quasi-representations with respect to the single family $\{\|\cdot\|\}$ are exactly ε-quasi-representations with respect to the metric $\|\cdot\|$. For brevity, they are simply called *ε-quasi-representations* in the corresponding normed space.

One of the main problems related to quasi-representations is in the study of conditions that guarantee the existence of an ordinary representation close to a given quasi-representation and in the investigation of properties of the collection of these representations (unless it is empty).

Definition 2.2.2 Let G be an amenable locally compact group, B is a Banach space. The group G is said to be *stably representable* in the Banach space B if for any $\varepsilon > 0$ and $C > 0$ there exists a $\delta > 0$ such that if T is a strongly continuous δ-quasi-representation of the group G in B by invertible operators such that $\|T(g)\| \leq C$ and $\|T(g)^{-1}\| \leq C$ for any $g \in G$ then there exists a representation R of the group G in B^* such that $\|T(g)^* - R(g)\| \leq \varepsilon$ for any $g \in G$ and the matrix elements $(R(g)F)(x)$ are continuous for $F \in B^*$, $x \in B$.

There are some natural operations, namely, the restriction to a subgroup, the direct sum of quasi-representations, tensor product and quasi-induction for bounded quasi-representations, integration and disintegration of bounded quasi-representations in Hilbert spaces.

2.2.2. *Almost multiplicative mappings and bounded*
ε-quasi-representations

Bounded ε-quasi-representations of a locally compact group G are related to the so-called almost multiplicative mappings (Johnson, 1986; Johnson, 1987; Johnson, 1988) of the group algebra $L^1(G)$ of the group G.

Namely, let T be an *essential ε-quasi-representation* of the locally compact group G in a normed space E (i.e., we have $\|T(g_1 g_2) - T(g_1)T(g_2)\| \leq \varepsilon$ for almost all $(g_1, g_2) \in G \times G$), and let T be essentially bounded (i.e., let T satisfy the condition $\|T(g)\| \leq C$ for some $C > 0$ and for almost all $g \in G$ with respect to the Haar measure on G); moreover, let $L(E)$ be the algebra of all bounded linear operators on the space E. Then the formula

$$T(f) = \int_G f(g)T(g)\, dg, \qquad f \in L^1(G),$$

where dg is a left Haar measure on G, defines an almost multiplicative mapping of the group algebra $L^1(G)$, where the linear mapping $T : L^1(G) \to L(E)$ is said to be almost multiplicative if

$$\|T(f_1 * f_2) - T(f_1)T(f_2)\| \leq \varepsilon\|f_1\|\|f_2\|$$

for any $f_1, f_2 \in L^1(G)$.

Conversely, suppose that S is an almost multiplicative mapping of the group algebra $L^1(G)$ into the algebra of bounded linear operators on a separable Banach space E conjugate to some Banach space E_* (which is automatically separable). Then there exists a mapping T of the group G into the space $L(E)$ such that the function $g \mapsto (T(g)\xi, \eta)$, $g \in G$, is measurable and essentially bounded for any $\xi, \eta \in E$; moreover, we have

$$(S(f)\xi, \eta) = \int_G f(g)(T(g)\xi, \eta)\, dg,$$

and the norm of the mapping T is equal to the essential supremum of the function $g \mapsto \|T(g)\|$, $g \in G$ (this is based upon (Dunford and Schwartz, 1958), Theorem VI.8.2, and upon the fact that the Banach space $L(E)$ is isometric to the conjugate space of a separable Banach space, which is the projective tensor product of E_* and E, cf. (Shaefer, 1966)).

2.2.3. ε-quasi-equivalence of quasi-representations and related topics

Definition 2.2.3 Quasi-representations R and T of the group G in F-spaces E_R and E_T, respectively, are said to be ε-quasi-equivalent if there exists a strongly continuous mapping A of the group G into the set of continuously invertible continuous linear operators from E_R into E_T such that for some pseudonorms $|\cdot|_R$ and $|\cdot|_T$ in E_R and E_T, respectively, the inequalities $|(A(g_1) - A(g_2))x| \leq \varepsilon|x|$ and $|(A^{-1}(g_1) - A^{-1}(g_2))y| \leq \varepsilon|y|$ hold for any $g_1, g_2 \in G$, $x \in E_R$ for which $|x|_R \leq 1$, and $y \in E_T$ for which $|x|_T \leq 1$, and we have $A(g)R(g) = T(g)A(g)$ for any $g \in G$. Quasi-representations R and T of the group G in F-spaces E_R and E_T, respectively, are said to be ε-equivalent if there exists a finite chain R_i, $i = 1, \ldots, n$, of pairwise ε-quasi-equivalent quasi-representations (in F-spaces) for which $R_1 = R$ and $R_n = T$.

Note that equivalence is 0-equivalence and 0-quasi-equivalence. For $\varepsilon > 0$, ε-quasi-equivalence can be not transitive.

Consider the following natural problem: Does there exist an $\varepsilon > 0$ for which (under additional assumptions) ε-equivalence of quasi-representations implies their ordinary equivalence? Even for ordinary representations of a given group G, this problem has different answers depending on whether G is amenable or not. For the notions related to amenability and related topics, see (Greenleaf, 1969; Paterson, 1988).

Proposition 2.2.1 (Shtern, 1991a) *Let G be an amenable locally compact group, let T and S be strongly continuous representations of this group in a Hilbert space H, and suppose that T is unitary, S is bounded, and representations T and S are ε-quasi-equivalent, whereas for any $g \in G$ the inequalities $\|A(g)\| \leq M$ and $\|A(g)^{-1}\| \leq M$ hold. Suppose that the representation S satisfies the condition $\|S(g)\| \leq C$ for any $g \in G$. Then the representations T and S are equivalent by a continuous linear operator B in the space H for which we have $\|B - A(g)\| \leq \varepsilon CM$ for any $g \in G$ and $\|B\| \leq M$.*

Theorem 2.2.1 (Shtern, 1991a) *Let E be a Fréchet space with barreled strong conjugate space E^*, let S be a left amenable semitopological semigroup, let R be a strongly continuous representation of the semigroup S in the space E by isometric operators mapping the space E onto itself, and let T be a weakly continuous representation of the semigroup S in the space E such that for some strictly monotone pseudonorm $|\cdot|$*

on E defining the topology on E and for some $q \in [0, 1)$ the condition $|T(s)x - R(s)x| \le q|x|$ holds for any $s \in S$ and any $x \in E$ with $|x| \le \varepsilon$. If the mapping $s \mapsto f(R(s)^{-1}T(s)x)$, $s \in S$, is continuous for all $f \in E^*$ and all $x \in E$, then the dual representations T^* and R^* are equivalent, and if $\|\cdot\|$ is a pseudonorm on E^{**} (where the stars denote the passage to the second strong conjugate space) defined by the condition that for any $x \in E^{**}$ the number $\|x\|$ is determined as the greatest lower bound of $c > 0$ for which x belongs to the bipolar of the set $\{y : |y| \le c\}$, then the equivalence of the representations T^* and R^* can be realized by an invertible continuous linear operator Z in E for which $\|Z^*x - 1_{E^{**}}x\| \le q\|x\|$ for any $x \in E^{**}$ such that $\|x\| \le \varepsilon$.

Theorem 2.2.1 means that the "ε-quasi-equivalence problem" for ordinary representations is related to a natural topology in the dual space of the group, namely, to a "uniform convergence topology on the group" (or semigroup) for representations in a given Hilbert space. It is worth comparing this situation with known compactness properties of the ordinary dual space (e.g., see (Shtern, 1971a; Shtern, 1971b)).

Example 2.2.1 (Shtern, 1991a) Consider the operators T and R in the space c of convergent complex sequences given as follows:

$$Rx = \{0, x_1, x_2, \ldots, x_{n-1}, \ldots\}, \qquad x = \{x_1, x_2, \ldots, x_n, \ldots\} \in c,$$

and

$$Tx = \{0, \lambda_1 x_1, \lambda_2 x_2, \ldots, \lambda_n x_{n-1}, \ldots\}, \qquad x = \{x_1, x_2, \ldots, x_n, \ldots\} \in c,$$

where $\lambda_n = e^{i\mu_n}$, $\mu_n > 0$, $\mu_n \to 0$, and the series $\sum_n \mu_n$ is divergent. Then the operators T^* and R^* are similar in $L(l^1)$, but T and R are not similar in the space $L(c)$ of continuous linear operators in c.

Thus, the passage to conjugate representations is meaningful.

Remark 2.2.1 Any ε-quasi-representation is ε-quasi-equivalent to an ε-quasi-representation that maps the identity element to the identity operator in the quasi-representation space.

Corollary 2.2.1 Let T and R be weakly continuous bounded representations of a left amenable topological group G in a Hilbert space H. If we have $\|T(g)\| \le C$ and $\|R(g)\| \le C$ for all $g \in G$ and there exists $q \in [0, 1)$ such that $\|T(g) - R(g)\| \le qC^{-1}$, then the representations T and R are equivalent by an operator Z for which we have $\|Z - 1_H\| \le q$.

Hence, for left amenable topological groups the dual space topology mentioned above is discrete. This is not true in general.

Example 2.2.2 (based upon (Kunze and Stein, 1960)) Let $G = \mathrm{SL}(2, \mathbb{R})$ be the real unimodular group of 2×2-matrices and let $g = \begin{pmatrix} a & b \\ c & d \end{pmatrix}$ for $g \in G$.

As is shown in (Kunze and Stein, 1960), there exist families of operator-valued functions $U^+(g, s)$ and $U^-(g, s)$, $g \in G$, $0 < \Re(s) < 1$, in a Hilbert space H such that $U^+(g, s)$ and $U^-(g, s)$ are continuous and weakly analytic with respect to s for fixed $g \in G$ and $U^+(g, s)$ and $U^-(g, s)$ are uniformly bounded (in the operator norm) in the strip $0 < \Re(s) < 1$, moreover, for $b = 0$ the operator families $U^+(g, s)$ and $U^-(g, s)$ do not depend on $s \in \mathbb{C}$. This implies that for the element $j \in G$ with $a = d = 0$ and $b = -d = 1$ the operators $U^+(j, s)$ and $U^-(j, s)$ depend on s continuously in the sense of the operator norm for $0 < \Re(s) < 1$, i.e., in this realization the formula $G = (jB) \cup (jBjB)$ implies that for small $|s - s_0|$ and for $\Re(s), \Re(s_0) \in (0, 1)$ the norms $\|U^+(g, s) - U^+(g, s_0)\|$ and $\|U^-(g, s) - U^-(g, s_0)\|$ are uniformly small on G, although the representations $U^+(g, s)$ and $U^+(g, s_0)$ (as well as $U^-(g, s)$ and $U^-(g, s_0)$) are equivalent only for $s = s_0$ or $s = 1 - s_0$; thus, for the group $SL(2, \mathbb{R})$ the "uniform convergence topology on the group" for representations of this group is not discrete.

Theorem 2.2.2 (Shtern, 1994b; Shtern, 1994c) *An almost connected locally compact group is amenable if and only if the "uniform convergence topology on the group" is discrete for the representations in any given Hilbert space.*

We conjecture that an arbitrary locally compact group is amenable if and only if the "uniform convergence topology on the group" is discrete for the representations in any given Hilbert space.

2.3. STRUCTURE OF QUASI-REPRESENTATIONS WHOSE ORBIT CLOSURES CONTAIN THE ORIGIN

Theorem 2.3.1 (Shtern, 1991b; Shtern, 1994a) *Let E be a topological vector space, let G be a group, and let T be a quasi-representation of the group G by continuous linear operators in the space E. Assume that the operators of the quasi-representation T are invertible and there exists an element $x \in E$ such that the closure $O_{T^{-1}}(x)^-$ of the T^{-1}-orbit $O_{T^{-1}}(x) = \{T(g)^{-1}x, g \in G\}$ of the element $x \in E$ contains the zero element of the space E. Then the close linear hull F of the orbit $O_{T^{-1}}(x)$ is invariant under T and the restriction of the mapping T to F is an ordinary representation of the group G.*

Remark 2.3.1 If E is finite-dimensional, then the condition that the quasi-representation operators are invertible holds automatically if the size of the set $U(T) = \{T(s_1 s_2) - T(s_1)T(s_2), s_1, s_2 \in S\}$ is sufficiently small (e.g., if E is endowed with some norm defining its topology and $U(T)$ is contained in the ball with center at the origin and with radius strictly lesser than one, because $\|T(g^{-1})T(g) - T(e)\|$ is less than one) and the operator $T(e)$ can be considered as the identity operator, as mentioned in Remark 2.2.1.

Thus, in finite-dimensional spaces, the statement of Theorem 2.3.1 holds for any quasi-representation whose orbit closure (for some nonzero vector) contains the origin. Theorem 2.3.1 generalizes the corresponding results of (Lawrence, 1985; Shtern, 1980; Shtern, 1982) related to one-dimensional and finite-dimensional quasi-representations.

Corollary 2.3.1 (Shtern, 1991b; Shtern, 1994a) *Let F be a finite-dimensional vector space, let G be a group, and let T be a quasi-representation of the group G in the space E. Suppose that there exists an element $x \in G$ for which the closure $O_{T^{-1}}(x)^-$ of the T^{-1}-orbit $O_{T^{-1}}(x) = \{T(g)^{-1}x, g \in G\}$ of the element $x \in E$ contains the zero element of the space E. Then the close linear hull F of the orbit $O_{T^{-1}}(x)$ is invariant under T and the restriction of the mapping T to F is an ordinary representation of the group G. Let F be a complementary subspace of F, and let the operator $T(g)$ be written in the block form*

$$T(g) \begin{pmatrix} f \\ w \end{pmatrix} = \begin{pmatrix} \lambda(g)f + \mu(g)w \\ \nu(g)w \end{pmatrix}, \qquad f \in F, \; w \in W, \; g \in G,$$

where λ, μ, and ν are some matrix-valued mappings. Let L be the linear hull of the images of the operators of the form $\mathrm{Im}(\nu(k)\nu(h) - \nu(kh))$, $k, h \in G$, and let N be a projection in W onto L. If $N = 1_W$, then T is similar to the direct sum of the mapping ν and the representation λ of the group G. If $N \neq 1_W$ and if M is the strong limit of the net $\lim_\alpha \lambda(g_\alpha^{-1})\mu(g_\alpha)N$, then the operator

$$R = \begin{pmatrix} 0 & -M \\ 0 & N \end{pmatrix}$$

is a projection satisfying the condition $TR = RTR$. If $S = 1 - R$, then $\mathrm{Im}(S)$ is not contained in F, and the mapping $g \to ST(g)|_{\mathrm{Im}\,S}$, $g \in G$, is a representation of the group G.

2.4. STRUCTURE OF QUASI-REPRESENTATIONS WHOSE ORBITS ARE UNBOUNDED

Theorem 2.4.1 (Shtern, 1997) *Let E be a Banach space, let G be a group, and let T be a quasi-representation of the group G by continuous linear operators in the space E. Assume that the operators of the quasi-representation T are invertible and that there exists an element $x \in E$ such that the T-orbit $O_T(x)$ is unbounded. Let F be the linear subspace of E formed by elements that determine bounded orbits.*

The subspace F is $T(g)$-invariant for all $g \in G$.

If the subspace F is closed and if there exists an ordinary representation R of G in F such that $I_{g \in G}(R(g^{-1}T(g)|_F)$ is invertible (this is the case if the constant that bounds the norms of both the restrictions of the operators

$T(g)$, $g \in G$, to the subspace F and the inverses of these restrictions, and if $T|_F$ is a δ-quasi-representation with sufficiently small δ, see Theorem 4.1.2 below), then the mapping T is ε-close to an ordinary representation, where ε, δ, and C are related as in Theorem 4.1.2.

Corollary 2.4.1 (Shtern, 1997) Let E be a Banach space, let G be a group, and let T be a quasi-representation of the group G by continuous linear operators in the space E. Assume that the operators of the quasi-representation T are invertible and that, for any nonzero $x \in E$, the T-orbit $O_T(x)$ is unbounded. Then the mapping T is an ordinary representation of the group G.

2.5. FINITE-DIMENSIONAL QUASI-REPRESENTATIONS: STRUCTURE AND RELATED QUASI-COCYCLES

Theorem 2.5.1 Let G be a group, and let T be a quasi-representation of G in a finite-dimensional vector space E_T. Let E_T^* be the conjugate space of E_T. Let L be the set of vectors $\xi \in E_T$ for which the orbit $\{T(g)\xi, g \in G\}$ is bounded; let M be the set of functionals $f \in E_T^*$ for which the orbit $\{T(g)^*f, g \in G\}$ is bounded in E_T^*; then L and the annihilator $M^\perp$ are vector subspaces in E_T invariant under T. Consider the collection of subspaces $\{0\}$, $L \cap M^\perp$, $M^\perp$, $L + M^\perp$, and $E = E_T$ (directed upwards) and write the matrix $t(g)$ of the operator $T(g)$, $g \in G$, in the block form (with respect to the decomposition of the space E into the direct sum of subspaces $L \cap M^\perp$, $M^\perp \setminus (L \cap M^\perp)$, $L \setminus (L \cap M^\perp)$, and $E \setminus (L + M^\perp)$, where "$\setminus$" denotes the choice of a complementary subspace):

$$t(g) = \begin{pmatrix} \alpha(g) & \varphi(g) & \sigma(g) & \tau(g) \\ 0 & \beta(g) & 0 & \rho(g) \\ 0 & 0 & \gamma(g) & \chi(g) \\ 0 & 0 & 0 & \delta(g) \end{pmatrix}, \qquad g \in G.$$

(Here we have $t_{23}(g) = 0$, because L is invariant under T.) Then the following assertions hold:

1) the mappings α, δ, γ, σ, and χ are bounded;
2) the mappings t_1 and t_2 defined by

$$t_1(g) = \begin{pmatrix} \alpha(g) & \varphi(g) \\ 0 & \beta(g) \end{pmatrix}, \qquad t_2(g) = \begin{pmatrix} \beta(g) & \rho(g) \\ 0 & \delta(g) \end{pmatrix},$$

are representations of the group G;

3) the mapping τ is a quasi-cocycle with respect to the representations t_1 and t_2, i.e., the mapping

$$(g, h) \to \tau(gh) - \alpha(g)\tau(h) - \varphi(g)\rho(h) - \tau(g)\delta(h), \qquad g, h \in G, \qquad (1)$$

is bounded.

Theorem 2.5.2 *For any amenable locally compact group, any its finite-dimensional unbounded quasi-representation is a bounded perturbation of an ordinary representation of this group in the same space, and any bounded finite-dimensional quasi-representation of the group is close to an ordinary finite-dimensional representation (which means that, for any $\varepsilon > 0$ and $C > 0$, there exists $\delta > 0$ such that if T is a continuous finite-dimensional δ-quasi-representation of the group G in B, and if the function (1) is bounded by a constant C for all $g \in G$, then there exists a continuous representation R of the group G, in the same space B, such that $\|T(g) - R(g)\| < \varepsilon$ for all $g \in G$). The perturbed representation is continuous provided that the original quasi-representation is continuous.*

Remark 2.5.1 Let G be a group. Consider the mapping of the form

$$T(g) = \begin{pmatrix} 1_E & \tau(g) \\ 0 & 1_F \end{pmatrix}, \qquad g \in G,$$

in the direct sum of linear spaces $E + F$, where $\tau(g)$ is a continuous linear operator from F to E. The mapping T is a quasi-representation if and only if the family of operators

$$\{\tau(g_1 g_2) - \tau(g_1) - \tau(g_2), \ g_1, g_2 \in G\}$$

is bounded; in this case we say that the mapping τ is an *additive quasi-representation* of the group G in the operator space $L(F, E)$. For the important special case in which F and E are one-dimensional, this is the main object from the point of view discussed in the next section.

3. One-dimensional additive quasi-symmetry

3.1. QUASI-CHARACTERS AND PSEUDOCHARACTERS. APPLICATIONS TO ADDITIVE QUASI-REPRESENTATIONS

Definition 3.1.1 Let S be a semigroup. A real function f on S is called a *(real) quasi-character* on S if the set $\{f(st) - f(s) - f(t), \ s, t \in S\}$ is bounded.

Thus, a (real) quasi-character is a special case of an additive quasi-representation (see Remark 2.4.1).

Definition 3.1.2 Let S be a semigroup. A quasi-character $f \, S \to \mathbb{R}$ is called a *(real) pseudocharacter* on S if $f(x^n) = nf(x)$ for all $x \in S$ and all $n \in \mathbb{N}$.

Theorem 3.1.1 (Shtern, 1983; Shtern, 1990); see also (Bouarich, 1995; Shtern, 1991b; Shtern, 1994a) *Let S be a semigroup and let f be a quasi-character on S.*

a) For any $s \in S$, the limit $\varphi(s) = \lim_{n \to \infty} n^{-1} f(s^n)$ exists, and if we have $|f(st) - f(s) - f(t)| \leq C$ for all $s, t \in S$, then $|f(s) - \varphi(s)| \leq C$ and $|\varphi(st) - \varphi(s) - \varphi(t)| \leq 4C$ for all $s, t \in S$.

b) If R is an (at least one-sided) amenable subsemigroup of S, then the restriction of φ to R is an ordinary (additive) character of R, i.e., $\varphi(st) = \varphi(s) + \varphi(t)$ for all $s, t \in S$. Moreover, the number $\varphi(s)$ can be defined as the common value of all invariant means of the semigroup on the bounded function $t \mapsto f(ts) - f(t)$, $t \in R$.

c) $\varphi(xy) = \varphi(yx)$ for all $x, y \in S$.

d) Let $S = G$ be a group, let N be a normal subgroup of G, and let π be the canonical epimorphism of G onto G/N. If a pseudocharacter φ vanishes on N, then there exists a pseudocharacter ψ on the group G/N such that $\varphi = \psi \circ \pi$.

Corollary 3.1.1 (Shtern, 1983; Shtern, 1990), see also (Shtern, 1991b; Shtern, 1994a) a) *A bounded pseudocharacter (in particular, a pseudocharacter on a semigroup with the zero element) is identically zero.*

b) *For any quasi-character f on a semigroup S there exists a unique pseudocharacter φ on S for which the difference $f - \varphi$ is bounded; moreover, for any $s \in S$ the limit*

$$\lim_{n \to \infty} n^{-1}(f(s^{n+k}) - f(s^k))$$

exists uniformly with respect to $k \in \mathbb{Z}$ and is equal to $\varphi(s)$. The mapping $f \to \varphi$, assigning to any quasi-character the related pseudocharacter, is a linear projection. If $\| \cdot \|$ is a quasi-norm in the linear space of real quasi-characters defined by the condition that $\|f\|$ is the radius of the smallest closed ball with center at the origin containing the set $\{f(st) - f(s) - f(t) \mid s, t \in S\}$, then the mapping $f \mapsto \varphi$ is continuous.

c) *For any pseudocharacter f on the semigroup S, the relation $f(xy) = f(yx)$ holds for all $x, y \in S$.*

d) *If the semigroup S is (one-sided) amenable, then any pseudocharacter on S is a homomorphism of S into $\mathbb{R}$.*

Assertion d) was also obtained by (Forti, 1987).

Example 3.1.1 (Shtern, 1983; Shtern, 1991b) (a nontrivial pseudocharacter) Let M be a group and let M contain subsets M^+ and M^- such that $M^+ \neq \emptyset$, $M^- = (M^+)^{-1}$, $M^+ \cap M^- = \emptyset$, and $M = M^+ \cup M^- \cup \{e\}$. Let us introduce the mapping sign : $M \to \mathbb{Z}$ by setting $\text{sign}(m)$ equal to 1 for $m \in M^+$, -1 for $m \in M^-$, and $\text{sign}(e) = 0$.

Let N be a group, $N \neq e$. We introduce the function f on the free product $G = M * N$ by setting $f(g) = \sum_{i=1}^{k+1} \text{sign}(m_i)$ for $g = (\prod_{i=1}^{k} m_i n_i) m_{k+1}$, where $m_i \in M$, $n_i \in N$, $n_i \neq e$ for all $i = 1, \ldots, k$ and $m_i \neq e$ for $i = 2, \ldots, k$. Then the function f is a quasi-character on G with values in $\mathbb{Z}$ and $|f(g_1 g_2) - f(g_1) - f(g_2)| \leq 3$ for all $g_1, g_2 \in G$.

Remark 3.1.1 The paper (Kazhdan, 1982) contains the example of a family of finite-dimensional quasi-representations of the fundamental group of a Riemann surface with arbitrarily small ε for which the distances from these quasi-representations to ordinary representations have positive lower bound. However, the dimensions of these quasi-representations are not arbitrary. In this connection, we note that the quasi-character constructed in Example 1.1 permits us to obtain an example of a one-dimensional quasi-representation which is not close to ordinary (unitary) characters. Indeed, if f is the quasi-character described above on the given group G for which the corresponding pseudocharacter is not an ordinary (additive) character, then, for any $\alpha \in G$, the function $\chi = \chi_\alpha$, where $\chi_\alpha(g) = \exp(i\alpha f(g))$, $g \in G$, is a one-dimensional quasi-representation of the group G. The distance between χ and the set of characters of the group G does not tend to zero as $\alpha \to 0$, $\alpha \neq 0$.

Example 3.1.2 (Shtern, 1990) (the pseudocharacter related to the Rademacher symbol) Let η be the Dedekind η-function on the upper half-plane C_+, i.e., $\eta(\tau) = e^{\pi i \tau/12} h(e^{2\pi i \tau})$, where

$$h(x) = \prod_{m=1}^{\infty}(1 - x^m), \qquad x = e^{2\pi i \tau}, \qquad \Im \tau > 0.$$

Let us use the well-known formulas for the logarithm of the η-function: if we choose a branch of the logarithm for which $\ln \eta(i) \in \mathbb{R}$, then

$$\log \eta(\tau') = \log \eta(\tau) + ((\text{sign}(c))^2 \log(c\tau+d))/2 - (\pi i/4)\text{sign}(c) + (\pi i/12)\Phi(g),$$

where

$$g = \begin{pmatrix} a & b \\ c & d \end{pmatrix} \in \text{SL}(2, \mathbb{Z}),$$

$\tau' = (\tau + b)/(c\tau + d)$, and Φ is the mapping of the group $G = \text{PSL}(2, \mathbb{Z})$ into $\mathbb{R}$ satisfying the relation

$$\Phi(gg') = \Phi(g) + \Phi(g') - 3\,\text{sign}(cc'c'')$$

for all $g, g' \in G$, where

$$g = \begin{pmatrix} a & b \\ c & d \end{pmatrix}, \qquad g' = \begin{pmatrix} a' & b' \\ c' & d' \end{pmatrix}, \qquad g'' = \begin{pmatrix} a'' & b'' \\ c'' & d'' \end{pmatrix},$$

and this means that Φ is a quasi-character on G taking its values in $\mathbb{Z}$ (cf. (Rademacher, 1932)). The pseudocharacter on G related to Φ is calculated (see (Shtern, 1990)).

3.2. APPLICATIONS, TRIVIALITY AND CONTINUITY

3.2.1. *Applications to linear operators*

Proposition 3.2.1 (Shtern, 1991b; Shtern, 1994a) (application to the Hyers theorem) *Let E be a weakly sequentially complete locally convex space, let G be a group such that any two elements of G belong to an amenable subgroup of G, and let f be a mapping of G to E for which the elements of the form $f(gh) - f(g) - f(h)$ belong to some bounded subset $O_f \subset E$ for all $g, h \in G$. Then the mapping f is a sum of a homomorphism $\Phi : G \to E$ and a mapping of G into the balanced convex hull of the set O_f. If E is a Baire space and f is continuous, then the both summands (Φ and $f - \Phi$) are continuous.*

This generalizes (Hyers, 1941) and (Johnson, 1988), Section 8.

3.2.2. *Continuous and Borel pseudocharacters on locally compact groups*

First we note the following additional property of continuous pseudocharacters on locally compact groups.

Proposition 3.2.2 (Shtern, 1993b) *Under the assumptions and notation of Theorem 3.1.1 d), assume that G is a locally compact group, N is closed, and φ is continuous. Then ψ is continuous.*

Now the following proposition shows that for a locally compact group, the notion of a continuous pseudocharacter is meaningful for disconnected groups only (see Theorem 3.1.1, d)).

Proposition 3.2.3 (cf. (Shtern, 1993b)) *Any continuous pseudocharacter on an almost connected locally compact group (not necessarily amenable) is an ordinary character.*

For continuous pseudocharacters on general locally compact groups we have the following description.

Proposition 3.2.4 (Shtern, 1997) *Any continuous pseudocharacter on a locally compact group G can be factored through either the disconnected quotient group of G (by the connected component G_0 of G) or through an extension of the additive group $\mathbb{R}$ of reals by the group $H = G/G_0$ that is defined by a bounded continuous system of factors on $H \times H$ (a bounded continuous 2-cocycle on H) related to the trivial action of H on $\mathbb{R}$.*

This is a corrected version of Corollary 3 in (Shtern, 1993b). I thank Professor Michael Grosser (Wien) for indicating a gap in my paper, see (Grosser, 1995).

However, continuous pseudocharacters form only a part of interesting pseudocharacters on topological groups. Namely, it follows from Theorem 3.1.1 a) that the pseudocharacter, on a topological group, related to a given continuous quasi-character belongs to the first Baire class on the group. For these pseudocharacters, the following assertion holds.

Proposition 3.2.5 (Shtern, 1997) *Any pseudocharacter on a locally compact group G that belongs to the first Baire class defines a continuous character on any amenable subgroup of G. This pseudocharacter can be factored (see Proposition 3.2.2) through either the disconnected quotient group of G (by the connected component G_0 of G), and in this case the original pseudocharacter is continuous, or through an extension of the additive group $\mathbb{R}$ of reals by the group $H = G/G_0$ that is defined by a bounded continuous system of factors on $H \times H$ (a bounded continuous 2-cocycle on H), or through an extension of the additive group $\mathbb{R}$ of reals by the group that is an extension of a semisimple Lie group by the totally disconnected group $H = G/G_0$, where both extensions are defined by bounded continuous systems of factors (bounded continuous 2-cocycles) related to the trivial action of H on $\mathbb{R}$.*

Thus, nontrivial extensions of $\mathbb{R}$, with trivial action, by semisimple groups, related to bounded continuous 2-cocycles, necessarily define discontinuous pseudocharacters on the extension that are continuous characters on any closed amenable subgroup.

The possibilities listed in Proposition 3.2.5 can occur indeed, as can be shown by examples related to the Guichardet–Wigner cocycle on the group $G = \mathrm{SL}(2, \mathbb{R})$ (Guichardet and Wigner, 1978), see also (Fuks, 1987), and to the restriction of this cocycle to $H = \mathrm{SL}(2, \mathbb{Z})$. For instance, let m be the mapping $m G \times G \to \mathbb{R}$ such that $m(g, h)$ is the oriented Lobachevski area of the triangle (on the upper half-plane $\mathbb{C}^+$ regarded as the Lobachevski plane) with the vertices z, gz, hgz, where $z \in \mathbb{C}^+$ (Fuks, 1987). Then the 2-cocycle m and its restriction to H are continuous, bounded and nontrivial.

Remark 3.2.1 It is obvious that any nontrivial pseudocharacter on a group G (or a semigroup S) defines a nontrivial pseudocharacter on a free group generated by generators of G (or S, respectively). This motivated the study of pseudocharacters on free objects (Faĭziev, 1987; Faĭziev, 1988), which was used by (Grigorchuk, 1995; Grigorchuk, 1996) in the problems related to bounded cohomology and weights on groups. It should be noted that in contrast to connected Lie groups and related Chevalley groups, Kac–Peterson groups (defined in (Kac and Peterson, 1985)) possess nontrivial characters (this is mentioned in (Shtern, 1991b)). We also note that the known example of a periodic nonamenable group (Ol'shanskiĭ, 1987) gives an explicit example of a discrete nonamenable group without nontrivial real characters; a description of groups without nontrivial real characters (in other terms related to the stability of quasi-representations) is given in (Forti, 1987).

4. Pseudorepresentations

4.1. QUASI-REPRESENTATIONS OF AMENABLE GROUPS. MAIN THEOREM

The study of quasi-representations of amenable groups was developed by many authors in different contexts with a natural result: in all particular cases quasi-representations are close (in a sense) to ordinary representations ((Forti, 1987; Grove e.a., 1974; de la Harpe and Karoubi, 1977; Kazhdan, 1982; Pap, 1988) etc.; a gap in the proof of the corresponding assertion in (Kazhdan, 1982) was pointed out in (Faĭziev, 1987)). For the sake of simplicity we discuss the proof of the main result in the Banach space setting (instead of more general Fréchet space setting, which is also available, cf. Theorem 2.2.1 and (Shtern, 1991a)).

We need the following part of the construction of B. E. Johnson.

Definition (cf. (Johnson, 1972a)) Let A be a Banach algebra, let $A \otimes A$ be its (projective) tensor square. A bounded net

$$d_\lambda = \sum_j a_{\lambda j} \otimes b_{\lambda j} \in A \otimes A, \qquad \lambda \in \Lambda,$$

is said to be an *approximate diagonal for A* if $\lim_\lambda (ad_\lambda - d_\lambda a) = 0$ for any $a \in A$ and the net $\{\pi(d_\lambda)\}$, $\lambda \in \Lambda$, is a bounded approximate unit in A, where $\pi : A \otimes A \to A$ is the canonical mapping defined by $(p \otimes q) = pq$, $p, q \in A$.

Theorem A (Johnson, 1972a; Johnson, 1972b) *The group algebra of an amenable locally compact group possesses an approximate diagonal.*

Definition (Johnson, 1972a; Johnson, 1972b) Let A and B be two Banach algebras. Denote by $\mathrm{Hom}(A, B)$ the set of continuous homomorphisms of A into B. The pair (A, B) is said to be an AMNM-*pair* (almost multiplicative maps are near multiplicative) if for any $\varepsilon > 0$ and $K > 0$ there exists $\delta > 0$ such that if $T \in L(A, B)$, $\|T\| \leq K$ and $\|T^\vee(a_1, a_2)\| \leq \|a_1\| \, \|a_2\|$, where

$$T^\vee(a_1, a_2) = T(a_1 a_2) - T(a_1)T(a_2), \qquad a_1, a_2 \in A,$$

then

$$\inf\{\|T - R\| : R \in \mathrm{Hom}(A, B)\} < \varepsilon.$$

Theorem B (Johnson, 1988) *Let A be a Banach algebra with an approximate diagonal. Suppose that B is a Banach algebra such that there exists a Banach B-bimodule B_* such that B is isomorphic as a B-bimodule with $(B_*)^*$. Then the pair (A, B) is AMNM.*

The following theorem unifies, generalizes and ameliorates results of (Grove e.a., 1974; de la Harpe and Karoubi, 1977; Kazhdan, 1982).

Theorem 4.1.1 *Let G be an amenable locally compact group, let B be a Banach space. Then the group G is stably representable in B, and we can take $\delta \le (4 + 8C^2)^{-1}\varepsilon$.*

The original proof of the corresponding theorem in (Johnson, 1988) is based on the formulas $T' = T + S$ and

$$S(a) = \lim_\lambda \sum_j T(a_{\lambda j}) T^\vee(b_{\lambda j}, a), \qquad a \in A,$$

which define the *Johnson's amelioration T'* of the mapping T. Under appropriate conditions we have $\lim T_n = R$, where $T_{n+1} = T'_n$ (Johnson, 1988). Certainly, the Johnson's amelioration depends on the choice of a approximate diagonal.

The next assertion gives an explicit formula for a Johnson's amelioration in the case of the group algebra of an amenable locally compact group G.

Theorem 4.1.2 *Let G be an amenable locally compact group, let B be a Banach space, let $\varepsilon \in [0, 1)$, and let T be an ε-quasi-representation of G in B that is measurable and bounded (together with the inverses) by a constant C such that $\delta = (4 + 8C^2)^{-1}\varepsilon$. Then the Johnson's amelioration process can be defined by the mapping T' given by the formula*

$$S(f) = I_{g \in G}(T(g)T(_g f)) - I_{g \in G}(T(g)T(g^{-1}))T(f), \qquad f \in L^1(G), \quad (2)$$

and $T' = T + S$, where $_g f(h) = f(gh)$, $g, h \in G$, $f \in L^1(G)$, and I is a left invariant mean on $L^\infty(G)$. For $T(g)^{-1} = T(g^{-1})$, $g \in G$, formula (2) becomes

$$T'(f) = I_{g \in G}(T(g)T(_g f)), \qquad f \in L^1(G). \quad (3)$$

This fact is discussed in Subsection 4.2 below. Note that formula (3) can be directly extended to the Fréchet setting.

The following theorem (Shtern, 1994b; Shtern, 1997) is a partial converse to Theorem 4.1.1. Its proof is based on the properties of induced representations and on general properties of the representations of $SL(2, \mathbb{R})$ and of its Borel subgroup.

Theorem 4.1.3 *Let G be an almost connected locally compact group. If G is stably representable in any Banach space, then G is amenable.*

We conjecture that an arbitrary locally compact group is amenable if and only if it is stably representable in any Hilbert space.

4.1.1. *Explicit formulas for representations close to left uniformly continuous quasi-representations*

In this subsection we exploit ideas of (de la Harpe and Karoubi, 1977). For close operator algebras, (Christensen, 1988) contains a good review.

A deficiency of Theorem 4.1.1 and Theorem 4.1.2 is as follows: the answer has an approximate form, and we have no explicit formula expressing the quasi-representation T to the representation R. For the case of left uniformly continuous quasi-representations, the situation is better.

Definition 4.1.1 Let G be a locally compact group and let B be a Banach space. The group G is said to be *uniformly weak stably representable* in the space B if, for any $\varepsilon > 0$ and $C > 0$, there exists $\delta > 0$ such that if T is a strongly left uniformly continuous δ-quasi-representation of the group G in the space B by invertible operators such that $\|T(g)\| \leq C$ and $\|T(g)^{-1}\| \leq C$ for any $g \in G$, then there exists a representation R of the group G in the space B^* such that $\|T(g)^* - R(g)\| \leq \varepsilon$ for any $g \in G$ and matrix elements $(R(g)F)(x)$ are continuous for $F \in B^*$ and $x \in B$.

Theorem 4.1.4 *Let G be an amenable locally compact group and let B be a Banach space. Then the group G is uniformly weak stably representable in B, and we can assume that $0 \leq \varepsilon \leq 1/64$ and $\delta \leq \varepsilon(2C)^{-9}$.*

4.1.2. Dependence of the approximating representation on the choice of the invariant mean

The constructions of representations close to a given quasi-representation used in the proofs of Theorems 4.1.2 and 4.1.4 involve invariant means. It turns out that different means necessarily lead to equivalent representations (and the equivalence operator can be taken close to the identity operator), and examples show that the representations S in Theorem 4.1.4 can differ for different means.

Theorem 4.1.5 *Let T be a strongly continuous quasi-representation of an amenable locally compact group G in a Banach space B. Let μ and ν be left invariant means on the space $CB(G)$ and let R_μ and R_ν be the corresponding representations of the group G, in the space B^*, defined in Theorem 4.1.2 or in Theorem 4.1.3. If either of these representations is strongly continuous and if $\varepsilon < (C + \varepsilon)^{-2}/2$, then the representations R_μ and R_ν are similar by an operator Q satisfying the inequality*

$$\|Q - 1|_{B^*}\| < 2\varepsilon(C + \varepsilon)^2.$$

Example 4.1.1 There exists a quasi-representation of an amenable discrete group such that representations R_μ and R_ν for some $\mu \neq \nu$ are distinct, namely, a certain 2-dimensional quasi-representation of the group $\mathbb{Z}$).

4.1.3. Definition of a pseudorepresentation

We give this definition in the F-space setting.

Definition 4.1.2 An ε-quasi-representation T of a semigroup S in an F-space E is called an *ε-σ-pseudorepresentation* (with respect to a given pseudonorm $|\cdot|$) if, for each $g \in S$ and each $n \in \mathbb{N}$ (if S is a group, then we assume that $n \in \mathbb{Z}$), there exists a linear operator $A(n, g)$ in the space $L(E^*)$ such that for each $x \in E^{**}$, $\|x\| \leq 1$, we have $\|A(n, g)^* x - x\| \leq \sigma \|x\|$ and $T(g^n)^* = A(n, g)T(g)^{*n} A(n, g)^{-1}$ for each $g \in S$ and $n \in \mathbb{N}$ (if S is a group, then we require this for each $n \in \mathbb{Z}$), where $\|\cdot\|$ is a pseudonorm in E^{**} introduced in Theorem 2.2.1.

Definition 4.1.3 Let G be a group and let B be a Banach space. The group G is said to be *stably pseudorepresentable in the Banach space B* if, for each $\varepsilon > 0$, $\sigma > 0$, and $C > 0$, there exists $\delta > 0$ such that if T is a δ-quasi-representation of the group G in B by invertible operators that satisfies the conditions $\|T(g)\| \leq C$ and $\|T(g)^{-1}\| \leq C$ for each $g \in G$, then there exists a ε-σ-pseudorepresentation R of the group G in the space B^* such that $\|T(g)^* - R(g)\| \leq \varepsilon$ for any $g \in G$.

Theorem 4.1.6 *Every group is stably pseudorepresentable in any Banach space.*

4.2. DISCUSSION

Remark 4.2.1 Operations over pseudorepresentations are not so comfortable as those over quasi-representations. For pseudorepresentations we have a natural definition of the direct sum (in a suitable direct sum of spaces). For bounded pseudorepresentations in normed spaces (i.e., for ε-δ-pseudorepresentations T which satisfy (for some $C > 0$) the conditions $\|T(s)\| \leq C$ and $\|T(s)^{-1}\| \leq C$ for all $s \in S$), there is a natural operation of tensor product. Integration and disintegration (into a direct integral) of pseudorepresentations is also well defined. However, seemingly, there is no natural operation of inducing (for which the result is necessarily a pseudorepresentation and needs no amelioration).

Remark 4.2.2 Merits of the notion of pseudorepresentation are evident, but the demerits are also essential.

First of all, it would be desirable to obtain an object of approximate nature as a fixed point of an ameliorating process, but even for the case in which T is a pseudorepresentation of a group G, and thus $T(h)^{-1} = T(h^{-1})$ for all $h \in G$, it is unclear whether there exists a fixed point, close to a given mapping T, of the Johnson's amelioration mapping $T \mapsto T'$ given by (3). Note that a representation is clearly a fixed point of the mapping $T \mapsto T'$.

Moreover, the set of natural operations seems to be tight and complicates the work with pseudorepresentations.

Finally, a pseudorepresentation of an amenable locally compact group can be not a representation (being "close" to a representation).

It is possible that pseudosymmetries are to be considered from a more integral viewpoint (cf. (Shtern, 1991c)), but the alliance of the notions of a pseudorepresentation and a pseudocharacter is very attractive.

Remark 4.2.3 If T is a bounded pseudorepresentation in a Banach space with sufficiently small positive ε and δ, then $T(ab)^*$ is similar to $T(ba)^*$.

Remark 4.2.4 In a more "category" way, we can define a U-V-pseudoho-momorphism (U is a given neighboring from the uniformity in H and V is a neighborhood of the unit in H) of a topological group G into a topological group H as a quasi-homomorphism (see the introduction to this section) of G into H such that the power of the image and the image of the power of any element in G are adjoint in H by an inner automorphism determined by an element from V. It can be shown that pseudohomomorphisms of solvable connected and simply connected Lie groups are ordinary homomorphisms.

4.3. EXAMPLES

4.3.1. *Nontrivial continuous pseudorepresentation of a Lie group*
Recall that continuous pseudocharacters of connected locally compact groups are ordinary characters (see Proposition 3.2.2). This is not the case for pseudorepresentations.

Example 4.3.1 The distance between a strongly continuous quasi-representation of the group $\mathrm{SL}(2,\mathbb{R})$ quasi-induced by any nontrivial one-dimensional pseudorepresentation χ_α of $\mathrm{SL}(2,\mathbb{Z})$ (see Remark 3.1.2) and the set of ordinary continuous unitary representations of $\mathrm{SL}(2,\mathbb{R})$ has positive lower bound for $\alpha \neq 0$.

4.3.2. *Swing*
The "swing" $A(n,g)$ in Definition 4.1.2 is essential.

Example 4.3.2 A quasi-representation of $\mathbb{Z}_2 * \mathbb{Z}_2 * \mathbb{Z}_2$ quasi-induced by a one-dimensional nontrivial pseudorepresentation of the subgroup $H \subset G$ isomorphic to $\mathbb{Z}_2 * \mathbb{Z}$) is not equivalent to any quasi-representation of G with $A(n,g) \equiv 1$ (see Definition 4.1.2).

References

Bouarich, A. (1995) Suites exactes en cohomologie bornée réelle des groups discrets, *C. R. Acad. Sci. Paris*, **320**, pp. 1355–1359.

Cerry, C. (1994) Non-commutative deformations of $C(T^2)$ and K-theory, Preprint.

Christensen, E. (1988) Close operator algebras, in: *Deformation Theory of Algebras and Structures and Applications*. Kluwer Academic Publishers, Dordrecht, pp. 537–556.

Dunford, N., and Schwartz, J. T. (1958) *Linear Operators. Part I: General Theory*. John Wiley & Sons, Interscience, New York–London.

Elliott, G. A., Exel, R., and Loring, T. (1995) The soft torus III: the flip, to appear.

Exel, R. (1993) The soft torus and applications to almost commuting matrices, *Pacific J. Math.* **160**, pp. 207–217.

Exel, R. (1994) The soft torus II: a variational analysis of commutator norms, Preprint.

Exel, R., and Loring, T. (1989) Almost commuting unitary matrices *Proc. Amer. Math. Soc.***106**, pp. 913–915.

Exel, R., and Loring, T. (1991) Invariants of almost commuting matrices *Proc. Amer. Math. Soc.***95**, pp. 364–376.

Farsi, C. (1994) Soft non-commutative tori, Preprint.

Faĭziev, V. A. (1987) Pseudocharacters on free products of semigroups *Funktsional. Analiz i Prilozhen.* **21**, No. 1, pp. 86–87.

Faĭziev, V. A. (1988) Pseudocharacters on free groups, free semigroups, and some group constructions *Uspekhi Mat. Nauk* **43** No. 5, pp. 225–226.

Forti, G. L. (1987) The stability of homomorphisms and amenability, with applications to functional equations *Abh. Math. Sem. Univ. Hamburg* **57**, pp. 215–226.

Fuks, D. B. (1987) Continuous cohomology of topological groups and characteristic classes, in: Brown, K. S., *Cohomology of Groups* [Russian translation], pp. 342–368.

Gel'fand, I. M. (1941) Zur Theorie der Charactere der Abelschen topologischen Gruppen *Mat. Sb., New ser.* **9 (51)**, pp. 49–50.

Greenleaf, F. (1969) *Invariant Means on Topological Groups and Their Applications*, Van Nostrand, New York e.a.

Grigorchuk, R. I. (1995) Some results on bounded cohomology, in: *Combinatorial and Geometric Group Theory. Edinburgh, 1993* (A. J. Duncan, N. D. Gilbert and J. Howie, editors), London Math. Soc. Lecture Notes Ser, **284**, Cambridge Univ. Press, Cambridge, pp. 111–163.

Grigorchuk, R. I. (1996) Weight functions on groups and criteria for the amenability of Beurling algebras *Mat. Zametki* **60**, No. 3, pp. 370–382.

Grosser M. (1995) The review of (Shtern, 1993b), *Math. Rev.*, 95g#22008.

Grove, K., Karcher, H., Roh, E. A. (1974) Jacobi fields and Finsler metrics on a compact Lie groups with an application to differential pinching problems *Math. Ann.* **211**, No. 1, pp. 7–21.

Guichardet A. and Wigner D. (1978) Sur la cohomologie réele des groupes de Lie simples réels *Ann. Sci. Éc. Norm. Sup.* **11**, pp. 277–292.

de la Harpe, P., Karoubi, M. (1977) Represéntations approchées d'un groupe dans une algébre de Banach *Manuscripta math.* **22**, No. 3, pp. 297–310.

Hyers, D. H. (1941) On the stability of the linear functional equation *Proc. Nat. Acad. Sci. USA* **27**, No. 2, pp. 222–224.

Hyers, D. H., and Ulam, S. M. (1945) On approximate isometries *Bull. Amer. Math. Soc.* **51**, No. 4, pp. 288–292.

Johnson, B. E. (1972a) Approximate diagonals and cohomology of certain annihilator Banach algebras *Amer. J. Math.* **94**, pp. 685–698.

Johnson, B. E. (1972b) Cohomology in Banach algebras *Memoirs Amer. Math. Soc.* **127** American Mathematical Society, Providence.

Johnson, B. E. (1986) Approximately multiplicative functionals *J. London Math. Soc.* **34**, pp. 489–510.

Johnson, B. E. (1987) Continuity of generalized homomorphisms *Bull. London Math. Soc.* **19**, No. 1, pp. 67–71.

Johnson, B. E. (1988) Approximately multiplicative maps between Banach algebras *J. London Math. Soc.* **37**, No. 2, pp. 294–316.

Kac, V. G. and Peterson, D. H. (1985) Defining relations of certain infinite-dimensional groups, in: *Proc. of the Cartan Conf.*, Lyon, 1984, *Astérisque, Numero hors série*, pp. 165–208.

Kazhdan, D. (1982) On ε-representations *Israel J. Math.* **43**, pp. 315–323.

Kunze, R. A., Stein, E. M. (1960) Uniformly bounded representations and harmonic analysis on the 2×2 real unimodular group *Amer. J. Math.* **82**, pp. 1–62.

Lawrence, J. M. (1985) The stability of multiplicative semigroup homomorphisms to real normed algebras *Aequationes Math.* **28**, pp. 94–101.

Loring, T. A. and Pedersen, G. K. (1994) Projectivity, transitivity and AF-telescopes, Preprint.

Lyubich, Yu. I. (1985) *Vvedenie v Teoriyu Banakhovykh Predstavleniĭ Grupp (Introduction to the Theory of Banach Representations of Groups)* , Khar'kov.

Naimark, M. A., Shtern, A. I. (1982) *Theory of Group Representations* Springer-Verlag, New York–Berlin.

Ol'shanskii, A. Yu. (1987) *Geometry of defining relations in groups*, Nauka, Moscow.

Pap, E. (1988) On approximately and bounded homomorphisms on uniform commutative semigroups *Univ. u Novom Sadu, Zb. Rad. Prirod.-Mat. Fak. Ser. Mat.* **18**, pp. 111–118.

Paterson, A. L. T. (1988) *Amenability*, Amer. Math. Soc., Providence, Rhode Island.

Rademacher, H. (1932) Zur Theorie der Modulfunktionen *J. Reine Angew. Math.* **167** pp. 312–336.

Shaefer, H. H. (1966) *Topological Vector Spaces*, The Macmillan Company, New York–London.

Shtern, A. I. (1971a) The connection between the topologies of a locally compact group and its dual space *Functional. Anal. Appl.* **5**, pp. 311–317.

Shtern, A. I. (1971b) Separable locally compact groups with discrete support for the regular representation *Soviet Math. Dokl.* **12**, pp. 994–998.

Shtern, A. I. (1973) Locally bicompact groups with finite-dimensional irreducible representations *Math. USSR Sbornik* **19**, No. 1, pp. 85–94.

Shtern, A. I. (1980) Roughness of positive characters *Uspekhi Mat. Nauk* **35**, No. 5, p. 218.

Shtern, A. I. (1982) Stability of homomorphisms into the group R^*, *Moscow Univ. Math. Bull.* **37**, No. 3, pp. 33–36.

Shtern, A. I. (1983) Stability of representations and pseudocharacters, Report on Lomonosov Readings 1983, Moscow State University.

Shtern, A. I. (1990) A pseudocharacter that is determined by the Rademacher symbol *Russ. Math. Surv.* **45**, No. 3, pp. 224–226.

Shtern, A. I. (1991a) On operators in Fréchet spaces that are similar to isometries *Moscow Univ. Math. Bull.* No. 4, pp. 67–70.

Shtern, A. I. (1991b) Quasirepresentations and pseudorepresentations *Funct. Anal. Appl.* **25**, No. 2, pp. 140–143.

Shtern, A. I. (1991c) Symmetry and its images, in: *A. F. Losev and XXth century culture*, Nauka, Moscow, pp. 79–82.

Shtern, A. I. (1993a) Almost convergence and its applications to the Fourier–Stieltjes localization *Russian Journal of Mathematical Physics* **1**, pp. 115–125.

Shtern, A. I. (1993b) Continuous pseudocharacters on connected locally compact groups are characters *Funktsional. Analiz i Prilozhen.* **27**, No. 4, pp. 94–96.

Shtern, A. I. (1994a) Quasi-symmetry. I. *Russian Journal of Mathematical Physics* **2**, pp. 353–382.

Shtern, A. I. (1994b) Characterizations of the amenability in the class of connected locally compact groups *Uspekhi Mat. Nauk* **49**, No. 2 (296), pp. 183–184.

Shtern, A. I. (1994c) Compact semitopological semigroups and reflexive representability of topological groups *Russian Journal of Mathematical Physics* **2**, pp. 131–132.

Shtern, A. I. (1997) Triviality and continuity of pseudocharacters and pseudorepresentations (to appear).

Ulam, S. M. (1960) *A collection of mathematical problems.* Interscience Tracts on Pure and Applied Mathematics, New York–London.

Zhelobenko, D. P. and Shtern, A. I. *Predstavleniya Grupp Li (Representations of Lie Groups).* Nauka, Moscow.

ORBITAL ISOMORPHISM BETWEEN TWO CLASSICAL INTEGRABLE SYSTEMS

The Euler case and the Jacobi problem

A.V. BOLSINOV AND A.T. FOMENKO

Department of Mechanics and Mathematics,
Moscow State University, 119899 Moscow, Russia

Abstract. We describe the orbital invariants of two famous integrable systems (the Euler case in rigid body dynamics and the Jacobi problem) and show that these systems are orbitally topologically equivalent.

Mathematics Subject Classification (1991): 58F05.

Key words: integrable Hamiltonian system, geodesic flow, rigid body dynamics.

1. Introduction

It is well known that the geodesic flow of the Riemannian metric on the ellipsoid in $\mathbb{R}^3$ is completely integrable in the Liouville sense, [12] (Jacobi). The corresponding dynamical Hamiltonian system is defined on the four-dimensional cotangent bundle of the 2-sphere. The isoenergy 3-manifolds (invariant under the geodesic flow) are all diffeomorphic to the three-dimensional projective space $\mathbb{RP}^3$. In the present paper we obtain the topological orbital classification of all these systems. We also consider another famous integrable system, namely the so-called Euler case in the rigid body dynamics in $\mathbb{R}^3$. This system describes the rotation of a rigid body fixed at its center of mass [4], [13], [19], [24]. It turns out that these two well-known systems are topologically orbitally equivalent.

Definition 1.1. Two smooth dynamical systems (v_1, Q_1) and (v_2, Q_2) are said to be *topologically orbitally equivalent* if there exists a homeomorphism from Q_1 into Q_2 which maps the trajectories of the first systems to those of

B. P. Komrakov et al. (eds.), Lie Groups and Lie Algebras, 359–382.
© 1998 *Kluwer Academic Publishers. Printed in the Netherlands.*

the second one whith preserving their orientation (here we do not assume that time on the trajectories is preserved).

The problem of the topological orbital classification for dynamical systems of different types has been discussed in many papers, in particular, [1] (Andronov A.A., Leontovich E.A., Gordon I.I., Mayer A.G.), [2] (Anosov D.V., Aranson S.Kh., Bronstein I.U., Grines V.Z.), [3] (Arnold V.I., Ilyashenko Yu.S.), [14] (Kozlov V.V.), [17] (Mozer U.). Let us also note the papers [16] (Leontovich E.A., Mayer A.G.), [23] (Umanskii Ya.L.) and [20] (Peixoto M.M.), where the orbital classification was obtained for Morse–Smale flows on two-dimensional surfaces and three-dimensional manifolds. Many interesting examples of orbitally equivalent systems can be obtained by using the Maupertuis principle.

Since three-dimensional constant-energy surfaces are invariant with respect to the flow, it is sufficient to investigate the topological orbital classification of integrable Hamiltonian systems (with two degrees of freedom) on three-dimensional energy level surfaces. The three main results of our paper are presented in the next section.

We are grateful to V.V. Kozlov, Yu.N. Fedorov, G. Paternain, O.E. Orel and V.V. Kalashnikov (Jr.) for discussions of integrability problems that were extremely valuable and useful for us.

2. Formulation of the main theorem

The Euler top is an arbitrary rigid body in three-dimensional space fixed at its center of mass and moving in the gravitational field. "Systems of Euler type" form a continuous family whose parameters are the principal momenta of inertia $1/A, 1/B, 1/C$ of the body (or their inverses A, B, C, we assume $A < B < C$), the value of the energy (i.e., Hamiltonian) h, and the value of the area constant g (here without loss of generality we put $g = 1$). Fixing the values of the so-called geometrical integral and area integral, we can assume a dynamical system of Euler type to be given on a four-dimensional symplectic manifold M^4.

Let us consider the domain in the real axis filled by admissible values of the Hamiltonian H in the Euler case. Marking the three following critical values $h_0 = A/2, h_1 = B/2, h_2 = C/2$, we divide this domain into three zones of energy $I = (h_0, h_1)$, $II = (h_1, h_2)$ and $III = (h_2, +\infty)$. According to this we shall speak of three types (zones) of energy I, II and III.

Let us consider two functions

$$S(h, A, B, C) = \frac{1}{\pi} \int_A^{2h} \frac{u\,du}{\sqrt{(2h - u)(C - u)(B - u)(u - A)}},$$

$$N(h, A, B, C) = -\frac{1}{\pi} \int_{2h}^C \frac{u\,du}{\sqrt{(2h - u)(C - u)(B - u)(u - A)}}.$$

Theorem A. *Consider two rigid bodies fixed at their centers of mass. Let* $(1/A, 1/B, 1/C)$ *and* $(1/A', 1/B', 1/C')$ *be their principal momenta of inertia and* h *and* h' *their energy values. Let* $v(A, B, C, h)$ *and* $v(A', B', C', h')$ *be the corresponding dynamical systems of Euler type on the three-dimensional isoenergy surfaces.*

1) *If* h *and* h' *belong to different zones, then the systems* $v(A, B, C, h)$ *and* $v(A', B', C', h')$ *are not orbitally equivalent.*

2) *If both* h *and* h' *belong to zone I (i.e.,* $h \in (A/2, B/2)$ *and* $h' \in (A'/2, B'/2)$*), then* $v(A, B, C, h)$ *and* $v(A', B', C', h')$ *are smoothly orbitally equivalent if and only if the conditions*

$$\frac{A^2}{(C - A)(B - A)} = \frac{A'^2}{(C' - A')(B' - A')},$$

and

$$S(h, A, B, C) = S(h', A', B', C')$$

hold.

3) *If both* h *and* h' *belong to zone* II*, then the systems* $v(A, B, C, h)$ *and* $v(A', B', C', h')$ *are topologically orbitally equivalent if and only if the two following conditions hold*

$$\frac{A^2}{(C - A)(B - A)} = \frac{A'^2}{(C' - A')(B' - A')},$$

and

$$N(h, A, B, C) = N(h', A', B', C').$$

4) *If both* h *and* h' *belong to zone III, then the systems* $v(A, B, C, h)$ *and* $v(A', B', C', h')$ *are topologically (and smoothly) orbitally equivalent if and only if the ellipsoids of inertia of the bodies are similar (i.e.,* $A/A' = B/B' = C/C'$*).*

Now let us consider the Jacobi problem. Define the ellipsoid in $\mathbb{R}^3$ by the equation

$$\frac{x^2}{a} + \frac{y^2}{b} + \frac{z^2}{c} = 1.$$

Theorem B. *The geodesic flows of two ellipsoids are orbitally topologically equivalent if and only if the ellipsoids are similar (i.e., their semiaxes are proportional).*

Finally, the last theorem establishes a relationship between the Euler case and Jacobi problem. Consider two dynamical systems on T^*S^2. The first one is the geodesic flow of the ellipsoid. The second is the dynamical system describing the integrable Euler case with zero area constant. Denote them by $v_J(a, b, c)$ and $v_E(A, B, C)$ respectively. It is easy to see that the dynamics of the systems does not depend on the energy level unless $H = 0$. The isoenergy surface $\{H = 0\}$ in both cases is singular and coincides with the zero section $S^2 \subset T^*S^2$. That is why it is natural to remove the zero section and consider the systems on $M^4 = T^*S^2 \backslash \{\text{zero section}\}$.

Theorem C. a) *The dynamical systems* v_J *(the Jacobi problem) and* v_E *(the Euler case) are orbitally topologically equivalent on* M^4 *in the following sense. For any* $a < b < c$ *there exist* $A < B < C$ *(uniquely defined up to proportionality) and for any* $A < B < C$ *there exist* $a < b < c$ *(also uniquely defined up to proportionality) such that the corresponding systems* $v_E(A, B.C)$ *and* $v_J(a, b, c)$ *are orbitally topologically equivalent.*

b) *The explicit formulas connecting the squares of semiaxes* a, b, c *of the ellipsoid and the principal moments of inertia* $1/A, 1/B, 1/C$ *of the corresponding rigid body in the Euler case are as follows*

$$-\frac{\int_{-b}^{-a} \Phi(u, c)du}{\int_0^{+\infty} \Phi(u, c)du} - 1 = \frac{-C}{\sqrt{(C - A)(C - B)}},$$

$$\frac{\int_{-c}^{-b} \Phi(u, a)du}{\int_0^{+\infty} \Phi(u, a)du} - 1 = \frac{A}{\sqrt{(C - A)(B - A)}},$$

where

$$\Phi(u, t) = \sqrt{\frac{u}{(u + a)(u + b)(u + c)(u + t)}}.$$

In other words, the systems $v_J(a, b, c)$ *and* $v_E(A, B, C)$ *are orbitally topologically equivalent if and only if the two above relations hold.*

Remark. In the classical theory of real rigid bodies, the principal momenta of inertia satisfy the triangle inequality. Nevertheless, the Euler–Poisson equations can be considered for arbitrary moments of inertia. That is why we do not confine "the store of rigid bodies" to the triangle inequalities and shall consider all kinds of number triples $1/A, 1/B, 1/C$ satisfying the only condition $1/A > 1/B > 1/C$. However, the same Euler–Poisson equations describe the geodesics on the Poisson sphere and, therefore, have meaning for any $1/A > 1/B > 1/C$.

All of these results follow from the general theory of orbital classification of nondegenerate integrable systems with two degrees of freedom on three-dimensional energy levels, which has been developed by the authors in [7]. The theory is based on the investigation of the new topological invariants of integrable Hamiltonian systems (rough molecules and marked molecules) discovered in the series of papers by A.T. Fomenko, H. Zieschang, S.V. Matveev, A.V. Bolsinov [6], [9–12], [22]. The main idea is to assign to every Hamiltonian system some invariant that is, in fact, a graph with numerical marks and is called a t-molecule. The point of the theory is the statement that two integrable nondegenerate systems are orbitally equivalent if and only if their t-molecules coincide. It turns out that in many concrete problems these new invariants can be effectively calculated, that is what leads to the above results. Throughout the paper we shall use the terminology and notation from [6], [9–11], [22].

3. Demonstrating the general theory of topological orbital classification in the case of Euler and Jacobi systems

3.1. FINE TOPOLOGICAL CLASSIFICATION OF THE EULER CASE AND JACOBI PROBLEM

Before answering the question whether two integrable systems are orbitally equivalent we should first find out if they are topologically equivalent. Let us recall that if a Hamiltonian system is Liouville integrable, then its phase manifold M^4 is foliated into orbits of the action of the Abelian group $\mathbb{R}^2$, generated by two commuting vector fields sgrad H and sgrad F, where H is the Hamiltonian and F is an integral of the system. All regular orbits of the action are 2-tori called *Liouville tori*. Singular orbits fill a set of measure zero. Thus the structure of a foliation (with singularities) called *the Liouville foliation of the system* appears on M^4

Definition 3.1. Two integrable Hamiltonian systems are called *finely topologically equivalent* if they have "the same Liouville foliation". This means that there exists a diffeomorphism (homeomorphism) mapping the Liouville foliation of the first system to that of the second one.

In the present paper we shall consider integrable systems satisfying the following natural conditions.

A1. The system is nondegenerate on the given isoenergy 3-surface, that is, it possesses a Bott integral.

A2. The system is topologically stable, that is, under small changes of the energy level, the system remains finely topologically equivalent to the initial one.

If we restrict the system to any of its regular isoenergy 3-surfaces then the fine topological type of the system can be completely described by means of the so-called *marked molecule W^**. The marked molecule W^* is a one-dimensional graph endowed with numerical marks. It consists of edges and vertices called "atoms". The edges of the molecule correspond to one-parametric families of regular Liouville tori. The atoms correspond to their bifurcations. Such bifurcations were classified in [6]. The numerical marks describe the rules of gluing the isoenergy surface from "the atoms".

The main result of the theory of topological classification ([6], [9–11]) states that two integrable nondegenerate Hamiltonian systems with two degrees of freedom are finely topologically equivalent (on their isoenergy surfaces) if and only if their marked molecules coincide.

The following statement describes the Liouville foliation structure for the Euler case and Jacobi problem.

Proposition 3.1. a) *The marked molecules for the systems $v(A, B, C, h)$ of Euler type are shown in Fig. 1-a, 1-b, 1-c. Two systems $v(A, B, C, H)$ and $v(A', B', C', h')$ are finely topologically equivalent if and only if the corresponding energy levels h and h' belong to zones of the same type.*

b) *The marked molecule of the geodesic flow $v_J(a, b, c)$ on an ellipsoid (Jacobi problem) is shown in Fig. 1-c. Here we consider the geodesic flow to be restricted to the isoenergy surface $Q^3 = \{H = 1\}$ (i.e., the unit covector bundle).*

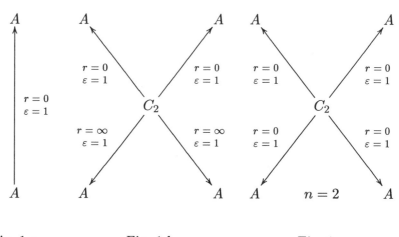

Fig. 1-a   Fig. 1-b   Fig. 1-c

Corollary. *The geodesic flow $v_J(a, b, c)$ is finely topologically equivalent to the system of Euler type $v(A, B, C, h)$ with $h \in III$.*

The marked molecules described in the above proposition have been calculated in [6] (Euler case) and [18] (Jacobi problem).

3.2. NEW ORBITAL INVARIANTS AND THE t-MOLECULE

The explicit description of the t-molecule in the general case is rather complicated (see [7]) and here, for the sake of simplicity, we examine the case of systems with the so-called simple molecules W.

We shall assume that the dynamical systems in question are sufficiently simple. Namely the corresponding molecules contain only four types of atoms: A, B, A^*, C_2. These are just the atoms which usually appear in specific physical and geometrical problems.

Definition 3.2. Let $v = \text{sgrad } H$ be an integrable Hamiltonian system with two degrees of freedom and $s_1, s_2, \varphi_1, \varphi_2$ be standard action-angle variables in a neighborhood of a Liouville torus T^2. According to the Liouville theorem on the torus we have $v = a\partial/\partial\varphi_1 + b\partial/\partial\varphi_2$. The ratio $\rho = a/b$ is called the *rotation number* of the system v on the torus T^2.

Let us remove from Q_h^3 all critical levels of the integral f. As a result, we obtain a finite set of one-parameter families e_i $(i = 1, \ldots, m)$ of Liouville tori. We can assume that the parameter on each family is the function f itself. Therefore for every family e_i the rotation function $\rho_i(f)$ appears. It is clear that $\rho_i(f)$ depends on the choice of basis on the corresponding Liouville torus. Let us recall that for every one-parameter family of Liouville tori in Q (represented as an edge of the molecule), we can define two admissible coordinate systems λ^-, μ^- and λ^+, μ^+ (see [6], [9]), corresponding to the atoms connected by this edge. Thus on every edge of the molecule there appears a transition matrix from one basis to another (the gluing matrix) $C = \begin{pmatrix} \alpha & \beta \\ \gamma & \delta \end{pmatrix}$ and the pair of rotation functions ρ^- and ρ^+. Let us note that the admissible coordinate systems are not uniquely defined. So the functions ρ^- and ρ^+ depend on the choice of admissible coordinate systems on the edge. However, if $\beta \neq 0$, then the function $\rho = \beta\rho^- - \alpha$ does not depend on the choice of λ^-, μ^- and λ^+, μ^+ and is an orbital invariant of the system, which is well defined (up to conjugacy, i.e., up to a monotone change of argument). It is easy to check that under some additional assumptions (see **A3** below) smooth rotation functions can be classified up to conjugacy by the so-called *rotation vector* R consisting of:

(1) local minima and maxima of ρ,

(2) two limits of ρ at the ends of the edge,

(3) symbols $+\infty$ and $-\infty$ showing the left and right limits of ρ at its poles.

Moving along the edge, we write out step by step all the components listed above. As the result, we obtain a sequence of numbers and symbols $\pm\infty$ called the *rotation vector* (or, simply, *R-vector*) of the integrable Hamiltonian system v on the edge. If $\beta = 0$, then we put $\rho = \rho^-$. Although here the function ρ depends of the choice of basis on a torus, the character of this nonuniqueness is very simple: if we change the basis, then ρ is changed by adding an *integer*. As in the case $\beta \neq 0$, here we can define the rotation vector R, but now we have to consider it modulo 1: $R \mod 1$.

By R^+ (respectively R^-) we shall denote the R-vector of the function ρ^+ (respectively ρ^-). We shall assume that the systems in question satisfy the following natural conditions (in addition to properties **A1** and **A2** above):

A3. the rotation function on every edge has only finite number of critical points and poles,

A4. the differential of the Poincaré map for any critical saddle circle is not an identity map.

Definition 3.3. We shall call an edge of the molecule W^*
(1) *finite*, if the corresponding mark r is finite,
(2) *infinite*, if r is infinite but the R-vector (on this edge) contains at least one finite component.
(3) *superinfinite*, if $r = \infty$ and the R-vector has no finite component.

Definition 3.4. Let us cut the molecule along all finite and infinite edges. As the result, the molecule is divided into several connected pieces (subgraphs). The pieces of the molecule different from atoms A will be called *radicals*.

Let U be an arbitrary radical. For every edge e_j incident to U (both interior and exterior) we define an integer $[\Theta]_j$ in the following way.

$$[\Theta]_j = \begin{cases} [\alpha_j/\beta_j], & \text{if} \quad e_j \text{ is finite and outgoing from } U \\ [-\delta_j/\beta_j], & \text{if} \quad e_j \text{ is finite and incoming to } U \\ [MR^+], & \text{if} \quad e_j \text{ is infinite and incoming to } U \\ -[-MR^-], & \text{if} \quad e_j \text{ is infinite and outgoing from } U \\ [-\gamma_j/\alpha_j], & \text{if} \quad e_j \text{ is an interior superinfinite edge.} \end{cases}$$

The expressions $[MR^+]$ and $[MR^-]$ in this formula denote the integer parts of the real numbers MR_j^+ and MR_j^- respectively, each of which is the average value of the finite components of the corresponding rotation vector.

Definition 3.5. The integer $b = \sum[\Theta]_j$ (where the sum is taken over all edges e_j incident to U) is said to be the *b-invariant* of the Hamiltonian system v on the radical U.

Finally, let us define another orbital invariant of an integrable Hamiltonian system, namely, the so-called Λ-invariant. For the class of systems we are working with, this invariant is assigned to the atom C_2 only. The atom C_2 contains two critical circles γ_1 and γ_2 of the integral f, which are simultaneously closed trajectories of the Hamiltonian system. For each of these trajectories, we can define its multiplicators (l_i, m_i). Recall that in this case the multiplicators are the eigenvalues of the differential of the Poincaré map defined on a two-dimensional transversal to the closed geodesic γ_i. Since the flow v is Hamiltonian, we have $l_i = -m_i$. Besides, if the trajectory γ_i is hyperbolic, then we may assume $l_i > 1$. Denote $\Lambda_i = \ln l_i$ and put this number on the corresponding vertex of the atom C_2 (that is, on the hyperbolic closed trajectory γ_i).

Definition 3.6. The pair $(\Lambda_1 : \Lambda_2)$ of real numbers considered up to proportionality is called the Λ-*invariant* of the Hamiltonian system v on the atom C_2.

It is well known that multiplicators (and, consequently, their logarithms Λ_i) are smooth (but not topological) orbital invariants of a dynamical system near its closed trajectory. It turns out that the Λ-invariant is still a topological orbital invariant of the system v in a neighborhood of C_2. Let us note that in the case of more complicated atoms containing several critical circles (i.e., hyperbolic closed trajectories) the Λ-invariant is defined in the same way.

Definition 3.7. The marked molecule W^* endowed with all its R-vectors, the Λ-invariants of all its atoms (of type C_2) and the b-invariants of all its radicals is called the *t-molecule* of the system and denoted by $W^{*t} = (W^*, R, \Lambda, b)$.

Remark. Here R denotes the R-vector for finite edges of the molecule and R-vector modulo 1 for its infinite edges.

Theorem 3.1. ([7]) *Let v be an integrable Hamiltonian system satisfying conditions* **A1**, **A2**, **A3**, **A4** *and such that its topology is sufficiently simple, that is, its molecule on a given isoenergy 3-manifold consists of the atoms* A, A^*, B *and* C_2 *only.*

a) *Then the corresponding t-molecule*

$$W^{*t} = (W^*, R, \Lambda, b)$$

is well defined, i.e., does not depend on the choice of admissible coordinates.

b) *the t-molecule is a complete orbital invariant of an integrable system. This means that two integrable systems are orbitally topologically equivalent if and only if their t-molecules coincide.*

4. Orbital classification of integrable systems of the Euler case

4.1. ROTATION FUNCTION IN THE EULER CASE

In this section we calculate the *orbital molecules (t-molecules)* for the well-known integrable Euler case in rigid body dynamics and obtain the complete orbital classification of all such Euler-type systems. Our analysis is based on analytical properties of this system extracted from the papers by V.V. Kozlov [13], Yu.A. Sadov [21] and Yu.A. Arkhangelskii [4]. The adaptation of these results for the purposes of our theory was carried out by O.E. Orel.

Let us consider in $\mathbb{R}^3$ a fixed basis $OXYZ$ and the moving coordinate system $oxyz$ connected with a moving rigid body. The axes x, y, z are directed along the principal axes of inertia of the body. Let K be the vector of kinetic moment of the rigid body and G the value of kinetic moment. It follows from the general theory describing the motion of a rigid body that $G^2 = \langle K, K \rangle$ is an additional integral if the body is fixed at its center of mass (the Euler case). We denote it by f, as usual. Then, f is a smooth function on the phase space M^4 and on the isoenergy 3-surface Q^3.

Let I_1, I_2, I_3 be the principal momenta of inertia of the body and A, B, C their inverse values (that is, the squares of the semiaxes for the inertia ellipsoid). As usual, we assume the momenta to be ordered, i.e., $I_1 > I_2 > I_3$, or $A < B < C$.

Let p, q, r be the projections of the vector Ω of angular velocity to the axes of inertia ox, oy, oz. Then $I_1 p, I_2 q, I_3 r$ are the projections of the kinetic moment on the same axes. In other words, $K = \text{diag}\,(I_1, I_2, I_3)\Omega$.

Let us denote by $\gamma_1, \gamma_2, \gamma_3$ the coordinates of the unit vertical vector γ with respect to the moving coordinate system. The classical Euler–Poisson equations, which describe the dynamics of a rigid body fixed at its center of mass, have the form:

$$\dot{K} = [K, \Omega], \quad \dot{\gamma} = [\gamma, \Omega].$$

Let us recall that the manifold M^4 is determined in $\mathbb{R}^6$ with coordinates $(p, q, r, \gamma_1, \gamma_2, \gamma_3)$, by the following two equations:

$$\langle \gamma, \gamma \rangle = \gamma_1^2 + \gamma_2^2 + \gamma_3^2 = 1,$$
$$\langle K, \gamma \rangle = I_1 p \gamma_1 + I_2 q \gamma_2 + I_3 r \gamma_3 = g = \text{const}.$$

The first function is called the *geometrical integral* and the second one is called the *area integral*.

Let us recall that the Hamiltonian H and the integral f of this system have the form

$$H = I_1 p^2 + I_2 q^2 + I_3 r^2, \quad f = I_1^2 p^2 + I_2^2 q^2 + I_3^2 r^2.$$

The *momentum mapping* $\pi = (H, f)$ takes M^4 to the 2-dimensional plane with the coordinates H and f. This mapping has some critical points. Their images are the *critical values* of π. The set of all the critical values is some 1-dimensional subset in the plane. This subset is called the *bifurcation diagram* (of the momentum mapping). In our case the bifurcation diagram has the form shown in Fig. 2. We assume here that the area constant g is not zero.

It turns out that an Euler-type system for large values of energy (i.e., for large values of H) is smoothly equivalent to Euler system with zero area constant.

Proposition 4.1. *Let us consider two different invariant isoenergy 3-surfaces* $Q_{h,g}$ *and* $Q_{h,0}$ *determined by the equations*

$$Q_{h,g} = (H = h, \langle K, \gamma \rangle = g, \langle \gamma, \gamma \rangle = 1),$$
$$Q_{h,0} = (H = h, \langle K, \gamma \rangle = 0, \langle \gamma, \gamma \rangle = 1).$$

Let h be a sufficiently large constant. Then the dynamical systems of Euler type on these two isoenergy 3-surfaces are smoothly equivalent (i.e., there exists a diffeomorphism between these 3-manifolds transforming one system into another).

Proof. Let us write the Euler–Poisson equations in the vector form: $\dot{K} = [K, \Omega]$, $\dot{\gamma} = [\gamma, \Omega]$. Consider the following change of coordinates: $\tilde{\gamma} = [\gamma, K]/\,|\,[\gamma, K]\,|$ and $\tilde{K} = K$. Direct calculation shows that the equations preserve their form after this transformation. On the other hand, it is clear that now the vectors $\tilde{K}$ and $\tilde{\gamma}$ turn out to be orthogonal and, consequently, $Q_{h,g}$ maps to $Q_{h,0}$. The only question is when does this coordinate transformation determine a diffeomorphism of the isoenergy 3-surfaces? It turns out that the answer depends on the value of h. If h is sufficiently large, then the coordinate change actually determines a diffeomorphism. Indeed, the transformation has an inverse one if $[\gamma, K]$ takes a nonzero value. But this is valid only for large h. Statement is proved. □

Remark. It is easy to see that in the above notation this statement means that the systems $v_E(A, B, C)$ and $v(A, B, C, h)$ are smoothly conjugate if $h \in III$.

Let us analyze the bifurcation diagram of Euler-type systems. The reader can find details, for example, in the paper by A.A. Oshemkov [19]. The infinite straight line segments of the diagram start at the coordinate origin. Let us denote these infinite rays by τ_1, τ_2, τ_3. Let τ_4 be the horizontal segment (Fig. 2) of the diagram. The equations of these lines are as follows:

$$\tau_1 = \{(H, f) : 2H/f = A\}, \tau_2 = \{(H, f) : 2H/f = B\},$$
$$\tau_3 = \{(H, f) : 2H/f = C\}.$$

The position of τ_4 depends on the choice of the area integral. More precisely, the segment τ_4 is the part of the horizontal line determined by the equation $f = g^2$, where g is the fixed value of the area integral. The image of the momentum mapping is divided into two regions X_1 and X_2 (Fig. 2).

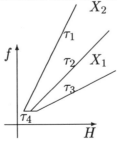

Fig. 2

Let us describe the basis cycles (λ, μ) necessary for the calculations (according to our general theory). Let us take a Liouville 2-torus corresponding to X_1, i.e., the preimage of an arbitrary inner point from X_1. Two natural basis cycles λ and μ are defined in a unique way on this torus. One of them, for example λ, is contracted into a point when the corresponding point in the domain X_1 (= projection of the torus in the domain X_1) tends to the ray τ_1. The second cycle μ annihilates when the point (= projection of the torus) tends to the segment τ_4. These two conditions determine both cycles uniquely (up to isotopy).

Let us consider a similar construction in the domain X_2. Let us denote by μ the cycle on a Liouville torus which transforms into the critical circle (= fiber of Seifert fibration) when the torus tends to the ray τ_2. Let λ' be the cycle contracted into a point when the torus tends to τ_3. Let us calculate the rotation function. This function can be calculated on the whole symplectic 4-manifold M^4 as a function of the Liouville 2-torus, i.e., as a function of two variables H and f. Considering the pair (H, f) as a point $y \in \mathbb{R}^2$ from the image of the momentum mapping, we can write the rotation function in the form $\rho = \rho(y)$. The two following statements can be extracted from [13].

Lemma 4.1 If $y \in X_1$, then the rotation function $\rho = \rho(y)$ corresponding to the cycles μ and λ has the form

$$\rho(y) = \frac{1}{\pi} \int_A^{2H/f} \frac{u\,du}{\sqrt{(\frac{2H}{f} - u)(C - u)(B - u)(u - A)}}.$$

If $y \in X_2$, then the rotation function $\rho = \rho(y)$ corresponding to the cycles μ and λ' has the form:

$$\rho(y) = -\frac{1}{\pi} \int_{2H/f}^C \frac{u\,du}{\sqrt{(\frac{2H}{f} - u)(C - u)(B - u)(u - A)}}.$$

Lemma 4.2 The limits of the rotation function $\rho(y)$ are:

a) $\displaystyle\lim_{y \to \tau_1} \rho(y) = K(A, B, C) = \frac{A}{\sqrt{(C - A)(B - A)}},$

b) $\displaystyle\lim_{y \to \tau_3} \rho(y) = L(A, B, C) = -\frac{C}{\sqrt{(C - A)(C - B)}},$

c) $\displaystyle\lim_{y \to \tau_2} \rho(y) = +\infty,$ if $y \in X_1$, $\displaystyle\lim_{y \to \tau_2} \rho(y) = -\infty,$ if $y \in X_2$,

d) $\displaystyle\lim_{y \to \tau_4} \rho(y) = S(H/g^2, A, B, C) =$

$$\frac{1}{\pi} \int_A^{2H/g^2} \frac{u\,du}{\sqrt{(\frac{2H}{g^2} - u)(C - u)(B - u)(u - A)}}, \text{ if } y \in X_1,$$

$$\lim_{y \to \tau_4} \rho(y) = N(H/g^2, A, B, C) =$$

$$-\frac{1}{\pi} \int_{2H/g^2}^C \frac{u\,du}{\sqrt{(\frac{2H}{g^2} - u)(C - u)(B - u)(u - A)}}, \text{ if } y \in X_1.$$

Now, using these lemmas and our general theory, we can easily calculate the t-molecules.

4.2. t-MOLECULES FOR ENERGY LEVELS OF TYPE I

The molecule W^* here has the simplest form:

$$A \xrightarrow[r\,=\,0]{} A.$$

The invariants Λ and b absent here (see the general theory above). Consequently, we must find the rotation vector R only. Let us assume that the edge of the molecule is oriented from τ_4 to τ_1. According to the definition of the rotation vector, we should consider the rotation function ρ with respect to a canonical basis (λ^-, λ^+) in the case of molecule $A \longrightarrow A$. Here the arrow shows the direction on the edge. The cycle λ^- is the disappearing (annihilating) cycle of the solid torus on the left atom, and λ^+ is the disappearing (annihilating) cycle of the other solid torus on the right atom A. Let us pass from the basis (λ^-, λ^+) to the basis (λ, μ), where the cycles λ and μ were described above. It is easily seen that in this case we can put $\mu = \lambda^-$ and $\lambda = \lambda^+$.

Thus, in our case the canonical basis has the form (μ, λ). Consequently, the rotation function ρ (with respect to this basis) on the line $H = h_0 = $ const in the domain X_1 is given by the formula of Lemma 4.1.

It is easy to prove (see the previous formula) that *the rotation function is strongly monotonous*. Consequently, to calculate the rotation vector on this edge of the molecule, we need only find the limits of the function $\rho(f)$ when f tends (as a parameter on the edge) to the end-values corresponding to the points of intersection of the line $H = h_0$ with the segment τ_4 and the ray τ_1. These limits were calculated above in Lemma 4.2 and are equal (respectively) to $S(h_0/g^2, A, B, C)$ and $K(A, B, C)$.

Proposition 4.2. *The orbital topological invariant (i.e., the t-molecule W^{*t}) for Euler-type systems $v(A, B, C, h_0)$ with $h_0 \in I$ has the form:*

$$A \xrightarrow[r\,=\,0]{} A.$$

with $R = (S(h_0/g^2, A, B, C), K(A, B, C))$.

We have discussed the area constant g above. Let us use a homothety transformation for all six variables in the initial phase space. As a result, we transform the initial Hamiltonian system into another one which is evidently topologically (and smoothly) orbitally equivalent to the initial system. On the other hand, the homothety changes the value of the area integral (area constant). Thus, we can assume that the value of the area integral is equal to 1 when $g \neq 0$. The case $g = 0$ corresponds to the energy values from the zone III and does not require a special analysis.

The first statement of the Theorem A follows from Proposition 4.2.

Let us fix the inertia momenta I_1, I_2, I_3 for the rigid body and the area constant g in the Euler case from the energy zone I (low energy = type I). When we change the energy value h_0, we change the isoenergy 3-manifold Q_{h_0} and, consequently, the integrable system v on it. Let us call all such systems *Euler-type systems from the zone I with fixed inertia ellipsoid.*

Corollary 4.1. *All integrable Euler-type systems from zone I (low energy) with a fixed inertia ellipsoid, corresponding to a different energy values h_0 and h'_0, are not orbitally topologically equivalent. In other words, two systems $v(A, B, C, h_0)$ and $v(A, B, C, h'_0)$ with $h_0, h'_0 \in I$ are orbitally topologically equivalent if and only if $h_0 = h'_0$.*

It is easy to see that conversely, now we can specify a lot of orbitally equivalent integrable systems. It is sufficient to choose the parameters characterizing the integrable systems (namely, semiaxes of inertia ellipsoids, values of area integral, values of energy) in such a way that the two rotation vectors will coincide. We can easily do this.

Corollary 4.2. *Let us consider two rigid bodies of Euler type with the following principal inertia momenta:*

$$\left(1, \frac{1}{4}, \frac{1}{5}\right) \quad \text{and} \quad \left(1, \frac{1}{3}, \frac{1}{7}\right).$$

Let h_0 be an arbitrary energy value from the zone I (low energy) for the first rigid body. Then for the second rigid body there always exists and is unique such an energy value h'_0 that the corresponding Euler-type systems $v(1, 4, 5, h_0)$ and $v(1, 3, 7, h'_0)$ are orbitally topologically equivalent.

Proof. It is easily seen that the second components of the rotation vectors for these two bodies coincide (because of the special choice of semiaxes for inertia ellipsoids). Therefore, it is sufficient to make the first components of the rotation vectors equal to obtain the coincidence of the corresponding t-molecules. Let us recall that $R = (a_1, a_2)$ and $R' = (a'_1, a'_2)$, where initially we have $a_2 = a'_2$. If a_1 and a'_1 become equal, then, according to our general theory, two systems become orbitally equivalent. Let us recall that both numbers a_1 and a'_1 are given as some integrals (see above). When we change the constant h_0', the first component a_1' of the rotation vector changes *monotonously*, and moreover, ranges over all real values from the value $a_2 = a_2'$ to infinity. Consequently, for some (and unique!) h_0' the component a_1' will coincide with a_1. The statement is proved. $\square$

Let us demonstrate even more effective corollary from the general theory. We take two rigid bodies T and T' of Euler type corresponding to the parameters (A, B, C) and (A', B', C') respectively.

Corollary 4.3. *Two rigid bodies T and T' of Euler type are orbitally topologically equivalent on the whole 4-dimensional domain in the symplectic phase manifold M^4 corresponding to the low energy values (i.e., for the energy values from zone I, see above) if and only if the following equality holds*

$$\frac{A^2}{(C - A)(B - A)} = \frac{A'^2}{(C' - A')(B' - A')}.$$

Proof. Let us consider two dynamical system of the above-mentioned type and the two corresponding 4-dimensional domains, namely two balls foliated into the union of 3-dimensional isoenergy spheres. As we know (see above) for each 3-sphere from the first 4-ball always there exists another (uniquely determined) 3-sphere from the second 4-ball, which is connected by some homeomorphism with the first 3-sphere and this homeomorphism transforms the trajectories of the first system into the trajectories of the second one. It is clear that the homeomorphism depends continuously on the energy level. Consequently, we can glue all these 3-dimensional homeomorphisms into one homeomorphism of a 4-ball onto another 4-ball and we obtain the proof of Corollary 4.3. $\square$

4.3. t-MOLECULES FOR ENERGY LEVELS OF TYPE II

In this case the molecule W^* has the form shown in Fig .1-b. We must find R-vectors on each of the four edges of the molecule so that they are Λ-invariant and b-invariant, because in this case there is a nontrivial radical U coinciding with the atom C_2. The numerical r-marks are equal to infinity for two edges and are equal to zero for the other two edges. Let us assume that the edges are oriented as shown in Fig. 1-b and are numbered by 1, 2, 3, 4.

Let us calculate the R-vectors on each of the four edges. This can be easily done because the rotation function was written above in the correct coordinate systems (on the tori). Indeed, on the upper finite edges of the molecule (in the domain X_1) the cycles λ^- and λ^+ coincide with the cycles μ and λ respectively. On the lower (bottom) infinite edges of the molecule (in the domain X_2) the pair of the cycles (μ, λ') can be considered as an admissible coordinate system.

Consequently, we find all four rotation vectors using the limits of rotation functions which were calculated above in Lemma 4.2. On the upper edges 1 and 2 of the molecule, we have

$$R_1 = R_2 = (+\infty, K(A, B, C)),$$

where K is (as above) the limit of ρ as the torus tends to τ_1.

On the lower edges 3 and 4, we must take the vectors R_3^- mod 1 and R_4^- mod 1 as the R-vectors. It follows from Lemma 4.2 that

$$R_3^- \text{ mod } 1 = R_4^- \text{ mod } 1 = (-\infty, N(\frac{h_0}{g^2}, A, B, C) \text{ mod } 1).$$

Now we calculate the b-invariant and Λ-invariant. It follows from Definition 3.5 (for b-invariant) in Section 3 that

$$b = [\frac{\alpha_1}{\beta_1}] + [\frac{\alpha_2}{\beta_2}] - [-MR_3^-] - [-MR_4^-].$$

Here the coefficients of the gluing matrices and the rotation vectors are calculated in some admissible coordinate system on the atom C_2. Let us specify such an admissible coordinate system. The first basis cycle λ^- on the boundary tori is the cycle μ (i.e., a fiber of the Seifert fibration). It can be checked that we may take the following cycles as the cycles $\mu_1^-, \mu_2^-, \mu_3^-, \mu_4^-$ (i.e., as the sections of the Seifert fibration):

μ_1^- and μ_2^- are the cycles λ,

μ_3^- and μ_4^- are cycles $\lambda' - \mu$.

Let us note that the four natural cycles $\lambda, \lambda, \lambda', \lambda'$ cannot be the boundary of a *global* section of the Seifert fibration. The obstacle for the existence of such a section is exactly the numerical mark n (see the general theory). Consequently, we need to correct this "natural set of 4 cycles". This "slight correction" was made above. As the result, we have annihilated the obstruction for the existence of the global section.

If we now consider the formula for b, we see that the numbers α_1 and α_2 are equal to zero. This follows from the fact that the first row (α_i, β_i) of the gluing matrix has the form $(0, 1)$. Indeed, the first row consists of the coefficients of the decomposition of the cycle λ_i^+ with respect to the cycles λ_i^- and μ_i^-. But on the upper edges of the molecule, we have $\lambda_i^+ = \mu_i^- = \lambda$, $i = 1, 2$.

The rotation functions ρ_3^- and ρ_4^- (with respect to the basis $\mu, \lambda' - \mu$) are connected with the rotation function ρ (with respect to the basis μ, λ') by the formula: $\rho_3^- = \rho_4^- = \rho + 1$. Hence $R_3^- = R_4^- = (-\infty, N + 1)$, where $N = N(h_0/g^2, A, B, C)$.

Consequently, $b = -2[N] + 2$, where $[\cdot]$ denotes the integer part of a number.

The last invariant that we must find is the Λ-invariant. In our case the system v has a symmetry determined by the multiplication of all six variables (of the initial space) by -1. Since the Hamiltonian and integral are quadratic, the system will transform into itself. Thus, both critical circles of the atom C_2 will interchange (see [18] for details). Because the symmetry preserves the system, both components of the Λ-invariant on two vertices of the atom C_2) are equal. Consequently, $\Lambda = (1 : 1)$.

We have proved the following statement.

Proposition 4.3. *The orbital topological invariant (i.e., the t-molecule W^{*t}) for Euler-type systems from the energy zone II (i.e., for mean values of energy) has the form shown in Fig. 3.*

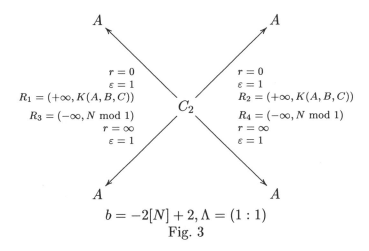

A A

$r = 0$
$\varepsilon = 1$
$R_1 = (+\infty, K(A, B, C))$

$R_3 = (-\infty, N \bmod 1)$
$r = \infty$
$\varepsilon = 1$

C_2

$r = 0$
$\varepsilon = 1$
$R_2 = (+\infty, K(A, B, C))$

$R_4 = (-\infty, N \bmod 1)$
$r = \infty$
$\varepsilon = 1$

A A

$b = -2[N] + 2, \Lambda = (1 : 1)$

Fig. 3

The statements that are analogues of Corollaries 4.1, 4.2, 4.3 and 4.4 are valid here too, i.e., for systems with energy values from the zone II. The only (but essential) difference is that the orbital topological equivalence in this case cannot be made smooth.

Thus, combining all these results, we obtain the following statement.

a. *For low energies, the Euler-type systems with appropriate values of inertia momenta are orbitally equivalent (for $h \in I$) both in the* smooth *and the* continuous *sense.*

b. *Let us increase the energy. When we come into the energy zone II (mean energy values), the* only *continuous orbital equivalence between these two systems is preserved (smooth equivalence disappears!).*

c. *When we come (by increasing the energy) into the energy zone III (high energy values), we see that the* continuous (topological) *equivalence also disappears and the two systems become orbitally nonequivalent (in topological sense!).*

4.4. t-MOLECULES FOR ENERGY LEVELS OF TYPE III

Proposition 4.4. *The orbital topological invariant (i.e., the t-molecule W^{*t}) for Euler-type systems from the energy zone III (high energy) has the form shown in Fig. 4.*

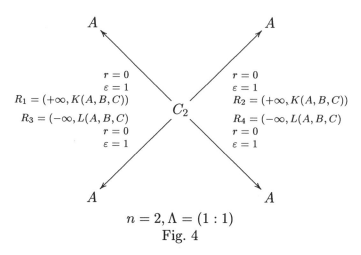

$$n = 2, \Lambda = (1 : 1)$$
Fig. 4

Proof. On the upper edges of the molecule the basis λ^-, λ^+ coincides with the basis μ, λ. On the lower (bottom) edges the basis λ^-, λ^+ coincides with μ, λ'. Thus, we can use Lemmas 4.1 and 4.2 to calculate the rotation vectors. Consequently, on the upper edges of the molecule, the rotation vectors have the form

$$R_1 = R_2 = (+\infty, K(A, B, C)),$$

and on the lower edges the rotation vectors are:

$$R_3 = R_4 = (-\infty, L(A, B, C)),$$

The Λ-invariant for this case is the same as for the Euler systems of energy type *II* (mean energy values), i.e., $\Lambda = (1 : 1)$. The arguments are the same.

Finally, the b-invariant for the unique radical U, coinciding here with the atom C_2 (as in the previous case), coincides with the numerical mark n, because the radical coincides with a family (see [7]). The proposition is proved. $\square$

Corollary 4.4. *Consider the integrable Euler-type systems $v(A, B, C, h_0)$ and $v(A', B', C', h_0')$ with h_0, h_0' from the zone III (high energy values). They are orbitally continuously equivalent if and only if the inertia ellipsoids of the bodies are similar (i.e., when $A/A' = B/B' = C/C'$).*

Proof. Since in this case the *t*-molecule does not depend on the energy value, we only need analyze the behavior of $K(A, B, C)$ and $L(A, B, C)$ as functions of the parameters A, B and C.

Consider the mapping

$$\Xi : (A, B, C) \longrightarrow (K(A, B, C), L(A, B, C)).$$

It maps the set of all "rigid bodies" (which are determined by their inertia ellipsoids) onto some region in Euclidean 2-plane. Corollary 4.4 follows from the following statement.

Proposition 4.5. *The image of the mapping Ξ is given by the two inequalities*

$$K > 0, \quad L < -1.$$

In addition, $\Xi(A, B, C) = \Xi(A', B', C')$ if and only if the triples of numbers (A, B, C) and (A', B', C') are proportional, i.e., the corresponding inertia ellipsoids (of two rigid bodies) are similar.

5. Calculation of t-molecules for the geodesic flow of the ellipsoid

5.1. ROTATION FUNCTIONS FOR THE JACOBI PROBLEM

Let us define the ellipsoid in $\mathbb{R}^3$ in the standard way:

$$\frac{x^2}{a} + \frac{y^2}{b} + \frac{z^2}{c} = 1.$$

The kinetic energy of a point moving along the ellipsoid is

$$H = \dot{x}^2 + \dot{y}^2 + \dot{z}^2.$$

An additional integral for the Jacobi problem can be written as follows:

$$f = abc \left(\frac{x^2}{a^2} + \frac{y^2}{b^2} + \frac{z^2}{c^2} \right) \left(\frac{\dot{x}^2}{a} + \frac{\dot{y}^2}{b} + \frac{\dot{z}^2}{c} \right).$$

It is clear that we can always reduce the geodesic flow of the ellipsoid from the total phase space to the 3-dimensional isoenergy manifold (3-surface of constant energy), which is diffeomorphic to the projective 3-space $\mathbb{RP}^3$. We can assume that $\{H = 1\}$ on this invariant 3-surface.

It is easy to see that the geodesic flow of the ellipsoid is a nondegenerate system on the whole 3-surfaces $H = h$, where $h \neq 0$. The segment of a real axis which is covered by the values of the integral f is $[a, c]$. The critical values of f are: $a = $ minimum, $b = $ saddle value, $c = $ maximum. Here $a < b < c$.

Let us consider one of the four edges of the molecule W in the Jacobi problem. We denote by t the parameter along the edge, i.e., $f = t$, where f is the integral of the geodesic flow (see above) and t is its value.

Let us calculate the rotation function $\rho(t)$ on the edge. To do this we introduce natural coordinates on the Liouville torus.

Definition 5.1. The *elliptic coordinates* of a point $P = (x, y, z)$ are defined as the three real numbers $\lambda_1 > \lambda_2 > \lambda_3$ that are the roots of the following equation:

$$\frac{x^2}{a + \lambda} + \frac{y^2}{b + \lambda} + \frac{z^2}{c + \lambda} = 1.$$

Let us notice that the initial ellipsoid can be written with respect to the elliptic coordinates $(\lambda_1, \lambda_2, \lambda_3)$ as $\lambda_1 = 0$.

Let us regard λ_2 and λ_3 as coordinates on the ellipsoid. They can be lifted (as functions) to the Liouville torus. As the result, we obtain two functions that we denote by the same symbols λ_2 and λ_3. These functions are in some sense "more natural" on the torus than on the ellipsoid. The fact is that the torus is projected on the ellipsoid in such a way that this projection almost everywhere is a 2-sheeted covering (except for the boundary of the projection G_t). Thus, in the region G_t, the equations $\lambda_2 = $ const and $\lambda_3 = $ const determine either circles or segments (i.e., the smooth arcs which are obtained when we glue a circle in a 2-sheeted way). The same equations on the Liouville torus always determine the circles only. After this remark, we can naturally define a basis on the Liouville torus. Namely, the two basic cycles on it can be determined by the equations

$$\lambda_2 = \text{const}, \quad \lambda_3 = \text{const}.$$

Thus, let us fix on the torus T_t^2 (which lies on the level $f = t$) the coordinates λ_2 and λ_3. Let us introduce the function

$$\Phi(u, t) = \sqrt{\frac{u}{(u + a)(u + b)(u + c)(u + t)}},$$

where a, b, c are the parameters of ellipsoid (see above).

Proposition 5.1. The rotation function $\rho'(t)$ on the edge of the molecule W with respect to the basis $(\lambda_2 = $ const, $\lambda_3 = $ const) (this is the second cycle) in Jacobi problem has the form:

$$\rho'(t) = \frac{\int_{-t}^{-a} \Phi(u, t)\, du}{\int_{-c}^{-b} \Phi(u, t)\, du}, \quad \text{where} \quad a < t < b,$$

$$\rho'(t) = \frac{\int_{-b}^{-a} \Phi(u, t)\, du}{\int_{-c}^{-t} \Phi(u, t)\, du}, \quad \text{where} \quad b < t < c.$$

Now we can rewrite the rotation function $\rho'(t)$ with respect to another "more suitable" basis. As a result we will obtain the final form of the function $\rho(t)$, which plays an important role in our classification theory. This "more correct" basis consists of two cycles (λ^-, λ^+) on a torus. Here λ^- is the cycle which appears from the critical saddle circle of the saddle atom C_2, and the cycle λ^+ is the annihilating cycle, i.e., the cycle which contracts to a point when the torus tends to another atom A on the second end of a given edge. It is easy to find the explicit formulas connecting the two bases (λ_2, λ_3) and (λ^-, λ^+). By rewriting the rotation function ρ with respect to the new basis (λ^-, λ^+), we get

$$\rho(t) = -\frac{\int_{-b}^{-a} \Phi(u, t)du}{\int_0^{+\infty} \Phi(u, t)du} - 1, \text{ where } b < t < c,$$

$$\rho(t) = \frac{\int_{-c}^{-b} \Phi(u, t)du}{\int_0^{+\infty} \Phi(u, t)du} - 1, \text{ where } a < t < b.$$

The two following lemmas describe the properties of ρ that we need to calculate the t-molecule.

Lemma 5.1. *The rotation function $\rho(t)$ is strongly monotonous on each of four edges of the molecule W. In particular, the rotation vector on each edge is completely determined by two numbers, namely, by the limits of the rotation function at the end-points of the edge.*

The proof follows from straightforward calculations.

Lemma 5.2. *The limits of the rotation function ρ (in Jacobi problem) are as follows:*

$$\lim_{t \to a} \rho(t) = k(a, b, c) = \frac{\int_{-c}^{-b} \Phi(u, a)du}{\int_0^{+\infty} \Phi(u, a)du} - 1,$$

$$\lim_{t \to c} \rho(t) = l(a, b, c) = -\frac{\int_{-b}^{-a} \Phi(u, c)du}{\int_0^{+\infty} \Phi(u, c)du} - 1,$$

$$\lim_{t \to b \mp 0} \rho(t) = \pm\infty.$$

5.2. THE t-MOLECULE FOR THE JACOBI PROBLEM

The statements proved above give us the rotation vectors for the Jacobi problem. As in the Euler case, the Λ-invariant takes the value $(1:1)$ (for the same reasons). Let us summarize all the obtained results in the form of the following theorem.

Theorem 5.1. *The topological orbital invariant, i.e., the t-molecule for the integrable Jacobi problem (geodesic flow on the ellipsoid) is shown in Fig. 5.*

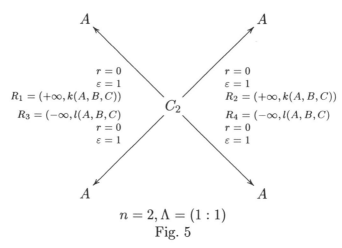

$$n = 2, \Lambda = (1 : 1)$$
Fig. 5

As we see, the geodesic flows of two different ellipsoids differ only by the values of the limits $k(a, b, c)$ and $l(a, b, c)$ for the rotation function. All others orbital topological invariants for the different ellipsoids coincide. Let us analyze the behavior of the functions $k(a, b, c)$ and $l(a, b, c)$. Consider the mapping $\xi : (a, b, c) \longrightarrow (k(a, b, c), l(a, b, c))$. The properties of ξ are similar to those of Ξ.

Proposition 5.2. *The image of the mapping ξ is given by the two inequalities $k > 0$, $l < -1$. In addition, $\xi(a, b, c) = \xi(a', b', c')$ if and only if the triples of numbers (a, b, c) and (a', b', c') are proportional, i.e., the corresponding ellipsoids are similar.*

The orbital classification of geodesic flows for ellipsoids (Theorem B) is a direct consequence of this statement and Theorem 5.1. In other words, the geodesic flows of two ellipsoids are orbitally equivalent if and only if the ellipsoids are similar (proportional).

6. Equivalence between the Jacobi problem and the Euler case

Let us consider two classical integrable systems:

1) the geodesic flow on the ellipsoid in $\mathbb{R}^3$ (Jacobi problem),

2) the dynamical system describing the motion of a rigid body in Euclidean 3-space around its center of mass (Euler integrable case).

The Euler equations include the so-called *area constant* g as a parameter. Let us consider the special case when $g = 0$. It was proved above (see Section 4), that Euler-type systems for large values of energy are smoothly conjugate to the Euler-type systems with zero area constant.

Let us fix a nonzero energy value h_0 for the function H and consider the dynamical system on the corresponding 3-manifold Q. The molecule W^* for the Euler case coincides with that for the Jacobi problem [5].

To prove Theorem C, it is sufficient to compare the t-molecules of the Euler case (type III) and of the Jacobi problem. They have the same structure and can only differ by the values of the limits for rotation functions on the end-atoms A (i.e., on the ends of the edges). In other words, the t-molecules coincide if and only if $K(A, B, C) = k(a, b, c)$ and $L(A, B, C) = l(a, b, c)$, i.e., $\Xi(A, B, C) = \xi(a, b, c)$. Hence, Theorem C follows immediately from Propositions 4.5 and 5.2.

All calculations have been done for 3-dimensional energy levels. But it turns out that for both systems the discovered topological orbital equivalence can be extended on 4-dimensional phase spaces. This follows from the observation that all 3-dimensional energy levels Q (for each problem) are homeomorphic, similar to each other and the flows on them are smoothly orbitally equivalent (the corresponding diffeomorphism is just the homothety on each cotangent space).

Let us recall that the Euler case and the Jacobi problem have different types of functions that integrate the problem, namely, *elliptic* functions for the Euler case and *hyperelliptic* ones for the Jacobi problem. But now we see from our theorem that the different character of the functions has no influence to the topology of these systems (because the systems are topologically orbitally equivalent).

We have considered the case of general position, i.e., when

$$1/A > 1/B > 1/C \text{ and } a < b < c.$$

An interesting problem is to analyze the case of symmetric ellipsoids (ellipsoids of revolution), when some of the inertia momenta (or semiaxes) coincide.

References

1. Andronov, A.A., Leontovich E.A., Gordon, I.I. and Mayer A.G.: *Quantitative Theory of Dynamical Systems of Second Order*, Nauka, Moscow, 1966 (in Russian).
2. Anosov, D.V., Aranson, S.H., Bronstein, I.U. and Grines, V.Z.: Smooth dynamical systems II, In: *Modern problems of mathematics*, **1** (1985) VINITI, Moscow, 151–242 (in Russian).
3. Arnold, V.I. and Ilyaschenko, Yu.S.: Ordinary differential equations, In: *Modern problems of mathematics*, **1** (1985) VINITI, Moscow, 7–149 (in Russian).
4. Arkhangelskii, Yu.A.: *Analytical Dynamics of a Rigid Body*, Nauka, Moscow, 1977 (in Russian).
5. Bolsinov, A.V.: Methods of calculation of the Fomenko–Zieschang invariant, In: *Topological Classification of Integrable System*, Adv. Soviet Math. **6** Amer. Math. Soc., Providence, RI, 147–183.

6. Bolsinov, A.V., Matveev, S.V. and Fomenko, A.T.: Topological classification of integrable Hamiltonian systems with two degrees of freedom. List of the systems of low complexity, *Russian Math. Surveys*, **45** (1990) no. 2, 59–94.

7. Bolsinov, A.V. and Fomenko A.T.: Orbital equivalence of integrable Hamiltonian systems with two degrees of freedom. A classification theorem I, *Russian Acad. Sci. Sb. Math.* **81** (1995) no. 2, 421–465; II, *Russian Acad. Sci. Sb. Math.* **82** (1995) no. 1, 21–63.

8. Bolsinov, A.V. and Fomenko A.T.: Orbital classification of integrable Euler-type systems in dynamics of a rigid body, *Uspekhi Matem. Nauk* **48** (1993) no. 5, 163–164 (in Russian).

9. Fomenko, A.T.: Topological classification of all Hamiltonian differential equations of general type with two degrees of freedom, In: *The Geometry of Hamiltonian Systems. Proceedings of a Workshop Held June 5-16, 1989*, Springer–Verlag, Berkeley – New York, 1991, 131–339.

10. Fomenko, A.T.: The topology of surfaces of constant energy in integrable Hamiltonian systems, and obstructions to integrability, *Math. USSR Izvestiya* **29** (1987) no. 3 pp. 629–658.

11. Fomenko, A.T. and Zieschang, H.: A topological invariant and a criterion for the equivalence of integrable Hamiltonian systems with two degrees of freedom, *Math. USSR Izvestiya* **36** (1991) no. 3, 567–596.

12. Jacobi, C.G.J.: *Lectures in Dynamics*, Moscow–Leningrad, 1936.

13. Kozlov, V.V.: *Methods of Quantitative Analysis in a Rigid Body Dynamics*, Moscow Univ. Press, Moscow, 1980.

14. Kozlov, V.V.: Two integrable problems of classical mechanics, *Vestnik MGU* **4** (1981), 80-83 (in Russian).

15. Kozlov, V.V. and Fedorov Yu.N.: *Memoir on Integrable Systems,* Springer–Verlag (to appear).

16. Leontovich, E.A. and Mayer A.G.: On the scheme determining the topological structure of a trajectories foliation, *Doklady AN SSSR* **103** (1955), no. 4 (in Russian).

17. Moser, J.: Three integrable Hamiltonian systems connected with isospectral deformations, *Advances in Math.* **16** (1975) no. 2, 197–220.

18. Nguen, T.Z. and Polyakova L.S.: A topological classification of integrable geodesic flows on the two-dimensional sphere with an additional integral quadratic in moments, *J. Nonlinear Sci.* **3** (1993) no. 1, 85–108.

19. Oshemkov, A.A.: Methods of calculation of the Fomenko–Zieschang invariant, In: *Topological Classification of Integrable System*, Adv. Soviet Math. **6** Amer. Math. Soc., Providence, RI, 67–146.

20. Peixoto, M.M.: On the classification of flows on manifolds, In: *Dynamical systems. Proc. Symp. Univ. of Bahia*, Acad. Press, New York–London, 1973, 389–419.

21. Sadov, Yu.A.: Action-angle variables in the Euler–Poinsot problem, *Priklad. Mat. i Mekh.* **34** (1970) no. 5, 962–964 (in Russian).

22. *Topological Classification of Integrable System,* Adv. Soviet Math. **6** Amer. Math. Soc., Providence, RI.

23. Umanskii, Ya.L.: Scheme of three-dimensional dynamical system of Morse–Smale type, *Doklady AN SSSR* **230** (1976) no. 6, 1286–1289.

24. Abraham, R. and Marsden, J.: *Foundations of Mechanics*, Benjamin–Cummings, New York, 1978.

NONCOMMUTATIVE DEFORMATION

OF THE KADOMTSEV–PETVIASHVILI HIERARCHY

AND THE UNIVERSAL GRASSMANN MANIFOLD

E.E. DEMIDOV

Department of Mechanics and Mathematics,
Moscow State University, 119899 Moscow, Russia

Abstract. We suggest a version of noncommutative analogue of the classical Kadomtsev–Petviashvili hierarchy based on replacing the classical infinite-dimensional time-space of the hierarchy by a noncommutative one in the spirit of the quantum group theory: $t_i t_j = q_{ij}^{-1} t_j t_i$. We prove that our analogue is an integrable hierarchy of noncommutative differential equations. Then this hierarchy is proved to be a (formal) dynamic system on the set of noncommutative points of the infinite-dimensional Grassmann manifold of Sato & Sato. We introduce a generalization of Mulase's Schur pairs and obtain a classification of the commutative subalgebras of noncommutative differential operators. Finally, all of this leads to a deformation of algebro-geometric data with a noncommutative base.

Mathematics Subject Classification (1991): 17B37, 16W35.

Key words: dressing transformation, noncommutative Sato–Wilson equation, PDO algebra, noncommutative de Rham complex, Schur pair, universal Grassmann manifold.

1. Introduction

The KP-hierarchy of nonlinear differential equations is known to be a fundamental mathematical object related to problems of mathematical physics,

B. P. Komrakov et al. (eds.), Lie Groups and Lie Algebras, 383–391.
© 1998 *Kluwer Academic Publishers. Printed in the Netherlands.*

algebraic geometry[1], representation theory, etc. It has a superanalog produced by a superization of its infinite-dimensional time-space (see e.g. [2]). In the present paper we define generalizations of the KP-hierarchy (NKP) and the Sato–Wilson (NSW) equation for a noncommutative time-space having "q-commuting" coordinates:

$$t_i t_j = q_{ij}^{-1} t_j t_i.$$

For this reason we need to modify the definitions of the pseudodifferential operator (PDO) and the differential calculus. Then we prove the unique solvability theorem for the analog of the Sato–Wilson equation. Further we show that the notion of universal Grassmann manifold **G** introduced by Sato & Sato [6] can be extended to the noncommutative case (i.e., one can speak of its noncommutative points). Moreover, reproducing the Sato & Sato correspondence in our noncommutative situation, we get, in a sense, a "formal dynamical system" of NSW-flows on the noncommutative points of **G**. It is natural, in this context, to investigate the modification of the so-called Schur pairs that classify commutative subalgebras of differential operators in the classical situation. We introduce a general version of the Schur pairs (cf. [4]). Next we describe a way that takes us from our Schur pairs to commutative subalgebras of differential operators with noncommutative coefficients and the converse way. Finally, we discuss the transformation law for Schur pairs under the action of NSW-flows.

   This work was supported in part by the grants RFFI 93-01-01542 and ISF M7N00.

2. The algebra of PDO

Fix a field k of characteristic zero. Let R be a k-algebra. A pair (∂, S), where $\partial : R \to R$ is a k-linear mapping, $S \in \mathrm{Aut}_k R$, is said to be a *skew derivation* of R if $\partial(a.b) = \partial a.b + S(a).\partial b$ for all $a, b \in R$ and $\partial \circ S = S \circ \partial$. The PDO algebra associated with (∂, S) is the set $\mathcal{E}_R$ of formal Laurent series in D^{-1} with coefficients in R and the following multiplication

$$D^n f = \sum_{k=0}^{\infty} \binom{n}{k} S^{n-k}(\partial^k f) D^{n-k}, \quad n \in \mathbb{Z}.$$

It is easy to see that $\mathcal{E}_R$ is an associative algebra, and the subset

$$\mathcal{V}_R = \{W \in \mathcal{E}_R \mid W = 1 + w_1 D^{-1} + \ldots\}$$

forms a multiplicative group.

---

[1]The remarkable solution of the Schottky problem by means of the soliton equation is presented in detail in [1].

Let v be a valuation on R, I be the valuation ideal and $\pi : R \to R/I$ be the canonical projection. Suppose that $v \circ S = v$ and $\partial(I) \subset I$. Consider the completed algebras

$$\hat{\mathcal{E}}_R = \left\{ P = \sum_{i \in \mathbb{Z}} p_i D^i \,\middle|\, \forall P \,\exists N, M, C > 0 \;:\; v(p_i) > Ci + M \text{ if } i > N \right\},$$

$$\hat{\mathcal{D}}_R = \{ P \in \hat{\mathcal{E}}_R \mid P_- = 0 \},$$

and the formal groups

$$\hat{\mathcal{E}}_R^\times = \{ P \in \hat{\mathcal{E}}_R \mid \pi(P) \in V_{R/I} \text{ and } \exists P^{-1} \in \hat{\mathcal{E}}_R \},$$

$$\hat{\mathcal{D}}_R^\times = \left\{ P \in \hat{\mathcal{D}}_R \mid \pi(P) = 1 \text{ and } \exists P^{-1} \in \hat{\mathcal{D}}_R \right\}.$$

Theorem 1 *There exists a unique group decomposition* $\hat{\mathcal{E}}_R^\times = V_R \hat{\mathcal{D}}_R^\times$.
The proof follows M. Mulase's work [3].

3. The noncommutative time-space

Consider the algebra $\mathcal{K}$ of noncommutative formal power series in $x, t_1, t_2,$ $\ldots$ that verify the following commutational relations:

$$t_i t_j = q_{ij}^{-1} t_j t_i, \quad x t_i = q_i^{-1} t_i x,$$

where $q_{ij}, q_i \in k^\times$ are the deformation parameters. Denote by T the subalgebra of $\mathcal{K}$ generated by t's. A skew derivation of $\mathcal{K}$ is defined by

$$\partial x = 1, \; \partial t_i = 0, \; S(x) = x, \; S(t_i) = q_i t_i.$$

Finally, fix a valuation v such that $v(x) = 0$, $v(t_i) = i$. Let $\mathcal{E}_\mathcal{K}$ be the PDO algebra associated with (∂, S).

4. The de Rham complex

Let us introduce the de Rham complex of the noncommutative time-space using an idea of J. Wess and B. Zumino [7]. Their construction is related to a finite-dimensional solution of the quantum Yang–Baxter equation, but one can show that it is possible to extend this construction to our infinite-dimensional case. Namely, our complex is defined to be an algebra Ω of noncommutative formal power series in $t_1, t_2, \ldots, \tau_1, \tau_2, \ldots$ subject to the following set of relations:

$$t_i t_j = q_{ij}^{-1} t_j t_i, \; \tau_i \tau_j = -c q_{ij}^{-1} \tau_j \tau_i, \; \tau_i^2 = 0,$$

$$t_i \tau_i = c \tau_i t_i,$$

$$t_i \tau_j = c q_{ij}^{-1} \tau_j t_i, \text{ if } i < j,$$

$$t_i \tau_j = (c - 1) \tau_i t_j + q_{ij}^{-1} \tau_j t_i, \text{ if } i > j.$$

Here $c \in k^\times$ is a new deformation parameter assumed not to be a root of 1. A map $d : \Omega \to \Omega$ taking t_i to τ_i and τ_i to 0 is a well-defined Leibnitz differential on Ω.

Let us construct the de Rham complex for the $\mathcal{K}$-module $\mathcal{E}_{\mathcal{K}}$ by putting $\Omega(\mathcal{E}_{\mathcal{K}}) = \Omega \otimes_T \mathcal{E}_{\mathcal{K}}$ as a vector space and by extending the differential d from Ω as $dx = 0$, $dD = 0$ with preserving the Leibnitz rule. Moreover, we set

$$x\tau_i = q_i^{-1}\tau_i x, \quad D\tau_i = q_i\tau_i D \text{ (or, in other words, } S(\tau_i) = q_i\tau_i).$$

5. The noncommutative KP-hierarchy

Define a $\mathcal{E}_{\mathcal{K}}$-valued form $\omega = \sum_{i>1} \tau_i L_i$ for a family $\{L_i = D^i + a_{i1}D^{i-1} + \ldots\}$. Put $Z_\pm = \pm\omega_\pm$. Here the subscript $\pm$ means the projections of a PDO to the spaces of purely differential and of "integral" operators respectively. The following system

$$\left\{ \begin{array}{rcl} d\omega &=& Z^+\omega + \omega Z^+ \\ \omega^2 &=& 0 \end{array} \right. \text{ or, equivalently, } \left\{ \begin{array}{rcl} d\omega &=& Z^-\omega + \omega Z^- \\ \omega^2 &=& 0 \end{array} \right.$$

$$\text{(NKP)}$$

is called the *noncommutative KP-hierarchy*. If one puts all the parameters equal to 1 and $L_i = L_1^i$, then one recover a consequence of the classical KP. The forms $Z^\pm$ play the role of the Zakharov–Shabat connections.

Let us find a solution of (NKP) in the form $\omega = W\omega_D W^{-1}$, where $\omega_D = \sum \tau_i D^i$.

Lemma 1 (i) $\omega_D^2 = 0$ *iff* $q_{ij} = cq_i^{-j}q_j^i$.

(ii) *If*

$$dW = Z^-W, \tag{NSW}$$

then $\omega = W\omega_D W^{-1}$ *is a solution of* (NKP).

We shall assume the conditions of (i) satisfied. Thus, the independent parameters of our deformation are $c, q_1, q_2, \ldots$. The *noncommutative Sato–Wilson equation* (NSW) has precisely the same form as in the classical and the super cases [3]. This was the reason M. Mulase called (NSW) "the generalized KP-hierarchy". In other words, (NSW) is a more primary object than (NKP).

6. Integrability of (NSW)

Theorem 2 *There exists a bijection between the solutions of* (NSW) *and the solutions of the equation*

$$dX = \omega_D X \tag{NL}$$

on $X \in \hat{\mathcal{E}}_{\mathcal{K}}^\times$.

The proof uses a nontrivial and very tedious calculation, which shows the existence of a unique operator $Y \in \hat{\mathcal{D}}_K^{\times}$ such that $dY = Z^+Y$. Technically, the proof of this fact follows M. Mulase's scheme [3]. In a sense, this differential equation is a complement to (NSW). Then, by an obvious using of the Leibnitz rule one derive that $X = W^{-1}Y$ satisfies (NL). Conversely, if X is a solution of (NL), then by the decomposition theorem $X = W^{-1}Y$, and (NL) implies (NSW).

The equation (NL) is linear, hence, its solution $X(t)$ with the initial condition X_0 is given by a c-exponent:

$$X(t) = \exp_c \left(\sum_{i \geq 1} t_i D^i \right) X_0,$$

where $\exp_c(u) = \sum_{n \geq 0} u^n/[n]_c!$ and $[n]_c = (c^n - 1)/(c - 1)$.

Corollary 1 *Equation* (NSW) *is uniquely solvable for every initial condition* $W_0 \in \mathcal{V}_{k[[x]]}$. *Moreover, the solution can be obtained from the decomposition*

$$W(t)^{-1}Y(t) = E(t)W_0^{-1}.$$

7. The infinite-dimensional Grassmannian

First, we define the set of noncommutative points of the universal Grassmann manifold of Sato & Sato [6]. Let R be a ring with 1. The semi-infinite index $I = (\ldots, i_{-3}, i_{-2}, i_{-1})$ is defined as an increasing sequence of integers such that $i_{-n} = -n$ for sufficiently large n. Below $\emptyset$ stands for $(\ldots, -3, -2, -1)$. A semi-infinite matrix M is the table (m_{ij}) of elements from R with $i \in \mathbb{Z}$ ánd $j \leq -1$. For a semi-infinite matrix M and an index I, let M_I be the submatrix $(m_{ij})_{i \in I, j \leq -1}$. Denote by Frame($R$) the whole set of semi-infinite matrices with entries from R satisfying the following conditions:

(1) $m_{-i,-i} = 1$ for all sufficiently large i,
(2) $m_{ij} = 0$ if $i < j$ and i is sufficiently small,
(3) M_I is an invertible matrix for some I.

Next consider a set $GL(R)$ of all invertible matrices $N = (n_{ij})_{i,j \leq -1}$ with entries from R such that (1) and (2) are satisfied. This set forms a group acting on Frame(R) by right multiplication.

By definition, R-points of the *infinite-dimensional Grassmannian* are the equivalence classes

$$\mathbf{G}(R) = \text{Frame}(R)/GL(R).$$

Denote by $\mathbf{G}^I(R)$ the set of classes $M(\text{mod}\, GL(R))$ from $\mathbf{G}(R)$ that have invertible submatrix M_I.

To get a more "invariant" description of $\mathbf{G}(R)$, let us introduce a free R-module $V = R((z))$ of the formal Laurent series, z being an auxiliary commutative variable. For every class $\xi = M(\mathrm{mod}\, GL(R))$ consider a left R-submodule U_ξ in V generated by the elements

$$u_i = \sum_{j \in \mathbb{Z}} m_{ji} z^j.$$

A basis of this type is called to be an *admissible basis*. It is evident, that the group $GL(R)$ acts transitively on the set of all admissible bases of our R-submodule U_ξ, therefore, U_ξ is well defined by $\xi \in \mathbf{G}(R)$.

So, we can define $\mathbf{G}(R)$ in an "invariant" manner. Namely, we say that a free left R-submodule $U \subset V$ is a point of $\mathbf{G}(R)$ if it has an admissible basis.

It is not hard to prove that in the case $R = k$ this description of $\mathbf{G}(k)$ is equivalent to the "Fredholm" one (see, e.g., [5]).

8. The Sato mapping

Note, that the map $X \mapsto (\tilde{X}_{ij})$ from $\mathcal{E}_R$ to the set of infinite R-valued matrices defined by the rule

$$D^i X = \sum_{j \in \mathbb{Z}} \tilde{X}_{ij} D^j$$

is a well-defined representation of $\mathcal{E}_R$.

Theorem 3 *The map* $\gamma : \mathcal{V}_\mathcal{K} \to \mathbf{G}^\emptyset(T)$ *taking an operator* W *to the point*

$$\gamma(W) = \left.\left(\widetilde{W}^{-1}\right|_{x=0}\right)_{i \in \mathbb{Z}, j \leq -1} (\mathrm{mod}\, GL(T)),$$

is a bijection.

Recall that T is the subalgebra of $\mathcal{K}$ generated by t's. Therefore, the action of (NSW)-flows can be interpreted as "formal dynamical system" on $\mathbf{G}^\emptyset$. This means that there exists a map $\mathbf{G}^\emptyset(k) \to \mathbf{G}^\emptyset(T)$ induced by the correspondence $W_0 \mapsto W(t)$. Formality is the price for dealing with formal power series.

In the rest of the paper the deformation parameters q_i's are assumed to be multiplicatively independent in $k^\times$.

9. The Schur pairs

Let F be a point of $\mathbf{G}(R)$, i.e., a certain submodule of V. Define the action of PDO's on $\mathbf{G}(R)$ as

$$aF = \tilde{a}F,$$

where F is the matrix representing the point F.

Definition 1 A pair (A, F) consisting of an algebra A of PDO's with coefficients in R (i.e., $A \subset \mathcal{E}_R$ or $A \subset \hat{\mathcal{E}}_R$) and a point F of $\mathbf{G}(R)$ is said to be an *R-Schur pair* if A stabilizes F, i.e., $AF \subset F$.

An R-Schur pair is called *maximal* if A is the stabilizer of F:

$$A = A_F =: \{a \in \hat{\mathcal{E}}_R \mid aF \subset F\}.$$

The notion of a Schur pair for PDO algebras with commutative coefficients was introduced in [4].

10. From Schur pairs to differential operators

Lemma 2 *Let* (A, F) *be a* T-*Schur pair with* $A \subset \mathcal{E}_T$. *Define the operator* $W \in \mathcal{V}_\mathcal{K}$ *as* $\gamma^{-1}(F)$. *Then for* $a \in A$ *the operator* WaW^{-1} *belongs to* $\mathcal{D}_\mathcal{K}$.

The proof of this lemma can be obtained from the matrix interpretation of the stability condition $aF \subset F$. The same result can be proved for a Schur pair with $A \subset \hat{\mathcal{E}}_T$. Note that the subalgebra of differential operators obtained is isomorphic to A. So, taking a T-Schur pair (A, F) with a commutative algebra A, we get a commutative subalgebra of differential operators.

11. From differential operators to Schur pairs

Consider a commutative algebra $B \hookrightarrow \mathcal{D}_\mathcal{K}$. We shall suppose that B contains an operator of the form

$$L = D^d + aD^{d-2} + \ldots \tag{L}$$

Moreover, the subalgebra B is assumed to have rank 1, i.e., for every sufficiently large d it contains an operator of order d.

We claim that all these algebras are classified by the Schur pairs of a special type. The proof of this claim closely follows the classical scheme and is based on the following three statements.

Lemma 3 *If* $L = D + u_1 D^{-1} + u_2 D^{-2} + \ldots \in \mathcal{E}_\mathcal{K}$, *then there exists an element* $W \in \mathcal{V}_\mathcal{K}$ *such that* $L = WDW^{-1}$. *This element is unique up to right multiplication by an operator* $C \in \mathcal{V}_k$.

Lemma 4 *For an operator* $X = D^d + a_{d-1}D^{d-1} + \ldots \in \mathcal{E}_\mathcal{K}$ *there exists a unique operator* $Y = D + b_0 + b_1 D^{-1} + \ldots \in \mathcal{E}_\mathcal{K}$ *such that* $Y^d = X$.

Lemma 5 *The centralizer* $\mathcal{Z}(L)$ *of an operator* $L = D^d + a_{d-1}D^{d-1} + \ldots$ *in* $\mathcal{E}_\mathcal{K}$ *coincides with* $k((L^{-1/d}))$.

Theorem 4 *For every commutative algebra* $B \hookrightarrow \mathcal{D}_K$ *that has rank 1 and satisfies the condition* (L) *there exists a unique subalgebra* $A \subset \mathcal{E}_k$ *and an operator* $W \in \mathcal{V}_K$ *which is unique up to a right multiplication by an element of* $\mathcal{V}_k$, *such that*

$$WAW^{-1} = B.$$

Finally, to get the Schur pair (A, F) corresponding to the given B, we put $F = \gamma(W)$ and $A \subset \mathcal{E}_k$ is precisely the same as in the theorem. The pair (A, F) is unique up to the left action of the group $\mathcal{V}_k$ on F.

12. Action of the NSW-flows

Let (A_0, F_0) be a k-Schur pair. Using the Sato correspondence, find the operator $W_0 \in \mathcal{V}_K$ such that $\gamma(W_0) = F_0$. Then solve (NSW) taking as the initial condition the operator W_0. So, we get the evolution of the point F_0, namely $F_t = \gamma(W(t))$. What is a natural transformation law for A_0? From general considerations it follows that A_t should be a subalgebra of $\mathcal{E}_T$ that stabilizes F_t.

Lemma 6 *If* (A_0, F_0) *is a Schur pair, then the subalgebra* $E(t)A_0E(t)^{-1}$ *of* $\hat{\mathcal{E}}_T$ *stabilizes* F_t.

The proof follows from the matrix interpretation.
Now *define* A_t as

$$E(t)A_0E(t)^{-1}.$$

This definition is compatible with the notion of maximality. In the commutative case the algebra A_t coincides with the given A_0. In our noncommutative situation they are only isomorphic and A_0 is contained in $\mathcal{E}_k$ while $A_t \subset \hat{\mathcal{E}}_T$. Therefore, according to our main lemma, the Schur pair (A_t, F_t) produces a commutative subalgebra in $\hat{\mathcal{D}}_K$.

13. Noncommutative deformations of the algebro-geometric data

Note that under the classical Burchnall–Chaundy–Krichever correspondence the Schur pair (A, F) gets a curve C such that $C - p = \operatorname{Spec} A$ and a sheaf $\mathcal{F}$ on it (as F is a A-module). So, a Schur pair (A, F) may be taken as the initial condition for two types of evolution: by the classical KP or by the noncommutative KP. Both types preserve the initial curve C. Hence, the sheaf $\mathcal{F}$ may be deformed with commutative and noncommutative bases as well. In the first case, this base is $\operatorname{Spec} k[[t_1, t_2, \ldots]]$, and in the second is "Spec" $k\langle\langle t_1, t_2, \ldots\rangle\rangle$. To shed some light to the last case, it is necessary to compute the effective deformation parameters as it was done in the classical case.

References

1. Demidov, E.E.: *The Kadomtsev–Petviashvili Hierarchy and the Schottky Problem*, Math. College of the Independent University of Moscow, 1995.
2. Manin, Yu.I. and Radul, A.O.: A supersymmetric extension of the Kadomtsev–Petviashvili hierarchy, *Comm. Math. Phys.* **98** (1985) 65–77.
3. Mulase, M.: Solvability of the super KP equations and a generalization of the Birkhoff decomposition, *Invent. Math.* **92** (1988) 1–46.
4. Mulase, M.: Cohomological structure in soliton equations and Jacobian varieties, *J.Diff. Geom.* **19** (1984) 403–430.
5. Mulase, M.: Normalization of the Krichever data, *Contemp. Math.* **136** (1992) 297–304.
6. Sato, M. and Sato, Y.: Soliton equations as dynamical systems on infinite dimensional Grassmann manifold, *Lect. Notes in Num. Appl. Anal.* **5** (1985) 259–271.
7. Wess, J. and Zumino B.: Covariant differential calculus on the quantum hyperplane, *Nucl. Phys. B (Proc. Suppl.)*, **18** (1990) 302.

SYMMETRIES OF COMPLETELY
INTEGRABLE DISTRIBUTIONS

BORIS DUBROV AND BORIS KOMRAKOV
International Sophus Lie Center,
P.B. 70, Minsk, 220123, Belarus

Abstract. This work is devoted to the integration problem of completely integrable distributions by means of symmetries. We reduce this problem to the well-known problem of integration of differential 1-forms with values in a Lie algebra that satisfy the Maurer–Cartan equation (see [4, 6, 5]). As an immediate applications we get a series of results generalizing the classical results of Sophus Lie. We show also that the same technique can be applied to the constructive equivalence of two local transitive actions.

Mathematics subject classification (1991): 58F07, 58F35.

Key words: completely integrable systems, symmetries, integration by quadrature.

1. Introduction

It is well known that the problem of solving systems of ordinary differential equations is equivalent to the integration of a certain completely integrable vector distribution (see, for example, [3]). The idea of using symmetries for this purpose is due to Sophus Lie, who constructed effective integration methods for solvable Lie algebras of symmetries [1]. In particular he showed that in this case the complete set of first integrals can be found by quadratures. Modern versions of his theory can be found, for example, in [2, 3, 7].

The standard technique of using symmetries requires a symmetry algebra of a dimension equal to the codimension of a distribution (or, what is the same, to the order of an ordinary differential equation). In Section 2 we show how extra symmetries can be used to derive first integrals even without the integration process. In Sections 3 and 4 we describe the language of

B. P. Komrakov et al. (eds.), Lie Groups and Lie Algebras, 393–405.

g-structures and describe methods for their integration. As an application, in Section 5 we generalize certain classical results of Sophus Lie. Finally, we explain in Section 6 how the same technique can be applied for finding a mapping that takes one transitive Lie algebra of vector fields to another (if possible).

Let M be a smooth manifold of dimension $n + m$, and let E be an m-dimensional completely integrable distribution on M. Using the Frobenius theorem, it is easy to show that the distribution E is determined uniquely by its algebra $I(E)$ of first integrals, namely

$$E_p = \bigcap_{f \in I(E)} \ker d_p f \quad \text{for all } p \in M.$$

An (infinitesimal) symmetry of E is a vector field X on M such that $[X, Y] \in \mathbb{D}(E)$ for all $Y \in \mathbb{D}(E)$. It is easily seen that the set of all symmetries of E is precisely the normalizer of the subalgebra $\mathbb{D}(E)$ in the Lie algebra $\mathbb{D}(M)$, and hence it is a subalgebra of $\mathbb{D}(M)$, called the symmetry algebra of the distribution E and denoted sym (E). We note that $\mathbb{D}(E)$ is, by definition, an ideal in sym (E). The elements of $\mathbb{D}(E)$ are called the characteristic symmetries of E.

It follows immediately from the definitions that

1. sym $(E)(I(E)) \subset I(E)$;
2. $I(E) \cdot \text{sym}(E) \subset \text{sym}(E)$.

Let $\mathfrak{g}$ be a subalgebra $\mathfrak{g}$ of sym (E). If p is a point in M, then let

$$\mathfrak{g}(p) = \{X_p \in T_p M \mid X \in \mathfrak{g}\};$$
$$\mathfrak{g}_p = \{X \in \mathfrak{g} \mid X_p \in E_p\}.$$

We point out that $\mathfrak{g}(p)$ is a subspace of $T_p(M)$, whereas $\mathfrak{g}_p$ is a subspace (and indeed a subalgebra) of $\mathfrak{g}$.

Let $\mathfrak{a} = \mathfrak{g} \cap \mathbb{D}(E)$. Since $\mathbb{D}(E)$ is an ideal in sym (E), we see that $\mathfrak{a}$ is an ideal in $\mathfrak{g}$. Notice that $\mathfrak{a}$ can also be defined as $\mathfrak{a} = \cap_{p \in M} \mathfrak{g}_p$

Definition. A Lie algebra $\mathfrak{g} \subset \text{sym}(E)$ is called a transitive symmetry algebra of E if $\mathfrak{g}(p) + E_p = T_p M$ for all $p \in M$. A transitive symmetry algebra $\mathfrak{g} \subset \text{sym}(E)$ is called simply transitive if $\mathfrak{g}_p = \mathfrak{a}$ for all $p \in M$.

Let $\mathfrak{g} \subset \text{sym}(E)$ be a symmetry algebra of E, and let G be the local Lie transformation group generated by $\mathfrak{g}$. Then G preserves the set $\mathbb{M}$ of all maximal integral manifolds of E, an moreover the following assertions are true:

1. The ideal $\mathfrak{a}$ is zero if and only if the action of G on $\mathbb{M}$ is locally effective.
2. The subalgebra $\mathfrak{g}_p$ is precisely the Lie algebra of the subgroup $G_p = \{g \in G \mid g.L_p \subset L_p\}$;

3. The Lie algebra $\mathfrak{g}$ is transitive if and only if the action of G on M is locally transitive.

4. The Lie algebra $\mathfrak{g}$ is simply transitive if and only if G_p does not depend on p and hence coincides with the ineffectiveness kernel of the action of G on M.

2. Normalizer theorem

Let $\mathfrak{g}$ be a transitive symmetry algebra of E. We fix a point a in M and consider the following set of points in M:

$$\{p \in M \mid \mathfrak{g}_p = \mathfrak{g}_a\}.$$

Let S_a denote the connected component of this set that contains a.

Since $\mathfrak{g}$ is a transitive symmetry algebra, there exist n vector fields $X_1, \ldots, X_n \in \mathfrak{g}$ such that the vectors

$$(X_1)_a, \ldots, (X_n)_a \in T_a M$$

form a basis for the complement of E_a in $T_a M$, and hence this will also be true in some neighborhood U of a:

$$\langle (X_1)_p, \ldots, (X_n)_p \rangle \oplus E_p = T_p M \text{ for all } p \in U.$$

In the neighborhood U, every vector field $Y \in \mathbb{D}(M)$ can be written uniquely in the form

$$Y = f_1 X_1 + \cdots + f_n X_n \quad (\text{mod } \mathbb{D}(E)). \tag{1}$$

Note that $Y \in \mathfrak{g}$ belongs to $\mathfrak{g}_p$ ($p \in U$) if and only if

$$f_1(p) = \cdots = f_n(p) = 0.$$

Let $\mathbb{F} = \{f_\alpha\}$ be the family of all functions that appear in the expansion (1) for all $Y \in \mathfrak{g}_a$. Note that, for every $p \in M$, the subalgebra $\mathfrak{g}_p$ has the same codimension in $\mathfrak{g}$, which is equal to codim (E). Therefore, the equality $\mathfrak{g}_p = \mathfrak{g}_a$ is equivalent to the inclusion $\mathfrak{g}_p \subset \mathfrak{g}_a$, which, for $p \in U$, can be written as

$$f(p) = 0 \quad \forall f \in \mathbb{F}. \tag{2}$$

Thus, in a neighborhood of a, the subset S_a is given by the simultaneous equations (2). Moreover, it is easy to show that S_a is a submanifold in M.

Theorem 1 Let $S = S_a$. Then

1. $T_p S \supset E_p$;

2. *the subspace*

$$\{X \in \mathfrak{g} \mid X_p \in T_pS\} \subset \mathfrak{g}$$

coincides with $N_{\mathfrak{g}}(\mathfrak{g}_p) = N_{\mathfrak{g}}(\mathfrak{g}_a),$

for all $p \in S.$

Proof. Since $\mathfrak{g}_p$ with $p \in S$ is independent of p, it will suffice to prove the theorem only for an arbitrary point $p \in S$, say $p = a$. Furthermore, since all assertions of the theorem are of local character, we can restrict our consideration to the neighborhood $U \subset M$, where S is given by equations (2).

1. It suffices to verify that all the functions $f \in \mathbb{F}$ are first integrals of E in U. If $Y \in \mathfrak{g}_a$ and $Z \in \mathbb{D}(E)$, then by (1) we have

$$[Z, Y - f_1X_1 - \cdots - f_nX_n] = [Z, Y] - f_1[Z, X_1] - \cdots - f_n[Z, X_n]$$
$$- Z(f_1)X_1 - \cdots - Z(f_n)X_n \in \mathbb{D}(E).$$

Now since $Y, X_1, \ldots, X_n \in \mathrm{sym}\,(E)$, we have $[Z, Y], [Z, X_1], \ldots, [Z, X_n] \in \mathbb{D}(E)$, and hence

$$Z(f_1)X_1 + \cdots + Z(f_n)X_n \in \mathbb{D}(E).$$

But this is possible only if $Z(f_1) = \cdots = Z(f_n) = 0$, so that $f_1, \ldots, f_n$ are indeed first integrals of E.

2. Let $Z \in \mathfrak{g}$. Since in a neighborhood of a, S can be given by (2), we see that $Z_a \in T_aS$ if and only if $d_af(Z_a) = Z(f)(a) = 0$ for all $f \in \mathbb{F}$.

Using (1) with $Y \in \mathfrak{g}_a$, we get

$$[Z, Y] = [Z, f_1X_1 + \cdots + f_nX_n] = Z(f_1)X_1 + \cdots + Z(f_n)X_n$$
$$+ f_1[Z, X_1] + \cdots + f_n[Z, X_n] \quad (\mathrm{mod}\ \mathbb{D}(E)).$$

This last equality, considered at the point a, gives

$$[Z, Y]_a = Z(f_1)(a)\,(X_1)_a + \cdots + Z(f_n)(a)\,(X_n)_a \quad (\mathrm{mod}\ E_a).$$

It follows that the condition $[Z, Y] \in \mathfrak{g}_a$ is equivalent to

$$Z(f_1)(a) = \cdots = Z(f_n)(a) = 0.$$

Thus Z_a lies in T_aS if and only if $[Z, Y] \in \mathfrak{g}_a$ for all $Y \in \mathfrak{g}_a$ that is if $Z \in N_{\mathfrak{g}}(\mathfrak{g}_a).$ $\square$

Corollary 1 *If the subalgebra* $\mathfrak{g}_a$ *coincides with its own normalizer, then* S *is a maximal integral manifold of the distribution* E.

Proof. Indeed, on the one hand, $T_pS \supset E_p$ for all $p \in S$, but on the other hand, $T_pS \cap \mathfrak{g}_p = E_p \cap \mathfrak{g}_p$ for all $p \in S$, which is possible only if $T_pS = E_p$, so that S is a maximal integral manifold of E. $\square$

Consider the restriction $\tilde{E}$ of the distribution E to S and also the set $\tilde{\mathfrak{g}} = N(\mathfrak{g}_a)$ restricted to S. Then $\tilde{\mathfrak{g}}$ is clearly a subalgebra of $\mathbb{D}(S)$ (which need not be isomorphic to $N(\mathfrak{g}_a)$), and as before, $\tilde{\mathfrak{g}} \subset \mathrm{sym}\,(\tilde{E})$.

Proposition 1 *The symmetry algebra $\tilde{\mathfrak{g}}$ of $\tilde{E}$ is simply transitive, and the ideal $\tilde{\mathfrak{a}} = \tilde{\mathfrak{g}} \cap \mathbb{D}(\tilde{E})$ coincides with $\mathfrak{g}_a$.*

Proof. The transitivity of $\tilde{\mathfrak{g}}$ follows from

$$
\begin{aligned}
T_pS &= T_pS \cap (E_p + \mathfrak{g}(p)) = E_p + T_pS \cap \mathfrak{g}(p) = E_p + N(\mathfrak{g}_p)(p) \\
&= E_p + N(\mathfrak{g}_a)(p) = E_p + \tilde{\mathfrak{g}}(p) \quad \text{for all } p \in S.
\end{aligned}
$$

Further, it is clear that $\tilde{\mathfrak{g}}_p = \mathfrak{g}_p = \mathfrak{g}_a$ for all $p \in S$, and therefore $\tilde{\mathfrak{a}} = \cap_{p \in S} \tilde{\mathfrak{g}}_p = \mathfrak{g}_a$, so that $\tilde{\mathfrak{g}}$ is a simply transitive symmetry algebra. $\square$

Thus, *the problem of integration of a distribution with the help of symmetries can be divided into the following two parts:*

1) the construction of the manifolds S_a;
2) the integration of the distributions $\tilde{E}$ on each of the manifolds S_a by means of the simply transitive symmetry algebras $\tilde{\mathfrak{g}}$.

3. Simply transitive symmetry algebras and $\mathfrak{g}$-structures

Let E be a completely integrable distribution on M, let $\mathfrak{h}$ be a simply transitive symmetry algebra of E, let $\mathfrak{a} = \mathfrak{h} \cap \mathbb{D}(E)$ be the ideal in $\mathfrak{h}$ consisting of characteristic symmetries, and finally let $\mathfrak{g} = \mathfrak{h}/\mathfrak{a}$.

We define a $\mathfrak{g}$-valued 1-form ω on M by requiring that

(i) $\omega(Y) = 0$ for all $Y \in \mathbb{D}(E)$;
(ii) $\omega(X) = X + \mathfrak{a}$ for all $X \in \mathfrak{h}$.

It is easy to verify that ω is well-defined.

Proposition 2 *The form ω has the following properties:*
1. $d\omega(X_1, X_2) = -[\omega(X_1), \omega(X_2)]$ *for all $X_1, X_2 \in \mathbb{D}(M)$;*
2. $\ker \omega_p = E_p$, $\mathrm{im}\,\omega_p = \mathfrak{g}$ *for all $p \in M$.*

Proof. 1. Since a vector field on M may be (uniquely) written in the form

$$
f_1 X_1 + \cdots + f_n X_n + Y, \qquad f_1, \ldots, f_n \in C^\infty(M),\ Y \in \mathbb{D}(E)
$$

and since both sides of the desired equality are $C^\infty(M)$-bilinear, we need to verify this equality only when

(i) $X_1, X_2 \in \mathbb{D}(E)$;
(ii) $X_1 \in \mathbb{D}(E)$, $X_2 \in \mathfrak{h}$;
(iii) $X_1, X_2 \in \mathfrak{h}$.

In case (i) both sides of our equality vanish identically. Note that $\omega(X) = $ const for all $X \in \mathfrak{h}$. We have

$$d\omega(X_1, X_2) = -\omega([X_1, X_2]) + X_1\omega(X_2) - X_2\omega(X_1)$$

for all $X_1, X_2 \in \mathbb{D}(M)$. It follows that in case (ii) both sides of the equality are also zero, because $[X_1, X_2] \in \mathbb{D}(E)$, $\omega(X_1) = 0$, and $\omega(X_2) = $ const. In case (iii) we have $[X_1, X_2] \in \mathfrak{h}$, so that

$$
\begin{aligned}
d\omega(X_1, X_2) &= -\omega([X_1, X_2]) = -[X_1, X_2] + \mathfrak{a} \\
&= -[X_1 + \mathfrak{a}, X_2 + \mathfrak{a}] = -[\omega(X_1), \omega(X_2)],
\end{aligned}
$$

as was to be proved.

2. The second statement follows immediately from the definition of ω.
□

Recall that a $\mathfrak{g}$-*structure on the manifold* M is a $\mathfrak{g}$-valued 1-form ω satisfying the first condition of Proposition 2 (see [4, 5]). We shall say that a $\mathfrak{g}$-structure ω is *nondegenerate*, if $\operatorname{im}\omega_p = \mathfrak{g}$ for all $p \in M$. We thus see that any simply transitive symmetry algebra of E defines a nondegenerate $\mathfrak{g}$-structure on M.

Conversely, a nondegenerate $\mathfrak{g}$-structure ω determines a distribution E on M by assigning to each point $p \in M$ the subspace $E_p = \ker \omega_p$. Moreover, for any $\overline{X} \in \mathfrak{g}$, the relation $\omega(X) = \overline{X}$ determines a unique (up to $\mathbb{D}(E)$) vector field X on M.

Proposition 3 1. *The distribution E given by a $\mathfrak{g}$-structure ω is completely integrable.*

2. *For any $\overline{X} \in \mathfrak{g}$, the vector field $X \in \mathbb{D}(E)$ is a symmetry of E.*

3. *The set $\mathfrak{h} = \{X \in \mathbb{D}(E) \mid \omega(X) = $ const$\}$ forms a simply transitive symmetry algebra of E with the following properties:*

(i) $\mathfrak{h} \supset \mathbb{D}(E)$ *and* $\mathfrak{h}/\mathbb{D}(E) \cong \mathfrak{g}$;

(ii) *any simply transitive symmetry algebra of E that determines the $\mathfrak{g}$-structure ω is contained in $\mathfrak{h}$.*

Proof. 1. Indeed, given $Y_1, Y_2 \in \mathbb{D}(E)$, we have

$$
\begin{aligned}
\omega([Y_1, Y_2]) &= -d\omega([Y_1, Y_2]) + Y_1\omega(Y_2) - Y_2\omega(Y_1) \\
&= [\omega(Y_1), \omega(Y_2)] + Y_1\omega(Y_2) - Y_2\omega(Y_1) = 0,
\end{aligned}
$$

since $\omega(Y_1) = \omega(Y_2) = 0$. Therefore $[Y_1, Y_2] \in \mathbb{D}(E)$, and the distribution E is completely integrable.

2. If $\overline{X} \in \mathfrak{g}$ and $Y \in \mathbb{D}(E)$, we have

$$\omega([X, Y]) = [\omega(X), \omega(Y)] + X\omega(Y) - Y\omega(X) = 0,$$

because $\omega(Y) = 0$ and $\omega(X) = \overline{X} = \text{const}$.

3. The inclusion $\mathfrak{h} \supset \mathbb{D}(E)$ is obvious. It is immediate from the definitions that the mapping $\mathfrak{h} \to \mathfrak{g}$ such that $X \mapsto \omega(X)$ is a surjective homomorphism of Lie algebras, and its kernel coincides with $\mathbb{D}(E)$. Hence $\mathfrak{h}/\mathbb{D}(E) \cong \mathfrak{g}$. If a simply transitive symmetry algebra $\mathfrak{h}'$ determines the same $\mathfrak{g}$-structure ω, then $\omega(X) = \text{const}$ for all $X \in \mathfrak{h}'$, so that $\mathfrak{h}' \subset \mathfrak{h}$. $\square$

We have thus proved that *completely integrable distributions on M with simply transitive symmetry algebras are in one-to-one correspondence with nondegenerate $\mathfrak{g}$-structures on M.*

4. Integration of $\mathfrak{g}$-structures

4.1. INTEGRALS

Let ω be a $\mathfrak{g}$-structure on M, not necessarily nondegenerate, and let G be a Lie group whose Lie algebra is isomorphic to $\mathfrak{g}$. We identify $\mathfrak{g}$ with $T_e G$ and, by means of *right* translations, with all the tangent spaces $T_g G$.

Definition. A mapping $f \colon M \to G$ is called an *integral of the $\mathfrak{g}$-structure* ω, if the differential $d_p f \colon T_p M \to T_{f(p)} G \equiv \mathfrak{g}$ of f coincides with ω_p for all $p \in M$.

A $\mathfrak{g}$-structure ω is said to be integrable, if there is an integral $f \colon M \to G$ of $\mathfrak{g}$.

Theorem 2 ([6]) *Any $\mathfrak{g}$-structure ω is locally integrable. Moreover, given $a \in M$ and $g_0 \in G$, there is a neighborhood U of a such that there exists a unique integral f of $\omega|_U$ satisfying the "initial" condition $f(a) = g_0$.*

Proof. Consider the distribution H on $M \times G$ defined by

$$H_{(p,g)} = \{X_p + \omega_p(X_p) \mid X_p \in T_p M\}, \quad p \in M, \ g \in G.$$

It is easy to check that H is completely integrable and that the dimension of H is equal to $\dim M$. If π is the natural projection of the direct product $M \times G$ onto M, then, as easily follows from the definition of H, the mapping $d_{(p,g)}\pi$ determines a homomorphism of $H_{(p,g)}$ onto $T_p M$ for all $(p, g) \in M \times G$.

It follows from the Frobenius theorem that there is a unique integral manifold L of the distribution H that passes through (a, g_0), and $\pi|_L$ is a local diffeomorphism of L and M at the point (a, g_0). Therefore, in some neighborhood U of a, there is a unique mapping $f \colon U \to M$ whose graph coincides with $L \cup (U \times G)$. It now follows that f is an integral of the $\mathfrak{g}$-structure $\omega|_U$. □

Example. Consider the trivial case in which $G = \mathbb{R}$ and $\mathfrak{g}$ is the one-dimensional commutative Lie algebra. The corresponding $\mathfrak{g}$-structure ω is then an ordinary closed 1-form on M, and its integrals are just integrals of closed forms, i.e., the functions $f: M \to G \equiv \mathbb{R}$ such that $df = \omega$. Thus the integration of $\mathfrak{g}$-structures may be viewed as the generalization of the integration of closed 1-forms.

Let H be the distribution on $M \times G$ defined by the $\mathfrak{g}$-structure ω. Then H may be regarded as a flat connection on the trivial principal G-bundle $\pi: M \times G \to G$. Since the connection H is flat, there is a natural homomorphism $\phi: \pi_1(M) \to \Phi$ of the fundamental group of M into the holonomy group of the connection H, and ω is globally integrable if and only if the group Φ is trivial. In the general case, ϕ is surjective, and any integrable manifold of the distribution H is a covering of M with respect to the projection π with fiber Φ. It follows in particular that any $\mathfrak{g}$-structure on a simply connected manifold is globally integrable.

Assume that the $\mathfrak{g}$-structure ω is globally integrable, and let f be an integral of ω. Then the mapping $g.f: M \to G$, $p \mapsto f(p)g$ is obviously an integral of ω too. Thus we have a right action of G on the set of all integrals of ω. If, in addition, M is connected, then by Theorem 2 this action is transitive. In the trivial case in which $G = \mathbb{R}$, this statement brings us to the well-known result that the integral of an exact 1-form is unique up to the addition of an arbitrary constant.

4.2. INTEGRALS OF $\mathfrak{g}$-STRUCTURES AND DISTRIBUTIONS

Let ω be a nondegenerate $\mathfrak{g}$-structure, and let E be the complete integrable distribution defined by ω. Then every (local) integral of ω is a submersion, and therefore $f^{-1}(g)$, for any $g \in G$, can be given a submanifold structure.

Proposition 4 *If f is an integral of the $\mathfrak{g}$-structure ω, then for any $g \in G$, the connected components of the submanifold $f^{-1}(g)$ will be maximal integral manifolds of the distribution E.*

Proof. If L is a connected component of f^{-1}, then

$$T_p L = \ker d_p f = \ker \omega_p = E_p$$

for all $p \in L$, so that L is an integral manifold of E. Assume that L' is an integral manifold of E containing L. Since L' is connected and $d_p f(T_p L') = d_p f(E_p) = \{0\}$ for all $p \in L'$, the mapping f is constant on L'. Thus $L' \subset L$, and therefore $L = L'$. □

Thus the integration of the distribution E reduces essentially to the integration of the corresponding $\mathfrak{g}$-structure.

4.3. REDUCTION

Let G_1 be a normal Lie subgroup of G, and let $\mathfrak{g}_1$ be the corresponding ideal in $\mathfrak{g}$. Consider the quotient group $G_2 = G/G_1$, whose Lie algebra $\mathfrak{g}_2$ is $\mathfrak{g}/\mathfrak{g}_1$. Suppose that $\pi: G \to G_2$ is the canonical surjection and $d\pi: \mathfrak{g} \to \mathfrak{g}_2$ the corresponding surjection of Lie algebras. The manifold M can be supplied with a natural $\mathfrak{g}_2$-structure ω_2:

$$(\omega_2)_p = d\pi \circ \omega_p \quad \text{for all } p \in M.$$

Let f be an integral of the $\mathfrak{g}$-structure ω. Then $\pi \circ f$ is obviously an integral of the $\mathfrak{g}_2$-structure ω_2.

Now assume that the $\mathfrak{g}$-structure ω is nondegenerate; then the $\mathfrak{g}_2$-structure ω_2 is also nondegenerate. If f_2 is an integral of ω_2 and $L = f_2^{-1}(g)$ is the inverse image of an arbitrary point $g \in G_2$, consider the restriction $\omega_1 = \omega|_L$. Then $(\omega_1)_p(T_pL) \subset \ker d\pi = \mathfrak{g}_1$ for all $p \in L$. Thus ω_1 may be considered as a $\mathfrak{g}_1$-structure on L. It is easy to show that ω_1 is also nondegenerate. Integrating ω_1 for all submanifolds $f_2^{-1}(g)$, $g \in G_2$, we obtain an integral of the $\mathfrak{g}$-structure ω.

Thus the problem of integrating a nondegenerate $\mathfrak{g}$-structure ω can be divided into smaller parts:

1) the construction of an integral f_2 of the $\mathfrak{g}_2$-structure ω_2;
2) the integration of the $\mathfrak{g}_1$-structure $\omega_1 = \omega|_{f_2^{-1}(g)}$ for $g \in G_2$.

Proposition 5 1. *The integration of any $\mathfrak{g}$-structure with a solvable Lie algebra $\mathfrak{g}$ reduces to the integration of closed 1-forms.*

2. *The integration of any $\mathfrak{g}$-structure reduces to the integration of $\mathfrak{g}$-structures with simple Lie algebras $\mathfrak{g}$ and to the integration of closed 1-forms.*

Proof. 1. We may assume without loss of generality that the Lie group G is connected and simply connected. Then G is diffeomorphic to $\mathbb{R}^k$ for some $k \in \mathbb{N}$, and moreover any normal virtual Lie subgroup of G is closed and also simply connected.

Since the Lie algebra $\mathfrak{g}$ is solvable, there is a chain

$$\mathfrak{g} = \mathfrak{g}_k \subset \mathfrak{g}_{k-1} \subset \ldots \subset \mathfrak{g}_1 \subset \mathfrak{g}_0 = \{0\}$$

of subalgebras of $\mathfrak{g}$ such that $\mathfrak{g}_{i-1}$ is an ideal of codimension 1 in $\mathfrak{g}_i$ ($i = 1, \ldots, k$). If G_i are the corresponding subgroups of G, then all quotient subgroups G_i/G_{i-1} are isomorphic to $\mathbb{R}$, and the problem of finding an integral of the $\mathfrak{g}$-structure ω reduces to the integration of k differential 1-forms.

2. The second statement of the theorem follows from the Levi theorem about the decomposition of a Lie group into the semidirect product of a semisimple Levi subalgebra and the radical, and from the decomposition of a simply connected semisimple Lie group into a direct product of simple Lie groups. □

4.4. INTEGRATION ALONG PATHS

In conclusion, we describe a procedure for finding an integral f for a given $\mathfrak{g}$-structure w that generalizes the process of integrating 1-forms along paths and coincides with it when $G = \mathbb{R}$. Recall that a $\mathfrak{g}$-structure ω may be regarded as a connection H on the principal G-bundle $\pi \colon M \times G \to M$.

Let (a, g_0) be a fixed point of the manifold $M \times G$; our goal is to construct an integral f of ω such that $f(a) = g_0$. Consider an arbitrary curve $\Gamma \colon [t_0, t_1] \to M$ with $\Gamma(t_0) = a$. There exists a unique curve $\widetilde{\Gamma} \colon [t_0, t_1] \to M \times G$ satisfying the following conditions:

1. $\pi \circ \widetilde{\Gamma} = \Gamma$.
2. $\widetilde{\Gamma}'(t) \in H_{\widetilde{\Gamma}(t)}$ for all $t \in (t_0, t_1)$.

This curve is precisely the horizontal lift of the curve Γ by means of the connection H.

The curve $\widetilde{\Gamma}$ may be equivalently described as follows: Consider the $\mathfrak{g}$-structure ω_1 defined on an interval $[t_0, t_1]$ by

$$\omega_1 = \omega \circ d\Gamma = X(t)dt,$$

where

$$X(t) = \omega_{\Gamma(t)}(\Gamma'(t)), \qquad t \in [t_0, t_1],$$

is a curve in $\mathfrak{g}$. Then the desired integral of ω_1 is a curve $g \colon [t_0, t_1] \to G$ in G satisfying the differential equation

$$g'(t) = X(t), \ t \in [t_0, t_1], \qquad g(t_0) = g_0. \tag{3}$$

It is not hard to show that the curve $\widetilde{\Gamma}$ on the manifold $M \times G$ has the form $\widetilde{\Gamma}(t) = (\Gamma(t), g(t))$.

Let L be the maximal integral manifold of the distribution H on $M \times G$ passing through (a, t_0). The desired integral f of the $\mathfrak{g}$-structure ω is a mapping $M \to G$ whose graph coincides with L. Since the curve $\widetilde{\Gamma}$ is tangent to H, it lies inside L, and therefore

$$f(\Gamma(t)) = g(t) \qquad \text{for all } t \in [t_0, t_1].$$

The method described above enables us to find the integral of the $\mathfrak{g}$-structure ω along any curve Γ on M. When $G = \mathbb{R}$, this method is the same as the usual procedure for the integration of 1-forms. The standard results of the theory of connections show that the point $f(\Gamma(t_1))$ depends only on the homotopy class of the curve $\Gamma(t)$. Thus our method allows to determine the integral f uniquely in any simply connected neighborhood of a.

5. Integration of distributions

The next theorem is a consequence of Proposition 5.

Theorem 3 *Let $\mathfrak{g}$ be a transitive symmetry algebra of the distribution E, and assume that there is a point $a \in M$ such that the Lie algebra $N(\mathfrak{g}_a)/\mathfrak{g}_a$ is solvable. Then E can be integrated by quadrature.*

In the most general case the problem of integration of a completely integrable distribution with a transitive symmetry algebra can be divided into the following three parts:

(1) the reduction of the problem to the integration of a distribution with a simply transitive symmetry algebra isomorphic to $N(\mathfrak{g}_a)/\mathfrak{g}_a$; in terms of coordinates, this is equivalent to the solution of simultaneous equations (which are, in general, transcendental);

(2) the integration of $\mathfrak{g}$-structures with simple Lie algebras $\mathfrak{g}$ from the decomposition of the Levi subalgebra of $N(\mathfrak{g}_a)/\mathfrak{g}_a$; in terms of coordinates, this is equivalent to the solution of finitely many ordinary differential equations of the form (3);

(3) the integration of a $\mathfrak{g}$-structure, where the Lie algebra $\mathfrak{g}$ is solvable and coincides with the radical of $N(\mathfrak{g}_a)/\mathfrak{g}_a$; this reduces to the integration of a finite number (equal to $\dim \mathfrak{g}$) of 1-forms, which, in terms of coordinates, is equivalent to ordinary integration.

6. Equivalence of local transitive actions

A Lie algebra $\bar{\mathfrak{g}}$ of vector fields on a manifold M is called *transitive*, if $\bar{\mathfrak{g}}(p) = T_p M$ for all points $p \in M$. Two transitive Lie algebras $\bar{\mathfrak{g}}_1$, $\bar{\mathfrak{g}}_2$ of vector fields on manifolds M_1, M_2 respectively are said to be *locally equivalent*, if there is a local diffeomorphism $\phi: M_1 \to M_2$ which takes $\bar{\mathfrak{g}}_1$ to $\bar{\mathfrak{g}}_2$ (in a neighborhood where ϕ is defined).

Fix two points $a_1 \in M_1$ and $a_2 \in M_2$, and let

$$\mathfrak{g}_i = \{X \in \bar{\mathfrak{g}}_i \mid X_{a_i} = 0\}.$$

The pairs $(\bar{\mathfrak{g}}_1, \mathfrak{g}_1)$ and $(\bar{\mathfrak{g}}_2, \mathfrak{g}_2)$ are said to be *isomorphic*, if there exists an isomorphism $\alpha: \bar{\mathfrak{g}}_1 \to \bar{\mathfrak{g}}_2$ of Lie algebras such that $\alpha(\mathfrak{g}_1) = \mathfrak{g}_2$; then α is said to be an isomorphism of the pairs $(\bar{\mathfrak{g}}_1, \mathfrak{g}_1)$ and $(\bar{\mathfrak{g}}_2, \mathfrak{g}_2)$. It is a well-known fact that $\bar{\mathfrak{g}}_1$ and $\bar{\mathfrak{g}}_2$ are locally equivalent if and only if the pairs $(\bar{\mathfrak{g}}_1, \mathfrak{g}_1)$ and $(\bar{\mathfrak{g}}_2, \mathfrak{g}_2)$ are isomorphic. A diffeomorphism $\phi \colon M_1 \to M_2$ establishing this equivalence is uniquely defined by the conditions:

(i) $\phi(a_1) = \phi(a_2)$;

(ii) $d\phi(X) = \alpha(X)$ for all $X \in \bar{\mathfrak{g}}_1$.

We shall now be concerned with the problem of finding a diffeomorphism ϕ for a given isomorphism α. Define a subset L of the manifold $M = M_1 \times M_2$ by

$$L = \{(p_1, p_2) \in M \mid \alpha((\bar{\mathfrak{g}}_1)_{p_1}) = (\bar{\mathfrak{g}}_2)_{p_2}\}.$$

Then $(a_1, a_2) \in L$. Let $\{X_1, \ldots, X_n\}$ ($n = \dim M_1 = \dim M_2$) be a basis of some complement of $\mathfrak{g}_1$ in $\bar{\mathfrak{g}}_1$. In a neighborhood of (a_1, a_2), every vector field $X \in \mathbb{D}(M_1 \times M_2)$ can be written uniquely in the form

$$X = f_1 X_1 + \cdots + f_n X_n + g_1 \alpha(X_1) + \cdots + g_n \alpha(X_n), \qquad (4)$$

where $f_1, \ldots, f_n \in C^\infty(M_1)$ and $g_1, \ldots, g_n \in C^\infty(M_2)$. The set L is given in this neighborhood by the equations $f_i = g_i$ for all functions f_i, g_i, $i = 1, \ldots, n$, that appear in the expansion (4) of all $X \in \bar{\mathfrak{g}}_1 \times \bar{\mathfrak{g}}_2$. In particular, it is not hard to show that L is a submanifold of M.

The proof of the next theorem is similar to those of Theorem 1 and Proposition 1.

Theorem 4

1. *If $X \in \bar{\mathfrak{g}}_1$, then the vector field $X + \alpha(X)$ is tangent to the submanifold L.*

2. *The distribution*

$$E_{(p_1, p_2)} = \{X_{p_1} + \alpha(X)_{p_2} \mid X \in \bar{\mathfrak{g}}_1\}, \quad (p_1, p_2) \in L,$$

on L is completely integrable, and its maximal integral manifold through (a_1, a_2) is the graph of the desired diffeomorphism ϕ.

3. *For $i = 1, 2$, the Lie algebra $\mathfrak{g} = Z_{\mathbb{D}(M_i)}(\bar{\mathfrak{g}}_i)$, viewed as a Lie algebra of vector fields on M, has the following properties:*

(i) *every vector field $X \in \mathfrak{g}$ is tangent to the submanifold L;*

(ii) *the restriction of $\mathfrak{g}$ to L is a simply transitive symmetry algebra of the distribution E;*

(iii) *$\mathfrak{g}$ is isomorphic to $N_{\bar{\mathfrak{g}}_i}(\mathfrak{g}_i)$.*

Remark. The Lie algebra $\mathfrak{g} = Z_{\mathbb{D}(M_i)}(\bar{\mathfrak{g}}_i)$ is called the *symmetry algebra* of the Lie algebra $\bar{\mathfrak{g}}_i$ of vector fields on M_i ($i = 1, 2$). To find it in the general case is not a trivial task. If, in particular, M_i is a Lie group and $\bar{\mathfrak{g}}_i$ is the Lie algebra of left-invariant (right-invariant) vector fields on M_i, then

$\mathfrak{g}$ is precisely the Lie algebra of right-invariant (left-invariant) vector fields, and to find it amounts to finding the multiplication in M_i for the given Lie algebra $\bar{\mathfrak{g}}_i$. Yet if we know the corresponding global transitive action, then the construction of $\mathfrak{g}$ presents no difficulties.

Indeed, let $\overline{G}_i$ be the global Lie transformation group acting on a manifold M_i and corresponding to a Lie algebra $\mathfrak{g}_i$ $(i = 1, 2)$. Let $G_i = (\overline{G}_i)_{a_i}$ be the stabilizer of a point $a_i \in M_i$. Then M_i may be identified with the set $\overline{G}_i/G_i$, which may be supplied with the following right action of the normalizer $N(G_i)$:

$$g.(\bar{g}G_i) = (\bar{g}g)G_i \quad \text{for all } g \in N(G_i), \, \bar{g} \in \overline{G}_i.$$

This action is easily seen to be well-defined, and its kernel of noneffectiveness is precisely G_i. The image of the corresponding mapping $N(\mathfrak{g}_i)/\mathfrak{g}_i \to \mathbb{D}(M_i)$ is precisely $Z_{\mathbb{D}(M_i)}(\bar{\mathfrak{g}}_i)$. For more details, see [8]

Thus the determination of a diffeomorphism ϕ establishing the equivalence of two locally transitive actions reduces to the integration of a completely integrable distribution with a given simply transitive symmetry algebra.

References

1. Lie, S.: Allgemeine Untersuchung über Differentialgleichnungen, die eine kontinuierliche, endliche Gruppe gestatten, *Math. Ann.* **25** (1885) no. 1, 71–151.

2. Duzhin, S.V. and Lychagin, V.V.: Symmetries of distributions and quadrature of ordinary differential equations, *Acta Appl. Math.* **24** (1991) 25–37.

3. Komrakov, B.P. and Lychagin, V.V.: Symmetries and integrals, Preprint Univ. Oslo, **15**, 1993.

4. Bernshtein, I.N. and Rosenfeld, I.B.: Homogeneous spaces of infinite-dimensional Lie algebras and characteristic classes of foliations, *Uspehi Mat. Nauk* **28** (1973) no. 5, 103–138 (in Russian).

5. Fuks, D.B.: *Cohomology of infinite-dimensional Lie algebras*, Nauka, Moscow, 1984 (in Russian).

6. Griffiths, P.: On Cartan's method of Lie groups and moving frames as applied to existence and uniqueness questions in differential geometry, *Duke J. Math.* **41** (1974) 775–814.

7. Olver, P.: *Applications of Lie groups to differential equations*, Springer, New-York, 1986.

8. Onishchik, A.L.: *Topology of transitive transformation groups*, Moscow, 1995 (in Russian).

ALGEBRAS WITH FLAT CONNECTIONS
AND SYMMETRIES OF DIFFERENTIAL EQUATIONS

I.S. KRASIL'SHCHIK
Moscow Institute for Municipal Economy
& the Diffiety Institute
1st Tverskoy-Yamskoy per. 14, Apt.45,
125047 Moscow, Russia.
E-mail: josephk@glas.apc.org

Abstract. In the category of algebras with flat, connections the concepts of symmetries and recursion operators are defined. Lie algebra structure of symmetries is described for the objects of this category possessing recursion operators. In particular, sufficient conditions for existence of infinite series of commuting symmetries are formulated. The results are applied to symmetries of differential equations.

Mathematics Subject Classification (1991): 58F07, 58F35, 58G05.

Key words: : differential equations, connections, integrable systems, symmetries, recursion operators, cohomologies, algebraic models.

1. Introduction

In [5, 8] we considered a cohomology theory arising on infinite prolongations of nonlinear differential equations. This theory gives rise to a series of invariants and in particular allows one to consider symmetries and recursion operators in a natural homological framework. In what follows, the theory is generalized to objects of a rather general algebraic nature — algebras with flat connections.

It is well known that differential equations with recursion operators possess infinite series of commuting symmetries (see [16] for example). Here we show that this fact is of a purely algebraic nature. We introduce a cat-

B. P. Komrakov et al. (eds.), Lie Groups and Lie Algebras, 407–424.
© 1998 *Kluwer Academic Publishers. Printed in the Netherlands.*

egory Conn(A) of algebras with flat connections and using Vinogradov's conceptual scheme of calculus in commutative algebras [19] construct an appropriate algebraic model of corresponding geometrical structures. Symmetries and recursion operators arise in this model as cohomological invariants of the order 0 and 1 respectively, cf. [7, 8]. If an object $\mathcal{O}$ of Conn(A) possesses a recursion operator $\mathcal{R}$, then Lie algebra structure of Sym($\mathcal{O}$) is determined by $[\![\mathcal{R}, \mathcal{R}]\!]$. Here $[\![\cdot, \cdot]\!]$ is the Frölicher–Nijenhuis bracket and $[\![\mathcal{R}, \mathcal{R}]\!]$ coincides in this case with the Nijenhuis torsion of $\mathcal{R}$ [10].

In what follows we consider an associative commutative unitary algebra A over a field $\mathbf{k}$ of nonzero characteristic. We use standard functors and constructions of calculus in the category Mod(A) of all A-modules (see [6, 19]). In particular, $D_r(P)$ denotes the A-module of P-valued r-derivations $P \to A$, while $\Lambda^r = \Lambda^r(A)$ is the A-module of differential forms of the degree r over A. We also use the notations $D_*(P) = \sum_{r \geq 0} D_r(P)$, $\Lambda^*(A) = \sum_{r \geq 0} \Lambda^r(A)$. In fact, the functors $D_r : \text{Mod}(A) \Rightarrow \text{Mod}(A)$ are representable and $\Lambda^r(A)$ are their representative objects:

$$D_r(P) = \hom_A(\Lambda^r(A), P) \tag{1}$$

for any A-module P. Moreover, for $r = 1$ any derivation $X \in D_1(A)$ is uniquely representable in the form $X = f_X \circ d$, where $f_X \in \hom_A(\Lambda^1, P)$ and

$$d : A \to \Lambda^1(A) \tag{2}$$

is the *first de Rham differential*. For any algebra A there exists a complex

$$0 \to A \xrightarrow{d} \Lambda^1(A) \to \ldots \to \Lambda^r(A) \xrightarrow{d} \Lambda^{r+1}(A) \to \ldots$$

which is called the *de Rham complex* of A and in which the first d coincides with (2).

We shall also need the following facts:

– any $\Lambda^r(A)$ is the r-th exterior power of $\Lambda^1(A)$: $\Lambda^r(A) = \Lambda^1(A) \wedge \ldots \wedge \Lambda^1(A)$ (r times);

– $\Lambda^1(A)$ as an A-module is generated over A by the elements $da = d(a), a \in A$, with the relations $d(a+b) = d(a) + d(b), d(ab) = ad(b) + bd(a)$;

– de Rham differentials are determined by the relations $d(d(a)) = 0$ and $d(\omega \wedge \theta) = d(\omega) \wedge \theta + (-1)^\omega \omega \wedge d(\theta)$ [1], where $\omega, \theta \in \Lambda^*$;

– any polyderivation $X \in D_r(P), r \geq 2$, can be considered as a derivation acting from A into $D_{r-1}(P)$, such that $X(a,b) + X(b,a) = 0$ for any $a, b \in A$, where $X(a,b) \stackrel{\text{def}}{=} (X(a))(b) \in D_{r-2}(P)$.

---

[1] Here and below we use the notation $(-1)^\omega$ instead of $(-1)^{\deg(\omega)}$

Algebras A below are supposed to be filtered:

$$A = \bigcup_{\alpha \in \mathbb{Z}} A_\alpha, \quad A_\alpha \subset A_{\alpha+1}, \quad A_\alpha \cdot A_\beta \subset A_{\max(\alpha,\beta)}.$$

We also consider filtered modules $P = \bigcup_\alpha P_\alpha$ over A and all maps $f : P \to Q$ under consideration are assumed to preserve filtrations: $f(P_\alpha) \subset Q_{\alpha+k}$ for all $\alpha \in \mathbb{Z}$ and some integer k. In this case k is filtration of f: $\deg(f) = k$. In particular, all derivations are filtered, while the modules $\Lambda^*(A)$ carry filtration

$$\deg(a_0 da_1 \wedge \ldots \wedge a_s) = \max(\deg(a_i)).$$

We restrict ourselves to the *smooth case* which means that all A_α-modules $\Lambda^1(A_\alpha)$ are projective and of finite type. These algebras are called *smooth*. One gets classical objects when taking $\mathbf{k} = \mathbb{R}$ and $A = C^\infty(M)$ for some smooth manifold M (here $A_\alpha = A$ for all α).

In what follows we use the algebraic setting exposed in [6, Chapter 1].

2. Basic structures

The basic structures needed below are: wedge products, contractions (internal product), the Richardson–Nijenhuis bracket (cf. [11, 13, 15]), Lie derivatives, the Frölicher–Nijenhuis bracket (cf. [2, 12, 14]). Here we briefly recall main definitions from [5] in a slightly modified form.

2.1. WEDGE PRODUCTS

A *wedge product*

$$\wedge : \Lambda^r(A) \otimes_A \Lambda^s(A) \to \Lambda^{r+s}(A), \quad \wedge : D_r(A) \otimes_A D_s(A) \to D_{r+s}(A)$$

is an operation defined both in $\Lambda^*(A)$ and in $D_*(A)$ with respect to which these modules acquire the structure of graded commutative algebras, i.e.,

$$v \wedge w = (-1)^{v \cdot w} v \wedge w,$$

where v, w are elements of either $\Lambda^*(A)$ or $D_*(A)$.

In the first case $\wedge$ is defined due to the decomposition $\Lambda^r(A) = \Lambda^1(A) \wedge \ldots \wedge \Lambda^1(A)$ (see above); for $D_*(A)$ we use an inductive definition. Namely, for $X, Y \in D_0(A) = A$ we set by induction

$$X \wedge Y = X \cdot Y,$$

where $\cdot$ is the multiplication in A. If $X \in D_r(A), Y \in D_s(A), r, s \geq 0$, and $a \in A$, then we set

$$(X \wedge Y)(a) = X \wedge Y(a) + (-1)^s X(a) \wedge Y.$$

This operation is well-defined (see [4]).

Below we also consider form-valued derivations $D_*(\Lambda^*) = \sum_{r,s \geq 0} D_r(\Lambda^s)$ and introduce the grading

$$\mathrm{gr}\,(\Omega) = r + s, \ \Omega \in D_r(\Lambda^s). \tag{3}$$

Define wedge product $\wedge : D_r(\Lambda^s) \otimes_A D_{r'}(\Lambda^{s'}) \to D_{r+r'}(\Lambda^{s+s'})$ by setting

$$(\omega \wedge \Theta)(a) = \omega \wedge \Theta(a), \ \Theta \wedge \omega = (-1)^{\Theta \cdot \omega} \omega \wedge \Theta, \tag{4}$$

where $a \in A, \Delta \in D_*(\Lambda^*)$ and $\omega \in \Lambda^s = D_0(\Lambda^s)$. For arbitrary $\Omega, \Theta \in D_*(\Lambda^*)$ we set by induction

$$(\Omega \wedge \Theta)a = \Omega \wedge \Theta(a) + (-1)^{\Theta}\Omega(a) \wedge \Theta, \ a \in A. \tag{5}$$

Proposition 2.1 *Wedge product* $\wedge : D_*(\Lambda^*) \otimes_A D_*(\Lambda^*) \to D_*(\Lambda^*)$ *is well-defined by* (4), (5) *and determines a commutative $\mathbb{Z}$-graded algebra structure in $D_*(\Lambda^*)$ with respect to grading* (3). *Restricted onto $\Lambda^*(A)$ and $D_*(A)$, this operation coincides with wedge product of differential forms and poly-derivations respectively.*

Let A be a smooth algebra and $A_\alpha \subset A$ be an element of its filtration. Then any A-module P is an A_α-module as well while any r-derivation $\Delta \in D_r(P)$ admits a restriction $\Delta_\alpha = \Delta|_{A_\alpha}$ which is an r-derivation over A_α and consequently can considered as an element of $P \otimes_{A_\alpha} D_r(A_\alpha)$. These restrictions are compatible with each other, $\Delta_{\alpha+1}|_{A_\alpha} = \Delta_\alpha$, and the following obvious fact is valid.

Lemma 2.1 *Let A be a smooth algebra and P be an A-module. Then any derivation $\Delta \in D_r(P)$ is completely determined by its restrictions $\Delta_\alpha \in P \otimes_{A_\alpha} D_r(A_\alpha)$. Conversely, if one has a series of mutually compatible derivations $\Delta_\alpha \in P \otimes_{A_\alpha} D_r(A_\alpha)$, then it determines an element $\Delta \in D_r(P)$.*

Hence, any $\Delta \in D_r(P)$ can be expressed in terms of decomposable elements of $P \otimes_{A_\alpha} D_r(A_\alpha)$.

In particular, if $\Omega = \omega \otimes X$, $\Theta = \theta \otimes Y$ are decomposable elements, then $\Omega \wedge \Theta = (-1)^{X \cdot \theta}(\omega \wedge \theta) \otimes (\Delta \wedge \nabla)$, where $\omega, \theta \in \Lambda^*(A), \Delta, \nabla \in D_*(A)$.

2.2. CONTRACTIONS

Contraction $i : D_*(\Lambda^*) \otimes_A D_*(\Lambda^*) \to D_*(\Lambda^*)$ generalizes usual internal product $i_X \omega$ of vector fields with differential forms and is defined as follows.

Let $\Omega \in D_r(\Lambda^s), \Theta \in D_{r'}(\Lambda^{s'})$. First consider the case $r' = 0$. Then for $r > s$, we set $i_\Omega \Theta = 0$ and $i_\Omega \Theta = \Omega \wedge \Theta$ for $r = 0$. In the case $r = s'$ we set $i_\Omega \Theta = \Omega(\Theta)$ using identification $D_r(\Lambda^s) = \mathrm{hom}_A(\Lambda^r, \Lambda^s)$ (see equality

(1)). Finally, for a decomposable form $da \wedge \omega \in \Lambda^{s'}$ its contraction with $\Theta \in D_*(\Lambda^*)$ is defined by

$$i_\Theta(da \wedge \omega) = i_{\Theta(a)}\omega + (-1)^\Theta da \wedge i_\Theta\omega,$$

where the grading of Θ is understood due to (3). For general Ω and Θ we define by induction

$$(i_\Omega\Theta)a = i_\Omega(\Theta(a)), \ a \in A.$$

Obviously, for decomposable elements one has

$$i_{\omega \otimes X}(\theta \otimes Y) = \omega \wedge (i_X\theta) \otimes Y,$$

where $\omega \in \Lambda^s, X \in D_r(A), \theta \in \Lambda^{s'}, Y \in D_{r'}(A)$.

Basic properties of contractions, using below, are expressed in the following

Proposition 2.2 *Let A be a commutative unitary smooth algebra. Then*

(i) *For any $X \in D_s(A)$ its contraction $i_X : \Lambda^*(A) \to \Lambda^*(A)$ is a graded differential operator of order s.*

(ii) *In particular, if $X \in D_1(A)$, then i_X is a graded derivation:*

$$i_X(\omega \wedge \theta) = i_X(\omega) \wedge \theta + (-1)^\omega \omega \wedge i_X\theta, \ \omega, \theta \in \Lambda^*.$$

(iii) *If $\Omega, \Theta \in D_1^v(A)$, then*

$$i_\Omega \circ i_\Theta = i_{\Omega\rfloor\Theta} + i_{\Omega\wedge\Theta}.$$

2.3. THE RICHARDSON-NIJENHUIS BRACKET

Consider two derivations $\Omega, \Theta \in D_1(\Lambda^*)$ and define

$$[\![\Omega, \Theta]\!]^\# = i_\Omega(\Theta) + (-1)^\Theta i_\Theta(\Omega).$$

The element $[\![\Omega, \Theta]\!]^\#$ is called the *Richardson–Nijenhuis bracket* of Ω and Θ.

Proposition 2.3 *Let A be a commutative unitary smooth algebra, and consider elements $\Omega, \Theta \in D_1(\Lambda^*)$. Then*

(i) $[i_\Omega, i_\Theta] = i_{[\![\Omega,\Theta]\!]^\#}$,

where $[\cdot, \cdot]$ denotes the graded commutator. It means that the Richardson–Nijenhuis bracket $[\![\cdot, \cdot]\!]^\#$ determines in $D_1(\Lambda^*)$ a graded Lie algebra structure with respect to grading (3). To be precise, one has

(ii) $[\![\Omega, \Theta]\!]^\# + (-1)^{(\Omega+1)(\Theta+1)}[\![\Theta, \Omega]\!]^\# = 0$

and

  (iii) *for any* $\Xi \in D_1(\Lambda^*)$

$$\oint (-1)^{(\Omega+\Xi)\cdot\Theta} [\![[\![\Omega,\Theta]\!]^\#,\Xi]\!]^\# = 0,$$

where $\oint$ denotes the sum of cyclic permutations.

  (iv) *If $\rho \in \Lambda^*(A)$, then*

$$[\![\Omega,\rho \wedge \Theta]\!]^\# = i_\Omega \rho \wedge \Theta + (-1)^{\Omega\cdot\rho}\rho \wedge [\![\Omega,\Theta]\!]^\#.$$

2.4. LIE DERIVATIVES

We now consider derivations $\Omega \in D_1(\Lambda^*)$ and introduce a new grading

$$\mathrm{gr}'(\Omega) = \mathrm{gr}(\Omega) - 1 \tag{6}$$

in the module $D_1(\Lambda^*)$. Define *Lie derivative* $L_\Omega : \Lambda^s(A) \to \Lambda^{s+\Omega}(A)$ by

$$L_\Omega = [i_\Omega, d] = i_\Omega \circ d + (-1)^\Omega d \circ i_\Omega, \tag{7}$$

cf. [1, 18].

Proposition 2.4 *Let A be a commutative unitary smooth algebra and $\Omega \in D_1(\Lambda^*)$. Then L_Ω defined by (7) possesses the following properties:*

  (i) *For any $\rho, \theta \in \Lambda^*$ one has*

$$L_\Omega(\rho \wedge \theta) = L_\Omega(\rho) \wedge \theta + (-1)^{(\Omega-1)\cdot\rho}\rho \wedge L_\Omega(\theta),$$

i.e., L_Ω is a derivation of $\Lambda^*(A)$ of the grading $\mathrm{gr}'(\Omega)$.

  (ii) *L_Ω commutes with the de Rham differential:*

$$[L_\Omega, d] = L_\Omega \circ d + (-1)^\Omega d \circ L_\Omega = 0.$$

  (iii) *If $\rho \in \Lambda^*(A)$ then*

$$L_{\rho\wedge\Omega} = \rho \wedge L_\Omega - (-1)^{\rho+\Omega} d\rho \wedge i_\Omega.$$

  (iv) *If $\mathrm{gr}'(\Omega) = 0$, i.e., $\Omega \in D_1(A)$, then L_Ω coincides with the usual Lie derivative while for decomposable Ω one has*

$$L_{\omega\otimes X} = \omega \wedge L_X + (-1)^\omega d\omega \wedge i_X,$$

where $\omega \in \Lambda^*(A)$ and $X \in D_1(A)$.

2.5. THE FRÖLICHER–NIJENHUIS BRACKET

Here we define the *Frölicher–Nijenhuis bracket* for elements of the module $D_1(\Lambda^*)$. This operation is the basic object in our considerations.

Proposition 2.5 *Let A be a commutative unitary smooth algebra. Take $\Omega, \Theta \in D_1(\Lambda^*)$. Then*

   (i) *The commutator of Lie derivatives $[L_\Omega, L_\Theta]$ is of the form L_Ξ for some $\Xi \in D_1(\Lambda^*)$.*

   (ii) *The correspondence $L : D_1(\Lambda^*) \to D_1^{\mathrm{gr}}(\Lambda^*)$, $\Omega \mapsto L_\Omega$, is injective and thus Ξ is uniquely defined by*

$$[L_\Omega, L_\Theta] = L_\Xi. \tag{8}$$

Here $D_1^{\mathrm{gr}}(\Lambda^*)$ denotes the module of graded derivations $\Lambda^* \to \Lambda^*$.

Definition 1 The element $\Xi \in D_1(\Lambda^*)$ defined by (8) is called the *Frölicher–Nijenhuis* bracket of Ω and Θ and is denoted by $\Xi = [\![\Omega, \Theta]\!]$.

The basic properties of the Frölicher–Nijenhuis bracket are described by

Proposition 2.6 *Let A be a commutative unitary smooth algebra. Then*

   (i) *If $\Omega = \omega \otimes X$ and $\Theta = \theta \otimes Y$ are decomposable elements, $\omega, \theta \in \Lambda^*, X, Y \in D_1(A)$, then*

$$\begin{aligned}
[\![\Omega, \Theta]\!] = (-1)^{X\cdot\theta}\omega \wedge \theta \otimes [X, Y] + \\
\omega \wedge L_X\theta \otimes Y + (-1)^\Omega d\omega \wedge i_X\theta \otimes Y - \\
(-1)^{\Omega\cdot\Theta}\theta \wedge L_Y\omega \otimes X - (-1)^{(\Omega+1)\cdot\Theta}d\theta \wedge i_Y\omega \otimes X = \\
= (-1)^{X\cdot\theta}\omega \wedge \theta \otimes [X, Y] + L_\Omega\theta \otimes Y - (-1)^{\Omega\cdot\Theta}L_\Theta\omega \otimes X.
\end{aligned}$$

   (ii) *For any $\Omega, \Theta, \Xi \in D_1(\Lambda^*)$ one has*

$$[\![\Omega, \Theta]\!] = (-1)^{\Omega\Theta}[\![\Theta, \Omega]\!]$$

and
   (iii)

$$\oint (-1)^{(\Omega+\Xi)\cdot\Theta}[\![\Omega, [\![\Theta, \Xi]\!]]\!] = 0, \tag{9}$$

i.e., $[\![\cdot, \cdot]\!]$ determines a graded Lie algebra structure in $D_1(\Lambda^*)$ with respect to grading (6).

   (iv) *An analog of the infinitesimal Stokes formula is valid:*

$$i_\Xi[\![\Omega, \Theta]\!] = [\![i_\Xi\Omega, \Theta]\!] +$$
$$(-1)^{\Omega\cdot(\Xi+1)}[\![\Omega, i_\Xi\Theta]\!] + (-1)^\Omega\, i_{[\![\Xi,\Omega]\!]}\Theta - (-1)^{(\Omega+1)\cdot\Theta}\, i_{[\![\Xi,\Theta]\!]}\Omega. \tag{10}$$

(v) *If $\rho \in \Lambda^*(A)$, then*

$$[\![\Omega, \rho \wedge \Theta]\!] =$$
$$L_\Omega(\rho) \wedge \Theta - (-1)^{(\Omega+1)\cdot(\Theta+\rho)} d\rho \wedge i_\Theta \Omega + (-1)^{\Omega\cdot\rho} \rho \wedge [\![\Omega, \Theta]\!].$$

(vi) *Finally,*

$$[L_\Omega, i_\Theta] = (-1)^\Omega L_{i_\Theta \Omega} + i_{[\![\Omega,\Theta]\!]}.$$

3. Algebras with connections and related cohomologies

Here we generalize the notion of connection and construct various invariants associated with this concept.

3.1. CONNECTIONS AND THEIR INVARIANTS

Let A, B be two **k**-algebras and an A-algebra structure $\varphi : A \to B$ be defined on B. Denote by $D_1(A, B)$ the module of B-valued derivations of A, i.e.,

$$D_1(A, B) = \{\Delta \in \hom_{\mathbf{k}}(A, B) \mid \Delta(aa') = \varphi(a)\Delta(a') + \varphi(a')\Delta(a)\}$$

for all $a, a' \in A$.

Definition 2 A B-homomorphism $\nabla : D_1(A, B) \to D_1(B)$ is called a *connection* in B, or B-connection, if for any $X \in D_1(A, B)$ one has

$$\nabla(X)|_A = X.$$

Two main invariants of connections are given by

Definition 3 Let ∇ be a B-connection.
 (i) The *connection form* U_∇ of ∇ is defined by

$$U_\nabla(X) = X - \nabla(X|_A), X \in D_1(B).$$

Thus, $U_\nabla \in \hom_B(D_1(B), D_1(B))$.
 (ii) Let $X, Y \in D_1(A, B)$. Then $\nabla(X) \circ Y - \nabla(Y) \circ X$ is evidently an element of $D_1(A, B)$ and one can define the *curvature form* R_∇ of ∇ by

$$R_\nabla(X, Y) = [\nabla(X), \nabla(Y)] - \nabla(\nabla(X) \circ Y - \nabla(Y) \circ X).$$

Hence, $R_\nabla \in \hom_B(D_1(A, B) \otimes_B D_1(A, B), D_1(B))$.

Consider the smooth case now.

Proposition 3.1 *Let A, B be smooth algebras such that $\varphi(A_\alpha) \subset B_{\alpha+k}$ for all α and some fixed k. Then for a B-connection ∇ one has*
(i) $U_\nabla \in D_1(\Lambda^1(B))$.
(ii) $R_\nabla \in D_1(B \otimes_A \Lambda^2(A))$.
(iii) *For any $X, Y \in D_1(B)$ one has*

$$[\![U_\nabla, U_\nabla]\!](X, Y) \overset{\text{def}}{=} i_Y(i_X[\![U_\nabla, U_\nabla]\!]) = 2R_\nabla(X|_A, Y|_A). \qquad (11)$$

Proof. Let C be an algebra with projective module $\Lambda^1(C)$ of finite type. Then for any $r, s \geq 0$ one has

$$\hom_C(D_r(C), D_s(C)) = \Lambda^r(C) \otimes_C D_s(C) \subset D_s(\Lambda^r(C)).$$

Therefore, the first two statements are consequences of Lemma 2.1.

To prove the third one, apply equality (10) to the element U_∇ twice. Then one obtains

$$\frac{1}{2}[\![U_\nabla, U_\nabla]\!](X, Y) = [U_\nabla(X), U_\nabla(Y)] - U_\nabla([U_\nabla(X), Y]) -$$
$$U_\nabla([X, U_\nabla(Y)]) + U_\nabla(U_\nabla([X, Y])).$$

Using the definition of U_∇, i.e., $U_\nabla(Z) = Z - \nabla(Z|_A)$, one gets

$$\frac{1}{2}[\![U_\nabla(X), U_\nabla(Y)]\!] = [X - \nabla(X|_A), Y - \nabla(Y|_A)] -$$
$$([X - \nabla(X|_A), Y] - \nabla([X - \nabla(X|_A), Y]|_A)) -$$
$$([X, Y - \nabla(Y|_A)] - \nabla([X, Y - \nabla(Y|_A)])|_A) +$$
$$[X, Y] - \nabla([X, Y]|_A) - \nabla(([X, Y] - \nabla([X, Y]|_A))|_A).$$

But due to identities $\nabla(Z|_A)|_A = Z|_A$ and

$$[X, Y]|_A = X \circ (Y|_A) - Y \circ (X|_A),$$

this equality is equivalent to (11). $\square$

Definition 4 A B-connection $\nabla \in \hom_B(D_1(A, B), D_1(B))$ is called *flat*, if $R_\nabla = 0$.

Now we introduce a category $\mathrm{Conn}(A)$ whose *objects* are pairs $\mathcal{O} = (B, \nabla)$, where B is an A-algebra with a flat connection ∇. Consider two objects $\mathcal{O} = (B, \nabla)$, $\mathcal{O}' = (B', \nabla')$ of $\mathrm{Conn}(A)$, then a *morphism* of $\mathcal{O}$ to $\mathcal{O}'$ is a homomorphism $\varphi : B \to B'$ of A-algebras such that the diagram

$$
\begin{array}{ccc}
B & \xrightarrow{\varphi} & B' \\
{\scriptstyle\nabla(X)}\downarrow & & \downarrow{\scriptstyle\nabla'(\varphi\circ X)} \\
B & \xrightarrow{\varphi} & B'
\end{array}
$$

is commutative for any $X \in D_1(A, B)$.

Let $\mathcal{O} = (B, \nabla)$ be an object of $\text{Conn}(A)$ and define a map $\partial_\nabla :$ $D_1(\Lambda^r) \to D_1(\Lambda^{r+1})$ by

$$\partial_\nabla(\Omega) \overset{\text{def}}{=} [\![U_\nabla, \Omega]\!], \ \Omega \in D_1(\Lambda^r),$$

where $\Lambda^i = \Lambda^i(B)$. Since ∇ is flat, from (11) it follows that $[\![U_\nabla, U_\nabla]\!] = 0$. Using (9) one can see then that

$$\partial_\nabla \circ \partial_\nabla = 0. \tag{12}$$

We also consider the map $L_\nabla \overset{\text{def}}{=} L_{U_\nabla} : \Lambda^r(B) \to \Lambda^{r+1}(B)$. Main properties of ∂_∇ and L_∇ follow from Propositions 2.4 and 2.6:

Proposition 3.2 *Let $\mathcal{O} = (B, \nabla)$ be an object of $\text{Conn}(A)$. Consider derivations $\Omega, \Theta \in D_1(\Lambda^*)$ and a form $\rho \in \Lambda^*$. Then*
(i) $\partial_\nabla(\rho \wedge \Omega) = L_\nabla(\rho) \wedge \Omega - d\rho \wedge i_\Omega U_\nabla + (-1)^\rho \rho \wedge \partial_\nabla \Omega$.
(ii) $[L_\nabla, i_\Omega] = i_{\partial_\nabla \Omega} + (-1)^\Omega L_{i_\Omega U}$.
(iii) $[i_\Omega, \partial_\nabla]\Theta + i_{[\![\Omega,\Theta]\!]} U_\nabla = [\![i_\Omega U_\nabla, \Theta]\!] + (-1)^\Omega \cdot i_{\partial_\nabla \Omega}\Theta$.
(iv)

$$\partial_\nabla[\![\Omega, \Theta]\!] = [\![\partial_\nabla \Omega, \Theta]\!] + (-1)^\Omega [\![\Omega, \partial_\nabla \Theta]\!]. \tag{13}$$

Due to (12) one can define cohomology **k**-modules

$$\tilde{H}_\nabla^r(B) = \frac{\ker\left(\partial_\nabla : D_1(\Lambda^r) \to D_1(\Lambda^{r+1})\right)}{\text{im}\left(\partial_\nabla : D_1(\Lambda^{r-1}) \to D_1(\Lambda^r)\right)}.$$

From (13) it follows that the module

$$\tilde{H}_\nabla^*(B) = \sum_{r \geq 0} \tilde{H}_\nabla^r(B)$$

inherits the Frölicher–Nijenhuis bracket and is a graded Lie algebra over **k** with respect to this bracket.

In what follows we shall need a subtheory of this cohomology theory.

3.2. VERTICAL COHOMOLOGIES

Definition 5 An element $\Omega \in D_1(\Lambda^*)$ is called *vertical*, if $L_\Omega(a) = 0$ for all $a \in \varphi(A) \subset B = \Lambda^0(B)$.

Denote by $D_1^v(B)$ the B-submodule formed by all vertical derivations $X \in D_1(B) = D_1(\Lambda^0)$.

Proposition 3.3 *Let $\mathcal{O} = (B, \nabla)$ be an object of* $\mathrm{Conn}(A)$. *Then*

(i) *The module* $D_1^v(\Lambda^*) \subset D_1(\Lambda^*)$ *is closed with respect to both the Frölicher–Nijenhuis bracket and the contraction operation:*

$$[\![D_1^v(\Lambda^r), D_1^v(\Lambda^s)]\!] \subset D_1^v(\Lambda^{r+s}), \quad i_{(D_1^v(\Lambda^r))}(D_1^v(\Lambda^s)) \subset D_1^v(\Lambda^{r+s-1}).$$

(ii) *An element* $\Omega \in D_1(\Lambda^*)$ *lies in* $D_1^v(\Lambda^*)$ *if and only if* $i_\Omega(U_\nabla) = \Omega$.

(iii) *The element* U_∇ *is vertical:* $U_\nabla \in D_1^v(\Lambda^1)$.

From the above said it follows that the differential ∂_∇ can be restricted onto $D_1^v(\Lambda^*)$, i.e., $\partial_\nabla : D_1^v(\Lambda^r) \to D_1^v(\Lambda^{r+1})$, and cohomology modules

$$H_\nabla^r(B) = \frac{\ker\left(\partial_\nabla : D_1^v(\Lambda^r) \to D_1^v(\Lambda^{r+1})\right)}{\mathrm{im}\left(\partial_\nabla : D_1^v(\Lambda^{r-1}) \to D_1^v(\Lambda^r)\right)}$$

are defined. Denote

$$H_\nabla^*(B) = \sum_{r \geq 0} H_\nabla^r(B).$$

From Propositions 3.2 and formula (12) one obtains

Proposition 3.4 *Let* $\mathcal{O} = (B, \nabla)$ *be an object of* $\mathrm{Conn}(A)$. *Then for any vertical derivations* $\Omega, \Theta \in D_1(\Lambda^*)$ *and a form* $\rho \in \Lambda^*(B)$ *one has:*

(i)

$$\partial_\nabla(\rho \wedge \Omega) = (L_\nabla \rho - d\rho) \wedge \Omega + (-1)^\rho \rho \wedge \partial_\nabla \Omega. \tag{14}$$

(ii) $[L_\nabla, i_\Omega] = i_{\partial_\nabla \Omega} + (-1)^\Omega L_\Omega.$

(iii).

$$[i_\Omega, \partial_\nabla] = (-1)^\Omega i_{\partial_\nabla \Omega}. \tag{15}$$

(iv) $\partial_\nabla [\![\Omega, \Theta]\!] = [\![\partial_\nabla \Omega, \Theta]\!] + (-1)^\Omega [\![\Omega, \partial_\nabla \Theta]\!].$

From Proposition 3.3 and the last two equalities it follows

Proposition 3.5 *For any object* $\mathcal{O} = (B, \nabla)$ *be an object of* $\mathrm{Conn}(A)$

(i) *The module* $H_\nabla^*(B)$ *is a graded Lie algebra with respect to the Frölicher–Nijenhuis bracket inherited from* $D_1^v(\Lambda^*)$.

(ii) *The module* $H_\nabla^*(B)$ *inherits the contraction operation from* $D_1^v(\Lambda^*)$:

$$i_{H_\nabla^r(B)} H_\nabla^s(B) \subset H_\nabla^{r+s-1}(B)$$

and $H_\nabla^*(B)$ *is a graded Lie algebra with respect to the inherited Richardson–Nijenhuis bracket, provided the grading be shifted by 1. In particular,* $H_\nabla^1(B)$ *is an associative algebra with respect to contraction.*

Other properties of $H_\nabla^*(B)$ are be considered in the next section.

4. Symmetries and recursion operators

Here we define symmetries and recursion operators for the objects of the category $\mathrm{Conn}(A)$ and describe a Lie algebra structure for the symmetry algebras of objects with nontrivial recursions.

4.1. SPLITTINGS

Consider equality (14) and define

$$d_h = d - L_\nabla. \tag{16}$$

Thus (14) can be rewritten as

$$\partial_\nabla(\rho \wedge \Omega) = -d_h\rho \wedge \Omega + (-1)^\rho \rho \wedge \partial_\nabla \Omega. \tag{17}$$

Proposition 4.1 *For any object $\mathcal{O}$ of* $\mathrm{Conn}(A)$ *one has*

$$L_\nabla \circ L_\nabla = 0, \quad d_h \circ d_h = 0, \quad [L_\nabla, d_h] = 0$$

and thus a spectral sequence is associated with the pair (d_h, L_∇). *Due to* $d = d_h + L_\nabla$, *this spectral sequence converges to the de Rham cohomology of B.*

The map $d_h : \Lambda^*(B) \to \Lambda^*(B)$ is called the *horizontal de Rham differential* of $\mathcal{O} = (B, \nabla)$ and its cohomologies

$$H_h^*(\mathcal{O}) = \sum_{r \geq 0} H_h^r(\mathcal{O})$$

are called the *horizontal cohomologies* of $\mathcal{O}$. From (17) it follows that $H_\nabla^*(B)$ is a graded module over $H_h^*(\mathcal{O})$.

Due to (16), the module $\Lambda^1(B)$ can represented as

$$\Lambda^1(B) = \mathcal{C}\Lambda^1(B) \oplus \Lambda_h^1(B), \tag{18}$$

where $\mathcal{C}\Lambda^1(B)$ and $\Lambda_h^1(B)$ are generated in $\Lambda^1(B)$ by images of $L_\nabla : B \to \Lambda^1(B)$ and $d_h : B \to \Lambda^1(B)$ respectively. From (18) it follows that

$$\Lambda^r(B) = \sum_{p+q=r} \mathcal{C}^p\Lambda(B) \wedge \Lambda_h^q(B), \tag{19}$$

where

$$\mathcal{C}^p\Lambda(B) = \underbrace{\mathcal{C}\Lambda^1(B) \wedge \ldots \wedge \mathcal{C}\Lambda^1(B)}_{p \text{ times}}$$

and

$$\Lambda_h^q(B) = \underbrace{\Lambda_h^1(B) \wedge \ldots \wedge \Lambda_h^1(B)}_{q \text{ times}}.$$

Due to (18), the module $D_1^v(\Lambda^r)$ splits into the direct sum

$$D_1^v(\Lambda^r) = \sum_{p+q=r} D_1^v(\mathcal{C}^p\Lambda(B) \wedge \Lambda_h^q(B)).$$

Proposition 4.2 *For any object* $\mathcal{O} = (B, \nabla)$ *one has*

$$\partial_\nabla (D_1^v(\mathcal{C}^p\Lambda(B) \wedge \Lambda_h^q(B)) \subset D_1^v(\mathcal{C}^p\Lambda(B) \wedge \Lambda_h^{q+1}(B)). \qquad (20)$$

Proof. To prove this fact note that $\Omega \in D_1^v(\Lambda^r)$ lies in $D_1^v(\mathcal{C}^p\Lambda(B) \wedge \Lambda_h^q(B))$ if and only if

$$(i_{\nabla(X_1)} \circ \ldots \circ i_{\nabla(X_p)} \circ i_{Y_1} \circ \ldots \circ i_{Y_q})\Omega = 0$$

for any $X_1, \ldots X_p \in D_1(A, B)$ and $Y_1, \ldots, Y_q \in D_1^v(B)$. Then the result is a consequence of (15). $\square$

From (19) and (20) it follows that $H_\nabla^r(B) = \sum_{p+q=r} H_\nabla^{p,q}(B)$, where

$$H_\nabla^{p,q}(B) = \frac{\ker(\partial_\nabla : D_1^v(\mathcal{C}^p\Lambda \wedge \Lambda_h^q) \to D_1^v(\mathcal{C}^p\Lambda \wedge \Lambda_h^{q+1}))}{\operatorname{im}(\partial_\nabla : D_1^v(\mathcal{C}^p\Lambda \wedge \Lambda_h^{q-1}) \to D_1^v(\mathcal{C}^p\Lambda \wedge \Lambda_h^q))}.$$

In particular, cohomology groups with trivial horizontal components are described by

Proposition 4.3 *If* $\mathcal{O} = (B, \nabla)$ *is an object of* $\operatorname{Conn}(A)$, *then*

$$H_\nabla^{p,0}(B) = \ker \left(\partial_\nabla \Big|_{D_1^v(\mathcal{C}^p\Lambda(B))}\right).$$

4.2. SYMMETRIES, RECURSION OPERATORS AND DESCRIPTION OF LIE ALGEBRA STRUCTURES

Detailed motivations for the following definition can be found in [5, 7, 8]. A brief discussion concerning applications to differential equations the reader will also find in concluding remarks below.

Definition 6 Let $\mathcal{O} = (B, \nabla)$ be an object of $\operatorname{Conn}(A)$.
  (i) The elements of $H_\nabla^{0,0}(B) = H_\nabla^0(B)$ are called *symmetries* of $\mathcal{O}$.
  (ii) The elements of $H_\nabla^{1,0}(B)$ are called *recursion operators* of $\mathcal{O}$.

We use the notations

$$\mathrm{Sym}\,(\mathcal{O}) \stackrel{\mathrm{def}}{=} \mathrm{H}_\nabla^{0,0}(\mathrm{B})$$

and

$$\mathrm{Rec}\,(\mathcal{O}) \stackrel{\mathrm{def}}{=} \mathrm{H}_\nabla^{1,0}(\mathrm{B}).$$

From Propositions 3.4, 3.5 and 4.3 one obtains

Theorem 4.4 *For any object $\mathcal{O} = (B, \nabla)$ of $\mathrm{Conn}(A)$ the following facts take place:*

(i) $\mathrm{Sym}\,(\mathcal{O})$ *is a Lie algebra with respect to commutator of derivations.*

(ii) $\mathrm{Rec}\,(\mathcal{O})$ *is an associative algebra with respect to contraction, U_∇ being the unit of this algebra.*

(iii) *The map* $\mathcal{R} : \mathrm{Rec}\,(\mathcal{O}) \to \mathrm{End}_{\mathbf{k}}(\mathrm{Sym}\,(\mathcal{O}))$, *where*

$$\mathcal{R}_\Omega(X) = i_X(\Omega), \ \Omega \in \mathrm{Rec}\,(\mathcal{O}), X \in \mathrm{Sym}\,(\mathcal{O})$$

is a representation of this algebra and hence

(iv) $i_{(\mathrm{Sym}\,(\mathcal{O}))}(\mathrm{Rec}\,(\mathcal{O})) \subset \mathrm{Sym}\,(\mathcal{O})$.

In what follows we shall need a simple consequence of basic definitions:

Proposition 4.5 *For any object $\mathcal{O} = (B, \nabla)$ of $\mathrm{Conn}(A)$*

$$[\![\mathrm{Sym}\,(\mathcal{O}), \mathrm{Rec}\,(\mathcal{O})]\!] \subset \mathrm{Rec}\,(\mathcal{O})$$

and

$$[\![\mathrm{Rec}\,(\mathcal{O}), \mathrm{Rec}\,(\mathcal{O})]\!] \subset \mathrm{H}_\nabla^{2,0}(\mathrm{B}).$$

Corollary 4.6 *If $\mathrm{H}_\nabla^{2,0}(B) = 0$, then all recursion operators of the object $\mathcal{O} = (B, \nabla)$ commute with each other with respect to the Frölicher–Nijenhuis bracket.*

We call the objects satisfying the conditions of the previous Corollary 2-*trivial*. To simplify notations we denote

$$\mathcal{R}_\Omega(X) = \Omega(X), \ \Omega \in \mathrm{Rec}\,(\mathcal{O}), X \in \mathrm{Sym}\,(\mathcal{O}).$$

From Proposition 4.5 and equality (10) one gets

Proposition 4.7 *Consider an object $\mathcal{O} = (B, \nabla)$ of $\mathrm{Conn}(A)$ and elements $X, Y \in \mathrm{Sym}\,(\mathcal{O})$, $\Omega, \Theta \in \mathrm{Rec}\,(\mathcal{O})$. Then*

$$[\![\Omega, \Theta]\!](X, Y) = [\Omega(X), \Theta(Y)] + [\Theta(X), \Omega(Y)] - \Omega([\Theta(X), Y] +$$
$$[X, \Theta(Y)]) - \Theta([\Omega(X), Y] + [X, \Omega(Y)]) + (\Omega \circ \Theta + \Theta \circ \Omega)\,[X, Y].$$

In particular, for $\Omega = \Theta$ one has

$$\frac{1}{2}[\![\Omega, \Omega]\!](X, Y) = [\Omega(X), \Omega(Y)] - \Omega([\Omega(X), Y]) - \Omega([X, \Omega(Y)]) +$$
$$\Omega(\Omega([X, Y])). \quad (21)$$

The proof of this statement is completely identical to that of Proposition 4.1. The right-hand side of (21) is called the *Nijenhuis torsion* of Ω (cf. [10]).

Corollary 4.8 *If $\mathcal{O}$ is a 2-trivial object, then*

$$[\Omega(X), \Omega(Y)] = \Omega\left([\Omega(X), Y] + [X, \Omega(Y)] - \Omega[X, Y]\right). \qquad (22)$$

Choose a recursion operator $\Omega \in \mathrm{Rec}\,(\mathcal{O})$ and for any symmetry $X \in \mathrm{Sym}\,(\mathcal{O})$ denote $\Omega^i(X) = \mathcal{R}_\Omega^i(X)$ by X_i. Then (22) can be rewritten as

$$[X_1, Y_1] = [X_1, Y]_1 + [X, Y_1]_1 - [X, Y]_2. \qquad (23)$$

Using (23) as the induction base one can prove the following

Proposition 4.9 *For any 2-trivial $\mathcal{O}$ and $m, n \geq 1$*

$$[X_m, Y_n] = [X_m, Y]_n + [X, Y_n]_m - [X, Y]_{m+n}.$$

Let, as before, X be a symmetry and Ω be a recursion operator. Then $\Omega_X \overset{\mathrm{def}}{=} [\![X, \Omega]\!]$ is a recursion operator again (Proposition 4.5). Due to (10), its action on $Y \in \mathrm{Sym}\,(\mathcal{O})$ can be expressed as

$$\Omega_X(Y) = [X, \Omega(Y)] - \Omega[X, Y]. \qquad (24)$$

From (24) one has

Proposition 4.10 *For any 2-trivial object $\mathcal{O}$, any two symmetries $X, Y \in \mathrm{Sym}\,(\mathcal{O})$, a recursion $\Omega \in \mathrm{Rec}\,(\mathcal{O})$ and integer $m, n \geq 1$ one has*

$$[X, Y_n] = [X, Y]_n + \sum_{i=0}^{n-1} (\Omega_X Y_i)_{n-i-1}$$

and

$$[X_m, Y] = [X, Y]_m - \sum_{j=0}^{m-1} (\Omega_Y X_j)_{m-j-1}.$$

From the last two results one obtains

Theorem 4.11 (the structure of a Lie algebra for $\mathrm{Sym}\,(\mathcal{O})$) *For any 2-trivial object $\mathcal{O}$, its symmetries $X, Y \in \mathrm{Sym}\,(\mathcal{O})$, a recursion operator $\Omega \in \mathrm{Rec}\,(\mathcal{O})$ and integer $m, n \geq 1$ one has*

$$[X_m, Y_n] = [X, Y]_{m+n} + \sum_{i=0}^{n-1} (\Omega_X Y_i)_{m+n-i-1} - \sum_{j=0}^{m-1} (\Omega_Y X_j)_{m+n-j-1}.$$

Corollary 4.12 *If* $X, Y \in \mathrm{Sym}\,(\mathcal{O})$ *are such that* Ω_X *and* Ω_Y *commute with* $\Omega \in \mathrm{Rec}\,(\mathcal{O})$ *with respect to the Richardson–Nijenhuis bracket, then*

$$[X_m, Y_n] = [X, Y]_{m+n} + n(\Omega_X Y)_{m+n-1} - m(\Omega_Y X)_{m+n-1}.$$

We say that a recursion operator $\Omega \in \mathrm{Rec}\,(\mathcal{O})$ is X-invariant, if $\Omega_X = 0$.

Corollary 4.13 (existence of infinite series of commuting symmetries) *If* $\mathcal{O}$ *is a 2-trivial object and if a recursion operator* $\Omega \in \mathrm{Rec}\,(\mathcal{O})$ *is* X-invariant, $X \in \mathrm{Sym}\,(\mathcal{O})$, *then a hierarchy* $\{X_n\}, n = 0, 1, \ldots$, *generated by* X *and* Ω *is commutative:*

$$[X_m, X_n] = 0$$

for all m, n.

4.3. CONCLUDING REMARKS

Here we briefly discuss relations of the above exposed algebraic scheme to partial differential equations.

First recall that correspondence between algebraic approach and geometrical picture is established by identifying the category of vector bundles over a smooth manifold M with the category of geometrical modules over $A = C^\infty(M)$, see [6]. In the case of differential equations M plays the role of the manifold of independent variables while $B = B_\infty = \bigcup_\alpha B_\alpha$ is the function algebra on the infinite prolongation of the equation $\mathcal{E}$ and $B_\alpha = C^\infty(\mathcal{E}^\alpha)$, where $\mathcal{E}^\alpha$, $\alpha = 0, 1, \ldots, \infty$, is the α-prolongation of $\mathcal{E}$. The map $\varphi : A \to B$ is dual to the natural projection $\pi_\infty : \mathcal{E}^\infty \to M$ and thus in applications to differential equations it suffices to consider the case $A = \bigcap_\alpha B_\alpha$.

If $\mathcal{E}$ is a formally integrable equation, the bundle $\pi_\infty : \mathcal{E}^\infty \to M$ possesses a natural connection (the Cartan connection $\mathcal{C}$) which sends a vector field X on M to corresponding total derivative on $\mathcal{E}^\infty$. Consequently, the category of differential equations [17] is embedded to the category of algebras with flat connections. Under this identification the spectral sequence defined in Proposition 4.1 coincides with A.M. Vinogradov's $\mathcal{C}$-spectral sequence [20] (or variational bicomplex), the module $\mathrm{Sym}\,(\mathcal{O})$, where $\mathcal{O} = (C^\infty(M), C^\infty(\mathcal{E}^\infty), \mathcal{C})$, is the Lie algebra of higher symmetries for the equation $\mathcal{E}$ and, in principle, $\mathrm{Rec}\,(\mathcal{O})$ consists of recursion operators for these symmetries. This last statement should be clarified.

In fact, if one tries to compute $\mathrm{Rec}\,(\mathcal{O})$ straightforwardly, the results will be trivial usually — even for equations really possessing recursion operators. The reason lies in nonlocal character of recursion operators for majority of interesting equations [16]. Thus extension of the algebra $C^\infty(\mathcal{E}^\infty)$ with nonlocal variables (see [9]) is the way to obtain nontrivial solutions — and actual computations show that all known (as well as new ones!) recursion operators can be obtained in such a way (see examples in [7, 8]). In practice, it usually suffices to extend $C^\infty(\mathcal{E}^\infty)$ by integrals of conservation laws.

The algorithm of computations becomes rather simple due to the following fact. For nonoverdetermined equations all cohomology groups $H_C^{p,q}(\mathcal{E})$ are trivial except for the cases $q = 0, 1$ while the differential

$$\partial_C : D_1^v(C^p\Lambda(\mathcal{E})) \to D_1^v(C^p\Lambda(\mathcal{E}) \wedge \Lambda_1^h(\mathcal{E}))$$

coincides with the universal linearization operator $\ell_\mathcal{E}$ of the equation $\mathcal{E}$, [3, 5]. Therefore, the modules $H_C^{p,0}(\mathcal{E})$ coincide with $\ker(\ell_\mathcal{E})$ (see Proposition 4.3) and thus can be computed efficiently.

In particular, it can be shown that for scalar evolution equations all cohomologies $H_C^{p,0}(\mathcal{E})$, $p \geq 2$, vanish and consequently equations of this type are 2-trivial and satisfy the conditions of Theorem 4.11.

References

1. Cabras, A. and Vinogradov, A.M.: Extension of the Poisson bracket to differential forms and multi-vector fields, *J. Geom. and Phys.* **9** (1992) no. 1, 75–100.
2. Frölicher, A. and Nijenhuis, A.: Theory of vector valued differential forms. Part I: Derivations in the graded ring of differential forms, *Indag. Math.* **18** (1956) 338–359.
3. Gessler, D.: On the Vinogradov C-spectral sequence for determined systems of differential equations, *J. Diff. Geom. Appl.* (to appear).
4. Krasil'shchik, I.S.: Schouten Brackets and Canonical Algebras, *Lecture Notes in Math.* **1334** (1988) Springer–Verlag, Berlin, 79–110.
5. Krasil'shchik, I.S.: Some new cohomological invariants for nonlinear differential equations, *J. Diff. Geom. Appl.* **2** (1992) 307-350.
6. Krasil'shchik, I.S., Lychagin, V.V., and Vinogradov, A.M.: *Geometry of Jet Spaces and Nonlinear Partial Differential Equations*, Gordon and Breach, New York, 1986.
7. Krasil'shchik, I.S. and Kersten, P.H.M.: Graded differential equations and their deformations: a computational theory for recursion operators, In: P.H.M. Kersten and I.S. Krasil'shchik I.S. (eds.), *Geometric and algebraic structures in differential equations*, Kluwer Acad. Publ., Dordrecht, 1995, 167–191.
8. Krasil'shchik, I.S. and Kersten, P.H.M.: Deformations of differential equations and recursion operators, In: A. Pràstaro and Th. M. Rassias (eds.), *Geometry in Partial Differential Equations*, World Scientific, Singapore, 1993, 114–154.
9. Krasil'shchik, I.S. and Vinogradov, A.M.: Nonlocal trends in the geometry of differential equations: symmetries, conservation laws, and Bäcklund transformations, In: A.M. Vinogradov (ed.), *Symmetries of Partial Differential Equations*, Kluwer Acad. Publ., Dordrecht, 1989, 161–209.
10. Kosmann-Schwarzbach, Y. and Magri, F.: Poisson–Nijenhuis structures, *Ann. Inst. Henri Poincaré* **53** (1990) 35–81.
11. Lecomte, P.A.B., Michor, P.W., and Schicketanz, H.: The multigraded Nijenhuis–Richardson algebra, its universal property and applications, *J. Pure Appl. Algebra* **77** (1992) 87–102
12. Michor, P.W.: Remarks on the Frölicher-Nijenhuis bracket, *Proc. Conf. in Differential Geometry and its Applications*, Brno, 1986, 197–220.

13. Michor, P.W.: Remarks on the Schouten-Nijenhuis bracket, *Proc. Winter School in Geometry and Physics, Srni, Suppl. Rend. Circ. Mat. Palermo, Ser. II* **16** (1987) 207–215.

14. Nijenhuis, A.: Jacobi type identities for bilinear differential concomitants of certain tensor fields. I, *Indag. Math.* **17** (1955) no. 3, 463–469.

15. Nijenhuis, A. and Richardson, R.W. Jr.: Deformations of Lie algebra structures, *J. Math. Mech.* **17** (1967) 89–105.

16. Olver, P.J.: Applications of Lie Groups to Differential Equations, *Graduate Texts in Mathematics* **107**, Springer–Verlag, New York, 1986.

17. Vinogradov, A.M.: Category of nonlinear differential equations, *Lecture Notes Math.* **1108** (1984) Springer–Verlag, Berlin, 77–102.

18. Vinogradov, A.M.: Unification of the Schouten–Nijenhuis and Frölicher–Nijenhuis brackets, cohomology, and superdifferential operators, *Mat. Zametki* **47** (1990) 138–140 (in Russian).

19. Vinogradov, A.M.: The logic algebra of linear differential operators, *Soviet Math. Dokl.* **13** (1972) no. 4, 1058–1062.

20. Vinogradov, A.M.: The C-spectral sequence, Lagrangian formalism, and conservation laws, *J. Math. Anal. Appl.* **100** (1984) 2–129.

ON THE GEOMETRY OF CURRENT GROUPS AND
A MODEL OF THE LANDAU–LIFSCHITZ EQUATION

A.M. LUKATSKY

Energy Research Institute, Russian Academy of Sciences,
31-2, Nagornaya str., 113186 Moscow, Russia

Abstract. This paper is devoted to the study of the Landau–Lifschitz equation (LL) as the Euler equations on a current group. The current group $G = G(M, SO(3))$ is the configuration space of the ferromagnet magnetization problem, which is described by LL. The Lie algebra of G is the current algebra $g = g(M, so(3))$. It is convenient to introduce a nonstandard Lie bracket in g, then the geodesics of a left-invariant metric on G become the solutions of LL. The expression of the curvature tensor on G is obtained. In the case $M = T^3$ (the 3-torus), examples of the calculation of the sectional curvature are presented.

Mathematics subject classification (1991): 81R10, 58B25.

Key words: current group, Euler equations, geodesics, left-invariant Riemannian metric, curvature tensor, sectional curvatures.

Let M be a 3-dimensional closed oriented Riemannian manifold. The current group $G = G(M, SO(3))$ may be regarded as the configuration space of the ferromagnet magnetization problem. Its Lie algebra is $g = g(M, so(3))$. If the manifold M is parallelizable (for example the 3-torus or 3-sphere), then we may identify the Lie algebra g with the space $V(M)$ of smooth vector fields on M with the pointwise vector product as the Lie bracket. Let us consider the ferromagnet magnetization problem [1]–[3] under the following assumptions:
  – there is no boundary effect;
  – the exchange effect function has a quadratic form;
  – the time is dimensionless (the hydromagnetic ratio is excluded).

B. P. Komrakov et al. (eds.), Lie Groups and Lie Algebras, 425–433.

Under these physical assumptions, the Landau–Lifschitz equation (LL) has the form:

$$\frac{dm}{dt} = m \times Pm. \tag{1}$$

Here $m(t)$ is a curve in the space $V(M)$; $a \times b$ is the pointwise vector product of vector fields; $P : V(M) \to V(M)$ is the operator which is the sum of an elliptic differential operator N and the pointwise orthogonal projection operator N_0 onto a one-dimensional subspace. The term N_0 expresses the anisotropic effect. The vector field m is called the field of density of the magnetic moment. Often the vector field $h = Pm$ is considered; it is called the effective magnetic field. Consider the standard scalar product on $V(M)$ given by

$$(m_1, m_2) = \int_M (m_1(x), m_2(x)) d\mu(x), \tag{2}$$

where $(m_1(x), m_2(x))$ is the pointwise scalar product. As is well known, this scalar product, regarded on the space of solutions of LL, is time-invariant not only globally, but at every point of M. In our model this fact is expressed by the property that LL has the current group G as its configuration space. The form (2) also can be interpreted in terms of Lie algebras. We can introduce the pointwise Killing form on the current group G as follows. At any point x from M, let us denote $k_x(u, v) = \text{tr}((adu)(adv))$, where u, v belong to $T_x(M)$ and ad is the adjoint operator in Lie algebra $so(3)$. It is easy to verify, that $(m_1(x), m_2(x)) = (-1/2)k_x(m_1(x), m_2(x))$. Below we suppose that P is symmetric operator and we consider another scalar product on $V(M)$, given by

$$\langle m_1, m_2 \rangle = -(m_1, Pm_2) = -\int_M (m_1(x), h_2(x)) d\mu(x). \tag{3}$$

As is well known, it is time-invariant for solutions of LL [3].

The metric (3) determines a left-invariant metric on the current group G, but its geodesics do not satisfy LL. To obtain the coincidence of LL solutions with geodesics, one can use a metric on the dual space g^*. Here we apply duality via the metric (3) but with an additional assumption. Note that the operator P is determined up to the additive term aI, where I is the identity operator. In general the operator P may be noninvertible. Instead of the operator P, let us consider the family of operators $P_a = P - aI$. Then there exists an $a_0 > 0$ such that for every $a > a_0$ the operator P_a is invertible. Let us fix a number $a > a_0$ and consider LL as an equation for the vector field $h_a = P_a(m)$. Physically, this corresponds to considering the representation of LL in the effective magnetic fields instead of the standard representation in the fields of density of the magnetic moment. Then LL has the form:

$$\frac{dh_a}{dt} = P_a(P_a^{-1}(h_a) \times h_a). \tag{4}$$

We also consider the metric

$$\langle h_{1a}, h_{2a} \rangle_1 = -(P_a^{-1} h_{1a}, h_{2a}). \tag{5}$$

It is easy to verify that the coadjoint operator $(\mathrm{ad}h)^*$ is expressed as follows:

$$(\mathrm{ad}h_1)^*(h_2) = -P_a(h_1 \times h_2). \tag{6}$$

Thus LL becomes the Euler equations on the current group G, [6].

In [6] an expression of the curvature of G for the representation of LL in the effective magnetic fields is obtained. But in this approach an additional problem appears: what relationship exists between this curvature and the properties of the solutions of LL for the standard representation. Therefore, it is more interesting to obtain LL as the Euler equations in the m-fields themselves. In order to achieve this, we ought to introduce a nonstandard Lie bracket in g. Namely, let us introduce the bracket:

$$[m_1, m_2] = P_a^{-1}(P_a(m_1) \times P_a(m_2)) = P_a^{-1}(h_1 \times h_2). \tag{7}$$

Here h_1, h_2 are the effective magnetic fields corresponding to m_1 and m_2 respectively. Let us denote by ad' the corresponding adjoint operator. Then we have:

$$(\mathrm{ad}'m)^* = -P_a \mathrm{ad}'m P_a^{-1}. \tag{8}$$

Thus we obtain LL as the Euler equations on G in the standard representation (in the fields of density of magnetic moment), but with the nonstandard Lie bracket $[\cdot, \cdot]$ instead of $\times$. It is easy to verify that these two Lie algebra structures in g are isomorphic.

Note that here we consider the left-invariant metric on G with the standard identification of $so(3)$ with the Lie algebra of left-invariant vector fields on $SO(3)$. We may also consider the Lie algebra of right-invariant vector fields on the Lie group $SO(3)$; then we must use the right-invariant metric on the current group G to obtain LL.

Below we obtain an expression for the curvature tensor of G and calculate the curvature. We use the formulas of Arnold [4] for the covariant derivative. Let us denote $B(u, v) = (\mathrm{ad}'v)^*(u)$. We have $B(u, v) = u \times Pv$. In particular, LL has the form:

$$\frac{du}{dt} = B(u, u).$$

Note that the Euler equations for solids have a similar form [5], while in the case of hydrodynamics of an ideal incompressible fluid the Euler equations have the form:

$$\frac{du}{dt} = -B(u, u).$$

Thus we see that LL has a closer analogy with solids rather than with ideal incompressible fluids.

Let a be a fixed real number and let us omit the subscript a in P_a. Set $Q = P^{-1}$. Let u, v, w be three vector fields from $V(M)$, denote $f = Pu, g = Pv, h = Pw$. According to [4], the expression of the covariant derivative has the form:

$$\nabla_u v = \frac{1}{2}(\mathrm{ad}'u(v) - B(u, v) - B(v, u)).$$

It follows that

$$\nabla_u v = \frac{1}{2}(Q(f \times g) + f \times v + g \times u). \tag{9}$$

Let us calculate the expressions of the components of the curvature tensor of G. We have:

$$\begin{aligned}
4\nabla_u \nabla_v w = {}& Q[f \times (g \times h) + f \times P(h \times v + g \times w)] + \\
& (g \times h) \times u + (P(h \times v + g \times w)) \times u + \\
& f \times Q(g \times h) + f \times (g \times w + h \times v), \\
4\nabla_v \nabla_u w = {}& Q[g \times (f \times h) + g \times P(h \times u + f \times w)] + \\
& (f \times h) \times v + P(h \times u + f \times w) \times v + \\
& g \times Q(f \times h) + g \times (f \times w + h \times u), \\
2\nabla_{[u,v]} w = {}& Q((f \times g) \times h) + h \times Q(f \times g) + (f \times g) \times w.
\end{aligned}$$

Now we can calculate the expression of the curvature tensor:

$$\begin{aligned}
R(u, v)w = {}& \frac{1}{4}\{Q[-(f \times g) \times h + f \times P(h \times v) - \\
& g \times P(h \times u) + f \times P(g \times w) - g \times P(f \times w)] - \\
& (f \times g) \times w - h \times (g \times u) + h \times (f \times v) + \\
& P(h \times v) \times u - P(h \times u) \times v + P(g \times w) \times u - P(f \times w) \times v + \\
& f \times Q(g \times h) - g \times Q(f \times h) - 2h \times Q(f \times g)\}. \tag{10}
\end{aligned}$$

Thus we obtain:

Theorem 1 *Let u, v be an orthogonal pair of vectors from $V(M)$. Then*

$$\langle R(u, v)v, u \rangle = -\frac{1}{4}(P(u \times g + v \times f), u \times g + v \times f)) - $$
$$\frac{1}{2}(u \times g - v \times f, f \times g)) + \frac{3}{4}(f \times g, Q(f \times g)) + (f \times u, P(g \times v)). \tag{11}$$

Below we consider the example of simple harmonics on the 3-torus. Let x_1, x_2, x_3 be the standard coordinates on T^3, denote $e_i = d/dx_i$ $(i = 1, 2, 3)$. We take as N the standard Laplace operator and choose the anisotropic term as $N_0(m)(x) = b * (m(x), e_3)e_3$, where b is a positive number. Then P_a is invertible for any $a > b$. Now we take the following pairs u, v, satisfying $(u(x), v(x)) = 0$ and $(u(x), P_a v(x)) = 0$ at every point $x \in T^3$:

$$u = \cos kx \, e_3, v = \cos lx \, (se_1 + re_2), k \neq 0, l \neq 0. \tag{12}$$

We will investigate the sign of the curvature $K(u, v)$ for such pairs (u, v). In the notations of Theorem 1 we have $f = (-k^2 - a + b)u, g = (-l^2 - a)v$. It follows

$$f \times u = v \times g = 0;$$
$$f \times g = (k^2 + a - b)(l^2 + a) \cos kx \cos lx(-re_1 + se_2);$$
$$u \times g + v \times f = (k^2 - l^2 - b) \cos kx \cos lx(-re_1 + se_2);$$
$$u \times g - v \times f = (-k^2 - l^2 - 2a + b) \cos kx \cos lx(-re_1 + se_2).$$

Let V be the volume of T^3. It is convenient to denote $T = 1/4(r^2+s^2)V, J = k^2 + a - b, H = l^2 + a, E = (k+l)^2 + a, G = (k-l)^2 + a$. Let us introduce the function

$$z(k) = \begin{cases} 1/2, & k \neq 0; \\ 1, & k = 0. \end{cases}$$

The necessary scalar products have the form

$$(f \times g, u \times g - v \times f) = -TJH(J + H)(z(k+l) + z(k-l));$$
$$(f \times q, Q(f \times g)) = -TJ^2H^2\left(\frac{z(k+l)}{E} + \frac{z(k-l)}{G}\right);$$
$$(u \times g + v \times f, P(u \times g + v \times f))$$
$$= -T(J - H)^2(z(k+l)E + z(k-l)G).$$

Substituting these expressions in (11), we obtain:

$$\langle R(u, v)v, u \rangle = \frac{T}{4}\{(z(k+l)E + z(k-l)G)(J - H)^2 +$$
$$2(J + H)JH(z(k+l) + z(k-l)) - 3J^2H^2(\frac{z(k+l)}{E} + \frac{z(k-l)}{G})\}. \tag{13}$$

Note that $\langle u, u \rangle = (k^2 + a - b)V/2 = JV/2, \langle v, v \rangle = (l^2 + a)(s^2 + r^2)V/2 = 2TH$. After the standard normalization procedure, we obtain the expression of the sectional curvature:

$$K(u, v) = \frac{1}{4V}\{z(k+l)E + z(k-l)G)(\frac{J}{H} + \frac{H}{J} - 2) +$$
$$2(J + H)(z(k+l) + z(k-l)) - 3JH(\frac{z(k+l)}{E} + \frac{z(k-l)}{G})\}. \tag{14}$$

Let us investigate the asymptotic behavior of $K(u, v)$. Note that for a pair of harmonics k, l we have

(i) If $k = l$ and $k \to \infty$, then $K(u, v) \to -\infty$. Namely, under this assumption we have $K(u, v) = -(3k^4)/(4aV) + o(k^4)$.

(ii) If k is fixed and $l \to \infty$, then $K(u, v) \to +\infty$. Namely, in this assumption we have $K(u,v) = l^4/(4(k^2 + a - b)V) + o(l^4)$.

Now let us construct the negative domain for the curvature $K(u, v)$.

Theorem 2 *Let the pair* (12) *of vector fields* (u, v) *have the harmonics* (k, l) *which satisfy the condition* $a \leq (k - l)^2 < k^2$. *Let us denote* $q^2 = (k - l)^2/k^2$ *and* $p^2 = a/(k^2)$ *(obviously* $0 < p \leq q < 1$ *). Then the following estimate of the curvature* $K(u, v)$ *holds:*

$$K(u, v) \leq \frac{C(k^2 + l^2 + a) + 2(a - b)}{4V}, \tag{15}$$

where $C = 4q^2(1 + q)^2/(1 - q)^2 + 2 - 3(1 - q)^2/((p^2 + q^2)(2 + q)^2)$. *In particular, if* $q = p = 1/5$, *then we have the estimate*

$$K(u, v) < \frac{-5(k^2 + l^2)}{8V}. \tag{16}$$

Proof. It is convenient to transform the expression of the curvature $K(u, v)$. Let us denote $F = (k^2 - l^2 - b)$. We have $K(u, v) = ((k^2 + l^2 + a)(F^2/(JH) + 2 - 3JH/(EG)) + 2(a - b))/(4V)$. Now we estimate the terms of this expression. It is easy to verify that: $F^2 \leq 4q^2(1 + q)^2(k^2)^2, JH \geq (1 - q)^2(k^2)^2$. It follows that $F^2/(JH) \leq 4q^2(1 + q)^2/(1 - q)^2$. Also we have $H/E \geq l^2/(k + l)^2 \geq (1 - q)^2/(2 + q)^2, J/G \geq 1/(p^2 + q^2), (JH)/(EG) \geq (1 - q)^2/((p^2 + q^2)(2 + q)^2)$ and we obtain (15). In the case $q = 1/5$ it is easy to verify that $C(q) < -2.59... < -5/2$. Since $aC(1/5) < -2a$, this implies the estimate (16). $\square$

Example. Let $b = 1/2, k = (3, 4, 5)$ and take $a = 1$; then the estimate (16) holds for harmonics l which belong to the set L of 18 elements. Below we give the list of harmonics $l \in L$ and the values $r = K(u, v)/((k^2 + l^2)/V)$, where $K(u, v)$ is the sectional curvature by the 2-dimensional space spanned by the vector fields (u, v) (12) having the simple harmonics (k, l). Note that according to (16), the value r must be less than -0.625.

$$
\begin{array}{lll}
(2,3,5) : -2.320, & (2,4,4) : -2.230, & (2,5,5) : -2.878 \\
(2,4,6) : -2.932, & (2,4,5) : -4.123, & (4,3,5) : -2.760 \\
(4,4,4) : -2.697, & (4,5,5) : -3.167, & (4,4,6) : -3.207 \\
(4,4,5) : -4.669, & (3,3,4) : -2.134, & (3,3,6) : -2.878 \\
(3,3,5) : -4.015, & (3,5,4) : -2.760, & (3,5,6) : -3.246 \\
(3,5,5) : -4.746, & (3,4,6) : -4.819, & (3,4,4) : -3.901
\end{array}
$$

We see that the estimate (16) holds by a large margin.

We also consider the orthogonal pair of vector fields

$$u' = \sin kx\, e_3, v' = \sin lx\, (se_1 + re_2), k \neq 0, l \neq 0. \tag{17}$$

It is easy to verify that $K(u,v) = K(u',v')$, where (u,v) is the pair (12). To calculate the curvatures $K(u',v)$ and $K(u,v')$, let us introduce the function

$$z'(k) = \left\{ \begin{array}{ll} \frac{1}{2}, & k \neq 0; \\ 0, & k = 0. \end{array} \right.$$

Now if in formula (14) we substitute z' for z, then we would calculate the curvature $K(u',v) = K(u,v')$. In particular we may estimate the values of $K(u',v)$ and $K(u,v')$. Note that under the assumptions of Theorem 2 we have $k \neq l$, which implies $z = z'$. Therefore the estimates (15),(16) for the values of $K(u',v)$ are done. Concerning the asymptotic behavior of $K(u',v)$, only the estimate (ii) is valid, while (i) is not valid, since under its assumptions $z \neq z'$.

Thus we have established that there exist pairs (u,v) of orthonormal vector fields on the 3-torus which give a negative value of the sectional curvature $K(u,v)$. On the other hand, in examples constructed the following property holds: if the vector field u is fixed, then the value of the curvature $K(u,v)$ may be negative only for vector fields v having simple harmonics l which belong to a finite set.

In conclusion let us investigate the constant vector field $u = e_3$ which corresponds to the homogeneous magnetizing torus. If v and v' are such as in (12) and (17), then by analogy with (13) and (14), it is easy to obtain that

$$R(e_3,v)v,e_3) = R(e_3,v')v',e_3) = \frac{1}{8}V(r^2 + s^2)(l^2 + a)^3,$$

$$K(e_3,v) = K(e_3,v') = \frac{(l^2 + a)^2}{4V(a - b)} \tag{18}$$

where $l \neq 0$ and

$$R(e_3,se_1 + re_2,)(se_1 + re_2),e_3) = \frac{1}{4}V(r^2 + s^2)a^3,$$

$$K(e_3,se_1 + re_2) = \frac{a^2}{4V(a - b)}. \tag{19}$$

Thus the sectional curvature of the homogeneous magnetizing torus is positive.

The investigation of the 3-dimensional case can be generalized to the current group $G = G(M,K)$, where K is a compact semi-simple Lie group,

M is a closed Riemannian oriented n-dimensional manifold and K is a subgroup of the orthogonal group $SO(n)$. The Lie algebra of G is the current algebra $g = g(M, k)$, where k is the Lie algebra of K. Let (E, p, M) be the principal bundle with structure group $SO(n)$ of orthonormal bases on the Riemannian manifold M. Note that there exists an action of the current group G on the fiber space E. Namely, let $i : K \rightarrow SO(n)$ be a monomorphism. For an element $g \in G$ and $e \in E$, we may put

$$ge = i(g(p(e)))(e). \qquad (20)$$

Note that the structure group acts by the same element in all fibers of the bundle E, while the current group acts by a different one, which depends on the point of the base M. Thus the current group G may be regarded as a subgroup of the group $\text{Diff}(E)$ of diffeomorphisms of E, its Lie algebra g being a subalgebra $V(E)$ of the Lie algebra of smooth vector fields on E. Generally speaking, we must consider the Landau–Lifschitz equation as a differential equation on the fiber space E. However, since the current group G acts only on fibers of the bundle E, we may reduce it to the differential equation on k-valued functions on the base M in the current algebra g. Namely, let us describe the generalized LL as follows. We have seen that the metric at a point for the LL model may be obtained as the pointwise Killing form of the current group $G(M, SO(3))$. Such an approach to the generalized LL gives the following. Let us consider the pointwise Killing form:

$$(m_1(x), m_2(x)) = -\text{tr}(\text{ad}(m_1(x))\text{ad}(m_2(x)), \qquad (21)$$

where $m_1, m_2 \in g$ and $x \in M$. Let $P : g \rightarrow g$ as above be the sum of an elliptic symmetrical differential operator and a pointwise orthogonal projection operator (in the sense of the natural metric). Let P_a be equal to $P - aI$. Using the Lie bracket $[\cdot, \cdot]$ in the Lie algebra k instead of the vector product in the 3-dimensional case, we can introduce the generalized LL as follows

$$\frac{dm}{dt} = [m, Pm]. \qquad (22)$$

Substituting (21) in the expressions of (2) and (3) instead of the scalar product, we will have similar forms for the current group $G(M, K)$. Now we can introduce the nonstandard Lie bracket in g by analogy with (7) and then obtain the expression of the coadjoint operator by analogy with (8). Then generalized LL (22) will have the form of the Euler equations of a solid body:

$$\frac{dm}{dt} = B(m, m),$$

where B is as above the coadjoint representation operator of the nonstandard Lie bracket. Moreover, the formalism of the preceding investigation of

LL is valid for the generalized LL. Thus we can obtain expressions for the covariant derivative similar to (9) and for the curvature tensor similar to (10) in the case of the current group $G(M, K)$. Theorem 1 (formula (11)) is also valid and allows us to investigate the stability of solutions of the generalized LL by calculating the curvatures.

In the special cases when $M = K$ or $M = T^n$, where $n = \dim K$, we may identify the Lie algebra g with the space $V(M)$ of smooth vector fields on M with pointwise Lie bracket. Then we obtain a situation similar to that of the 3-sphere or with the 3-torus.

Acknowledgments

The research described in this publication was made possible in part by grants RO4000 of ISF, RO4300 of ISF and Russian Government, FF60 of RFFI.

References

1. Broun, W.F.: *Micromagnetics*, New-York – London, 1963.
2. Kotsevitch, A.M., Ivanov, B.A. and Kovalev, A.S.: *Nonlinear waves of magnetization. Dynamical and topological solitons*, Kiev, 1983 (in Russian).
3. Barouch, E., Fokas, A.S. and Papageorgiou, V.G.: The bi-Hamiltonian formulation of the Landau–Lifschitz equation, *J.Math.Physics* 29 (1988) no. 12, 2628–2633.
4. Arnold, V.I.: Sur la géometrie différentielle des groupes de Lie de dimension infinie and ses applications à l'hydrodynamique des fluides parfais, *Ann. Inst. Fourier* **16** (1966) Grenoble, 319–361.
5. Arnold, V.I.: *Mathematical methods of classic mechanics*, Springer–Verlag, 1989.
6. Aleksovsky, V.A. and Lukatsky, A.M.: Nonlinear dynamics of ferromagnetic magnetization and motion of generalized body with the current group, *Theor. and Math. Physics* **85** (1990) no. 1, 115–123 (in Russian).

CHANGE OF VARIABLE FORMULAS

FOR GAUSSIAN INTEGRALS OVER SPACES OF PATHS

IN COMPACT RIEMANNIAN MANIFOLDS

O.G. SMOLYANOV
Department of Mechanics and Mathematics,
Moscow State University, 119899 Moscow, Russia

Abstract. A method for constructing measures on spaces of paths in compact Riemannian manifolds embedded in Euclidean space is developed. For measures obtained by this way, a Cameron–Martin–Girsanov–Maruyama–Ramer transformation formula is obtained. A particular case of this formula is the formula for transformations of the measures describing Brownian motion on compact Riemannian manifolds.

Mathematics Subject Classification (1991): 60H07, 60H05, 60H10, 60H15, 60J25.

Key words: Gaussian integrals, Riemannian manifolds, spaces of paths, Cameron–Martin–Girsanov–Maruyama–Ramer formula, logarithmic derivative, differentiable measure, Radon measure, surface measure.

We discuss certain formulas for transformations of measures over spaces of functions of a real variable ranging in compact Riemannian manifolds. Such measures are called Gaussian if they are — in a sense — surface measures corresponding to Gaussian measures over functions taking values in a Euclidean space in which the Riemannian manifold is embedded. The reduction of some problems related to measures over spaces of functions taking values in a (compact) Riemannian manifold to similar problems related to measures over spaces of functions taking values in Euclidean spaces is the main idea of the present paper.

In particular, the transformation formulas obtained below are deduced from Cameron–Martin–Girsanov–Maruyama–Ramer transformation formulas that are obtained for the vector space case (cf. [1]–[7]). Our main tool

B. P. Komrakov et al. (eds.), Lie Groups and Lie Algebras, 435–442.
© 1998 *Kluwer Academic Publishers. Printed in the Netherlands.*

is the notion of logarithmic derivative of a measure; the logarithmic derivative of a surface measure ν generated by a measure η is (under suitable assumptions) the restriction of the logarithmic derivative of η; having a logarithmic derivative, one can immediately obtain transformation formulas ([2], [3], [4]).

For the sake of simplicity, some definitions are formulated below for arbitrary "locally convex manifolds", not only for manifolds of mappings.

1. Terminology, notations, and preliminaries

The terminology and some notations from [9] and [10] are used without explanation. One assumes that the field of scalars for all considered vector spaces is the field of real numbers and that all topological spaces are Hausdorff. For any two topological spaces F and G denote by $C(F, G)$ the space of all continuous mappings of F into G, equipped with the compact-open topology. If the space G is equipped with some supplementary structure (that of vector space, normed space, group, algebra), then we assume that the space $C(F, G)$ is endowed with a (naturally defined) similar structure.

For any topological space Q, the symbols $C_b(Q)$ and $M(Q)$ denote respectively the vector space of bounded continuous real functions on the topological space Q and the vector space of all (real-valued Borel) Radon measures [9] on Q, the symbol $B(Q)$ denotes the σ-algebra of all Borel subsets of Q.

A mapping f of a locally convex space (LCS) E into a LCS G is called differentiable, if f is Gateaux-differentiable at each point $x \in E$ and if f and its Gateaux derivative are continuous as mappings of E into G and into $L(E, G)$ respectively. Here $L(E, G)$ denotes the space of all continuous linear mappings from E into G equipped with the topology of compact convergence. The derivatives of higher order are defined by induction. The mapping $f : E \to G$ is called smooth, if it is differentiable infinitely many times at each point.

2. Differentiable measures

Let Q be a topological space and τ be a topology on $M(Q)$ compatible with the vector space structure.

Definition 1 ([2], [3]) A function f defined on an open interval S of the real axis and taking values in the space $M(Q)$ is called τ-*differentiable* at the point t_0 if f is differentiable at this point as a mapping from S into $(M(Q), \tau)$; in this case the τ-*derivative* of the function f at the point t_0 is defined as the derivative of the latter mapping at t_0 and is denoted by $f'(t_0)$ (one does not use the symbol τ in the notation of the τ-derivative, since for any two comparable topologies τ_1, τ_2 the corresponding derivatives at a point coincide if they exist).

If f is τ-differentiable at t_0 and the measure $f'(t_0)$ is absolutely continuous with respect to $f(t_0)$, $f'(t_0) \ll f(t_0)$, then the Radon–Nicodym derivative $d(f'(t_0))/d(f(t_0))$ is called the logarithmic derivative of f [3] at t_0 and is denoted by $\rho(t, f)$.

In particular, if $\nu \in M(Q)$ and $\mathcal{F} = \{\mathcal{F}(t) : -\varepsilon < t < \varepsilon\}$ is a family of $(B(Q), B(Q))$-measurable transformations of Q with $\mathcal{F}(0) = \mathrm{id}$, then ν is called τ-differentiable along the family $\mathcal{F}$ iff the map

$$f : t \mapsto \nu_t = \nu \circ T_t^{-1}$$

is τ-differentiable at $t_0 = 0$, [2]. In this case the derivative $f'(0)$ is called the derivative of ν along (the family) $\mathcal{F}$ and is denoted by $\nu'_{\mathcal{F}}$ and the logarithmic derivative $\rho(0, f)$ (if it exists) is called the τ-logarithmic derivative of ν along $\mathcal{F}$ and is denoted by $\beta^\nu_{\mathcal{F}}$.

Let now V_E be a locally convex C_∞-manifold (LCM), modelled by a locally convex space E (the definition of locally convex manifold is quite similar to the definition of a Banach C_∞-manifold [11], the only difference is that one uses the notion of differentiability introduced above instead of the notion of Fréchet differentiability). The tangent bundle $T(V_E)$ of V_E is defined in a natural way. A vector field h on V_E is a mapping of V_E into the tangent bundle $T(V_E)$ such that for any $x \in V_E$ one has $\pi h(x) = x$, where π is the natural mapping of $T(V_E)$ onto V_E. The vector space of all vector fields on V_E is denoted by $\mathrm{vect}\, V_E$. The derivatives of mappings between two LCM and smooth functions on a LCM are defined in a natural way. Let us assume that there exists $\varepsilon > 0$ and a family

$$\mathcal{F}_h = \{\mathcal{F}_h(t) : -\varepsilon < t < \varepsilon\}$$

of mappings of V_E into itself such that for each $x \in V_E$ the function $t \mapsto \mathcal{F}_h(t)(x)$ is differentiable and $\mathcal{F}'(0)(x) = -h(x)$ ($h \in \mathrm{vect}\, V_E$). Then one calls $\nu \in M(V_E)$ τ-differentiable along the vector field h if it is τ-differentiable along the family $\mathcal{F}_h$; the corresponding derivative and logarithmic derivative (if it exists) are denoted respectively by $\nu'h$ and by $\beta^\nu_h(\cdot)$ and are called the derivative and the logarithmic derivative of ν along the vector field h. If $V_E = E$ and $h(x) = h_0 \in E$ for each $x \in E$, then instead of $\beta^\nu_h(\cdot)$ one uses the symbol $\beta^\nu(h_0, \cdot)$.

A function $f : V_E \to \mathbb{R}^1$ is called smooth if it is differentiable infinitely many times.

Let now C be a (vector) space of smooth functions on V_E that is norm defining for $M(V_E)$, i.e., $\|\nu\|_1 = \sup\{\int \phi d\nu : \phi \in C, \|\phi\|_\infty \leq 1\}$, where $\|\cdot\|_1$ is the total variation norm and $\|\cdot\|_\infty$ is the sup-norm. Let $h \in \mathrm{vect}\, V_E$ be such that, for any $\phi \in C$, $\phi' h \in C$.

A measure $\nu \in M(V_E)$ is called C-differentiable along the vector field $h \in \operatorname{vect} V_E$, if there exists a measure $\nu'h \in M(V_E)$, called the C-derivative of ν along h, such that

$$\int \psi'h d\nu = -\int \psi d(\nu'h)$$

for all $\psi \in C$.

As above, if $\nu'h \ll \nu$, then the corresponding Radon–Nicodym derivative is called the logarithmic derivative of ν along h and is again denoted by β_h^ν. The following proposition (in particular) asserts that β_h^ν is well defined.

Proposition 1 (cf. [2]) *Let $\mathcal{F}_h = \{\mathcal{F}_h(t) : t \in \mathbb{R}^1\}$ be a family of mappings of V_E into V_E satisfying the preceding assumptions. Let τ_c be the topology $\tau(M(V_E), C)$. Then the measure $\nu \in M(V_E)$ is τ_c-differentiable along $\mathcal{F}_h$ iff it is C-differentiable along h; in this case τ_c- and C-derivatives of ν and the corresponding logarithmic derivatives (if they exist) coincide.*

The proof is quite similar to the proof of an analogous statement in [2].

3. Surface measures and logarithmic derivatives

Everywhere below we assume that E is a Fréchet space. Let G be a complementable topological vector subspace of E and V_G be a submanifold of E modelled by G; the definition of a submanifold of a LCM, in particular, of a LCS, is the same as the definition of a submanifold of a Banach manifold; this notion is well defined because the chain rule is valid for (n times) differentiable mapping of LCS (for any n) (cf. [11], Ch. 2, § 2, Lemma 1).

Suppose that $\nu \in M(E)$, $\eta \in M(V_E)$, and for any open subset V of E one has $\nu(V) > 0$. We shall say that η is the surface measure on V_G generated by ν, if the following statement holds. For any $f \in C_b(E)$ and for every $\varepsilon > 0$ there exists a neighborhood W_0 of zero in E such that for each neighborhood W of zero in E satisfying $W \subset W_0$ one has

$$\left| \int_{V_G} f(x)\eta(dx) - (\nu(V_G + W))^{-1} \int_{V_G+W} f(x)\nu(dx) \right| < \varepsilon.$$

Remark 1 If $g \in L_1(E, \nu)$, then a function $g_0 \in L_1(V_G, \eta)$ is called the (L_1)-restriction of g to V_G, if for any $\varepsilon > 0$ there exists a neighborhood W of zero in E such that for each neighborhood of zero $W \subset W_0$,

$$\left| \int_{V_G} g_0(x)\eta(dx) - (\nu(V_G + W))^{-1} \int_{V_G+W} g(x)\nu(dx) \right| < \varepsilon.$$

Theorem 1 *Let η be the surface measure on V_G generated by a Radon measure ν on E (which is positive on each open subset V of E). Let $h \in \operatorname{vect} E$ have the following properties:*

(1) h is (everywhere) differentiable and, together with its derivative, is uniformly continuous;

(2) $h'(x)(T_x(V_G)) \subset T_x(V_G)$ for each $x \in V_G$ ($T_x(V_G)$ is the tangent space of V_G at x, i.e., is regarded as be embedded into E,

and let h_0 be the restriction of h to V_G.

If ν is C-differentiable along h with its logarithmic derivative $\beta\nu_h$ that has an L_1-restriction to V_G, then η is C-differentiable along h_0 with the logarithmic derivative $\beta^\eta_{h_0}$ which is an L_1-restriction to V_G of β^ν_h.

This statement can be deduced from the definitions of a surface measure ν, of a logarithmic derivative along the vector field, and of an L_1-restriction.

Now we consider some typical examples which the introduced definitions and statements can be applied to.

For any topological space T, any $a \in T$ and any $\tau \in (0, \infty]$ let the symbol $C_a([0, \tau), T)$ denote the subspace of $C([0, \tau), T)$ consisting of all functions taking the value a at $t = 0$.

Let $E = C([0, \infty), \mathbb{R}^n)$ and let ν_a be the (probabilistic) measure on E generated by a homogeneous Markov process ξ on $\mathbb{R}^n$, $\xi(0) = a$, that is governed by the stochastic differential equation $d\xi(t) = \alpha\, dt + b\, dw$ whose drift coefficient α and diffusion coefficient b have the following properties: α is a smooth function bounded on each bounded subset, and b is a bounded smooth function, $b > \varepsilon > 0$ everywhere.

Let K be a compact Riemannian (C_∞-) submanifold in $\mathbb{R}^n$ (one assumes that the metric on K is induced by the metric on $\mathbb{R}^n$), and let $G = C_0([0, \infty), \mathbb{R}^k)$, where $k = \dim K$.

Finally let η_a be the measure on $V_G^a = C_a([0, \infty), K)$, $a \in K$, generated by the (homogeneous) Markov process ξ_K on K, $\xi_K(0) = a$, governed by the stochastic differential equation whose coefficients are the restrictions on K of the coefficients of the preceding stochastic differential equation.

Theorem 2 *The surface measure on V_G^a corresponding to ν_a is η_a.*

A particular case of this theorem can be found in [1]; an essential part of the proof is based on the following statement.

For every $t \in [0, \infty)$ and every sequence γ of real numbers $0 = t_0 < t_1 < t_2 < \ldots < t_n = t$, let the function $p_\gamma(t, \cdot, \cdot) : K \times K \to \mathbb{R}^1$ and the number $\delta(\gamma)$ be defined by the following relations:

$$\delta(\gamma) = \max\{|t_j - t_{j-1}|; \; j = 1, 2, \ldots, n\};$$

$$p_\gamma(t, x, z) = \int_K \cdots \int_K p(t_1, x, x_1) \cdot p(t_2 - t_1, x, x_1) \cdots$$

$$\cdots p(t - t_n, x_n, z) dx_1 dx_2 \cdots dx_n \,,$$

where $p(\cdot, \cdot, \cdot)$ is the transition probability for the Markov process ξ in $\mathbb{R}^n$. Finally for each $t \in [0, \infty)$, let $p^K(t, \cdot, \cdot) : K \times K \to \mathbb{R}^1$ be the transition probability for the Markov process in K defined above.

Theorem 3 *If $\delta(\gamma) \to 0$, then for each $t \in [0, \infty)$, $x, z \in K$ one has*

$$\frac{p_\gamma(t, x, z)}{\int_K p_\gamma(t, x, z_1) dz_1} \to p^K(t, x, z),$$

where $\int_K p_\gamma(t, x, z_1) dz_1$ is the integral over the Lebesgue measure on K induced by the Riemannian structure in K.

Remark 2 For the particular case in which $\alpha = 0$, $b = 1$, $K = U(r)$ (the unitary group), $n = 2r^2$, and $\mathbb{R}^n$ is realized as the space of all complex-valued matrices, this statement was formulated in [1]; the detailed proof can be found in [12]; in the general case the proof is similar.

Let us recall that if ν is a Gaussian (cylindrical) measure on a (separable) Hilbert space H whose correlation operator is B $(\in L(H, H))$ and mean value is a $(\in H)$, then the subspace $D(\nu)$ of differentiability of ν (i.e., the space of all vectors along which ν is differentiable) is $\mathrm{Im}\, B^{1/2}$ and for each $h \in \mathrm{Im}\, B$ one has

$$\beta^\nu(h, x) = -(B^{-1}h, a - x).$$

Let us recall also that this fact implies that if w_a is the Wiener measure on the space $C([0, t), \mathbb{R}^n)$ of all continuous $\mathbb{R}^n$-valued functions defined on $[0, t)$ concentrated on the subspace $C_a([0, t), \mathbb{R}^n)$, then the subspace $D(w_a)$ of differentiability of w_a is equal to the (Sobolev) space $W_2^1([0, t), \mathbb{R}^n)$ consisting of all absolutely continuous function vanishing at 0 and having square summable derivatives, and for $h \in W_2^1([0, t), \mathbb{R}^n)$ we have

$$\beta^\nu(h, w) = -\int\limits_0^t (h'(\tau), d(w(\tau) - a)),$$

where the symbol $(\, , \,)$ denotes the scalar product in $\mathbb{R}^n$.

4. Transformations of measures

We keep here all the preceding assumptions about measures and spaces.

Let $\mathcal{F}$ be a family of twice differentiable transformations of E having twice differentiable inverse transformations and preserving V_G. Let $\nu \in M(E), \eta \in M(V_G)$ and η be the surface measure generated by ν.

Our next purpose is to derive the formula for the density of $\eta(\mathcal{F}_{V_G}(t, \cdot))$ with respect to η (here $\mathcal{F}_{V_G}$ denotes the restriction of $\mathcal{F}(t, \cdot)$ to V_G and

$\eta(\mathcal{F}_{V_G}(t,\cdot))$ denotes the image of η with respect to the map $(\mathcal{F}_{V_G}(t,\cdot))^{-1})$, under the assumptions of differentiability of ν having its logarithmic derivative with respect to $\mathcal{F}$).

This can be done in two ways. First, one can calculate (as shown in [3]) the logarithmic derivative of η and afterwards use the general formula for transformations of a measure from [3]. Second, one can calculate the transformation of ν directly and afterwards find the surface measure of the transformed measure.

In any of these ways, we obtain the following statement, in which one assumes that the measure-valued function $\tau \mapsto \nu\mathcal{F}(\tau, \cdot)$, $[0, t) \to M(E)$ satisfies the assumptions of Theorem 3.3 from [3].

Proposition 2 *The following relation holds:*

$$\eta\mathcal{F}(t, \cdot) = \left(\exp\{\int_0^t \beta_{\mathcal{F}}^{\nu}(\mathcal{F}(\tau, \cdot))d\tau\}\right) \cdot \eta .$$

Now let $a \in K \subset \mathbb{R}^n$ and w_a^K be the measure generated on $C([0, t), K)$ by the Wiener measure on $C([0, t), \mathbb{R}^n)$ concentrated on $C_a([0, t), \mathbb{R}^n)$.

Let h ($\in$ vect $C([0, t), \mathbb{R}^n)$ have the following property: for every $\tau \in [0, t)$ the function $x \mapsto h(x)(\tau)$ is measurable with respect to the algebra of subsets of $C([0, t), \mathbb{R}^n)$ generated by the functions $x \mapsto x(s)$, $s \leq \tau$. Let, moreover, the vector field have the integral flow $\mathcal{F}(\cdot, \cdot) : \mathbb{R}^1 \times E \to E$.

Theorem 4 *The shift of η along integral curves [3] of the flow $\mathcal{F}$, restricted to V_G, is defined by*

$$\eta\mathcal{F}(\tau, \cdot) = \left(\exp\{-\frac{\tau^2}{2}\int_0^t ((h(x))'(\alpha))^2 d\alpha + \tau \int_0^t (h(x))'(\alpha)dw(\alpha - a)\}\right) \cdot \eta.$$

A special case of this formula, related to the situation described in Remark 2, was obtained in [1].

References

1. Smolyanov, O.G.: Smooth measures on loop groups, *Doklady Mathematics,* **52** (1995) no.3, 408–411.
2. Smolyanov, O.G. and van Weizsacker, H.: The calculus of differentiable measures (to appear).
3. Smolyanov, O.G. and van Weizsacker, H.: Differentiable families of measures, *J. of Functional Analysis,* **118** (1993) no. 2, 455–476.
4. Smolyanov, O.G. and van Weizsacker, H.: Change of measures and their logarithmic derivatives under smooth transformations, *C.R. Acad. sci, Paris, Sér. 1,* **321** (1995) 103–108.
5. Daletski, Yu.L. and Sohadze, G.: Absolute continuity of smooth measures, *Funct. Anal. Appl.* **22** (1988) no. 2, 77–78.

6. Nualart, D.: *The Malliavin calculus and related topics*, Springer–Verlag, 1995.
7. Bell, D.: A quasi-invariant theorem for measures in Banach spaces, *Trans. Amer. Math. Soc.* **290** (1985) 851–855.
8. Pressley, A. and Segal, G.: *Loop groups*, Oxford, Clarendon, 1986.
9. Smolyanov, O.G. and Fomin, S.V.: Measures on topological linear spaces, *Russian Math. Surveys*, **3** (1976) 3–56.
10. Schaefer, H.: *Topological vector spaces*, New York–London, 1966.
11. Lang, S.: *Introductions to differentiable manifolds*, New York–London, 1962.
12. Smirnova, M.G.: A construction of the Brownian motion on compact Lie groups, (to appear).

Other *Mathematics and Its Applications* titles of interest:

P.H. Sellers: *Combinatorial Complexes. A Mathematical Theory of Algorithms.*
1979, 200 pp.   ISBN 90-277-1000-7

P.M. Cohn: *Universal Algebra.* 1981, 432 pp.
                    ISBN 90-277-1213- 1 (hb), ISBN 90-277-1254-9 (pb,

J. Mockor: *Groups of Divisibility.* 1983, 192 pp.   ISBN 90-277-1539-4

A. Wwarynczyk: *Group Representations and Special Functions.* 1986, 704 pp.
                    ISBN 90-277-2294-3 (pb), ISBN 90-277-1269-7 (hb)

I. Bucur: *Selected Topics in Algebra and its Interrelations with Logic, Number
Theory and Algebraic Geometry.* 1984, 416 pp.   ISBN 90-277-1671-4

H. Walther: *Ten Applications of Graph Theory.* 1985, 264 pp.
                                                      ISBN 90-277-1599-8

L. Beran: *Orthomodular Lattices. Algebraic Approach.* 1985, 416 pp.
                                                      ISBN 90-277-1715-X

A. Pazman: *Foundations of Optimum Experimental Design.* 1986, 248 pp.
                                                      ISBN 90-277-1865-2

K. Wagner and G. Wechsung: *Computational Complexity.* 1986, 552 pp.
                                                      ISBN 90-277-2146-7

A.N. Philippou, G.E. Bergum and A.F. Horodam (eds.): *Fibonacci Numbers and
Their Applications.* 1986, 328 pp.   ISBN 90-277-2234-X

C. Nastasescu and F. van Oystaeyen: *Dimensions of Ring Theory.* 1987, 372 pp.
                                                      ISBN 90-277-2461-X

Shang-Ching Chou: *Mechanical Geometry Theorem Proving.* 1987, 376 pp.
                                                      ISBN 90-277-2650-7

D. Przeworska-Rolewicz: *Algebraic Analysis.* 1988, 640 pp.   ISBN 90-277-2443-1

C.T.J. Dodson: *Categories, Bundles and Spacetime Topology.* 1988, 264 pp.
                                                      ISBN 90-277-2771-6

V.D. Goppa: *Geometry and Codes.* 1988, 168 pp.   ISBN 90-277-2776-7

A.A. Markov and N.M. Nagorny: *The Theory of Algorithms.* 1988, 396 pp.
                                                      ISBN 90-277-2773-2

E. Kratzel: *Lattice Points.* 1989, 322 pp.   ISBN 90-277-2733-3

A.M.W. Glass and W.Ch. Holland (eds.): *Lattice-Ordered Groups. Advances and
Techniques.* 1989, 400 pp.   ISBN 0-7923-0116-1

N.E. Hurt: *Phase Retrieval and Zero Crossings: Mathematical Methods in Image
Reconstruction.* 1989, 320 pp.   ISBN 0-7923-0210-9

Du Dingzhu and Hu Guoding (eds.): *Combinatorics, Computing and Complexity.*
1989, 248 pp.   ISBN 0-7923-0308-3

Other *Mathematics and Its Applications* titles of interest:

A.Ya. Helemskii: *The Homology of Banach and Topological Algebras*. 1989, 356 pp.   ISBN 0-7923-0217-6

J. Martinez (ed.): *Ordered Algebraic Structures*. 1989, 304 pp.
ISBN 0-7923-0489-6

V.I. Varshavsky: *Self-Timed Control of Concurrent Processes. The Design of Aperiodic Logical Circuits in Computers and Discrete Systems*. 1989, 428 pp.
ISBN 0-7923-0525-6

E. Goles and S. Martinez: *Neural and Automata Networks. Dynamical Behavior and Applications*. 1990, 264 pp.   ISBN 0-7923-0632-5

A. Crumeyrolle: *Orthogonal and Symplectic Clifford Algebras. Spinor Structures*. 1990, 364 pp.   ISBN 0-7923-0541-8

S. Albeverio, Ph. Blanchard and D. Testard (eds.): *Stochastics, Algebra and Analysis in Classical and Quantum Dynamics*. 1990, 264 pp. ISBN 0-7923-0637-6

G. Karpilovsky: *Symmetric and G-Algebras. With Applications to Group Representations*. 1990, 384 pp.   ISBN 0-7923-0761-5

J. Bosak: *Decomposition of Graphs*. 1990, 268 pp.   ISBN 0-7923-0747-X

J. Adamek and V. Trnkova: *Automata and Algebras in Categories*. 1990, 488 pp.
ISBN 0-7923-0010-6

A.B. Venkov: *Spectral Theory of Automorphic Functions and Its Applications*. 1991, 280 pp.   ISBN 0-7923-0487-X

M.A. Tsfasman and S.G. Vladuts: *Algebraic Geometric Codes*. 1991, 668 pp.
ISBN 0-7923-0727-5

H.J. Voss: *Cycles and Bridges in Graphs*. 1991, 288 pp.   ISBN 0-7923-0899-9

V.K. Kharchenko: *Automorphisms and Derivations of Associative Rings*. 1991, 386 pp.   ISBN 0-7923-1382-8

A.Yu. Olshanskii: *Geometry of Defining Relations in Groups*. 1991, 513 pp.
ISBN 0-7923-1394-1

F. Brackx and D. Constales: *Computer Algebra with LISP and REDUCE. An Introduction to Computer-Aided Pure Mathematics*. 1992, 286 pp.
ISBN 0-7923-1441-7

N.M. Korobov: *Exponential Sums and their Applications*. 1992, 210 pp.
ISBN 0-7923-1647-9

D.G. Skordev: *Computability in Combinatory Spaces. An Algebraic Generalization of Abstract First Order Computability*. 1992, 320 pp.   ISBN 0-7923-1576-6

E. Goles and S. Martinez: *Statistical Physics, Automata Networks and Dynamical Systems*. 1992, 208 pp.   ISBN 0-7923-1595-2

Other *Mathematics and Its Applications* titles of interest:

M.A. Frumkin: *Systolic Computations.* 1992, 320 pp.   ISBN 0-7923-1708-4

J. Alajbegovic and J. Mockor: *Approximation Theorems in Commutative Algebra.* 1992, 330 pp.   ISBN 0-7923-1948-6

I.A. Faradzev, A.A. Ivanov, M.M. Klin and A.J. Woldar: *Investigations in Algebraic Theory of Combinatorial Objects.* 1993, 516 pp.   ISBN 0-7923-1927-3

I.E. Shparlinski: *Computational and Algorithmic Problems in Finite Fields.* 1992, 266 pp.   ISBN 0-7923-2057-3

P. Feinsilver and R. Schott: *Algebraic Structures and Operator Calculus.* Vol. I. Representations and Probability Theory. 1993, 224 pp.   ISBN 0-7923-2116-2

A.G. Pinus: *Boolean Constructions in Universal Algebras.* 1993, 350 pp.
ISBN 0-7923-2117-0

V.V. Alexandrov and N.D. Gorsky: *Image Representation and Processing. A Recursive Approach.* 1993, 200 pp.   ISBN 0-7923-2136-7

L.A. Bokut' and G.P. Kukin: *Algorithmic and Combinatorial Algebra.* 1994, 384 pp.   ISBN 0-7923-2313-0

Y. Bahturin: *Basic Structures of Modern Algebra.* 1993, 419 pp.
ISBN 0-7923-2459-5

R. Krichevsky: *Universal Compression and Retrieval.* 1994, 219 pp.
ISBN 0-7923-2672-5

A. Elduque and H.C. Myung: *Mutations of Alternative Algebras.* 1994, 226 pp.
ISBN 0-7923-2735-7

E. Goles and S. Martínez (eds.): *Cellular Automata, Dynamical Systems and Neural Networks.* 1994, 189 pp.   ISBN 0-7923-2772-1

A.G. Kusraev and S.S. Kutateladze: *Nonstandard Methods of Analysis.* 1994, 444 pp.   ISBN 0-7923-2892-2

P. Feinsilver and R. Schott: *Algebraic Structures and Operator Calculus.* Vol. II. Special Functions and Computer Science. 1994, 148 pp.   ISBN 0-7923-2921-X

V.M. Kopytov and N. Ya. Medvedev: *The Theory of Lattice-Ordered Groups.* 1994, 400 pp.   ISBN 0-7923-3169-9

H. Inassaridze: *Algebraic K-Theory.* 1995, 438 pp.   ISBN 0-7923-3185-0

C. Mortensen: *Inconsistent Mathematics.* 1995, 155 pp.   ISBN 0-7923-3186-9

R. Abłamowicz and P. Lounesto (eds.): *Clifford Algebras and Spinor Structures.* A Special Volume Dedicated to the Memory of Albert Crumeyrolle (1919–1992). 1995, 421 pp.   ISBN 0-7923-3366-7

W. Bosma and A. van der Poorten (eds.), *Computational Algebra and Number Theory.* 1995, 336 pp.   ISBN 0-7923-3501-5

Other *Mathematics and Its Applications* titles of interest:

A.L. Rosenberg: *Noncommutative Algebraic Geometry and Representations of Quantized Algebras*. 1995, 316 pp.   ISBN 0-7923-3575-9

L. Yanpei: *Embeddability in Graphs*. 1995, 400 pp.   ISBN 0-7923-3648-8

B.S. Stechkin and V.I. Baranov: *Extremal Combinatorial Problems and Their Applications*. 1995, 205 pp.   ISBN 0-7923-3631-3

Y. Fong, H.E. Bell, W.-F. Ke, G. Mason and G. Pilz (eds.): *Near-Rings and Near-Fields*. 1995, 278 pp.   ISBN 0-7923-3635-6

A. Facchini and C. Menini (eds.): *Abelian Groups and Modules*. (Proceedings of the Padova Conference, Padova, Italy, June 23–July 1, 1994). 1995, 537 pp.   ISBN 0-7923-3756-5

D. Dikranjan and W. Tholen: *Categorical Structure of Closure Operators*. With Applications to Topology, Algebra and Discrete Mathematics. 1995, 376 pp.   ISBN 0-7923-3772-7

A.D. Korshunov (ed.): *Discrete Analysis and Operations Research*. 1996, 351 pp.   ISBN 0-7923-3866-9

P. Feinsilver and R. Schott: *Algebraic Structures and Operator Calculus*. Vol. III: Representations of Lie Groups. 1996, 238 pp.   ISBN 0-7923-3834-0

M. Gasca and C.A. Micchelli (eds.): *Total Positivity and Its Applications*. 1996, 528 pp.   ISBN 0-7923-3924-X

W.D. Wallis (ed.): *Computational and Constructive Design Theory*. 1996, 368 pp.   ISBN 0-7923-4015-9

F. Cacace and G. Lamperti: *Advanced Relational Programming*. 1996, 410 pp.   ISBN 0-7923-4081-7

N.M. Martin and S. Pollard: *Closure Spaces and Logic*. 1996, 248 pp.   ISBN 0-7923-4110-4

A.D. Korshunov (ed.): *Operations Research and Discrete Analysis*. 1997, 340 pp.   ISBN 0-7923-4334-4

W.D. Wallis: *One-Factorizations*. 1997, 256 pp.   ISBN 0-7923-4323-9

G. Weaver: *Henkin–Keisler Models*. 1997, 266 pp.   ISBN 0-7923-4366-2

V.N. Kolokoltsov and V.P. Maslov: *Idempotent Analysis and Its Applications*. 1997, 318 pp.   ISBN 0-7923-4509-6

J.P. Ward: *Quaternions and Cayley Numbers*. Algebra and Applications. 1997, 250 pp.   ISBN 0-7923-4513-4

E.S. Ljapin and A.E. Evseev: *The Theory of Partial Algebraic Operations*. 1997, 245 pp.   ISBN 0-7923-4609-2

Other *Mathematics and Its Applications* titles of interest:

S. Ayupov, A. Rakhimov and S. Usmanov: *Jordan, Real and Lie Structures in Operator Algebras*. 1997, 235 pp. ISBN 0-7923-4684-X

A. Khrennikov: *Non-Archimedean Analysis: Quantum Paradoxes, Dynamical Systems and Biological Models*. 1997, 389 pp. ISBN 0-7923-4800-1

G. Saad and M.J. Thomsen (eds.): *Nearrings, Nearfields and K-Loops*. (Proceedings of the Conference on Nearrings and Nearfields, Hamburg, Germany. July 30–August 6, 1995). 1997, 458 pp. ISBN 0-7923-4799-4

L.A. Lambe and D.E. Radford: *Introduction to the Quantum Yang–Baxter Equation and Quantum Groups: An Algebraic Approach*. 1997, 314 pp.
ISBN 0-7923-4721-8

H. Inassaridze: *Non-Abelian Homological Algebra and Its Applications*. 1997, 271 pp. ISBN 0-7923-4718-8

B.P. Komrakov, I.S. Krasil'shchik, G.L. Litvinov and A.B. Sossinsky (eds.): *Lie Groups and Lie Algebras*. Their Representations, Generalisations and Applications. 1998, 358 pp. ISBN 0-7923-4916-4